Lecture Notes in Computer Science 4588

Commenced Publication in 1973
Founding and Former Series Editors:
Gerhard Goos, Juris Hartmanis, and Jan van Leeuwen

T0223144

Tero Harju   Juhani Karhumäki
Arto Lepistö (Eds.)

Developments in Language Theory

11th International Conference, DLT 2007
Turku, Finland, July 3-6, 2007
Proceedings

 Springer

Volume Editors

Tero Harju
Department of Mathematics
University of Turku
20014 Turku, Finland
E-mail: tero.harju@utu.fi

Juhani Karhumäki
Department of Mathematics
University of Turku
20014 Turku, Finland
E-mail: karhumak@utu.fi

Arto Lepistö
Department of Mathematics
University of Turku
20014 Turku, Finland
E-mail: alepisto@utu.fi

Library of Congress Control Number: 2007929029

CR Subject Classification (1998): F.4.3, F.4.2, F.4, F.3, F.1, G.2

LNCS Sublibrary: SL 1 – Theoretical Computer Science and General Issues

ISSN   0302-9743
ISBN-10   3-540-73207-1 Springer Berlin Heidelberg New York
ISBN-13   978-3-540-73207-5 Springer Berlin Heidelberg New York

Springer is a part of Springer Science+Business Media

springer.com

© Springer-Verlag Berlin Heidelberg 2007
Printed in Germany

Typesetting: Camera-ready by author, data conversion by Scientific Publishing Services, Chennai, India
Printed on acid-free paper   SPIN: 12080298   06/3180   5 4 3 2 1 0

Preface

The 11th International Conference on Developments in Language Theory (DLT 2007) was held at the University of Turku, Finland, July 3–6, 2007. This was the second time DLT took place in Turku. Indeed, the very first meeting was organized in Turku in 1993. Consequent meetings were held in Magdeburg (1995), Thessaloniki (1997), Aachen (1999), Vienna (2001), Kyoto (2002), Szeged (2003), Auckland (2004), Palermo (2005), and Santa Barbara (2006). The conference series is held under the auspices of the European Association for Theoretical Computer Science.

The DLT meeting can be viewed as the main conference on automata and formal language theory. The current topics of the conference include the following: grammars, acceptors and transducers for strings, trees, graphs and arrays, efficient text algorithms, algebraic theories for automata and languages, combinatorial and algebraic properties of words and languages, variable-length codes, symbolic dynamics, decision problems, relations to complexity theory and logic, picture description and analysis, polyominoes and bidimensional patterns, cryptography, concurrency, bio-inspired computing, quantum computing. This volume of *Lecture Notes in Computer Science* contains the papers that were presented at DLT 2007, including the abstracts or full papers of the six invited speakers Volker Diekert (Stuttgart), Thomas Henzinger (Lausanne), Michal Kunc (Brno), Ming Li (Waterloo), Jacques Sakarovitch (Paris), and Kai Salomaa (Kingston)

For the conference, 32 contributed papers were selected from a record-breaking 74 submissions. We warmly thank the authors of the papers, the members of the Program Committee, who faced many hard decisions, and the reviewers of the submitted papers for their valuable work. All these efforts were the basis of the success of the conference, In particular, we are very thankful to the invited speakers of the conference. Finally, we thank the Organizing Committee for its splendid work and also the members of the Steering Committee.

Finally, we wish to thank the support of the conference sponsors: The Academy of Finland, The Finnish Cultural Foundation, the Finnish Academy of Science and Letters / Vilho, Yrjö and Kalle Väisälä Foundation, the City of Turku, the University of Turku, the Turku Centre for Computer Science, and Centro Hotel.

April 2007

Tero Harju
Juhani Karhumäki
Arto Lepistö

Organization

DLT 2007 was organized by the Department of Mathematics and Centre for Fundamentals of Computing and Discrete Mathamatics (FUNDIM), University of Turku.

Invited Speakers

Volker Diekert (Stuttgart, Germany)
Thomas Henzinger (Lausanne, Switzerland)
Michal Kunc (Brno, Czech Republic)
Ming Li (Waterloo, Canada)
Jacques Sakarovitch (Paris, France)
Kai Salomaa (Kingston, Canada)

Program Committee

Cristian Calude (Auckland, New Zealand)
Julien Cassaigne (Marseille, France)
Christian Choffrut (Paris, France)
Jürgen Dassow (Magdeburg, Germany)
Bruno Durand (Marseille, France)
Massimiliano Goldwurm (Milan, Italy)
Tero Harju, Co-chair (Turku, Finland)
Juraj Hromkovič (Zurich, Switzerland)
Oscar Ibarra (Santa Barbara, USA)
Lucian Ilie (London, Canada)
Masami Ito (Kyoto, Japan)
Nataša Jonoska (Tampa Bay, USA)
Juhani Karhumäki, Co-chair (Turku, Finland)
Wojciech Rytter (Warsaw, Poland)
Mikhail Volkov (Ekaterinburg, Russia)
Klaus Wagner (Würzburg, Germany)

Organizing Committee

Juhani Karhumäki (Co-chair)
Vesa Halava (Co-chair)
Elena Czeizler
Eugen Czeizler
Eero Lehtonen
Arto Lepistö
Alexander Okhotin
Petri Salmela

Referees

F. Ablayev	I. Honkala	F. Otto
R. Alur	J. Honkala	B. Palano
A. Angeleska	J. Hromkovič	G. Paun
M. Babenko	O. Ibarra	D. Perrin
A. Ballier	L. Ilie	G. Pighizzini
L. Bienvenu	M. Ito	J.-E. Pin
H.-J. Böckenhauer	E. Jeandel	J. Pirnot
P. Boldi	N. Jonoska	W. Plandowski
L. Breveglieri	J. Justin	R. Radicioni
H. Buhrman	C. Kapoutsis	B. Ravikumar
C. Calude	J. Karhumäki	C. Reutenauer
O. Carton	J. Kari	G. Richomme
J. Cassaigne	L. Kari	A.E. Romashchenko
A. Cherubini	D. Kephart	W. Rytter
C. Choffrut	R. Kralovic	K. Saari
M. Chrobak	D. Krob	P. Salmela
B. Courcelle	M. Kutrib	A. Salomaa
M. Crochemore	S. Lasota	K. Salomaa
F. D'Alessandro	M. Latteux	N. Santean
J. Dassow	A. Lepistö	M. Santha
F. Drewes	F. Levé	S. Seibert
E. Duchi	G. Lischke	V. Selivanov
B. Durand	K. Lodya	A. Shen
Z. Fülöp	M. Lohrey	A. Staninska
E. Formenti	S. Lombardy	J.-M. Talbot
A. Frid	V. Lonati	D. Tamascelli
P. Gastin	P. Madhusudan	N. Tran
M. Goldwurm	P. Massazza	S. Travers
P. Goralčík	B. Masson	R. Treinen
J. Gruska	C. Mereghetti	G. Vaszil
Y. Guesnet	T. Momke	M. Volkov
V. Halava	B. Nagy	K. Wagner
T. Harju	E. Ochmański	H.-C. Yen
M. Hirvensalo	A. Okhotin	S. Yu
M. Holzer	N. Ollinger	

Sponsoring Institutions

The Academy of Finland
The Finnish Cultural Foundation
Finnish Academy of Science and Letters / Vilho, Yrj and Kalle Visl Fund
Nokia Foundation
Turku University Foundation
Centro Hotel

Table of Contents

On First-Order Fragments for Words and Mazurkiewicz Traces

A Survey

Volker Diekert[1] and Manfred Kufleitner[2]

[1] Universität Stuttgart, FMI, Germany
diekert@fmi.uni-stuttgart.de
[2] Université Bordeaux 1, LaBRI, France
manfred.kufleitner@labri.fr

Abstract. We summarize several characterizations, inclusions, and separations on fragments of first-order logic over words and Mazurkiewicz traces. The results concerning Mazurkiewicz traces can be seen as generalizations of those for words. It turns out that over traces it is crucial, how easy concurrency can be expressed. Since there is no concurrency in words, this distinction does not occur there. In general, the possibility of expressing concurrency also increases the complexity of the satisfiability problem.

In the last section we prove an algebraic and a language theoretic characterization of the fragment $\Sigma_2[E]$ over traces. Over words the relation E is simply the order of the positions. The algebraic characterization yields decidability of the membership problem for this fragment. For words this result is well-known, but although our proof works in a more general setting it is quite simple and direct. An essential step in the proof consists of showing that every homomorphism from a free monoid to a finite aperiodic monoid M admits a factorization forest of finite height. We include a simple proof that the height is bounded by $3\,|M|$.

1   Introduction

The concept of partially commutative free monoids has first been considered by Cartier and Foata [1]. Later Keller and Mazurkiewicz used them as a model for concurrent systems and Mazurkiewicz established the notion of *trace* monoids for these structures [16,19,20]. Since then the elements of partially commutative monoids are called *Mazurkiewicz traces*. Many aspects of traces and trace languages have been researched, see *The Book of Traces* [7] for an overview.

Over words it has turned out that finite monoids are a powerful technique to refine the class of recognizable languages [9]. For fragments of first-order logic, in many cases it is a characterization in terms of algebra which leads to decidability of the membership problem. For example, on the algebraic side first-order logics as well as temporal logics corresponds to aperiodic monoids, see e.g. [12]. The probably most interesting fragment of them is given by the variety **DA**. It admits many different characterizations, which led to the title *Diamonds are Forever* in

T. Harju, J. Karhumäki, and A. Lepistö (Eds.): DLT 2007, LNCS 4588, pp. 1–19, 2007.

[30]. One of the purposes of this paper is to survey the situation over words and Mazurkiewicz traces.

Words can be seen as a special case of Mazurkiewicz traces and the corresponding results for words have been known before their generalizations to traces. Since over words we do not have any concurrency the situation is more complex for traces, and therefore not all word results remain valid for traces. It turns out that for traces the distinction between so-called dependence graphs and partial orders is rather crucial. Over words, both notions coincide.

The paper is organized as follows. In Section 2 we introduce Mazurkiewicz traces using a graph theoretic approach since this directly translates into the logic setting. After that we present further notions used in this paper which include the definition of fragments of first-order logic and temporal logic, some language operations, and the connections to finite monoids. In Section 3 we give several characterizations of languages whose syntactic monoid is aperiodic or in the variety **DA**. In a second part of this section we describe the alternation hierarchy of first-order logic using language operations. Section 4 contains some ideas and approaches revealing how concurrency increases the expressive power of logical fragments and in Section 5 we present some results showing that in general, concurrency also increases the complexity of the satisfiability problem.

Finally, in Section 6 we give a self-contained proof of a language theoretic and an algebraic characterization of the fragment Σ_2 over traces. The algebraic characterization yields decidability of the membership problem for this fragment. For words this result is well-known, but although our proof works in a more general setting it is quite simple and direct. A main tool in this proof are factorization forests. We give a simple and essentially self-contained proof for Simon's theorem on factorization forests in the special case of finite aperiodic monoids M. Our proof can be generalized to arbitrary monoids and still yields that the height of the factorization forests is bounded by $3|M|$. The previously published bound was $7|M|$, see [2]. After having completed our paper we learned that the bound $3|M|$ has been stated in the Technical Report [3], too.

2 Preliminaries

Words and Mazurkiewicz Traces

A *dependence alphabet* is a pair (Γ, D) where the alphabet Γ is a finite set (of actions) and the *dependence relation* $D \subseteq \Gamma \times \Gamma$ is reflexive and symmetric. The *independence relation* I is the complement of D. A *Mazurkiewicz trace* is an isomorphism class of a node-labeled directed acyclic graph $t = [V, E, \lambda]$, where V is a finite set of vertices labeled by $\lambda : V \to \Gamma$ and $E \subseteq (V \times V) \setminus \mathrm{id}_V$ is the edge relation such that for any two different vertices $x, y \in V$ we have either $(x, y) \in E$ or $(y, x) \in E$.

We call $[V, E, \lambda]$ a *dependence graph*. By $<$ we mean the transitive closure of E. We write $x \parallel y$ if $x \neq y$ and the vertices x and y are incomparable with respect to $<$. In this case we say that x and y are *independent* or *concurrent*. Node labeled graphs (V, E, λ) and (V', E', λ') are isomorphic if and only if the corresponding labeled *partial orders* $(V, <, \lambda)$ and $(V', <', \lambda')$ are isomorphic. The transitive reduction of a trace is called the *Hasse diagram*.

For $D = \Gamma \times \Gamma$ we obtain *words*. The vertices in words are linearly ordered and the relations E and $<$ are identical. Let $t_1 = [V_1, E_1, \lambda_1]$ and $t_2 = [V_2, E_2, \lambda_2]$ be traces. Then we define the concatenation of t_1 and t_2 to be $t_1 \cdot t_2 = [V, \leq, \lambda]$ where $V = V_1 \cup V_2$ is a disjoint union, $\lambda = \lambda_1 \cup \lambda_2$, and $E = E_1 \cup E_2 \cup \{(x, y) \in V_1 \times V_2 \mid (\lambda(x), \lambda(y)) \in D\}$. The set $\mathbb{M}$ of traces becomes a monoid with the empty trace $1 = (\emptyset, \emptyset, \emptyset)$ as unit. It is generated by Γ, where a letter a is viewed as a graph with a single vertex labeled by a. Thus, we obtain a canonical surjective homomorphism $\pi : \Gamma^* \to \mathbb{M}$. The effect of the mapping π can be made explicit as follows. We start with a word $w = a_1 \cdots a_n$ where all a_x are letters in Γ. Each x is viewed as an element in $\{1, \ldots, n\}$ with label $\lambda(x) = a_x$. We draw an arc from x to y if and only if both, $x < y$ and $(a_x, a_y) \in D$. This dependence graph is $\pi(w)$. Note that $\mathbb{M}$ is also canonically isomorphic to the quotient monoid $\Gamma^* / \{ab = ba \mid (a, b) \in I\}$. By abuse of notation we often identify a trace t and its word representatives $w \in \pi^{-1}(t)$.

Example 1. Let $(\Gamma, D) = a \text{—} b \text{—} c \text{—} d$ where self-loops are omitted. Consider the trace $t = acdbca$. We have $acdbca = cabadc$ in $\mathbb{M}$. The trace t has the following graphical presentations:

Hasse diagram: Dependence graph E: Partial order $<$:

In t, the node labeled with d is concurrent to all nodes labeled with a or b. □

There is a basic observation which holds for all $t \in \mathbb{M}$ and all vertices x, y of t:

$$(x, y) \in E \iff (x, y) \in E^+ \wedge (\lambda(x), \lambda(y)) \in D \tag{1}$$

$$(x, y) \in E^+ \iff \exists x_1 \cdots \exists x_{|\Gamma|} : \left\{ \begin{array}{l} x_{|\Gamma|} = y \wedge (x, x_1) \in E \wedge \\ \bigwedge_{1 \leq i < |\Gamma|} (x_i, x_{i+1}) \in E \cup \mathrm{id}_V \end{array} \right\} \tag{2}$$

This shows that traces can be either represented by their dependence graphs or as a partial order without losing any information. There are some standard notations we adopt here. By $\mathrm{alph}(t)$ we denote the *alphabet* of a trace t, i.e., the set of letters occurring as labels of some position. By $|t|$ we denote the *length* of a trace, i.e., the number of vertices of t. A *trace language* L is a subset of $\mathbb{M}$.

First-Order Logic and Temporal Logic

The syntax of first-order logic formulas $\mathrm{FO}[E]$ is built upon atomic formulas of type

$$\top, \quad \lambda(x) = a, \quad \text{and} \quad (x, y) \in E,$$

where $\top$ means *true*, x, y are variables and $a \in \Gamma$ is a letter. If φ, ψ are first-order formulas, then $\neg\varphi$, $\varphi \vee \psi$, $\exists x\, \varphi$ are first-order formulas, too. We use the usual shortcuts as $\bot = \neg\top$ meaning *false*, $\varphi \wedge \psi = \neg(\neg\varphi \vee \neg\psi)$, and $\forall x\, \varphi = \neg\exists x\, \neg\varphi$. Note that $x = y$ can be expressed by

$$\bigvee_{a \in \Gamma} (\lambda(x) = a \,\wedge\, \lambda(y) = a) \,\wedge\, (x,y) \notin E \,\wedge\, (y,x) \notin E$$

We let $\mathrm{FO}^m[E]$ be the set of all formulas with at most m different names for variables. There are completely analogous definitions for the first-order logic $\mathrm{FO}[<]$. The only difference is that instead of $(x,y) \in E$ we have an atomic predicate $x < y$.

Given $\varphi \in \mathrm{FO}[E] \cup \mathrm{FO}[<]$ the semantics is defined as usual [32]. In particular, if all free variables in φ belong to a set $\{x_1, \ldots, x_m\}$, then for all $t \in \mathbb{M}$ and all $x_1, \ldots, x_m \in t$ we write $t, x_1, \ldots, x_m \models \varphi$ if t satisfies $\varphi(x_1, \ldots, x_m)$. We identify formulas by semantic equivalence (over finite traces). Hence, if φ and ψ are formulas with m free variables, then we write $\varphi = \psi$ as soon as $t, x_1, \ldots, x_m \models (\varphi \leftrightarrow \psi)$ for all $t \in \mathbb{M}$ and all $x_1, \ldots, x_m \in t$. Due to (1) we have that $\mathrm{FO}^m[E]$ is a fragment of $\mathrm{FO}^m[<]$. A *first-order sentence* is a formula in $\mathrm{FO}[E]$ or $\mathrm{FO}[<]$ without free variables. For a first-order sentence φ we define $L(\varphi) = \{t \in \mathbb{M} \mid t \models \varphi\}$. A trace language $L \subseteq \mathbb{M}$ is called *first-order definable* if $L = L(\varphi)$ for some first-order sentence φ and we let $\mathrm{FO}(\mathbb{M}) = \{L(\varphi) \mid \varphi \in \mathrm{FO}[E]\}$. We do not write $\mathrm{FO}[E](\mathbb{M})$, because $\mathrm{FO}(\mathbb{M}) = \{L(\varphi) \mid \varphi \in \mathrm{FO}[<]\}$ as well, due to (2). So, in first-order it is not necessary to distinguish between E and $<$. However, for subclasses of FO we need this distinction. We define the following classes for $E' = E$ and $E' = <$, respectively.

The fragment $\Sigma_n[E']$ contains all formulas in prenex normal form with n blocks of alternating quantifiers starting with a block of existential quantifiers whereas in $\Pi_n[E']$ formulas start with a block of universal quantifiers. According to our convention to identify equivalent formulas, it makes sense to write e.g. $\varphi \in \Sigma_n[E'] \Leftrightarrow \neg\varphi \in \Pi_n[E']$. Although in general the transitive closure of binary relations is not expressible in first-order logic, we have $\bigcup_{0 \leq n} \Sigma_n[E] = \mathrm{FO}[<]$ due to the following observation obtained from (1) and (2):

$$\Sigma_n[E] \subseteq \Sigma_n[<] \subseteq \Sigma_{n+1}[E]$$

For $E' = E$ and $E' = <$ we define the following language classes:

- $\mathrm{FO}^m[E'](\mathbb{M}) = \{L(\varphi) \mid \varphi \in \mathrm{FO}^m[E']\}$.
- $\Sigma_n[E'](\mathbb{M}) = \{L(\varphi) \mid \varphi \in \Sigma_n[E']\}$.
- $\Pi_n[E'](\mathbb{M}) = \{L(\varphi) \mid \varphi \in \Pi_n[E']\}$.
- $\Delta_n[E'](\mathbb{M}) = \Sigma_n[E'](\mathbb{M}) \cap \Pi_n[E'](\mathbb{M})$.

Now, $\mathrm{FO}^m[E'](\mathbb{M})$ and $\Delta_n[E'](\mathbb{M})$ are Boolean algebras and $\Sigma_n[E'](\mathbb{M})$ and $\Pi_n[E'](\mathbb{M})$ are closed under union and intersection.

Local temporal logic formulas are defined by first-order formulas having at most one free variable. In this paper we focus on unary operators and local semantics. In temporal logic we write $a(x)$ for the atomic formula $\lambda(x) = a$.

Inductively, we define $\mathsf{SF}\varphi(x)$ (*Strict Future*), $\mathsf{SP}\varphi(x)$ (*Strict Past*), $\mathsf{M}\varphi(x)$ (*soMewhere*), $\mathsf{Eco}\varphi(x)$ (*Exists concurrently*) as follows.

$$\mathsf{SF}\varphi(x) = \exists y : x < y \land \varphi(y)$$
$$\mathsf{SP}\varphi(x) = \exists y : y < x \land \varphi(y)$$
$$\mathsf{M}\varphi(x) = \exists y : \varphi(y)$$
$$\mathsf{Eco}\varphi(x) = \exists y : x \parallel y \land \varphi(y)$$

It is common to write φ instead of $\varphi(x)$. Let $\mathcal{C}$ be a subset of temporal operators from the set above, then $\mathrm{TL}[\mathcal{C}]$ means the formulas where all operators are from $\mathcal{C}$. In order to pass to languages we would like to define $L(\varphi) \subseteq \mathbb{M}$, even if φ has a free variable. There is however no canonical choice, so we use an existential variant; and we define here:

$$L_\exists(\varphi) = \{\, t \in \mathbb{M} \mid \exists x \in t : t, x \models \varphi \,\} = L(\mathsf{M}\varphi).$$

Define $\mathrm{TL}[\mathcal{C}](\mathbb{M})$ as the Boolean closure of languages defined by $L_\exists(\varphi)$ with $\varphi \in \mathrm{TL}[\mathcal{C}]$.

Languages and Language Operations

We now define some operations on classes of languages that are used to describe the expressive power of logical fragments. Let $\mathcal{V}$ be a class of trace languages. By $\mathbb{B}(\mathcal{V})$ we denote the Boolean closure of $\mathcal{V}$. A language L is a *monomial* over $\mathcal{V}$ of *degree* m if there exist $n \leq m$, $a_i \in \Gamma$ and $L_i \in \mathcal{V}$ with

$$L = L_0 a_1 L_1 \cdots a_n L_n$$

Note that the degree of a monomial is not unique. A finite union of monomials over $\mathcal{V}$ is called a *polynomial* over $\mathcal{V}$. A polynomial has *degree* m if it can be written as a union of monomials of degree m. The class of all polynomials over $\mathcal{V}$ is denoted by $\mathrm{Pol}(\mathcal{V})$. The class $\mathrm{Pol}(\mathcal{V})$ is often called the *polynomial closure* or the *closure under product and union* of the class $\mathcal{V}$. By $\mathrm{co\text{-}Pol}(\mathcal{V})$ we denote the class of languages L such that $\mathbb{M} \setminus L \in \mathrm{Pol}(\mathcal{V})$. If we speak of monomials and polynomials without referring to some class $\mathcal{V}$ then we mean monomials and polynomials over $\mathcal{A} = \{\, A^* \mid A \subseteq \Gamma \,\}$, respectively. In particular, $\mathrm{Pol} = \mathrm{Pol}(\mathcal{A})$ and $\mathrm{co\text{-}Pol} = \mathrm{co\text{-}Pol}(\mathcal{A})$. For example, if $A, B \subseteq \Gamma$ then $A^* B^* \in \mathrm{Pol}$ since

$$A^* B^* = A^* \cup \bigcup_{b \in B} A^* b B^*$$

The class of *star-free* languages SF is the closure of the empty set under Boolean operations and polynomials. If $\mathcal{V}$ is a class of word languages then $\mathrm{UPol}(\mathcal{V})$ consists of the word languages that are disjoint finite unions of unambiguous monomials. A monomial $L_0 a_1 L_1 \cdots a_n L_n$ is *unambiguous* if every $w \in L_0 a_1 L_1 \cdots a_n L_n$ has a unique factorization $w = w_0 a_0 w_1 \cdots a_n w_n$ with $w_i \in L_i$. A similar language operation is $\mathbb{B}\text{-UPol}$. By $\mathbb{B}\text{-UPol}(\mathcal{V})$ we denote the closure of $\mathcal{V}$ under Boolean operations and unambiguous products. An *unambiguous product* is an unambiguous monomial of the form $L_0 a_1 L_1$. We set UPol = UPol($\mathcal{A}$) and $\mathbb{B}$-UPol = $\mathbb{B}$-UPol($\mathcal{A}$). For example, the word language $\{a,b\}^* ab \{a,b\}^*$ is in UPol since

$$\{a,b\}^* ab \{a,b\}^* = \{b\}^* a \{a\}^* b \{a,b\}^*$$

whereas the polynomials $\{a,b\}^* aa \{a,b\}^*$ and $\{a,b,c\}^* ab \{a,b,c\}^*$ are not in UPol. See [23] for more information on the language operations UPol($\mathcal{V}$) and $\mathbb{B}$-UPol($\mathcal{V}$). The operation $\mathbb{B}$-UPol($\mathcal{V}$) has been extended to classes of trace languages [18].

Algebraic Descriptions

Finite monoids are an elementary tool in the description and classification of recognizable languages. Remember that a *monoid M* is a set equipped with an associative binary operation and a neutral element 1. An *ordered monoid* is a monoid M equipped with a partial order relation $\leq$ such that $a \leq b$ implies $ca \leq cb$ and $ac \leq bc$ for all $a, b, c \in M$. Every monoid M forms an ordered monoid $(M, =)$. For homomorphisms $h : (M, \leq) \to (N, \preceq)$ between ordered monoids we additionally require that $a \leq b$ implies $h(a) \preceq h(b)$ for all $a, b \in M$. If a is an element of an ordered monoid $(M, \leq)$ then we define $\lfloor a \rfloor = \{b \in M \mid b \leq a\}$. More details on ordered monoids can be found in [22]. An element e of a monoid is called *idempotent* if $e^2 = e$. For every finite monoid M there exists a number $\omega \in \mathbb{N}$ such that a^ω is idempotent for every $a \in M$. The element a^ω is the unique idempotent generated by a. Therefore we use the ω-notation also if the finite monoid M is not fixed to denote the idempotent generated by some element. A language L is called *recognizable* if $L = h^{-1}h(L)$ for some homomorphism $h : \mathbb{M} \to M$, where M is a finite monoid. In this case we say that M recognizes L. The minimal monoid recognizing L is its syntactic monoid. For a language $L \subseteq \mathbb{M}$ we define its *syntactic pre-order* $\leq_L$ by

$$s \leq_L t \iff (\forall p, q \in \mathbb{M} : ptq \in L \Rightarrow psq \in L)$$

and its *syntactic congruence* $\sim_L$ by $s \sim_L t$ if and only if $s \leq_L t$ and $t \leq_L s$. The natural homomorphism $\mu_L : \mathbb{M} \to \mathbb{M}/\sim_L : t \mapsto [t]_{\sim_L}$ is called the *syntactic homomorphism* of L and the monoid $M(L) = \mathbb{M}/\sim_L$ is called the *syntactic monoid* of L. A language L is recognizable if and only if $M(L)$ is finite. The *syntactic pre-order* $\leq_L$ of L induces a partial order on $M(L)$ such that $(M(L), \leq_L)$ forms an ordered monoid. It is called the *syntactic ordered monoid* of L. For $\mu_L : (\mathbb{M}, =) \to (M(L), \leq_L)$ we have

$$L = \bigcup_{a \in \mu_L(L)} \mu_L^{-1}(\lfloor a \rfloor)$$

A class of recognizable languages $\mathcal{V}$ is a *language variety* if it is closed under Boolean operations, left and right quotients, and inverse homomorphic images. A class of finite monoids $\mathbf{V}$ is called a *variety* if it is closed under taking finite products, submonoids and homomorphic images [21]. Eilenberg has shown that language varieties of word languages and varieties of finite monoids are in a one-to-one correspondence [9]. Ordered monoids are designed to serve as a similar tool for classes of languages which are not closed under complementation. Syntactic (ordered) monoids play a crucial role in these correspondences. This yields to the observation that properties of classes of languages can be expressed in terms of properties of syntactic monoids. In a lot of cases, a description of the variety generated by the syntactic monoids $M(L)$ for $L \in \mathcal{V}$ yields decidability of the membership problem for this language variety $\mathcal{V}$. An important tool to describe the structure of monoids are *Green's relations*. For $a, b \in M$ we define

$$a \mathrel{\mathcal{J}} b \iff MaM = MbM \qquad a \leq_{\mathcal{J}} b \iff MaM \subseteq MbM$$
$$a \mathrel{\mathcal{R}} b \iff aM = bM \qquad\quad a \leq_{\mathcal{R}} b \iff aM \subseteq bM$$
$$a \mathrel{\mathcal{L}} b \iff Ma = Mb \qquad\quad a \leq_{\mathcal{L}} b \iff Ma \subseteq Mb$$
$$a \mathrel{\mathcal{H}} b \iff a \mathrel{\mathcal{R}} b \text{ and } a \mathrel{\mathcal{L}} b$$

Note that $\mathcal{J}$, $\mathcal{R}$, $\mathcal{L}$, and $\mathcal{H}$ are equivalence relations, whereas $\leq_{\mathcal{J}}$, $\leq_{\mathcal{R}}$, and $\leq_{\mathcal{L}}$ are pre-orders. Equations are another tool to describe properties of finite monoids. Let Ω be a finite set and let $v, w \in \Omega^*$. A monoid M *satisfies* the equation $v = w$, if for all homomorphisms $h : \Omega^* \to M$ we have $h(v) = h(w)$. For example, commutative monoids satisfy $xy = yx$. We also allow the ω-operator in equations and define $h(v^\omega) = h(v)^\omega$. By $[\![\, v = w \,]\!]$ we denote the class of finite monoids satisfying $v = w$. The class of all monoids satisfying an equation forms a variety. We define the variety of *aperiodic monoids* $\mathbf{A}$ by $\mathbf{A} = [\![\, x^\omega = x^{\omega+1} \,]\!]$. Another important variety is $\mathbf{DA} = [\![\, (xy)^\omega x(xy)^\omega = (xy)^\omega \,]\!]$. By mapping y to 1 we see that $\mathbf{DA} \subseteq \mathbf{A}$. In the following we summarize some basic properties of these varieties.

Proposition 1 ([21]). *For every finite monoid M the following are equivalent:*

1. $M \in \mathbf{A}$.
2. M is $\mathcal{H}$-trivial, i.e., every $\mathcal{H}$-class contains exactly one element.
3. All groups in M are trivial, i.e., if a subsemigroup of M is a group then it contains only one element.

Proposition 2 ([17]). *For every finite monoid M the following are equivalent:*

1. $M \in \mathbf{DA}$.
2. $M \in [\![\, (xy)^\omega y(xy)^\omega = (xy)^\omega \,]\!]$.
3. $M \in [\![\, (xyz)^\omega y(xyz)^\omega = (xyz)^\omega \,]\!]$.
4. $M \in \mathbf{A}$ and $\forall a, b, e \in M : e = e^2$ and $a \mathrel{\mathcal{J}} b \mathrel{\mathcal{J}} e$ implies $ab \mathrel{\mathcal{J}} e$.
5. $\forall e, f \in M : e = e^2$ and $e \mathrel{\mathcal{J}} f$ implies $f = f^2$.

3   Expressivity Results

In the following two theorems we summarize characterizations of trace languages whose syntactic monoid is aperiodic or in $\mathbf{DA}$. Note that this includes the special case of word languages. The results are using some temporal operators which we did not introduce yet. The operator X is an existential next-operator, i.e., $\mathsf{X}\varphi$ is true at a position x if at some minimal position in the future of x the formula φ holds. Over words, this position is unique. The until-operator U is a binary operator. The formula $\varphi\,\mathsf{U}\,\psi$ is true at a position x if there exists a position $y \geq x$ at which ψ holds and all positions between x and y (i.e., all positions from the current position x "*until*" y) satisfy φ. The formula $\mathsf{X}_a\,\varphi$ for $a \in \Gamma$ is true at a position x if there exists a position $y > x$ labeled by a and if at the first of these a-labeled positions in the future of x the formula φ holds. The operator Y_a is left-right symmetric to X_a. With $\mathrm{TL}[\mathsf{X}_a,\mathsf{Y}_a]$ we mean that we have X_a and Y_a operators for every $a \in \Gamma$. The definition of the languages generated by formulas in $\mathrm{TL}[\mathsf{X},\mathsf{U}]$ and $\mathrm{TL}[\mathsf{X}_a,\mathsf{Y}_a]$ is slightly different from the one that we propose above for unary temporal logic.

Theorem 1 ([5,8,14,15]). *Let $L \subseteq \mathbb{M}$. Then the following are equivalent:*

1. $M(L) \in \mathbf{A}$.
2. $L \in \mathrm{SF}$.
3. L is expressible in $\mathrm{FO}^3[<]$.
4. L is expressible in $\mathrm{FO}[<]$.
5. L is expressible in $\mathrm{FO}[E]$.
6. L is expressible in $\mathrm{TL}[\mathsf{X},\mathsf{U}]$.

Theorem 2 ([6,18]). *Let $L \subseteq \mathbb{M}$. Then the following are equivalent:*

1. $M(L) \in \mathbf{DA}$.
2. $L \in \mathrm{Pol} \cap \mathrm{co\text{-}Pol}$.
3. $L \in \mathbb{B}\text{-}\mathrm{UPol}$.
4. L is expressible in $\mathrm{FO}^2[E]$.
5. L is expressible in $\Delta_2[E]$.
6. L is expressible in $\mathrm{TL}[\mathsf{X}_a,\mathsf{Y}_a]$.
7. L is expressible in $\mathrm{TL}[\mathsf{SF},\mathsf{SP}]$.
8. L is expressible in $\mathrm{TL}[\mathsf{SF},\mathsf{SP},\mathsf{M}]$.

For word languages $L \subseteq \Gamma^*$ we additionally have $M(L) \in \mathbf{DA}$ if and only if $L \in \mathrm{UPol}$, see [25]. In particular, UPol is closed under complementation. Since membership in both varieties $\mathbf{A}$ and $\mathbf{DA}$ is decidable, membership for all characterizations in Theorem 1 and Theorem 2 is decidable.

Theorem 3 ([6,11]). *Let $L \subseteq \mathbb{M}$. Then the following are equivalent:*

1. L is expressible in $\mathrm{FO}^2[<]$.
2. L is expressible in $\mathrm{TL}[\mathsf{SF},\mathsf{SP},\mathsf{Eco}]$.

The following theorem gives a language theoretic characterization of the alternation hierarchy for first-order logic over words. It is the connection to the *Straubing-Thérien hierarchy* in which one describes classes of word languages by alternating Boolean closure and polynomial closure starting with the empty set. By definition, the limit of this process is the class of star-free languages. In the following we use $\mathbb{B}\Sigma_n$ as a shortcut for $\mathbb{B}\big(\Sigma_n[<](\Gamma^*)\big)$. Note that $\mathbb{B}\Sigma_n = \mathbb{B}\Pi_n$.

Theorem 4 ([24]). *Over words we have the following*

1. $\Sigma_0[<](\Gamma^*) = \mathbb{B}(\Sigma_0) = \{\emptyset, \Gamma^*\}$.
2. $\Sigma_{n+1}[<](\Gamma^*) = \text{Pol}(\mathbb{B}\Sigma_n)$.
3. $\Pi_{n+1}[<](\Gamma^*) = \text{co-Pol}(\mathbb{B}\Sigma_n)$.
4. $\Delta_{n+1}[<](\Gamma^*) = \text{UPol}(\mathbb{B}\Sigma_n)$.

A basis for the last part of this theorem is the more general fact that $\text{UPol}(\mathcal{V}) = \text{Pol}(\mathcal{V}) \cap \text{co-Pol}(\mathcal{V})$ if $\mathcal{V}$ is a variety of word languages. This follows from an algebraic description in terms of Mal'cev products [24]. Another language theoretic characterization of Σ_2 is $\Sigma_2[<](\Gamma^*) = \text{Pol}$. We give a detailed proof of this characterization in the more general setting of traces over dependence graphs in Section 6. It is well-known that the alternation hierarchy for first-order logic is strict [29], i.e.:

- For $n \geq 1$ the classes $\Sigma_n[<](\Gamma^*)$ and $\Pi_n[<](\Gamma^*)$ are incomparable.
- For $n \geq 1$ the class $\Sigma_n[<](\Gamma^*)$ is strictly contained in $\Delta_{n+1}[<](\Gamma^*)$.
- For $n \geq 1$ the class $\Delta_n[<](\Gamma^*)$ is strictly contained in the class $\Sigma_n[<](\Gamma^*)$.

Recently, Weis and Immerman have shown that the alternation hierarchy for FO^2 on words is strict [33]. In the next section we consider the alternation hierarchy for first-order logic over traces. The distinction between partial orders $<$ and dependence graphs E turns out to be crucial. Using (2) we can express $<$ in terms of E, but this requires variables and it requires quantifiers, but in FO^2 the number of variables is restricted whereas in Σ_n the number of quantifier alternations is bounded.

4   Separation Results

We start this section with a simple observation. Let $(\Gamma, D) = a\,—\,b\,—\,c$ and consider the traces $x = abc$ and $y = b$. Then for all $n \in \mathbb{N}$ the trace $(xy)^n$ is a sequence in which all positions are totally ordered whereas in the trace $(xy)^n x (xy)^n$ we have a factor xx whose Hasse diagram is

$$\begin{array}{ccc} & a \longrightarrow b \longrightarrow c \\ \nearrow \nearrow \nearrow \\ a \longrightarrow b \longrightarrow c \end{array}$$

In particular, in xx there exist two concurrent actions. Consider the formula $\varphi = \exists z_1 \exists z_2 \colon z_1 \| z_2 \in \text{FO}^2[<] \cap \Sigma_1[<]$ where $z_1 \| z_2$ is a macro for $\neg(z_1 = z_2 \vee z_1 < z_2 \vee z_2 < z_1)$. Then for all $n \geq 1$ we have

$$(xy)^n x (xy)^n \models \varphi \qquad \text{and} \qquad (xy)^n \not\models \varphi$$

This shows that the syntactic monoid of the trace language $L(\varphi)$ is not in $\mathbf{DA} = [\![\,(xy)^\omega x (xy)^\omega = (xy)^\omega\,]\!]$. Now, whenever the dependence relation is not transitive we find some letters a, b and c with the dependencies $a\,—\,b\,—\,c$. On the other hand, if the dependence relation is transitive then the partial order $<$ and the edge relation E of the dependence graph are identical. Together with $\Sigma_1[<](\mathbb{M}) \subseteq \Delta_2[<](\mathbb{M})$ we obtain the following theorem.

Theorem 5. *Let* $\mathbb{M}$ *be the trace monoid generated by the dependence alphabet* (Γ, D). *The following are equivalent:*

1. *The dependence relation* D *is transitive.*
2. *For every trace, the relations* $<$ *and* E *are identical.*
3. $\mathrm{FO}^2[E](\mathbb{M}) = \mathrm{FO}^2[<](\mathbb{M})$.
4. $\Delta_2[E](\mathbb{M}) = \Delta_2[<](\mathbb{M})$.

The main technique in the proofs of the following theorems are Ehrenfeucht-Fraïssé games, see e.g. [29,32]. Let $\mathbb{M}$ be a trace monoid over the following dependence alphabet:

$$(\Gamma, D) \ = \ \begin{array}{c} a \\ f \diagdown \ | \diagup b \\ | \quad \# \quad | \\ e \diagup \ | \diagdown c \\ d \end{array}$$

Theorem 6 ([6]). *For the above trace monoid* $\mathbb{M}$ *the trace language*

$$L = \{\, t \in \mathbb{M} \mid \exists x, y, z \in t \colon (x \parallel y \wedge y \parallel z \wedge z \parallel x) \,\}$$

consisting of all traces with three pairwise concurrent actions is expressible in $\Sigma_1[<]$ *but not in* $\mathrm{FO}^2[<]$.

The main idea in the proof of this theorem is to consider the traces $\#q\#$ and $p = \#r^n\#$ in which every action has the same set of concurrent actions, but in p there are at most two pairwise independent actions. The Hasse diagram of $\#q\#$ is:

$$\begin{array}{c} a \longrightarrow b \\ \diagup \quad \diagdown \\ \# \longrightarrow c \diagdown \quad \diagup d \longrightarrow \# \\ \diagdown \quad \diagup \\ e \longrightarrow f \end{array}$$

and the Hasse diagram of the trace $p = \#r^n\#$ is sketched below:

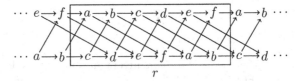

For every formula $\varphi \in \mathrm{FO}^2[<]$ we can find a sufficiently large number n such that the two traces $p^n q p^n \in L$ and $p^{2n} \notin L$ either both are models of φ or none of them is a model. Therefore, $L \notin \mathrm{FO}^2[<](\mathbb{M})$. The previous two results can be summarized as follows: "two concurrent actions" is in $\mathrm{FO}^2[<]$ and $\Delta_2[<]$ but not in $\mathrm{FO}^2[E] = \Delta_2[E]$ and "three concurrent actions" is in $\Delta_2[<]$ but not in $\mathrm{FO}^2[<]$. The next theorem implies that in general $\mathrm{FO}^2[<]$ and $\Delta_2[<]$ are incomparable. It is open whether membership is decidable for $\mathrm{FO}^2[<](\mathbb{M})$ or $\Delta_2[<](\mathbb{M})$. Also note that the following result is rather unexpected since $\mathrm{FO}^2[<] \subseteq \mathrm{FO}[<] = \bigcup_n \Sigma_n[<]$.

Theorem 7 ([6]). *For every $n \geq 0$ there exists a trace monoid $\mathbb{M}$ and a trace language $L \subseteq \mathbb{M}$ such that $L \in \mathrm{FO}^2[<](\mathbb{M})$ but $L \notin \Sigma_n[<](\mathbb{M})$.*

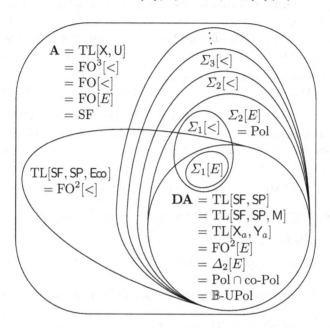

Remember $\Sigma_n[<] \subseteq \Sigma_{n+1}[E] \subseteq \Sigma_{n+1}[<]$. We already know from the word case that the inclusion $\Sigma_n[<] \subseteq \Sigma_{n+1}[E]$ is strict. The following theorem says that in general the second inclusion is also strict and that the fragments $\Pi_{n-1}[<]$ and $\Sigma_n[E]$ are incomparable.

Theorem 8 ([6]). *Let $\mathbb{M}$ be the trace monoid generated by the dependence alphabet (Γ, D). The following are equivalent:*

1. *The dependence relation D is transitive.*
2. $\exists n \geq 1 : \Sigma_n[E](\mathbb{M}) = \Sigma_n[<](\mathbb{M})$.
3. $\exists n \geq 2 : \Pi_{n-1}[<](\mathbb{M}) \subseteq \Sigma_n[E](\mathbb{M})$.

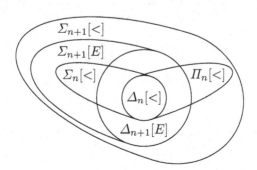

5   Complexity of Satisfiability

The possibility of being able to speak about concurrency increases the expressivity of most first-order fragments. In this section we will see how it also increases the complexity of the satisfiability problem. The (uniform) *satisfiability problem* (SAT) for some class of logical formulas $\mathcal{C}$ is the following:

Input: A dependence alphabet (Γ, D) and a formula $\varphi \in \mathcal{C}$.
Question: Does there exists $t \in \mathbb{M} = \mathbb{M}(\Gamma, D)$ such that $t \in L(\varphi)$?

and the *non-uniform satisfiability problem* for $\mathcal{C}$ over a dependence alphabet (Γ, D) is the satisfiability problem where the dependence alphabet (Γ, D) is fixed and not part of the input:

Input: A formula $\varphi \in \mathcal{C}$.
Question: Does there exists $t \in \mathbb{M}$ such that $t \in L(\varphi)$?

We summarize some complexity results in the following theorem.

Theorem 9 ([6,10,11,13,27,28])

1. *SAT for temporal logics is* PSPACE-*complete.*
2. *SAT for* FO[<] *is not elementary.*
3. *The non-uniform satisfiability problem for* TL[X, F] *over* $\{a, \overline{a}\}^*$ *is* PSPACE-*hard.*
4. *SAT for* TL[SF, SP, M] *is* NP-*complete.*
5. *The non-uniform satisfiability problem for* TL[SF, SP, Eco] *over some dependence alphabet is* PSPACE-*hard. In fact, non-uniform satisfiability for the stutter-invariant fragment* TL[F, Eco] *is already* PSPACE-*hard.*
6. *SAT for* $FO^2[E]$ *is in* NEXPTIME.
7. *The satisfiability problem for* $FO^2[<]$ *is in* EXPSPACE *and* NEXPTIME-*hard. In fact, satisfiability for* $FO^2[||]$ *in which* $||$ *is the only binary relation is already* NEXPTIME-*hard.*

The parts "*4.*" and "*5.*" in Theorem 9 show that allowing the Eco operator increases the complexity of the satisfiability problem (unless NP = PSPACE). Part "*4.*" is proved by giving a *small model property* for TL[SF, SP, M], i.e., if there exists a model then there also exists a model whose size is polynomially bounded. For part "*5.*" a reduction of "*3.*" is used. In the following we sketch the idea of how to simulate the X-operator using the Eco-operator over the following *independence* alphabet:

$$(\Gamma, I) \quad = \quad \begin{array}{c} a \\ e \overset{\overline{a}}{\diagup \diagdown} b \\ \diagdown \quad \diagup \\ d \,\text{---}\, c \end{array}$$

For a word $w = a_1 \cdots a_n \in \{a, \overline{a}\}^+$ we define a trace $\widetilde{w} = a_1(bcde) \cdots a_n(bcde) \in \mathbb{M} = \mathbb{M}(\Gamma, I)$.

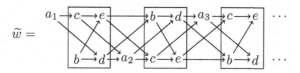

$\widetilde{w} =$

For the trace $\widetilde{w}$ we can use Eco to simulate X on the positions with label $a, \overline{a}$. The transformation of X ψ is given by

$$\widetilde{X\psi} = \text{Eco}\,(b \wedge \text{Eco}\,(c \wedge \text{Eco}\,(d \wedge \text{Eco}\,(e \wedge \text{Eco}\,((a \vee \overline{a}) \wedge \widetilde{\psi})))))$$

It is easy to verify that $\widetilde{X\psi}$ indeed reaches the next a or $\overline{a}$ position.

6   The Fragment $\Sigma_2[E]$

In this section we give a self-contained proof of the following theorem. An important tool in the proof are factorization forests.

Theorem 10. *Let* $L \subseteq \mathbb{M} = \mathbb{M}(\Gamma, I)$ *be a recognizable trace language and let* $\mu : \mathbb{M} \to (M(L), \leq) : t \mapsto [t]$ *be the syntactic homomorphism onto its syntactic ordered monoid. The following are equivalent:*

1. *For all* $e, s \in \mathbb{M}$: $[e] = [e^2]$ *and* $\text{alph}(s) \subseteq \text{alph}(e)$ *implies* $[ese] \leq [e]$.
2. L *is a polynomial.*
3. L *is expressible in* $\Sigma_2[E]$.

The syntactic ordered monoid of a recognizable trace language (given in any reasonable presentation) is effectively computable. Since property "1." in Theorem 10 can be effectively verified we obtain the following corollary.

Corollary 1. *It is decidable if* $L \subseteq \mathbb{M}$ *is definable in* $\Sigma_2[E]$.

6.1   Factorization Forests

Let M be a finite monoid. A *factorization forest* of a homomorphism $\varphi : \Gamma^* \to M$ is a function d which maps every word w with length $|w| \geq 2$ to a factorization $d(w) = (w_1, \ldots, w_n)$ of $w = w_1 \cdots w_n$ such that $n \geq 2$ and w_i is not empty for all $i \in \{1, \ldots, n\}$ and such that $n \geq 3$ implies that $\varphi(w_1) = \ldots = \varphi(w_n)$ is idempotent in M. The *height* h of a word w is defined as

$$h(w) = \begin{cases} 0 & \text{if } |w| \leq 1 \\ 1 + \max\{\,h(w_1), \ldots, h(w_n)\,\} & \text{if } d(w) = (w_1, \ldots, w_n) \end{cases}$$

We call the tree defined by the "branching" d for the word w the *factorization tree* of w. The height $h(w)$ is the height of this tree. The *height* of d is defined as $\sup\{\,h(w) \mid w \in \Gamma^*\,\}$. A famous theorem of Simon says that every homomorphism $\varphi : \Gamma^* \to M$ has a factorization forest of height $\leq 9|M|$, see [26]. By generalizing techniques of [2] we can improve this bound to $3|M|$. Using another approach, this bound has been shown independently in [3]. Below we present a simple proof of this fact in the special case of aperiodic monoids. The proof requires only basic facts from the theory of finite semigroups such as:

– The intersection of an $\mathcal{R}$-class and an $\mathcal{L}$-class within the same $\mathcal{J}$-class yields a unique $\mathcal{H}$-class within that $\mathcal{J}$-class.
– $x \leq_{\mathcal{L}} y$ and $x \mathcal{J} y$ implies $x \mathcal{L} y$; $x \leq_{\mathcal{R}} y$ and $x \mathcal{J} y$ implies $x \mathcal{R} y$.
– In aperiodic monoids every $\mathcal{H}$-class consists of only one element.

Theorem 11. *Let M be a finite aperiodic monoid. Every homomorphism $\varphi : \Gamma^* \to M : w \mapsto [w]$ has a factorization forest of height $< 3\,|M|$.*

Proof. We show that for every $w \in \Gamma^*$ there exists a factorization tree of height $h(w) < 3\,|\{\, x \in M \mid [w] \leq_{\mathcal{J}} x \,\}|$. The $\mathcal{J}$-class of 1 in aperiodic monoids is trivial. Let $w \in \Gamma^*$ with $|w| \geq 2$. If $[w] = 1$ then for all $b \in \mathrm{alph}(w)$ we have $[b] = 1$. Hence $d(w) = (b_1, \ldots, b_n)$ yields a factorization tree of height 1 for $w = b_1 \ldots b_n$. Now let $[w] <_{\mathcal{J}} 1$. Then w has a unique factorization

$$w = w_0 a_1 w_1 \cdots a_m w_m$$

with $a_i \in \Gamma$ and $w_i \in \Gamma^*$ satisfying the following two conditions:

$$\forall\, 1 \leq i \leq m \colon [a_i w_i] \, \mathcal{J} \, [w] \quad \text{and} \quad \forall\, 0 \leq i \leq m \colon [w] <_{\mathcal{J}} [w_i]$$

Let $w_i' = a_i w_i$ for $1 \leq i \leq m$. For each $1 \leq i < m$ define a pair (L_i, R_i) where L_i is the $\mathcal{L}$-class of $[w_i']$ and R_i is the $\mathcal{R}$-class of $[w_{i+1}']$. Every such pair represents an $\mathcal{H}$-class within the $\mathcal{J}$-class of $[w]$. Therefore, the number of different such pairs does not exceed $|\{\, x \mid [w] \, \mathcal{J} \, x \,\}|$. For the above factorization of w we perform an induction on the cardinality of the set $\{\, (L_i, R_i) \mid 1 \leq i < m \,\}$ to show that w has a factorization tree of height

$$h(w) \; < \; 3\,|\{\, (L_i, R_i) \mid 1 \leq i < m \,\}| \; + \; 3\,|\{\, x \mid [w] <_{\mathcal{J}} x \,\}|$$

Note that the number on the right-hand side of this inequality does not exceed $3\,|\{\, x \in M \mid [w] \leq_{\mathcal{J}} x \,\}|$. If every pair (L, R) occurs at most twice then we have $m - 1 \leq 2\,|\{\, (L_i, R_i) \mid 1 \leq i < m \,\}|$. We define a factorization tree for w by $d(w) = (w_0 w_1', w_2' \cdots w_m')$, $d(w_0 w_1') = (w_0, w_1')$, $d(w_i' \cdots w_m') = (w_i', w_{i+1}' \cdots w_m')$ for $2 \leq i < m$ and $d(w_i') = (a_i, w_i)$ for $1 \leq i \leq m$. Since $[w] <_{\mathcal{J}} [w_i]$, by induction every w_i has a factorization tree of height $h(w_i) \; < \; 3\,|\{\, x \mid [w_i] \leq_{\mathcal{J}} x \,\}| \; \leq \; 3\,|\{\, x \mid [w] <_{\mathcal{J}} x \,\}|$. This yields:

$$\begin{aligned} h(w) \; &< \; m + 3\,|\{\, x \mid [w] <_{\mathcal{J}} x \,\}| \\ &\leq \; 3\,|\{\, (L_i, R_i) \mid 1 \leq i < m \,\}| + 3\,|\{\, x \mid [w] <_{\mathcal{J}} x \,\}| \end{aligned}$$

Note that the height might decrease if some of the w_i are empty. Now suppose that there exists a pair $(L, R) \in \{\, (L_i, R_i) \mid 1 \leq i < m \,\}$ occurring (at least) three times. Let $i_0 < \cdots < i_k$ be the sequence of all positions with $(L, R) = (L_{i_j}, R_{i_j})$. Let $\widehat{w_j} = w_{i_{j-1}+1}' \cdots w_{i_j}'$ for $1 \leq j \leq k$. For all $1 \leq j \leq \ell \leq k$ we have

– $[\widehat{w_j} \cdots \widehat{w_\ell}] \leq_{\mathcal{L}} [w_{i_\ell}'] \, \mathcal{L} \, [w_{i_0}']$.
– $[\widehat{w_j} \cdots \widehat{w_\ell}] \leq_{\mathcal{R}} [w_{i_{j-1}+1}'] \, \mathcal{R} \, [w_{i_0+1}']$.
– $[w_{i_\ell}'] \leq_{\mathcal{J}} [\widehat{w_j} \cdots \widehat{w_\ell}] \leq_{\mathcal{J}} [w] \, \mathcal{J} \, [w_{i_\ell}'] \, \mathcal{J} \, [w_{i_0+1}']$ by assumption on the factorization.

Thus for all $1 \leq j \leq \ell \leq k$ and $1 \leq j' \leq \ell' \leq k$ we get

- $[\widehat{w_j} \cdots \widehat{w_\ell}] \; \mathcal{L} \; [w'_{i_1}] \; \mathcal{L} \; [\widehat{w_{j'}} \cdots \widehat{w_{\ell'}}]$ and
- $[\widehat{w_j} \cdots \widehat{w_\ell}] \; \mathcal{R} \; [w'_{i_1+1}] \; \mathcal{R} \; [\widehat{w_{j'}} \cdots \widehat{w_{\ell'}}]$ and therefore
- $[\widehat{w_j} \cdots \widehat{w_\ell}] \; \mathcal{H} \; [\widehat{w_{j'}} \cdots \widehat{w_{\ell'}}]$ and since M is aperiodic we find
- $[\widehat{w_j} \cdots \widehat{w_\ell}] = [\widehat{w_{j'}} \cdots \widehat{w_{\ell'}}]$.

Therefore, all $[\widehat{w_j} \cdots \widehat{w_\ell}]$ denote the same element in M and since $k \geq 2$ this element is idempotent. In particular, we have $[\widehat{w_j}]^2 = [\widehat{w_j}] = [\widehat{w_\ell}]$ for all $1 \leq j, \ell \leq k$. We construct a factorization tree of w by

$$d(w) = (w_0 w'_1 \cdots w'_{i_0}, w'_{i_0+1} \cdots w'_m)$$
$$d(w'_{i_0+1} \cdots w'_m) = (\widehat{w_1} \cdots \widehat{w_k}, w'_{i_k+1} w'_m)$$
$$d(\widehat{w_1} \cdots \widehat{w_k}) = (\widehat{w_1}, \ldots, \widehat{w_k})$$

Now, the pair (L, R) does not occur in any of the words $w_0 w'_1 \cdots w'_{i_0}, w'_{i_k+1} w'_m$ and $\widehat{w_j}$. By induction on the number of pairs (L_i, R_i) there exist factorization trees for them whose height is bounded by

$$3 \, | \{ (L_i, R_i) \mid 1 \leq i < m \} \setminus \{ (L, R) \} | \; + \; 3 \, | \{ x \mid [w] <_{\mathcal{J}} x \} |$$

Hence the height of the factorization tree of w satisfies the desired bound. $\quad\square$

6.2   Proof of Theorem 10

Lemma 1. *Let $\mu : \mathbb{M} \to (M, \leq) : t \mapsto [t]$ be a homomorphism into an ordered monoid. If M is finite and satisfies the following property for all $e, s \in \mathbb{M}$:*

$$[e] = [e^2] \text{ and } \mathrm{alph}(s) \subseteq \mathrm{alph}(e) \text{ implies } [ese] \leq [e] \tag{3}$$

then for every $p \in M$ the language $\mu^{-1}(\lfloor p \rfloor)$ is a polynomial.

Proof. By considering the case $s^\omega = e$ the property (3) implies $[s^\omega s s^\omega] = [s^\omega s] \leq [s^\omega]$ and furthermore

$$[s^\omega] = [s^\omega s^\omega] \leq [s^\omega s^{\omega-1}] \leq [s^\omega s^{\omega-2}] \leq \cdots \leq [s^\omega s]$$

Hence $[s^\omega s] = [s^\omega]$ for all $s \in \mathbb{M}$ and therefore M is aperiodic. By Theorem 11 there exists a factorization forest d of height $< 3 \, |M|$ for the homomorphism $\Gamma^* \to M : w \mapsto [\pi(w)]$ where $\pi : \Gamma^* \to \mathbb{M}$ is the natural projection. We define the height $h(t)$ of a trace t with respect to this factorization forest as the minimal height of one of its word representatives $w \in \pi^{-1}(t)$ and set $d(t) = \big(\pi(w_1), \ldots, \pi(w_n)\big)$ where $d(w) = (w_1, \ldots, w_n)$. We show that for every $t \in \mathbb{M}$ there exists a monomial L_t of the form

$$a_1 A_1^* a_2 \cdots A_n^* a_{n+1}$$

whose (minimal) degree is bounded by (a sufficiently large function in) the height $h(t)$ of the factorization tree of t and that has the property $t \in L_t \subseteq \mu^{-1}(\lfloor [t] \rfloor)$.

Since $h(t) < 3|M|$ there exist only finitely many such languages and therefore the following union

$$\bigcup_{t \in \mu^{-1}(\lfloor p \rfloor)} L_t$$

is finite and gives a polynomial representation for $\mu^{-1}(\lfloor p \rfloor)$.

If $|t| \leq 1$ then $L_t = \{t\}$ is a monomial with constant degree. Now let $|t| > 1$. The first case is $d(t) = (t_1, t_2)$. Then by induction on the height there exist monomials for t_1 and t_2 with $t_i \in L_{t_i} \subseteq \mu^{-1}(\lfloor [t_i] \rfloor)$ for $i = 1, 2$ whose degree is bounded by a function in $h(t) - 1$. We define the monomial $L_t = L_{t_1} \emptyset^* L_{t_2}$. Clearly, we have $t \in L_t$. It remains to verify $L_t \subseteq \mu^{-1}(\lfloor [t] \rfloor)$. Let $t_1' t_2' \in L_t$ with $t_1' \in L_{t_1}$ and $t_2' \in L_{t_2}$. Then

$$[t_1' t_2'] = [t_1'][t_2'] \leq [t_1][t_2] = [t_1 t_2] = [t]$$

The second case is $d(t) = (t_1, \ldots, t_n)$ with $[t_1]^2 = [t_1] = [t_2] = \ldots = [t_n] = [t]$. By induction there exist languages L_i with $t_i \in L_{t_i} \subseteq \mu^{-1}(\lfloor [t_i] \rfloor)$ for $i = 1, n$ whose degree is bounded by a function in $h(t) - 1$. We define the monomial $L_t = L_{t_1}(\text{alph}(t))^* L_{t_n}$. Again, $t \in L_t$ is clear. It remains to verify $L_t \subseteq \mu^{-1}(\lfloor [t] \rfloor)$. Let $t_1' s t_n' \in L_t$ with $t_1' \in L_{t_1}$, $t_n' \in L_{t_n}$ and $\text{alph}(s) \subseteq \text{alph}(t)$. Then

$$[t_1' s t_n'] = [t_1'][s][t_n'] \leq [t_1][s][t_n] = [t][s][t] \leq [t]$$

where the last inequality follows by (3).   □

Lemma 2. *Every monomial $A_0^* a_1 A_1^* \cdots a_m A_m^*$ is expressible in $\Sigma_2[E]$.*

Proof. We show that for every trace $t = t_0 a_1 t_1 \cdots a_m t_m$ with $\text{alph}(t_i) \subseteq A_i$ there exists a $\Sigma_2[E]$-sentence φ_t whose size is bounded by a function in m and the size of the alphabet Γ (and not by $|t|$) such that

$$t \in L(\varphi_t) \subseteq A_0^* a_1 A_1^* \cdots a_m A_m^*$$

Since there are only finitely many such sentences the following disjunction is finite

$$\bigvee_{t \in A_0^* a_1 A_1^* \cdots a_m A_m^*} \varphi_t$$

and it describes exactly the monomial $A_0^* a_1 A_1^* \cdots a_m A_m^*$. The lemma then follows since $\Sigma_2[E]$ is closed under finite disjunctions.

Using the convention that a_0 is the empty trace we define $B_i = \text{alph}(a_i t_i)$ for $0 \leq i \leq m$. For each i and each letter $b \in B_i$ fix a first position $x_{f,i,b}$ with label b in the factor $a_i t_i$ and a last position $x_{\ell,i,b}$ with label b in the factor $a_i t_i$. There is a $\Sigma_2[E]$-formula $\psi_t(\overline{x})$ with free variables $\overline{x} = (x_{f,i,b}, x_{\ell,i,b})_{0 \leq i \leq m, b \in B_i}$ which reflects exactly the labeling and the partial ordering (i.e., not only the edge relation in the dependence graph) of the chosen positions in t. Furthermore the size of $\psi_t(\overline{x})$ does only depend on m and Γ. The formula φ_t we are looking for can be specified as follows:

$$\varphi_t = \exists \overline{x}: \psi_t(\overline{x}) \wedge \forall y: \bigvee_{b \in B_i, 0 \leq i \leq m} \lambda(y) = b \wedge x_{f,i,b} \leq y \leq x_{\ell,i,b}$$

Note that it is allowed to write $x_{f,i,b} \leq y \leq x_{\ell,i,b}$ also over dependence graphs because $\psi_t(\overline{x})$ specifies the labels such that $\lambda(x_{f,i,b}) = \lambda(x_{\ell,i,b}) = b$. $\qquad\qquad\square$

Lemma 3. *Let $L \subseteq \mathbb{M}$ be a trace language and let $\mu : \mathbb{M} \to (M(L), \leq)$ be its syntactic ordered homomorphism. If L is definable in $\Sigma_2[E]$ then $M(L)$ has the property that $[e] = [e^2]$ and $\mathrm{alph}(s) \subseteq \mathrm{alph}(e)$ implies $[ese] \leq [e]$ for all $e, s \in \mathbb{M}$.*

Proof. Let $\varphi = \exists\overline{x}\,\forall\overline{y}\colon \psi(\overline{x}, \overline{y}) \in \Sigma_2[E]$ where $\overline{x} = (x_1, \ldots, x_n)$, $\overline{y} = (y_1, \ldots, y_n)$, and ψ is a propositional formula. Let $p, q, s, t \in \mathbb{M}$ and assume $\mathrm{alph}(s) \subseteq \mathrm{alph}(t)$. We show that for all $k \geq (n+1)^2$ we have

$$pt^k q \models \varphi \;\Rightarrow\; pt^k st^k q \models \varphi \qquad\qquad (4)$$

If $u = pt^k q$ models φ then there exist positions $X_1, \ldots, X_n$ in the trace u such that

$$u, \overline{X} \models \forall\overline{y}\colon \psi(\overline{X}, \overline{y}) \qquad\qquad (5)$$

where $X = (X_1, \ldots, X_n)$. We refer to the k copies of the factor t in u as *blocks* numbered by 1 to k from left to right. By choice of k there exist n consecutive blocks such that no X_i is a position within these blocks, i.e.,

$$u = pt^{k_1} \cdot t^n \cdot t^{k_2} q$$

and all X_i are positions either in the prefix pt^{k_1} or in the suffix $t^{k_2} q$ of u. Consider the following factorization of $v = pt^k st^k q$:

$$v = pt^{k_1} \cdot t^{k_1'} st^{k_2'} \cdot t^{k_2} q$$

Since the prefix and suffix in this factorization are equal to that in the factorization of u and since all X_i correspond to positions in these parts of u we can choose the corresponding positions $X_1', \ldots, X_n'$ in the identical parts of v. We claim that for $\overline{X'} = (X_1', \ldots, X_n')$ we have

$$v, \overline{X'} \models \forall\overline{y}\colon \psi(\overline{X'}, \overline{y})$$

By contradiction, suppose there exist positions $Y_1', \ldots, Y_n'$ in v such that for $\overline{Y'} = (Y_1', \ldots, Y_n')$ we have

$$v, \overline{X'}, \overline{Y'} \models \neg\psi(\overline{X'}, \overline{Y'})$$

We show that this contradicts (5). If Y_i' is a position in the prefix pt^{k_1} or in the suffix $t^{k_2} q$ of v we can choose an analogous position Y_i in u. W.l.o.g. we assume that all Y_i are positions in the middle factor $t^{k_1'} st^{k_2'}$ and that $i < j$ implies $(Y_j', Y_i') \notin E$, i.e., $Y_1', \ldots, Y_n'$ is a linearization of the positions in $\overline{Y'}$. We now let Y_i be any position in the block $k_1 + i$ of u with the same label as Y_i'. This is possible since $\mathrm{alph}(s) \subseteq \mathrm{alph}(t)$. Now, all Y_i are positions in the middle factor t^n of u. By construction, we have

$$(X_i, X_j) \in E \;\Leftrightarrow\; (X_i', X_j') \in E$$
$$(Y_i, Y_j) \in E \;\Leftrightarrow\; (Y_i', Y_j') \in E$$
$$(X_i, Y_j) \in E \;\Leftrightarrow\; (X_i', Y_j') \in E$$
$$(Y_i, X_j) \in E \;\Leftrightarrow\; (Y_i', X_j') \in E$$

Note that this would not be true for partial orders instead of dependence graphs. From $v, \overline{X'}, \overline{Y'} \models \neg\psi(\overline{X'}, \overline{Y'})$ it now follows

$$u, \overline{X}, \overline{Y} \models \neg\psi(\overline{X}, \overline{Y})$$

in contradiction to (5). This proves (4). For $L = L(\varphi)$ it follows that $[t^k s t^k] \leq [t^k]$ holds in the syntactic ordered monoid $(M(L), \leq)$ of L. The lemma now follows since $[t^k] = [t]$ if $[t] = [e]$ is idempotent.   □

Proof (Theorem 10). The implication "*1.* ⇒ *2.*" follows by Lemma 1 since L is the union of languages of the form $\mu^{-1}(\lfloor p \rfloor)$ with $p \in M(L)$. "*2.* ⇒ *3.*" follows from Lemma 2 since $\Sigma_2[E]$ is closed under finite disjunctions. Finally, the implication "*3.* ⇒ *1.*" is Lemma 3.   □

References

1. Cartier, P., Foata, D. (eds.): Problèmes combinatoires de commutation et réarrangements. Lecture Notes in Mathematics, vol. 85. Springer, Heidelberg (1969)
2. Chalopin, J., Leung, H.: On factorization forests of finite height. Theoretical Computer Science 310(1-3), 489–499 (2004)
3. Colcombet, T.: On Factorization Forests. Technical report, number hal-00125047, Irisa, Rennes (2007)
4. Diekert, V., Gastin, P.: LTL is expressively complete for Mazurkiewicz traces. Journal of Computer and System Sciences 64, 396–418 (2002)
5. Diekert, V., Gastin, P.: Pure future local temporal logics are expressively complete for Mazurkiewicz trace. Information and Computation 204, 1597–1619 (2006)
6. Diekert, V., Horsch, M., Kufleitner, M.: On first-order fragments for Mazurkiewicz traces. In: Fundamenta Informaticae (to appear)
7. Diekert, V., Rozenberg, G. (eds.): The Book of Traces. World Scientific, Singapore (1995)
8. Ebinger, W., Muscholl, A.: Logical definability on infinite traces. Theoretical Computer Science 154, 67–84 (1996)
9. Eilenberg, S.: Automata, Languages, and Machines, vol. B. Academic Press, New York and London (1976)
10. Emerson, E.A.: Temporal and modal logic. In: van Leeuwen, J. (ed.) Handbook of Theoretical Computer Science, vol. B, ch. 16, pp. 995–1072. Elsevier Science Publisher B.V, Amsterdam (1990)
11. Etessami, K., Vardi, M.Y., Wilke, T.: First-order logic with two variables and unary temporal logic. Information and Computation 179(2), 279–295 (2002)
12. Gabbay, D., Hodkinson, I., Reynolds, M.: Temporal Logic: Mathematical Foundations and Computational Aspects. Clarendon Press, Oxford (1994)
13. Gastin, P., Kuske, D.: Satisfiability and model checking for MSO-definable temporal logics are in PSPACE. In: Amadio, R.M., Lugiez, D. (eds.) CONCUR 2003. LNCS, vol. 2761, Springer, Heidelberg (2003)
14. Thomas, W.: Languages, automata and logic. In: Salomaa, A., Rozenberg, G. (eds.) Handbook of Formal Languages. Beyond Words, vol. 3, Springer, Berlin Heidelberg (1997)

15. Guaiana, G., Restivo, A., Salemi, S.: Star-free trace languages. Theoretical Computer Science 97, 301–311 (1992)
16. Keller, R.M.: Parallel program schemata and maximal parallelism I. Fundamental results. Journal of the Association for Computing Machinery 20(3), 514–537 (1973)
17. Kufleitner, M.: Logical Fragments for Mazurkiewicz Traces: Expressive Power and Algebraic Characterizations. Universität Stuttgart, Dissertation (2006)
18. Kufleitner, M.: Polynomials, fragments of temporal logic and the variety DA over traces. In: Ibarra, O.H., Dang, Z. (eds.) DLT 2006. LNCS, vol. 4036, Springer, Heidelberg (2006)
19. Mazurkiewicz, A.: Concurrent program schemes and their interpretations. DAIMI Rep. PB 78, Aarhus University, Aarhus (1977)
20. Mazurkiewicz, A.: Trace theory. In: Brauer, W., Reisig, W., Rozenberg, G. (eds.) Petri Nets, Applications and Relationship to other Models of Concurrency. LNCS, vol. 255, pp. 279–324. Springer, Heidelberg (1987)
21. Pin, J.É.: Varieties of Formal Languages. North Oxford Academic, London (1986)
22. Pin, J.-É.: A variety theorem without complementation. In: Russian Mathematics (Izvestija vuzov.Matematika), vol. 39, pp. 80–90 (1995)
23. Pin, J.É., Straubing, H., Thérien, D.: Locally trivial categories and unambiguous concatenation. Journal of Pure. and Applied Algebra 52, 297–311 (1988)
24. Pin, J.É., Weil, P.: Polynominal closure and unambiguous product. Theory Comput. Syst. 30(4), 383–422 (1997)
25. Schützenberger, M.P.: Sur le produit de concatenation non ambigu. Semigroup Forum 13, 47–75 (1976)
26. Simon, I.: Factorization forests of finite height. Theoretical Computer Science 72(1), 65–94 (1990)
27. Sistla, A.P., Clarke, E.: The complexity of propositional linear time logic. Journal of the Association for Computing Machinery 32, 733–749 (1985)
28. Stockmeyer, L.: The complexity of decision problems in automata theory and logic. PhD thesis, TR 133, MIT, Cambridge (1974)
29. Straubing, H.: Finite Automata, Formal Logic, and Circuit Complexity. Birkhäuser, Boston, Basel and Berlin (1994)
30. Tesson, P., Thérien, D.: Diamonds are Forever: The Variety DA. In: dos Gomes Moreira da Cunha, G.M., da Silva, P.V.A., Pin, J.É. (eds.) Semigroups, Algorithms, Automata and Languages, Coimbra (Portugal) 2001, pp. 475–500. World Scientific, Singapore (2002)
31. Thiagarajan, P.S., Walukiewicz, I.: An expressively complete linear time temporal logic for Mazurkiewicz traces. In: Proc. of LICS'97, pp. 183–194 (1997)
32. Thomas, W.: Languages, automata and logic. In: Salomaa, A., Rozenberg, G. (eds.) Handbook of Formal Languages. Beyond Words, vol. 3, Springer, Heidelberg (1997)
33. Weis, P., Immerman, N.: Structure theorem and strict alternation hierarchy for FO^2 on words. Technical report, Department of Computer Science University of Massachusetts, Amherst (2006)

Quantitative Generalizations of Languages[*]

Thomas A. Henzinger

EPFL, Switzerland, and University of California, Berkeley

In the traditional view, a language is a set of words, i.e., a function from words to boolean values. We call this view "qualitative," because each word either belongs to or does not belong to a language. Let Σ be an alphabet, and let us consider infinite words over Σ. Formally, a *qualitative language* over Σ is a function A: $\Sigma^\omega \to \mathbb{B}$. There are many applications of qualitative languages. For example, qualitative languages are used to specify the legal behaviors of systems, and zero-sum objectives of games played on graphs. In the former case, each behavior of a system is either legal or illegal; in the latter case, each outcome of a game is either winning or losing. For defining languages, it is convenient to use finite acceptors (or generators). In particular, qualitative languages are often defined using finite-state machines (so-called ω-automata) whose transitions are labeled by letters from Σ. For example, the states of an ω-automaton may represent states of a system, and the transition labels may represent atomic observables of a behavior. There is a rich and well-studied theory of finite-state acceptors of qualitative languages, namely, the theory of the ω-regular languages.

There are two common, orthogonal quantitative generalizations of languages. In the first quantitative view, a language is a set of probability distributions on words, i.e., a set of functions from words to the real interval $[0, 1]$. We call this view "probabilistic." A *probabilistic word* over the alphabet Σ is a probability distribution on Σ^ω. We write $\mathcal{D}(\Sigma^\omega)$ for the set of probabilistic words. A *probabilistic language* over Σ is a function B: $\mathcal{D}(\Sigma^\omega) \to \mathbb{B}$. Probabilistic languages can be defined by Markov decision processes (MDPs) whose transitions are labeled by letters from Σ. MDPs generalize ω-automata by distinguishing between controllable states, where an outgoing transition is chosen according to a policy[1] (or strategy), and probabilistic states, where an outgoing transition is chosen according to a given probability distribution. Given an MDP, and a policy for resolving all controllable decisions, the outcome is a probabilistic word. By collecting the outcomes of all policies in a set, we obtain a probabilistic language. Many basic questions about such finite-state generators of probabilistic languages are unsolved. For example, the language-inclusion problem for MDPs is central to the algorithmic verification of probabilistic systems: it asks, given two finite-state MDPs M_1 and M_2, if the probabilistic language defined by M_1 is

---

[*] This research was supported in part by the Swiss National Science Foundation and by the NSF grant CCR-0225610.

[1] Policies may in general be probabilistic. A *policy* is a function mapping each finite state sequence (representing the history of a behavior) to a probability distribution on the possible next states.

T. Harju, J. Karhumäki, and A. Lepistö (Eds.): DLT 2007, LNCS 4588, pp. 20–22, 2007.
© Springer-Verlag Berlin Heidelberg 2007

a subset of the probabilistic language defined by M_2; in other words, if for every policy p_1 of M_1, there is a policy p_2 of M_2 such that the outcome of applying p_1 in M_1 is equal to the outcome of applying p_2 in M_2. To our knowledge, it is an open problem if this question can be decided.

In the second quantitative view, a language is a function from words to real values. These values may represent rewards or costs. We call this view "numerical." Formally, a *numerical language* over the alphabet Σ is a function C: $\Sigma^\omega \to \mathbb{R}$. We refer to $C(w)$ as the *value* of a word w in the language C. There are several ways of generating numerical languages. One mechanism for obtaining finite numerical values for infinite words is discounting, which gives geometrically less weight to letters that occur later in a word. Let M be a state machine whose transitions are labeled by letters from Σ. Given a real-valued discount factor $\lambda \in (0,1)$, the value $M(w)$ of each word $w \in \Sigma^\omega$ can be defined as $1 - \lambda^n$, where n is the number of letters in the longest prefix of w that is accepted by M (if all prefixes of w are accepted by M, then $M(w) = 1$). Numerical languages are also generated by weighted state machines, whose transitions are labeled both with letters from Σ and with real values. The numerical label (or weight) of a transition may represent a reward obtained or a cost incurred by traversing the transition. Let M be a weighted state machine, and let r be a run of M over a word $w \in \Sigma^\omega$. The run r can be defined to assign to w either the supremal transition value occurring in r, or the limsup of all transition values in r, or their limit average, or their discounted sum (for some discount factor λ). The weighted state machine M, then, assigns to each word w as value the supremum of all values assigned to w by accepting runs of M over w. In game theory, objectives that try to maximize a numerical value are common and well-studied; in system modeling, the numerical value of a run may represent a resource requirement of a behavior, such as power consumption.

The probabilistic and numerical views can be combined, resulting in quantitative languages of the type $[\mathcal{D}(\Sigma^\omega) \to \mathbb{R}]$. In this talk, we survey some theoretical results about such quantitative generalizations of languages, and review some of their applications in system design and verification [1–6].

References

1. Chakrabarti, A., Chatterjee, K., Henzinger, T.A., Kupferman, O., Majumdar, R.: Verifying quantitative properties using bound functions. In: Borrione, D., Paul, W. (eds.) CHARME 2005. LNCS, vol. 3725, pp. 50–64. Springer, Heidelberg (2005)
2. Chakrabarti, A., de Alfaro, L., Henzinger, T.A., Stoelinga, M.: Resource interfaces. In: Alur, R., Lee, I. (eds.) EMSOFT 2003. LNCS, vol. 2855, pp. 117–133. Springer, Heidelberg (2003)
3. Chatterjee, K., de Alfaro, L., Faella, M., Henzinger, T.A., Majumdar, R., Stoelinga, M.: Compositional quantitative reasoning. In: Proceedings of the Third Annual Conference on Quantitative Evaluation of Systems, IEEE Computer Society Press, Los Alamitos (2006)
4. Chatterjee, K., Henzinger, T.A.: Value iteration. In: 25 Years of Model Checking. LNCS, Springer, Heidelberg (2007)

5. de Alfaro, L., Henzinger, T.A., Jhala, R.: Compositional methods for probabilistic systems. In: Larsen, K.G., Nielsen, M. (eds.) CONCUR 2001. LNCS, vol. 2154, pp. 351–365. Springer, Heidelberg (2001)
6. de Alfaro, L., Henzinger, T.A., Majumdar, R.: Discounting the future in systems theory. In: Baeten, J.C.M., Lenstra, J.K., Parrow, J., Woeginger, G.J. (eds.) ICALP 2003. LNCS, vol. 2719, pp. 1022–1037. Springer, Heidelberg (2003)

What Do We Know About Language Equations?

Michal Kunc*

Department of Mathematics and Statistics, Masaryk University,
Janáčkovo nám. 2a, 602 00 Brno, Czech Republic
kunc@math.muni.cz, http://www.math.muni.cz/~kunc/

In the talk we give an overview of recent developments in the area of language equations, with an emphasis on methods for dealing with non-classical types of equations whose theory has not been successfully developed already in the previous decades, and on results forming the current borderline of our knowledge. This abstract is in particular meant to provide the interested listener with references to the material discussed in the talk.

Motivations for studying equations over languages come from several sources (e.g. formal grammars, automata constructions, word equations, set constraints, games or natural computing) and most of the results on these equations are related to one of these topics.

Language equations were first applied in [14] to elegantly define semantics for context-free grammars by means of explicit systems of equations with the operations of union and concatenation. Some interesting examples of using these systems can be found in [52]. By allowing in these systems also intersection, one obtains the notion of conjunctive languages [36,37], which are more general than context-free ones even over a unary alphabet [17]. The special case of linear conjunctive languages was studied in [39].

The theory of explicit systems of language equations with concatenation and all Boolean operations was developed in [46], and even one-variable systems were proved computationally universal [43]. The appropriate restriction of these systems to define Boolean grammars was described in [38]. Several basic open problems about conjunctive and Boolean languages are proposed in [45]. The classes of languages obtained by allowing in explicit systems additionally to concatenation all possible clones of Boolean operations were also determined [47,44]. Explicit systems were further shown to naturally define arithmetical hierarchy [40]. Solutions of explicit systems with some language operations other than concatenation were also described, e.g. equations employing homomorphisms are related to ET0L languages [53].

Implicit language equations where concatenation is the only operation naturally appear as a generalization of equations over words to sets of words. Existence of solutions of word equations with constants was proved decidable by Makanin [34]. Currently best algorithms for solving word equations can be found in [49]. It is also well known that solvability of word equations is decidable even for infinite rational systems of equations [10,2,15].

---

* Supported by the Grant no. 201/06/0936 of the Grant Agency of the Czech Republic.

T. Harju, J. Karhumäki, and A. Lepistö (Eds.): DLT 2007, LNCS 4588, pp. 23–27, 2007.

For equations over languages, the situation is completely different. In [48] existence of arbitrary solutions was proved undecidable for equations with finite constants employing union and concatenation. When regular constant languages are allowed, the problem is undecidable already for one-variable systems using only the operation of concatenation [29]. But there is no such result about equations with only finite constants, and we also have virtually no knowledge about solvability of finite systems over finite or regular languages. On the other hand, it is known that already for a very simple rational system of equations with only concatenation we cannot algorithmically decide whether given finite languages form its solution [32,20,30].

Most of the results about implicit language equations and inequalities concern inequalities of particular forms, often related to important classes of formal grammars or basic automata constructions like those of minimal and universal automata (see [54]). Results of this kind were surveyed in [26]. General treatment of systems of implicit equations was initiated by Okhotin [41], who considered also strict inequalities [42].

General systems of equations and inequalities with constant right-hand sides were studied by Conway [9], and the exact complexity of determining their solvability was established in [6]. The study of such equations was also extended to the simplest equations with more general operations than concatenation based on shuffle and deletion along trajectories [22,23,12].

Some generalizations of standard systems of right-linear equations were considered by Leiss [31]. For general systems of right-linear inequalities, basic problems can be solved using Rabin's results on MSO logic on infinite trees [50]; the complexity of these problems has been determined in [1,8,4,3,5]. Regularity of largest solutions in the case of inequalities with non-regular left-hand sides was established in [28].

The method of proving regularity by means of well quasi-orders was developed by Ehrenfeucht et al. [13]; a number of results on regularity of languages based on well quasi-orders can be found in [11]. Well quasi-orders were used to show regularity of largest solutions of systems of inequalities with certain restrictions on constant languages [25].

The borderline between equations with algorithmically constructible regular largest solutions and those having universal expressive power appears to be formed by semi-commutation inequalities $XK \subseteq LX$. For any regular language L, the largest solution of such an inequality is always regular [25], but the only known proof of this fact is non-constructive, based on Kruskal's tree theorem [24], and we know how to algorithmically find the largest solution only in a very special case [33]. However, systems of two semi-commutation inequalities possess universal expressive power [27]. A prominent role among these systems is played by commutation equations, which were first considered by Conway [9] and later studied in many papers (see [21] for a survey). Basic results were achieved and conjectures formulated in [51]; regularity of the largest solutions was proved for three-element languages [19] (a more general result based on lexicographic ordering can be found in [35]) and regular codes [18]. The expressive universality of commutation equations over finite languages was established in [29] (a more intuitive incremental construction for this result is described in [16]). Some partial results were proved also for equations expressing conjugacy of languages [7]; an undecidability result for these equations can be found in [29].

References

1. Aiken, A., Kozen, D., Vardi, M., Wimmers, E.: The complexity of set constraints. In: Meinke, K., Börger, E., Gurevich, Y. (eds.) CSL 1993. LNCS, vol. 832, pp. 1–17. Springer, Heidelberg (1994)
2. Albert, M.H., Lawrence, J.: A proof of Ehrenfeucht's conjecture. Theoret. Comput. Sci. 41(1), 121–123 (1985)
3. Baader, F., Küsters, R.: Unification in a description logic with transitive closure of roles. In: Nieuwenhuis, R., Voronkov, A. (eds.) LPAR 2001. LNCS (LNAI), vol. 2250, pp. 217–232. Springer, Heidelberg (2001)
4. Baader, F., Narendran, P.: Unification of concept terms in description logics. J. Symbolic Comput. 31(3), 277–305 (2001)
5. Baader, F., Okhotin, A.: Complexity of language equations with one-sided concatenation and all Boolean operations. In: Proc. UNIF'06, pp. 59–73 (2006)
6. Bala, S.: Complexity of regular language matching and other decidable cases of the satisfiability problem for constraints between regular open terms. Theory Comput. Syst. 39(1), 137–163 (2006)
7. Cassaigne, J., Karhumäki, J., Maňuch, J.: On conjugacy of languages. Theor. Inform. Appl. 35(6), 535–550 (2001)
8. Charatonik, W., Podelski, A.: Co-definite set constraints. In: Nipkow, T. (ed.) Rewriting Techniques and Applications. LNCS, vol. 1379, pp. 211–225. Springer, Heidelberg (1998)
9. Conway, J.H.: Regular Algebra and Finite Machines. Chapman and Hall, London (1971)
10. Culik II, K., Karhumäki, J.: Systems of equations over a free monoid and Ehrenfeucht's conjecture. Discrete Math. 43(2–3), 139–153 (1983)
11. de Luca, A., Varricchio, S.: Finiteness and Regularity in Semigroups and Formal Languages. Springer, Heidelberg (1999)
12. Domaratzki, M., Salomaa, K.: Decidability of trajectory-based equations. Theoret. Comput. Sci. 345(2–3), 304–330 (2005)
13. Ehrenfeucht, A., Haussler, D., Rozenberg, G.: On regularity of context-free languages. Theoret. Comput. Sci. 27(3), 311–332 (1983)
14. Ginsburg, S., Rice, H.G.: Two families of languages related to ALGOL. J. ACM 9(3), 350–371 (1962)
15. Guba, V.S.: Equivalence of infinite systems of equations in free groups and semigroups to finite subsystems. Mat. Zametki 40(3), 321–324 (1986)
16. Jeandel, E., Ollinger, N.: Playing with Conway's problem. Technical report ccsd-00013788, Laboratoire d'Informatique Fondamentale de Marseille (2005). Available at http://hal.archives-ouvertes.fr/hal-00013788
17. Jeż, A.: Conjunctive grammars can generate non-regular unary languages. In this volume
18. Karhumäki, J., Latteux, M., Petre, I.: Commutation with codes. Theoret. Comput. Sci. 340(2), 322–333 (2005)

19. Karhumäki, J., Latteux, M., Petre, I.: Commutation with ternary sets of words. Theory Comput. Syst. 38(2), 161–169 (2005)
20. Karhumäki, J., Lisovik, L.P.: The equivalence problem of finite substitutions on ab^*c, with applications. Internat. J. Found. Comput. Sci. 14(4), 699–710 (2003)
21. Karhumäki, J., Petre, I.: Two problems on commutation of languages. In: Current Trends in Theoretical Computer Science, The Challenge of the New Century, vol. 2, pp. 477–494. World Scientific, Singapore (2004)
22. Kari, L.: On language equations with invertible operations. Theoret. Comput. Sci. 132(1–2), 129–150 (1994)
23. Kari, L., Sosík, P.: On language equations with deletion. Bull. EATCS 83, 173–180 (2004)
24. Kruskal, J.B.: Well-quasi-ordering, the tree theorem, and Vazsonyi's conjecture. Trans. Amer. Math. Soc. 95(2), 210–225 (1960)
25. Kunc, M.: Regular solutions of language inequalities and well quasi-orders. Theoret. Comput. Sci. 348(2–3), 277–293 (2005)
26. Kunc, M.: Simple language equations. Bull. EATCS 85, 81–102 (2005)
27. Kunc, M.: On language inequalities $XK \subseteq LX$. In: De Felice, C., Restivo, A. (eds.) DLT 2005. LNCS, vol. 3572, pp. 327–337. Springer, Heidelberg (2005)
28. Kunc, M.: Largest solutions of left-linear language inequalities. In: Proc. AFL 2005, University of Szeged, pp. 178–186 (2005), Also available at http://www.math.muni.cz/~kunc/math/left_linear.ps.
29. Kunc, M.: The power of commuting with finite sets of words. Theory Comput. Syst. 40(4), 521–551 (2007)
30. Kunc, M.: The simplest language where equivalence of finite substitutions is undecidable. Preprint available at http://www.math.muni.cz/~kunc/math/finite_substitutions.pdf.
31. Leiss, E.L.: Language Equations. Springer, Heidelberg (1999)
32. Lisovik, L.P.: The equivalence problem for finite substitutions on regular languages. Dokl. Akad. Nauk 357(3), 299–301 (1997)
33. Ly, O.: A constructive solution of the language inequation $XA \subseteq BX$. Preprint available at http://www.labri.fr/perso/ly/publications/LanguageEquation.pdf.
34. Makanin, G.S.: The problem of solvability of equations in a free semigroup. Mat. Sb. 103(2), 147–236 (1977)
35. Massazza, P., Salmela, P.: On the simplest centralizer of a language. Theor. Inform. Appl. 40(2), 295–301 (2006)
36. Okhotin, A.: Conjunctive grammars. J. Autom. Lang. Comb. 6(4), 519–535 (2001)
37. Okhotin, A.: Conjunctive grammars and systems of language equations. Program. Comput. Software 28(5), 243–249 (2002)
38. Okhotin, A.: Boolean grammars. Inform. and Comput. 194(1), 19–48 (2004)
39. Okhotin, A.: On the equivalence of linear conjunctive grammars and trellis automata. Theor. Inform. Appl. 38(1), 69–88 (2004)
40. Okhotin, A.: A characterization of the arithmetical hierarchy by language equations. Internat. J. Found. Comput. Sci. 16(5), 985–998 (2005)
41. Okhotin, A.: Unresolved systems of language equations: Expressive power and decision problems. Theoret. Comput. Sci. 349(3), 283–308 (2005)
42. Okhotin, A.: Strict language inequalities and their decision problems. In: Jedrzejowicz, J., Szepietowski, A. (eds.) MFCS 2005. LNCS, vol. 3618, pp. 708–719. Springer, Heidelberg (2005)
43. Okhotin, A.: Computational universality in one-variable language equations. Fund. Inform. 74(4), 563–578 (2006)

44. Okhotin, A.: Language equations with symmetric difference. In: Grigoriev, D., Harrison, J., Hirsch, E.A. (eds.) CSR 2006. LNCS, vol. 3967, pp. 292–303. Springer, Heidelberg (2006)
45. Okhotin, A.: Nine open problems for conjunctive and Boolean grammars. Bull. EATCS 91 (to appear, 2007)
46. Okhotin, A.: Decision problems for language equations. Submitted for publication. Preliminary version. In: Baeten, J.C.M., Lenstra, J.K., Parrow, J., Woeginger, G.J. (eds.) ICALP 2003. LNCS, vol. 2719, pp. 239–251. Springer, Heidelberg (2003)
47. Okhotin, A., Yakimova, O.: Language equations with complementation: Decision problems. Theoret. Comput. Sci. (to appear)
48. Parikh, R., Chandra, A., Halpern, J., Meyer, A.: Equations between regular terms and an application to process logic. SIAM J. Comput. 14(4), 935–942 (1985)
49. Plandowski, W.: An efficient algorithm for solving word equations. In: Proc. STOC'06, pp. 467–476. ACM, New York (2006)
50. Rabin, M.O.: Decidability of second-order theories and automata on infinite trees. Trans. Amer. Math. Soc. 141, 1–35 (1969)
51. Ratoandromanana, B.: Codes et motifs. RAIRO Inform. Théor. Appl. 23(4), 425–444 (1989)
52. Rozenberg, G., Salomaa, A. (eds.): Handbook of Formal Languages. Springer, Heidelberg (1997)
53. Ruohonen, K.: A note on language equations involving morphisms. Inform. Process. Lett. 7(5), 209–212 (1978)
54. Sakarovitch, J.: Elements of Automata Theory. Cambridge University Press, Cambridge (2007)

Information Distance and Applications

Ming Li

Canada Research Chair in Bioinformatics
University of Waterloo

We know how to measure distance from Turku to Toronto. However, do you know how to measure the distance between two information carrying entities? For example: two genomes, two music scores, two programs, two articles, two emails, two concepts, or from a question to an answer? Furthermore, such a distance measure must be application-independent, must be universal in the sense it is provably better than all other distances, and must be applicable.

From a simple and accepted assumption in thermodynamics, we have developed such a theory. I will present this theory and will talk about some new applications of this theory, including a question answering system.

Reference

1. Li, M., Vitanyi, P.: An introduction to Kolmogorov complexity and its applications, 3rd edn., Springer, Heidelberg (to appear in 2007/2008)

T. Harju, J. Karhumäki, and A. Lepistö (Eds.): DLT 2007, LNCS 4588, p. 28, 2007.
© Springer-Verlag Berlin Heidelberg 2007

Finite Automata and the Writing of Numbers

Jacques Sakarovitch

LTCI, ENST/CNRS
46, rue Barrault, 75013 Paris, France
sakarovitch@enst.fr

Abstract. Numbers do exist, independently of the way we represent them, of the way we write them. And there are many ways to write them: integers as finite sequence of digits once a base is fixed, rational numbers as a pair of integer or as an ultimately periodic infinite sequence of digits, or reals as an infinite sequence of digits but also as a continued fraction, just to quote a few. Operations on numbers are defined, independently of the way they are computed. But when they are computed they amounts to be algorithms that work on the representations of numbers.

Here, numbers will be represented by their development in a base, hence by *words* over an alphabet of digits and the algorithms we shall consider are those that can be performed by finite state machines, that is, by the simplest machines one can think of. Which operations can be thus defined? which set of numbers can be thus described? how this is related to the chosen base? how the choice of the alphabet of digits may influence the way the operations may be computed? These are the questions that will be asked and, hopefully and to a certain extent, answered in this conference.

We shall begin with the example of divisibility by a given integer in a given base p, the generalization — due to Blaise Pascal — of the casting out nines and, more seriously, with the beautiful Cobham's Theorem [1,2,3]. This result leads to the distinction between *recognizable* and *p-recognizable* sets of integers that generalizes to set of tuples of integers and sets the problem of the decidability of the former among the latter, answered positively by Honkala, Muchnik and Leroux [4,5,6].

Another obvious appearance of finite automata, of finite transducers indeed, in the processing of written numbers occurs when signed digits are used, as has been popularized in the field of computer arithmetics by Avizienis for instance [7]. In this framework arises the interesting problem of the trade-off between the *redundancy* of a number system and the "compexity" of the operations performed on numbers written in that system.

The next case that will retain our attention is the one of *non standard number systems*; here, a non integer real β is taken as a base and the (real) numbers are written in this base. We put into correspondance the so-called *arithmetic* properties of β — that is, which kind of algebraic integer β is — the rationality of the set of expansions in such a base and the possibility of defining a linear recurrence that yields a system for representing the integers (cf. [8, Ch. VII]). A striking result is the fact that the addition is realized by a finite transducer if, and only if, β is a

T. Harju, J. Karhumäki, and A. Lepistö (Eds.): DLT 2007, LNCS 4588, pp. 29–30, 2007.

Pisot number [9,10]. In all these systems, the algorithm for computing the expansions is the *greedy* algorithm and produces the most significant digit first.

In a last part we shall touch on a more recent topic: rational base number systems (cf. [11]). In these systems, every integer has a unique finite expansion, which is not computed by a greedy algorithm but by a right to left algorithm, that is, by an algorithm which computes the least significant digit first. The set of all expansions is not a rational language, a very intriguing set of words indeed, but a finite transducer exists which converts a representation written on any alphabet of digits into a representation of the same number written on the canonical alphabet.

References

1. Cobham, A.: On the base-dependance of the sets of numbers recognizable by finite automata. Math. Systems Theory 3, 186–192 (1969)
2. Eilenberg, S.: Automata, Languages and Machines, vol. A. Academic Press (1974)
3. Bruyère, V., Hansel, G., Michaux, Ch., Villemaire, R.: Logic and p-recognizable sets of integers. Bull. Belg. Soc. Math. 1, 191–238 (1994)
4. Honkala, J.: A decision method for the recognizability of sets defined by number systems. RAIRO Informatique Théor. 20, 395–403 (1986)
5. Muchnik, A.: The definable criterion for definability in Presburger arithmetic and its applications. Theoret. Computer Sci. 290, 1433–1444 (2003) Late publication in a journal of a preprint (in russian) issued in 1991
6. Leroux, J.: A polynomial time Presburger criterion and synthesis for number decision diagrams. In: Logic in Computer Science 2005 (LICS'2005), pp. 147–156. IEEE Computer Society Press, Los Alamitos (2005) New version at arXiv:cs/0612037v1
7. Avizienis, A.: Signed-digit number representations for fast parallel arithmetic. IRE Transactions on electronic computers 10, 389–400 (1961)
8. Lothaire, M.: Algebraic Combinatorics on Words. Cambridge University Press, Cambridge (2002)
9. Frougny, Ch.: Representation of numbers and finite automata. Math. Systems Theory 25, 37–60 (1992)
10. Berend, D., Frougny, Ch.: Computability by finite automata and Pisot bases. Math. Systems Theory 27, 275–282 (1994)
11. Akiyama, S., Frougny, Ch., Sakarovitch, J.: Powers of rationals modulo 1 and rational base number systems. Israël J. Math (to appear)

Descriptional Complexity of Nondeterministic Finite Automata*

Kai Salomaa

School of Computing, Queen's University, Kingston, Ontario K7L 3N6, Canada
ksalomaa@cs.queensu.ca

Abstract. In this talk, I will survey recent results and discuss open problems on the state and transition complexity of nondeterministic finite automata.

Finite-state automata are one of the simplest models of computation and a basis for the study of fundamental questions in complexity of computing. During the last ten years, motivated by new applications of regular languages that require automata of very large size, descriptional complexity of finite automata has received increased attention [4,14,22]. The majority of the work centers on the state complexity of deterministic finite automata (DFAs). An interesting aspect of the work is that it often combines experiments with purely theoretical work. When dealing with the state complexity of more involved combined operations [18,19], the worst-case examples are, typically, found experimentally using software tools such as Grail+ [21].

While DFAs can be efficiently minimized, the minimization of nondeterministic finite automata (NFAs) is known to be PSPACE-complete [15], and moreover the minimal NFA cannot be efficiently approximated [6,11]. Further results in this direction can be found in [8,10,17].

The number of transitions gives for NFAs a more realistic descriptional complexity measure than the number of states because the number of transitions determines the size of a complete description of an NFA. There has been much work on the transition complexity of converting regular expressions to NFAs and [20] has established a tight lower bound for the transformation. More references can be found in [6,20].

Here our focus is on questions relating the nondeterministic transition complexity and state complexity, and on questions on operational transition complexity [2], that is, how does the (minimal) number of transitions change when applying various regularity preserving operations to NFAs. Also, we can study trade-offs between the number of states and the number of transitions. There are examples where the number of transitions in state minimal NFAs may be significantly reduced already by allowing one additional state.

---

* Supported by the Natural Sciences and Engineering Research Council of Canada Grant OGP0147224.

T. Harju, J. Karhumäki, and A. Lepistö (Eds.): DLT 2007, LNCS 4588, pp. 31–35, 2007.
© Springer-Verlag Berlin Heidelberg 2007

1 Definitions

A nondeterministic finite automaton is a tuple $A = (\Sigma, Q, q_0, Q_F, \delta)$ where Σ is the input alphabet, Q is the finite set of states, $q_0 \in Q$ is the start state, $Q_F \subseteq Q$ is the set of accepting states and $\delta \subseteq Q \times \Sigma \times Q$ gives the set of transitions.

Let L be a regular language. The *nondeterministic state complexity* of L is the smallest number of states of any NFA recognizing L and it is denoted as $\mathrm{nsc}(L)$. The (nondeterministic) *transition complexity* of L, $\mathrm{tc}(L)$, is the smallest number of transitions of any NFA that recognizes L.

For $k \geq 0$, the *k-strict transition complexity* of L, $\mathrm{stc}_k(L)$, is the smallest number of transitions of any NFA A for L such that A has at most $\mathrm{nsc}(L) + k$ states. For any regular language L and $k \geq 0$, the following relations follow directly from the definitions

$$\mathrm{nsc}(L) - 1 \leq \mathrm{tc}(L) \leq \mathrm{stc}_{k+1}(L) \leq \mathrm{stc}_k(L).$$

To describe the transition complexity of operations on regular languages, we need the following notions dealing with numbers of transitions originating from the start state or entering the accepting states in transition minimal NFAs for the language. For a regular language L we denote by $\mathcal{M}(L)$ the family of all NFAs for L where the number of transitions is exactly $\mathrm{tc}(L)$. Now we define

$$s(L) = \min_{A \in \mathcal{M}(L)} \{|\delta \cap (\{q_0\} \times \Sigma \times Q)| : A = (\Sigma, Q, q_0, Q_F, \delta)\},$$

$$f(L) = \min_{A \in \mathcal{M}(L)} \{(|\delta \cap Q \times \Sigma \times Q_F)| : A = (\Sigma, Q, q_0, Q_F, \delta)\},$$

$$fs(L) = \min_{A \in \mathcal{M}(L)} \{|\delta \cap ((\{q_0\} \times \Sigma \times Q) \cup (Q \times \Sigma \times Q_F))| : A = (\Sigma, Q, q_0, Q_F, \delta)\}.$$

2 Transition Complexity and State Complexity

Recently nondeterministic state complexity has been used to provide estimations for the deterministic state complexity of combined operations [19]. In many cases the composition of nondeterministic state complexities of basic operations turns out to be fairly close to the nondeterministic state complexity of the combined operation, while the same is not true for the deterministic state complexity of combined operations [18].

Table 1 summarizes the results for nondeterministic state complexity [12] and transition complexity [2] of basic operations. The lower bound for state complexity of complementation is from [1].

In the table L_i, $i = 1, 2$, are regular languages where $\mathrm{nsc}(L_i) = n_i$ and $\mathrm{tc}(L_i) = m_i$. The cardinality of the alphabet is denoted by k.

When the upper and lower bounds do not coincide, in the table the row element for that operation is divided into two parts. The entry (†) refers to the case where the alphabet has two letters and a transition minimal NFA for L_i has the same number of transitions for both symbols, $i = 1, 2$. In the general case, the upper bound for transition complexity of intersection depends on the numbers of transitions for each symbol. The entries (‡) refer to the case where L_1

Table 1. Nondeterministic state and transition complexity

	State complexity	Transition complexity
Union	$n_1 + n_2 + 1$	$m_1 + m_2 + s(L_1) + s(L_2)$
Intersection	$n_1 n_2$	$\frac{1}{2} m_1 m_2^{(\dagger)}$
Complement (u.b.)	2^{n_1}	$k 2^{m_1+1}$
(l.b.)		$2^{m_1/2-2} - 1$
Catenation	$n_1 + n_2$	$m_1 + m_2 + f(L_1)$
Kleene star (u.b.)	$n_1 + 1$	$m_1 + k + fs(L_1)^{(\ddagger)}$
(l.b.)		$m_1 + fs(L_1)^{(\ddagger)}$
Reversal	$n_1 + 1$	$m_1 + f(L_1)$

does not contain the empty word, in the other case the upper and lower bound are both $m_1 + f(L_1)$.

In the results for transition complexity, further work is needed to determine how the measures $s(L_i)$, $f(L_i)$ and $fs(L_i)$ interact with $\mathrm{tc}(L_i)$. For example, in the worst-case examples for catenation one could try to determine what values $(\leq \mathrm{tc}(L_1))$ the term $f(L_1)$ may have.

Problem 2.1. Which range of values the measures $s(\cdot)$, $f(\cdot)$, $fs(\cdot)$ may have in worst-case examples for the transition complexity of basic operations given in Table 1?

The state complexity of morphisms and inverse morphisms is usually not examined because the constructions yield easily tight bounds, in particular, the standard construction for inverse morphism does not increase the number of states of an NFA. On the other hand, only a quadratic upper bound, and no matching lower bound, is known for transition complexity of inverse morphisms [2].

Problem 2.2. What is the transition complexity of inverse morphism?

In a worst-case comparison of nondeterministic state complexity and transition complexity, it has been established using counting arguments that there exist finite languages L_n, $n \geq 1$, with $\mathrm{tc}(L_n) \in \Omega(\frac{\mathrm{nsc}(L_n)^2}{\log(\mathrm{nsc}(L_n))})$ [9,16]. However, the counting arguments do not yield efficiently constructible languages having a corresponding transition complexity lower bound. An explicit construction of finite languages L_n, $n \geq 1$, with $\mathrm{tc}(L_n) \in \Omega(\mathrm{nsc}(L_n) \cdot \sqrt{\mathrm{nsc}(L_n)})$ is given in [3].

It seems difficult to obtain useful general purpose tools for proving lower bounds for the transition complexity of particular regular languages, similar as the fooling set methods used to prove lower bounds for the number of states of an NFA [5,13]. The lack of such tools makes it hard to obtain tight lower bounds when considering operational transition complexity.

Problem 2.3. Develop general purpose tools for proving transition complexity lower bounds (in the spirit of the techniques [5,13] used for nondeterministic state complexity).

If the number of states is fixed, it is much easier to prove lower bounds for the number of transitions. For suitably constructed NFAs it is relatively straightforward to establish lower bounds for the k-strict transition complexity and, for any $k \geq 0$, one can find families of regular languages $L_{n,k}$ ($n \geq 1$) for which $\mathrm{stc}_k(L_{n,k})$ is of a different order than $\mathrm{tc}(L_{n,k})$.

Proposition 2.1. [3] There exist regular languages L_n, $n \geq 1$, such that $\mathrm{stc}_1(L_n) \in O(\mathrm{nsc}(L_n))$ and $\mathrm{stc}_0(L_n) \in \Omega((\mathrm{nsc}(L_n))^2)$.

The result of Proposition 2.1 represents a maximal trade-off between the number of states and the number of transitions in any NFAs recognizing the languages L_n. If the NFAs are restricted to be state minimal, the number of transitions has to be quadratic in $\mathrm{nsc}(L_n)$ but by allowing one additional state in the NFA it is possible to have a number of transitions that is linear in $\mathrm{nsc}(L_n)$. Proposition 2.1 can be generalized to establish an analogous maximal gap in transition complexity when comparing NFAs that, for an arbitrary $k \geq 1$, allow respectively, $k - 1$ and k additional states compared to the size of a state-minimal NFA [3]. Earlier it was shown in [7] that by allowing a non-constant number of additional states, the number of transitions can be decreased from quadratic to linear.

When considering the reverse trade-off, one can construct a family of regular languages L_n, $n \geq 1$, such that for a constant $c > 1$,

$$\text{any transition minimal NFA for } L_n \text{ needs at least } c \cdot \mathrm{nsc}(L_n) \text{ states.} \tag{1}$$

Above the constant c depends on the alphabet, but it is not clear how large the gap between the number of states of a transition minimal NFA and the nondeterministic state complexity of the corresponding language can become. We conjecture that the number of states in transition minimal NFAs for any regular languages L_n, $n \geq 1$, is $O(\mathrm{nsc}(L_n))$.

Problem 2.4. Determine an upper bound, depending on the alphabet, for the constant c in (1).

References

1. Birget, J.-C.: Partial orders on words, minimal elements of regular languages and state complexity. Theoret. Comput. Sci. 119, 267–291 (1993)
2. Domaratzki, M., Salomaa, K.: Transition complexity of language operations. In: Leung, H., Pighizzini, G. (eds.): Proc. of Descriptional Complexity of Formal Systems, DCFS 2006, Las Cruces, NM, pp. 141–152 (2006)
3. Domaratzki, M., Salomaa, K.: Lower bounds for the transition complexity of NFAs. In: Královič, R., Urzyczyn, P. (eds.) MFCS 2006. LNCS, vol. 4162, pp. 315–326. Springer, Heidelberg (2006)
4. Goldstine, J., Kappes, M., Kintala, C.M.R., Leung, H., Malcher, A., Wotschke, D.: Descriptional complexity of machines with limited resources. J. Universal Computer Science 8, 193–234 (2002)
5. Glaister, I., Shallit, J.: A lower bound technique for the size of nondeterministic finite automata. Inf. Proc. Letters 59, 75–77 (1996)

6. Gramlich, G., Schnitger, G.: Minimizing NFA's and regular expressions. In: Diekert, V., Durand, B. (eds.) STACS 2005. LNCS, vol. 3404, pp. 399–411. Springer, Heidelberg (2005)
7. Gruber, H., Holzer, M.: A note on the number of transitions of nondeterministic finite automata. In: Fernau, H. (ed.) 15. Theorietag der GI-Fachgruppe 0.1.5 Automaten und Formale Sprachen, pp. 24–25 (2005)
8. Gruber, H., Holzer, M.: Finding lower bounds for nondeterministic state complexity is hard. In: Ibarra, O.H., Dang, Z. (eds.) DLT 2006. LNCS, vol. 4036, pp. 363–374. Springer, Heidelberg (2006)
9. Gruber, H., Holzer, M.: Results on the average state complexity of finite automata accepting finite languages. In: Leung, H., Pighizzini, G. (eds.) Proc. of Descriptional Complexity of Formal Systems, DCFS 2006, Las Cruces, NM, pp. 267–275 (2006)
10. Gruber, H., Holzer, M.: Computational complexity of NFA minimization for finite and unary languages. In: Proc. of 1st International Conference on Language Theory and Applications, LATA (2007)
11. Gruber, H., Holzer, M.: Inapproximability of nondeterministic state and transition complexity assuming P $\neq$ NP. In: Proc. of 11th International Conference Developments in Language Theory, DLT (2007)
12. Holzer, M., Kutrib, M.: Nondeterministic descriptional complexity of regular languages. Internat. J. Foundations of Computer Science 14, 1087–1102 (2003)
13. Hromkovič, J.: Communication Complexity and Parallel Computing. Springer, Heidelberg (1997)
14. Hromkovič, J.: Descriptional complexity of finite automata: Concepts and open problems. J. Automata, Languages and Combinatorics 7, 519–531 (2002)
15. Jiang, T., Ravikumar, B.: Minimal NFA problems are hard. SIAM J. Computing 22, 1117–1141 (1993)
16. Kari, J.: Personal communication (2006)
17. Malcher, A.: Minimizing finite automata is computationally hard. Theoret. Comput. Sci. 327, 375–390 (2004)
18. Salomaa, A., Salomaa, K., Yu, S.: State complexity of combined operations. Theoret. Comput. Sci. (to appear)
19. Salomaa, K., Yu, S.: On the state complexity of combined operations and their estimation. Internat. J. Foundations of Computer Science (to appear)
20. Schnitger, G.: Regular expressions and NFAs without ε-transitions. In: Durand, B., Thomas, W. (eds.) STACS 2006. LNCS, vol. 3884, pp. 432–443. Springer, Heidelberg (2006)
21. Yu, S.: Grail+: A symbolic computation environment for finite-state machines, regular expressions and finite languages. (2002),
http://www.csd.uwo.ca/Research/grail
22. Yu, S.: State complexity of regular languages. J. Automata, Languages and Combinatorics 6, 221–234 (2001)

From Determinism to Non-determinism in Recognizable Two-Dimensional Languages[*]

Marcella Anselmo[1], Dora Giammarresi[2], and Maria Madonia[3]

[1] Dipartimento di Informatica ed Applicazioni,
Università di Salerno I-84084 Fisciano (SA) Italy
anselmo@dia.unisa.it
[2] Dipartimento di Matematica. Università di Roma "Tor Vergata",
via della Ricerca Scientifica, 00133 Roma, Italy
giammarr@mat.uniroma2.it
[3] Dip. Matematica e Informatica, Università di Catania,
Viale Andrea Doria 6/a, 95125 Catania, Italy
madonia@dmi.unict.it

Abstract. Tiling systems that recognize two-dimensional languages are intrinsically non-deterministic models. We introduce the notion of deterministic tiling system that generalizes deterministic automata for strings. The corresponding family of languages matches all the requirements of a robust deterministic class. Furthermore we show that, differently from the one-dimensional case, there exist many classes between deterministic and non-deterministic families that we separate by means of examples and decidability properties.

Keywords: Automata and Formal Languages. Unambiguity, Determinism. Two-dimensional languages.

1 Introduction

Two-dimensional languages are sets of pictures or two-dimensional arrays of symbols chosen in a finite alphabet. The increasing interest for pattern recognition and image processing has motivated the research on two-dimensional (2D for short) languages, and nowadays this is a research field of great interest. Since the sixties, many approaches have been presented in the literature in order to find in 2D a counterpart of what regular languages are in one dimension (1D): finite automata, grammars, logics and regular expressions. In 1991, an unifying point of view was presented by A. Restivo and D. Giammarresi who defined the family REC of *recognizable picture languages* (see [6] and [7]). This definition takes as starting point a characterization of recognizable string languages in terms of local languages and projections (cf. [5]): the pair of a local picture language and a projection is called *tiling system*.

---

[*] This work was partially supported by PRIN project *Linguaggi Formali e Automi: aspetti matematici e applicativi.*

T. Harju, J. Karhumäki, and A. Lepistö (Eds.): DLT 2007, LNCS 4588, pp. 36–47, 2007.
© Springer-Verlag Berlin Heidelberg 2007

REC family inherits several properties from the class of regular string languages. A crucial difference lies in the fact that the definition of recognizability by tiling systems is intrinsically non-deterministic. Deterministic machine models to recognize two-dimensional languages have been considered in the literature: they always accept classes of languages smaller than the corresponding non-deterministic ones (see for example, [3,8,13]). This seems to be unavoidable when jumping from one to two dimensions. Further REC family is not closed under complement and therefore the definition of any constraint to force determinism in tiling systems should necessary result in a class smaller than REC.

In this paper we provide a definition of deterministic recognizable picture languages based on the formalism of tiling system, that generalizes 1D case. We first observe that a tiling system is not an effective computation device: given a tiling system and a picture, if we want to decide whether the picture belongs to the language recognized by the tiling system, we have to try to cover the picture with the given tiles, in a way that they match each others and the local symbols project to underlying symbols of the picture. All the attempts can be done following any scanning strategy: we could either start in the top-left corner and going row by row (from top to bottom) or by columns or in a spiral-like way or in many other more or less natural or strange ways of proceeding. Then in a sense, a set of tiles is the set of undirected transitions for a sort of automaton that reads the given picture along a fixed scanning strategy. Moreover, in general, such recognition process is non-deterministic: at each step of a recognition process for a picture of size (m, n), one can have a backtracking on all already scanned positions, i.e. a backtracking of $O(m \times n)$ steps. Further recall that parsing for 2D languages is a NP-complete problem [11]. The complexity of unary tiling-recognizable picture languages was recently considered in [2].

The definition of determinism we introduce consists of a property on the tiling system (i.e. the undirected transitions of the automata in the 1D case) that leads to no backtracking in any reasonable associated "computation". Furthermore determinism is a decidable property that implies unambiguity and polynomial parsing. More in details we will define four types of determinism, one for each corner-to-corner direction of reading of a picture. Observe that this is also the case for string languages. The notion of determinism on strings is somehow an "oriented" notion. When a set of undirected transitions is given for strings, there are two notions of determinism according to the reading direction: determinism (from left-to-right) and co-determinism (from right-to-left). *Deterministic Recognizable Languages* are defined as languages that admit a deterministic tiling system along one of the four corner-to-corner directions: $DREC$ denotes the class of all deterministic recognizable languages. As one would expect DREC class results to be closed under complement. In [4,14] it is given a different definition of determinism for tiling systems based on the way a tiling system is used to recognize pictures. Such definition is conceptually different and it does not reduce to conventional determinism on strings when restricting to one-row pictures.

In formal language theory, an intermediate notion between determinism and non-determinism is the notion of unambiguity. In an unambiguous model, we require that each accepted object admits only one successful computation. Both determinism and unambiguity correspond to the existence of a *unique* process of computation, but while determinism is a "local" notion, unambiguity is a fully "global" one. Unambiguous recognizable two-dimensional languages have been introduced in [6], and their family is referred to as UREC. Informally, a picture language belongs to UREC when it admits an unambiguous tiling system, that is if every picture has a unique counter-image in its corresponding local language; and this is an "orientation-free" notion. In [1], the proper inclusion of UREC in REC is proved. We show here that also DREC is properly included in UREC.

Hence DREC $\subset$ UREC $\subset$ REC, differently from the 1D case where all the corresponding classes collapse. Then we further strengthen this result and show that there is a very rich hierarchy of classes between determinism and non -determinism in 2D. We exhibit some classes, denoted *Col-UREC* and *Row-UREC*, that strictly separate DREC from UREC. Recall that DREC is the class of languages that can be accepted with backtracking zero at each step of the computation while UREC languages may require backtracking linear in the size of the pictures during computation. As intermediate classes, Col-UREC and Row-UREC are defined in such a way to have backtracking at most linear in one dimension of the picture at each step of its computation: they are defined by means of column-unambiguous and row-unambiguous tiling systems, respectively.

We conclude the paper by considering a decidability issue: it is easy to prove that it is decidable whether a given tiling system is deterministic while in [1] it is shown that it is undecidable whether it is unambiguous. Here we prove that for those intermediate notions of row-/ column-unambiguous tiling system such problem is still decidable.

2   Preliminaries

We introduce some definitions about two-dimensional languages. The notations used and more details can be mainly found in [7].

A *two-dimensional string* (or a *picture*) over a finite alphabet Σ is a two-dimensional rectangular array of elements of Σ. The set of all pictures over Σ is denoted by Σ^{**} and a *two-dimensional language* over Σ is a subset of Σ^{**}.

Given a picture $p \in \Sigma^{**}$, let $p_{(i,j)}$ denote the symbol in p with coordinates (i,j), $\ell_1(p) = m$, the number of rows and $\ell_2(p) = n$ the number of columns; the pair (m,n) is the *size* of p. The set of all pictures over Σ of size (m,n) is denoted by $\Sigma^{m,n}$. It will be needed to identify the symbols on the boundary of a given picture: for any picture p of size (m,n), we consider the *bordered picture* $\hat{p}$ of size $(m+2, n+2)$ obtained by surrounding p with a special *boundary symbol* $\# \notin \Sigma$: positions of $\hat{p}$ will be indexed in $\{0, 1, \cdots, m+1\} \times \{0, 1, \cdots, n+1\}$.

A *tile* is a picture of dimension $(2,2)$ and $B_{2,2}(p)$ is the set of all sub-blocks of size $(2,2)$ of a picture p. Given an alphabet Γ, a two-dimensional language $L \subseteq \Gamma^{**}$ is *local* if there exists a finite set Θ of tiles over $\Gamma \cup \{\#\}$ such that $L = \{p \in \Gamma^{**} | B_{2,2}(\hat{p}) \subseteq \Theta\}$ and we will write $L = L(\Theta)$.

A *tiling system* is a quadruple $(\Sigma, \Gamma, \Theta, \pi)$ where Σ and Γ are finite alphabets, Θ is a finite set of tiles over $\Gamma \cup \{\#\}$ and $\pi : \Gamma \to \Sigma$ is a projection. A two-dimensional language $L \subseteq \Sigma^{**}$ is *tiling recognizable* if there exists a tiling system $(\Sigma, \Gamma, \Theta, \pi)$ such that $L = \pi(L(\Theta))$ (extending π in the usual way). We denote by REC the family of all *tiling recognizable* picture languages.

The family REC is closed with respect to different types of operations (see [7] for all the proofs). The *column concatenation* of p and q (denoted by $p \oslash q$) and the *row concatenation* of p and q (denoted by $p \ominus q$) are partial operations, defined only if $\ell_1(p) = \ell_1(q)$ and if $\ell_2(p) = \ell_2(q)$, respectively and are given by:

$$p \oslash q = \boxed{\begin{array}{c|c} p & q \end{array}} \qquad\qquad p \ominus q = \boxed{\begin{array}{c} p \\ \hline q \end{array}}.$$

REC family is closed under row and column concatenation and their closures, under union, intersection and under rotation. All those closure properties confirm the close analogy with the one-dimensional case. The big difference regards the complement operation. In [7] and, in a different set-up, in [9], it is shown that the family REC is *not* closed under complement.

Let us give some examples to which we will refer later.

Example 1. Let $L_{fc=lc}$ be the language of pictures over $\Sigma = \{a, b\}$ whose the first column is equal to the last one. Language $L_{fc=lc} \in REC$. Informally we can define a local language where information about first column symbols of a picture p is brought along horizontal direction, by means of subscripts, to match the last column of p. Tiles are defined to have always same subscripts within a row while, in the right-border tiles, subscripts and main symbols should match. Below it is an example of a picture $p \in L_{fc=lc}$ together with a corresponding local picture p'.

$$p = \begin{array}{|c|c|c|c|c|} \hline b & b & a & b & b \\ \hline a & a & b & a & a \\ \hline b & a & a & a & b \\ \hline a & b & b & b & a \\ \hline \end{array} \qquad p' = \begin{array}{|c|c|c|c|c|} \hline b_b & b_b & a_b & b_b & b_b \\ \hline a_a & a_a & b_a & a_a & a_a \\ \hline b_b & a_b & a_b & a_b & b_b \\ \hline a_a & b_a & b_a & b_a & a_a \\ \hline \end{array}.$$

Let $L_{fc=c'}$ be the language of pictures such that the first column is equal to some i-th column, $i \neq 1$. Note that $L_{fc=c'} = L_{fc=lc} \oslash \Sigma^{**}$ and thus $L_{fc=c'} \in REC$. Similarly we can show that the languages $L_{c'=lc} = \Sigma^{**} \oslash L_{fc=lc}$, and $L_{c=c'} = \Sigma^{**} \oslash L_{fc=lc} \oslash \Sigma^{**}$ are in REC.   □

An interesting model of 2D automaton to recognize picture languages is the *two-dimensional on-line tessellation acceptor* (OTA) introduced in [8]. In a sense the OTA is an infinite array of identical finite-state automata in a two dimensional space. The computation goes by counter-diagonals starting from top-left towards bottom-right corner of the picture. A run of a OTA on a picture consists in associating a state to each position of the picture. The state for some position (i, j) is given by the transition function and depends on the symbol in that position and on the states already associated to positions $(i, j-1)$, $(i-1, j-1)$ and $(i-1, j)$ (note that an equivalent definition is possible with the state not depending on the state in the top-left corner, $(i-1, j-1)$). A deterministic version of this model is referred to as DOTA. The family of languages recognized by the two versions of the model ($\mathcal{L}(OTA)$, $\mathcal{L}(DOTA)$) are different. Although this kind of automaton is quite difficult to manage, this is actually the machine counterpart of a tiling system: in [10], it is proved that $REC = \mathcal{L}$(OTA).

3   Deterministic Tiling Systems

In this section we focus on the well accepted model of tiling system to discuss on the question of defining a corresponding deterministic model and to establish a "robust" definition for the class of *Deterministic Recognizable Two-dimensional Languages*. The main property for a deterministic model should be that a recognition process does *not* have *any* backtracking at each step of the computation. Moreover, as tiling systems generalize finite automata for strings to two dimensions (and in fact they coincide with finite automata in the special case of one-row pictures), we require the same for deterministic tiling systems.

To better fix these ideas we jump for a while to the one dimensional case. Recall that a string language L is accepted by a finite automaton if and only if it is the projection of a local language (given by a finite set of length-two strings on a local alphabet). In fact, using the given set of length-two strings, one can easily define a transition function of a conventional automaton for L. By conventional automaton here we mean an automaton that reads any input string starting from the leftmost position and going from left to right (the conventional reading direction, at least for occidental people!). It is easy to verify that the same set of length-two strings can be also used to define an automaton that recognizes strings in L by starting from the rightmost position and then proceeding from right to left (probably more natural for Arabian people!). The two automata can be obtained one from the other by exchanging initial and final states and reversing arrow directions. In the string case we have two notions of determinism: (conventional) determinism and co-determinism. In fact, if the right-to-left automaton is deterministic we say that conventional automaton is co-deterministic. This implies that a "deterministic property" on the set of length-two strings needs to be given according to a fixed direction. Moreover recall that not all regular string languages admit automata that are both deterministic and co-deterministic.

We now extend such considerations to the two dimensional case. In 2D there are 4 possible starting positions (the four corners) and therefore 4 possible main scanning directions (one from each corner). For a while let us focus on the direction from the top-left corner towards the bottom-right one, denoted by *tl2br-direction*: any reading of a picture along this direction has the property that we can read position (x, y) only if we have already read all the positions that are above and to the left of (x, y) that is all the positions (i, j) with $i \leq x$ and $j \leq y$. Similarly we can define all the others *corner-to-corner directions* in the set $C2C = \{tl2br, tr2bl, bl2tr, br2tl\}$.

Remark that, unlike the 1D case, once fixed a scanning direction there can be several reading paths on the picture p that are "compatible with" that direction.

For example, if we take the tl2br-direction, we can have the path that visits p column by column from left to right and each column from top to bottom, or another path that goes row by row from top to bottom and each row from left to right or another path that starts from top-left corner and then explores p by counter-diagonals, each one from top to bottom and so on... Observe that, we could also consider scanning processes that do not follow a fixed direction but this would not reduce to the conventional reading of a string when we restrict to one-row pictures: therefore they are not interesting for our purposes.

We are now ready to introduce *deterministic tiling systems*. As in 1D case, determinism will be defined as a property of the tiling system referred to a direction (one of the 4 main directions from the corners). Then any computation that follows a scanning path compatible with that scanning direction will be a deterministic computation (next step is determined with backtracking 0).

Definition 1. *A tiling system* $(\Sigma, \Gamma, \Theta, \pi)$ *is* tl2br-*deterministic if for any* γ_1, γ_2, $\gamma_3 \in \Gamma \cup \{\#\}$ *and* $\sigma \in \Sigma$ *there exists at most one tile* $\begin{array}{|c|c|} \hline \gamma_1 & \gamma_2 \\ \hline \gamma_3 & \gamma_4 \\ \hline \end{array} \in \Theta$, *with* $\pi(\gamma_4) = \sigma$.

Similarly we define d-*deterministic tiling systems for any corner-to-corner direction* $d \in C2C$.

Example 2. Let $L_{fr=fc}$ be the language of squares over a two-letters alphabet $\Sigma = \{a, b\}$ with the first row equal to the first column. $L_{fr=fc} \in REC$: indeed we will exhibit a tiling system $\mathcal{T} = (\Sigma, \Gamma, \Theta, \pi)$ recognizing L. The tiling system $\mathcal{T}$ is such that, for any picture p, the information on each letter of the first row is brought down till the diagonal and then left towards the first column. More precisely, we use a local alphabet $\Gamma = \{x_y^z$ with $x, y \in \{a, b\}$, $z \in \{0, 1, 2\}\}$ and define $\pi(x_y^z) = x$. The superscript symbol 0 occurs only in positions below the diagonal, the symbol 1 occurs only on the diagonal and symbol 2 occurs only above the diagonal, while the subscript symbols correspond to information we are bringing from the first row to the first column (making a turn at the diagonal). Here below it is given an example of a picture $p \in L_{fr=fc}$ together with the corresponding local picture p' (i.e. $\pi(p') = p$).

$$p = \begin{array}{|c|c|c|c|c|} \hline a & a & b & b & a \\ \hline a & b & b & a & a \\ \hline b & b & a & a & b \\ \hline b & b & a & a & a \\ \hline a & a & a & a & b \\ \hline \end{array} \qquad p' = \begin{array}{|c|c|c|c|c|} \hline a_a^1 & a_a^2 & b_b^2 & b_b^2 & a_a^2 \\ \hline a_a^0 & b_a^1 & b_b^2 & a_b^2 & a_a^2 \\ \hline b_b^0 & b_b^0 & a_b^1 & a_b^2 & b_a^2 \\ \hline b_b^0 & b_b^0 & a_a^0 & a_b^1 & a_a^2 \\ \hline a_a^0 & a_a^0 & a_a^0 & a_a^0 & b_a^1 \\ \hline \end{array}$$

It is easy to see that the tiling system $\mathcal{T}$ is tl2br-deterministic. Remark that it is not br2tl-deterministic: tiles $\begin{array}{|c|c|} \hline a_a^1 & a_a^2 \\ \hline a_a^0 & b_a^1 \\ \hline \end{array}$, $\begin{array}{|c|c|} \hline a_b^1 & a_a^2 \\ \hline a_a^0 & b_a^1 \\ \hline \end{array} \in \Theta$ with $\pi(a_a^1) = \pi(a_b^1) = a$.

Another important property of determinism should be the decidability. We show that it is decidable whether a given tiling system is deterministic.

Proposition 1. *It is decidable whether a given tiling system is d-deterministic for a given corner-to-corner direction d ∈ C2C.*

Proof. Given a tiling system $\mathcal{T} = (\Sigma, \Gamma, \Theta, \pi)$, in order to test, for example, whether it is tl2br-deterministic it suffices to verify whether there exist in Θ two tiles $\begin{array}{|c|c|}\hline \gamma_1 & \gamma_2 \\\hline \gamma_3 & \gamma_4 \\\hline\end{array}$, $\begin{array}{|c|c|}\hline \gamma_1 & \gamma_2 \\\hline \gamma_3 & \gamma_4' \\\hline\end{array} \in \Theta$, with $\gamma_4 \neq \gamma_4'$ and $\pi(\gamma_4) = \pi(\gamma_4')$. □

A recognizable two-dimensional language L is *deterministic*, if it admits a d-deterministic tiling system for some corner-to-corner direction d. Moreover, we denote by *DREC*, the class of *Deterministic Recognizable Two-dimensional Languages.*

We first observe that, as one would expect, deterministic recognizable languages are unambiguous (i.e. DREC⊆UREC). In fact, if at each step of recognition of a given picture we have only one possible local symbol to choose, then we have only one possible local counter-image for the input picture. We will prove that there are unambiguous recognizable languages that are not deterministic. Moreover, in the next section we will stress this result by exhibiting some other classes between DREC and UREC.

Remark that DREC is closed under rotation. Indeed if L is recognized by a d-deterministic tiling system, say a tl2br-deterministic one, then its (clockwise) 90°-rotation is accepted by a tr2bl-deterministic tiling system obtained by rotation of tiles. We now show that DREC family has a natural counterpart in the formalism of 2OTA.

Proposition 2. *The class DREC is equal to the closure by rotation of $\mathcal{L}(DOTA)$.*

Proof. Let $L \in DREC$ and let $\mathcal{T}$ be a d-deterministic tiling system for L. If $\mathcal{T}$ is tl2br-deterministic, then the OTA simulating the tiling system $\mathcal{T}$ and accepting the language L, as in the proof of [10], results to be deterministic. Then $L \in \mathcal{L}(DOTA)$. If $\mathcal{T}$ is d-deterministic, for some $d \in C2C$, then a proper rotation of $\mathcal{T}$ will be tl2br-deterministic and the proof follows.

Now, let $L \in \mathcal{L}(DOTA)$. The tiling system for L, obtained as in [10], is tl2br-deterministic. The proof is completed by the closure by rotation of DREC. □

Proposition 3. *DREC is properly included in UREC*

Proof. We will exhibit a language $L_{frames} \in$ UREC \ DREC using the characterization in Proposition 2. Consider language $L_{fr=fc}$ as defined in Example 2. One can easily show that $L_{fr=fc} \in \mathcal{L}(DOTA)$; while its 180° rotation, say $L_{lr=lc}$, the language of all square pictures with the last row equal to the last column is not in $\mathcal{L}(DOTA)$ (cf. [8]). Hence define L_{frames} as the intersection of four languages over $\Sigma = \{a, b\}$: $L_{fr=fc}$, $L_{lr=lc}$, L', the language of all square pictures with the second row equal to the reverse of the second-last column, and L'', the language of all square pictures with the second-last row equal to the reverse of the second column. Formally, let $L_{frames} = \{p \in \Sigma^{**} | l_1(p) = l_2(p) = n,\ p_{(n,i)} = p_{(i,n)},$

$p_{(2,i)} = p_{(n-i+1,n-1)}$, $p_{(i,1)} = p_{(1,i)}$ and $p_{(n-1,i)} = p_{(n-i+1,2)} \forall i = 1, \ldots, n\}$. The proof that $L_{frames} \notin$ DREC is rather involved and we omit it for lack of space.

Moreover, it can be shown that each one of the four languages defining L_{frames} is in UREC (for example the tiling system for $L_{fr=fc}$ given in Example 2 is tl2br-deterministic and then unambiguous) and, since UREC is closed with respect to intersection (cf. [1]), we have $L_{frames} \in$ UREC.   □

We conclude this section by stating that the class of deterministic recognizable languages is closed under complement and therefore it shares such important property with any other 'deterministic' model. (Remember that the whole REC family is not closed under complement.) The proof follows from Proposition 2 and the closure of $\mathcal{L}(DOTA)$ under complement [8].

Proposition 4. *DREC is closed under complement.*

4 Between DREC and UREC Classes

In this section we show that differently from one dimensional case, there is a very rich hierarchy of classes between determinism and non determinism by exhibiting some classes, we denote *Col-UREC* and *Row-UREC*, that strictly separate DREC from UREC. DREC is the class of languages that can be accepted with backtracking 0 in their computations; while UREC languages may require backtracking linear in the size of the pictures during computation. Col-UREC and Row-UREC are defined in such a way to have backtracking at most linear in one dimension of the picture. They correspond to an intermediate notion between determinism and unambiguity, and hence they lie between DREC and UREC. Note that the situation is extremely more complex than in 1D where all the corresponding classes collapse. Finally we prove some decidability results regarding those new definitions (Proposition 6).

We now define *column-* and *row-unambiguous* languages. For this, we use a different point of view for two-dimensional scanning directions: we somehow consider one dimension at each time and therefore move only along that direction. More precisely, we consider four side-to-side scanning directions namely left-to-right and vice versa, top-to-bottom and vice versa. In particular any reading of a picture p along the side-to-side direction for left-to-right, denoted by *l2r-direction*, has the property that we can read position (x, y) only if we have already read *all* the positions in the columns to the left, that is *all* the positions (i, j) with $j < y$. In other words the scanning of p proceeds column by column (despite we do not pay attention to the order of reading inside a given column). Similarly we can define all the others *side-to-side directions* in the set $S2S = \{l2r, r2l, t2b, b2t\}$.

We are now ready to give the definition of l2r-unambiguous tiling systems. Informally, a tiling system is l2r-unambiguous if, when used to recognize a picture by reading it along a l2r direction, there is only one possible next local column.

Definition 2. *A tiling system* $(\Sigma, \Gamma, \Theta, \pi)$ *is* l2r-unambiguous *if for any column* $col' \in \Gamma^{m,1} \cup \{\#\}^{m,1}$, *and picture* $p \in \Sigma^{m,1}$, *there exists at most one* *local column* $col'' \in \Gamma^{m,1}$, *such that* $\pi(col'') = p$ *and* $B_{2,2}(p') \subseteq \Theta$ *where* $p' = \{\#\}^{1,2} \ominus (col' \oslash col'') \ominus \{\#\}^{1,2}$.

Similar properties define d-unambiguous tiling systems, for any side-to-side direction $d \in S2S$.

We say that a language is *column-unambiguous* if it is recognized by a d-unambiguous tiling system for some $d \in \{l2r, r2l\}$ and it is *row-unambiguous* if it is recognized by a d-unambiguous tiling system for some $d \in \{t2b, b2t\}$. Finally, we denote by *Col-UREC* the class of column-unambiguous languages and by *Row-UREC* the class of row-unambiguous languages.

Remark 1. A column-unambiguous tiling system is such that, during the computation of a picture of size (m, n), the backtracking at each step is at most m. This is because the next local column is uniquely determined without ambiguity after backtracking of m steps at most. Same remarks hold for row-unambiguity.

Remark 2. It is interesting to note that we could similarly define diagonal unambiguity, requiring that the next diagonal of local symbols is uniquely determined from the previous one (for example, the counter-diagonals like OTA's transitions waves). In this case, such a diagonal unambiguity would coincide with determinism, since the local symbol in a position on the diagonal does not depend on the other local symbols on the diagonal.

Example 3. Let $L_{fr=fc}$ be the language of squares over $\Sigma = \{a, b\}$ with first row equal to the first column and $\mathcal{T} = (\Sigma, \Gamma, \Theta, \pi)$, as introduced in Example 2. We show that $\mathcal{T}$ is $l2r$-unambiguous and hence $L_{fr=fc} \in$ Col-UREC.

Informally, for any local column $col' \in \Gamma^{m,1}$, and picture $p \in \Sigma^{m,1}$, the local column (if any) $col'' \in \Gamma^{m,1}$ (in Definition 2), is univocally determined as follows. The position of col'' on the diagonal is determined from the position of a symbol with superscript 1 in col'; we have x_y^1 when x is the underlying symbol of p and y is the matching symbol from the first row and first column. Above this position, we have $col''_{(i,1)} = x_y^2$ iff $p_{(i,1)} = x$ and $p_{(1,1)} = y$; below diagonal position, we have $col''_{(i,1)} = x_y^0$ iff $p_{(i,1)} = x$ and $col'_{(i,1)} = x_y'^0$.

Moreover, by similar argument one can show that $\mathcal{T}$ is also $r2l$-, $t2b$-, and $b2t$-unambiguous. Then $L_{fr=fc} \in$ Row-UREC.

The following proposition compares all the classes DREC, Col-UREC, Row-UREC, UREC and REC. The proof is almost trivial and it is omitted. In the sequel we will be able to show that all these inclusions are strict.

Proposition 5. *DREC* $\subseteq$ *(Col-UREC* $\cap$ *Row-UREC)* $\subseteq$ *(Col-UREC* $\cup$ *Row-UREC)* $\subseteq$ *UREC* $\subseteq$ *REC*.

Now we state some necessary condition for Col-UREC (and Row-UREC) family. We will associate the string language over the alphabet of the columns (or rows) with a two-dimensional language, in a way similar to [1,12]. More precisely, let $L \subseteq \Sigma^{**}$ be a picture language. For any $m \geq 1$, we consider the subset $L_h(m) \subseteq L$ containing all pictures with exactly m rows. Such language $L_h(m)$

can be viewed as a string language over the alphabet $\Sigma^{m,1}$ of the columns, i.e. words in $L_h(m)$ have a "fixed height m". In an analogous way one can define the language $L_w(n)$ of pictures with fixed width n.

Furthermore if L is in REC and $\mathcal{T}$ is a tiling system that recognizes it, we can construct an automaton $\overrightarrow{\mathcal{A}}_m\,(\mathcal{T})$ that recognizes $L_h(m)$, as follows. The states of $\overrightarrow{\mathcal{A}}_m\,(\mathcal{T})$ are the local columns of $\Gamma^{m,1}$ plus the initial state that is the column $\{\#\}^{m,1}$; for each pair of states col', col'', we add a transition labelled $\pi(col'')$, from col' to col'', iff $B_{2,2}(\{\#\}^{1,2}\ominus(col'\oplus col'')\ominus\{\#\}^{1,2}) \subseteq \Theta$. The final states are columns $col' \in \Gamma^{m,1}$ such that $B_{2,2}(\{\#\}^{1,2}\ominus(col'\oplus\{\#\}^{m,1})\ominus\{\#\}^{1,2}) \subseteq \Theta$. Note that the number of states of $\overrightarrow{\mathcal{A}}_m\,(\mathcal{T})$ is $|\Gamma|^m+1$, at most, and thus upper limited by k^m for some k. In an analogous way, we can construct the automata $\overleftarrow{\mathcal{A}}_m\,(\mathcal{T})$, $\mathcal{A}_n^{\downarrow}(\mathcal{T})$ and $\mathcal{A}_n^{\uparrow}(\mathcal{T})$ for $L_h(m)^{Rev}$, $L_w(n)$ and $L_w(n)^{Rev}$, respectively, where Rev denotes the reverse of a (string) language.

Finally for any $L \subseteq \Sigma^*$ denote by M_L the infinite boolean matrix $M_L = \|a_{\alpha\beta}\|_{\alpha\in\Sigma^*,\beta\in\Sigma^*}$ where $a_{\alpha\beta} = 1$ iff $\alpha\beta \in L$.

Theorem 1. *Let $L \subseteq \Sigma^{**}$.*
If $L \in Col\text{-}UREC$, then there is a k such that, for all $m \geq 1$, the number of different rows of either $M_{L_h(m)}$ or $M_{L_h(m)^{Rev}}$ is less than or equal to k^m.
If $L \in Row\text{-}UREC$, then there is a k such that, for all $n \geq 1$, the number of different rows of either $M_{L_w(n)}$ or $M_{L_w(n)^{Rev}}$ is less than or equal to k^n.

Proof. Let $\mathcal{T} = (\Sigma, \Gamma, \Theta, \pi)$ be a tiling system recognizing L. The main observation is now that if $\mathcal{T}$ is d-unambiguous with $d = l2r$ ($r2l$, $t2b$, or $b2t$, resp.) then the automaton $\overrightarrow{\mathcal{A}}_m\,(\mathcal{T})$ ($\overleftarrow{\mathcal{A}}_m\,(\mathcal{T})$, $\mathcal{A}_n^{\downarrow}(\mathcal{T})$, or $\mathcal{A}_n^{\uparrow}(\mathcal{T})$, resp.) will result deterministic. Consider the automaton $\overrightarrow{\mathcal{A}}_m\,(\mathcal{T})$. For any state $col' \in \Gamma^{m,1} \cup \{\#\}^{m,1}$, and symbol $\sigma \in \Sigma^{m,1}$, the arriving state (if any) is col'' uniquely determined by the Definition 2. So $\overrightarrow{\mathcal{A}}_m\,(\mathcal{T})$ is deterministic. Therefore there exists k such that, for all $m \geq 1$ the string language $L_h(m)$ is accepted by a *deterministic* (string) automaton with k^m states at most.

From Myhill-Nerode Theorem, we also know that the number of states of the minimal deterministic automaton accepting $L_h(m)$ is equal to the number of different rows of $M_{L_h(m)}$. Therefore the number of different rows of $M_{L_h(m)}$ is less than or equal to k^m. The proof is analogous in the other cases. □

As an application of Theorem 1, let us show a language not in Col-UREC.

Example 4. Consider the language $L = L_{fc=c'} \cap L_{c'=lc}$ and, for any $m > 1$, consider language $L_h(m)$ of pictures of fixed height m in L. This is the string language over the alphabet $A = \Sigma^{m,1}$ with at least two occurrences of the first and of the last symbol. If Σ has σ symbols, then $A = \Sigma^{m,1}$ has σ^m elements. One can show that $M_{L_h(m)}$ has at least 2^{σ^m} different rows. Indeed the rows corresponding to two pictures with different sets of columns are distinct. Since the different subsets of columns in $\Sigma^{m,1}$ are 2^{σ^m}, then $M_{L_h(m)}$ has at least 2^{σ^m} different rows. The same holds for $M_{L_h(m)^{Rev}}$ since $L^{Rev} = L$ and then, $L \notin Col\text{-}UREC$.

Next theorem shows that the inclusions in Proposition 5 are all strict, by exhibiting languages that separate the classes. First we state the following lemma. Its proof easily follows by observing that the tiling system for the intersection of two languages, as constructed in [7], preserves side-to-side unambiguity.

Lemma 1. *If $L_1, L_2 \subseteq \Sigma^{**}$ are recognized by a d-unambiguous tiling system for some $d \in S2S$, then so is $L_1 \cap L_2$.*

Theorem 2. *DREC$\subset$ (Col-UREC$\cap$ Row-UREC)$\subset$ (Col-UREC$\cup$ Row-UREC) $\subset$ UREC $\subset$ REC, with all strict inclusions.*

Proof. We exhibit three languages L_1, L_2, L_3 showing the three first strict inclusions. A language in REC\ UREC is shown in [1] (namely $L_{c=c'}$).

Language L_1 is L_{frames} as introduced in the proof of Proposition 3. We have $L_1 \in$ (Col-UREC $\cap$ Row-UREC)\$DREC$. Indeed L_1 is defined as the intersection of four languages; each one can be recognized by a tiling system that is d-unambiguous for any $d \in S2S$ (see Example 3 for part of the proof) and Lemma 1 holds. The proof that $L_{frames} \notin$ DREC is rather involved and we omit it for lack of space.

Language L_2 is $L_2 = L_{fc=c'} \cap L_{c'=lc} \cap S$, where S is the language of squares pictures. $L_2 \in$ Row-UREC\ Col-UREC. A t2b-unambiguous tiling system can be constructed as here sketched. The idea is, starting from the first row of a picture, to mark both all the columns candidates to be equal to the first one and all the columns candidates to be equal to the last one and to propagate this information downwards. In the last row, we can check whether two entire columns were found, one equal to the first one and another equal to the last one. The condition that the picture is a square is needed to detect when the *last* row is reached. Hence $L_2 \in$ Row-UREC. On the contrary $L_2 \notin$ Col-UREC, using Theorem 1. In fact $M_{L_h(m)}$ has at least $(\sigma^m/m)^m$ different rows, by calculations similar to the ones done in Example 4.

Language L_3, is obtained by intersection of $L_{fc=c'}$ (see Example 1) with its three 90°-rotations. $L_3 \in$UREC since $L_{fc=c'} \in UREC$ and UREC is closed by rotations and intersection. On the other hand $L_3 \notin$ (Col-UREC $\cup$ Row-UREC). One can show that $M_{L_h(m)}$ has at least $2^{\sigma^{m-2}}$ different rows, similarly as in Example 4. So applying Theorem 1, $L_3 \notin$Col-UREC; $L_3 \notin$Row-UREC since it coincides with its rotations. $\square$

We conclude with some decidability issues. Proposition 1 shows that it is decidable whether a given tiling system is corner-to-corner deterministic. On the contrary in [1] it was shown that it is undecidable whether a given tiling system is unambiguous. Here we show that it is still decidable whether a tiling system is column-/ row-unambiguous.

Proposition 6. *It is decidable whether a tiling system is d-unambiguous for a given side-to-side direction $d \in S2S$.*

Proof. Consider a tiling system $\mathcal{T} = (\Sigma, \Gamma, \Theta, \pi)$ and direction $l2r$. Denote P the cardinality of the set $\{(\alpha, \beta) |\ \alpha, \beta \in \Gamma\}$. Trivially, P is the upper bound on the length of shortest (i.e. with the minimal number of rows) pictures col' and p for which there is no col'' as in the Definition 2. Then, it suffices to verify whether, for any $n \leq P$, there are no pictures $col_1 \in \Gamma^{n,1} \cup \{\#\}^{n,1}$, $col_2, col_3 \in \Gamma^{n,1}$, such that $col_2 \neq col_3$, $\pi(col_2) = \pi(col_3)$, $B_{2,2}(p_1) \subseteq \Theta$, $B_{2,2}(p_2) \subseteq \Theta$ where $p_1 = \{\#\}^{1,2} \ominus (col_1 \oplus col_2) \ominus \{\#\}^{1,2}$ and $p_2 = \{\#\}^{1,2} \ominus (col_1 \oplus col_3) \ominus \{\#\}^{1,2}$. The proof is similar for $d \in \{r2l, t2b, b2t\}$. $\square$

Acknowledgments

We are grateful to Antonio Restivo for all our fruitfully discussions.

References

1. Anselmo, M., Giammarresi, D., Madonia, M., Restivo, A.: Unambiguous Recognizable Two-dimensional Languages. RAIRO: Theoretical Informatics and Applications EDP Sciences 2006 40(2), 227–294 (2006)
2. Bertoni, A., Goldwurm, M., Lonati, V.: On the complexity of unary tiling-recognizable picture languages. In: Thomas, W., Weil, P. (eds.) STACS 2007. LNCS, vol. 4393, Springer-Verlag, Heidelberg (To appear, 2007)
3. Blum, M., Hewitt, C.: Automata on a two-dimensional tape. IEEE Symposium on Switching and Automata Theory, pp. 155–160 (1967)
4. B. Borchert, K. Reinhardt. Deterministically and Sudoku-Deterministically Recognizable Picture Languages. http://www-fs.informatik.uni-tuebingen.de/~borchert/papers/Borchert-Reinhardt_2006_Sudoku.pdf
5. Eilenberg, S.: Automata, Languages and Machines, vol. A. Academic Press, San Diego (1974)
6. Giammarresi, D., Restivo, A.: Recognizable picture languages. Int. Journal Pattern Recognition and Artificial Intelligence 6(2,3), 241–256 (1992)
7. Giammarresi, D., Restivo, A.: Two-dimensional languages. In: Rozenberg, G., et al. (ed.) Handbook of Formal Languages, vol. III, pp. 215–268. Springer Verlag, Heidelberg (1997)
8. Inoue, K., Nakamura, A.: Some properties of two-dimensional on-line tessellation acceptors. Information Sciences 13, 95–121 (1977)
9. Inoue, K., Nakamura, A.: Nonclosure properties of two-dimensional on-line tessellation acceptors and one-way parallel/sequential array acceptors. Transaction of IECE of Japan 6, 475–476 (1977)
10. Inoue, K., Takanami, I.: A characterization of recognizable picture languages. In: Nakamura, A., Saoudi, A., Inoue, K., Wang, P.S.P., Nivat, M. (eds.) ICPIA 1992. LNCS, vol. 654, Springer-Verlag, Berlin Heidelberg (1993)
11. Lindgren, K., Moore, C., Nordahl, M.: Complexity of two-dimensional patterns. Journal of Statistical Physics 91(5-6), 909–951 (1998)
12. Matz, O.: On piecewise testable, starfree, and recognizable picture languages. In: Nivat, M. (ed.) ETAPS 1998 and FOSSACS 1998. LNCS, vol. 1378, Springer, Heidelberg (1998)
13. Potthoff, A., Seibert, S., Thomas, W.: Nondeterminism versus determinism of finite automata over directed acyclic graphs. Bull. Belgian Math. Soc. 1, 285–298 (1994)
14. Reinhardt, K.: On some recognizable picture-languages. In: Brim, L., Gruska, J., Zlatuška, J. (eds.) MFCS 1998. LNCS, vol. 1450, pp. 760–770. Springer, Heidelberg (1998)

Coding Partitions: Regularity, Maximality and Global Ambiguity⋆

Marie-Pierre Béal[1], Fabio Burderi[2], and Antonio Restivo[2]

[1] Institut Gaspard-Monge,
Laboratoire d'informatique UMR 8049,
Université de Marne-la-Vallée, 77454 Marne-la-Vallée Cedex 2, France
beal@univ-mlv.fr
[2] Dipartimento di Matematica ed Applicazioni,
Università degli studi di Palermo,
Via Archirafi 34, 90123 Palermo, Italy
{Burderi,Restivo}@math.unipa.it

Abstract. The canonical coding partition of a set of words is the finest partition such that the words contained in at least two factorizations of a same sequence belong to a same class. In the case the set is not uniquely decipherable, it partitions the set into one unambiguous class and other parts that localize the ambiguities in the factorizations of finite sequences.

We firstly prove that the canonical coding partition of a regular set contains a finite number of regular classes. We give an algorithm for computing this partition. We then investigate maximality conditions in a coding partition and we prove, in the regular case, the equivalence between two different notions of maximality. As an application, we finally derive some new properties of maximal uniquely decipherable codes.

1 Introduction

In this paper, we call code a set of finite words. An important class of codes is the class of Uniquely Decipherable (UD) codes. This property allows the decoding of a sequence of concatenated codewords. Nevertheless, some classes of codes are used in information theory although they are not uniquely decipherable (see for instance [7], [9] and [10]). The condition of unique decipherability can also be weakened by considering that it applies only to codes with constraints (see [1]) or to codes with a constraint source (see [4], [6]). In [6], the classification of ambiguities of codes is investigated in the study of natural languages. From a combinatorial point of view, the study of ambiguities helps to understand the structure of a code.

To this purpose, the notions of coding partition and canonical coding partition of a code were introduced in [3] to study some decipherability conditions for codes that are weaker than the unique decipherability. The notion of coding partition

---

⋆ Partially supported by Italian MURST Project of National Relevance "Linguaggi Formali e Automi: Metodi, Modelli e Applicazioni".

T. Harju, J. Karhumäki, and A. Lepistö (Eds.): DLT 2007, LNCS 4588, pp. 48–59, 2007.
© Springer-Verlag Berlin Heidelberg 2007

generalizes that of UD code: indeed UD codes correspond to the extremal case in which each class contains exactly one element. In general, for codes that are not UD, the notion of coding partition allows to recover "unique decipherability" at the level of classes of the partition. In other words, such a notion gives a tool to *localize* the ambiguities for a code that is not UD: indeed the ambiguities are localized inside each class of the partition and a kind of mutual unambiguity holds between the different classes.

By taking into account the natural ordering between the partitions of a set X, where finer is higher, we have that the coding partitions form a complete lattice. As a consequence, given a code X, we can define the finest coding partition P of X. It is called the *characteristic* partition of X and it is denoted by $P(X)$.

The structure of $P(X)$ gives useful information about coding properties of X. In particular, an extremal case (each class of $P(X)$ is a singleton) corresponds to UD codes. The opposite extremal case ($P(X)$ contains only one class) gives rise to the definition of *Globally Ambiguous (GA)* codes. Such considerations lead to define a *canonical decomposition* of a code in at most one unambiguous component and in a set (possibly empty) of GA components.

Remark that the notion of coding partition is related to some special cases of the notion of $\mathfrak{F}$-factorization, introduced in [8].

In [3] it is given a Sardinas-Patterson like algorithm for computing the canonical coding partition of a finite code.

In this paper, we firstly prove that the canonical coding partition of a regular code has a finite number of classes, each one being regular. This result was conjectured in [3]. We give an exponential time algorithm for computing all classes of the partition which is based on automata constructions.

We then introduce the notion of maximality of coding partition with respect to a component and, in the regular case, we prove that if a coding partition is maximal with respect to one component, then it is maximal with respect to all the components. As an application, we prove in the last section that, if a regular UD code X is maximal, then any code containing strictly X is GA.

2   Partitions of a Code

Let A be a finite alphabet. We denote by A^* the set of finite words over the alphabet A, and by A^+ the set of nonempty finite words. A *code* X is here a subset of A^+. Its elements are called *code words*, the elements of X^* *messages* .

Let X be a code and let

$$P = \{X_1, X_2, \dots\},$$

be a partition of X i.e. : $\bigcup_{i \geq 1} X_i = X$ and $X_i \cap X_j = \emptyset$, for $i \neq j$.

A *P-factorization* of an element $w \in X^+$ is a factorization $w = z_1 z_2 \cdots z_t$, where

- $\forall i \ z_i \in X_k^+$,   for some $k \geq 1$
- if $t > 1$, $z_i \in X_k^+ \Rightarrow z_{i+1} \notin X_k^+$,   for all $1 \leq i \leq t - 1$.

The partition P is called a *coding partition* if any element $w \in X^+$ has a *unique P-factorization*, i.e. if

$$w = z_1 z_2 \cdots z_s = u_1 u_2 \cdots u_t,$$

where $z_1 z_2 \cdots z_s$, $u_1 u_2 \cdots u_t$ are *P-factorizations* of w, then $s = t$ and $z_i = u_i$ for $i = 1, \ldots, s$.

We say that a partition P is *concatenatively independent* if, for $i \neq j$,

$$X_i^+ \cap X_j^+ = \emptyset.$$

Then a necessary condition for a partition P to be a coding partition, is that P is concatenatively independent.

Let X be a code and let $x_1 x_2 \cdots x_s = y_1 y_2 \cdots y_t$ be two factorizations into code words of a message $w \in X^+$. We say that the relation $x_1 x_2 \cdots x_s = y_1 y_2 \cdots y_t$ is *prime* if for all $i < s$ and for all $j < t$ one has $x_1 x_2 \cdots x_i \neq y_1 y_2 \cdots y_j$.

In [3] it is proved that P is a coding partition of a code X iff for every prime relation $x_1 x_2 \cdots x_s = y_1 y_2 \cdots y_t$ these code words belong to the same component of the partition.

Recall that there is a natural order between the partitions of a set X: if P_1 and P_2 are two partitions of X, $P_1 \leq P_2$ if the elements of P_1 are unions of elements of P_2. In [3] is proved the next theorem.

Theorem 1. *The set of the coding partitions of a code X is a complete lattice.*

As a consequence of previous theorem we can give the next definition.

Given a code X, the finest coding partition P of X is called the *characteristic partition* of X and it is denoted by $P(X)$.

A code X is called *ambiguous* if it is not UD. It is called *globally ambiguous* (GA) if $|X| > 1$ and $P(X)$ is the trivial partition.

So UD codes and GA codes correspond to the two extremal cases: a code is UD if $|P(X)| = |X|$ and a code is GA if $|P(X)| = 1$.

Let X be a code and let $P(X)$ be the characteristic partition of X. Let X_0 be the union of all classes of $P(X)$ having only one element, i.e. of all classes $Z \in P(X)$ such that $|Z| = 1$. The code X_0 is a UD code and is called the *unambiguous component* of X. From $P(X)$ one then derives another partition of X

$$P_C(X) = \{X_0, X_1, \ldots\},$$

where $|X_i| > 1$, for $i \geq 1$. The sets X_i, with $i \geq 1$, are (see[3]) GA. They are called the GA *components* of X. The partition $P_C(X)$ is called the *canonical partition* of X: it defines a *canonical decomposition* of a code X in at most one unambiguous component and a (possibly empty) set of GA components. Roughly speaking, if a code X is not UD, then its canonical decomposition, on one hand separates the unambiguous component of the code (if any), and, on the other, localizes the ambiguities inside the GA components of the code. On the contrary, if X is UD, then its canonical decomposition contains only the unambiguous component X_0. Moreover if X is UD then every partition of X is a coding partition.

In [3] is given a Sardinas-Patterson like algorithm for computing the canonical coding partition of a finite code X and is also proved the next result.

Theorem 2. *Given a partition $P = \{X_1, X_2, \ldots, X_n\}$ such that X_i, for $i = 1, 2, \ldots, n$, is a regular set, then it is decidable whether P is a coding partition.*

In the same paper it was conjectured that *if X is regular, the number of classes of $P_C(X)$ is finite and each class of $P_C(X)$ is a regular set.*

The conjecture will be proved in the next section so the restrictive conditions considered in the Theorem 2 are not actually a restriction for regular codes.

3 Coding Partition of a Regular Code

In this section, we consider a regular code X.

We say that a coding partition of a code is *finite* if is has a finite number of components. We say that a coding partition of a code is *regular* if all the components of the partitions are regular. The following theorem gives a positive answer to previous conjecture.

Theorem 3. *The canonical partition of a regular code is finite and regular.*

Remark 1. Given a coding partition $P = \{X_1, X_2, \ldots\}$ of a code $X \subseteq A^+$, the condition that every word $w \in X^+$ admits a unique P-factorization has the following algebraic interpretation: the submonoid X^* is isomorphic to the *free product* of the submonoids X_i^*. We say that a submonoid $M \subseteq A^*$ is *indecomposable* if M is not factorizable in the free product of others submonoids. Then the previous theorem can be restated in the following algebraic setting.

Theorem 4. *Any regular monoid admits a canonical decomposition into a free product of at most one regular free monoid and a finite number (possibly zero) of regular indecomposable monoids.*

In order to prove Theorem 3, we give an algorithm for computing the finite automata accepting the components of the partition from a finite automaton accepting the code X.

A finite *automaton* $\mathcal{A} = (Q, I, E, T)$ is made of a finite set of states Q, a set of edges E labelled on an alphabet A, a set of initial states I and a set of final states T. We shall also consider automata labelled in A^*. A *successful path* is a path going from a state of I to a state of T. The set of labels of successful paths is the *language accepted* by the automaton.

An automaton is *unambiguous* if for any word z, any states p, q, there is at most one path going from p to q and labelled by z.

Let $\mathcal{A} = (Q, I, E, T)$ be a finite automaton. We define the automaton $\mathcal{A} \times \mathcal{A} = (Q', I', E', T')$ called the *square* of $\mathcal{A}$, where $Q' = Q \times Q$, $E' = \{(p, q) \xrightarrow{a} (p', q') \mid p \xrightarrow{a} p'$ and $q \xrightarrow{a} q' \in E\}$. The set of initial states I' and the set of final states T' will be specified later. A state (p, q) will be also denoted by $\begin{bmatrix} p \\ q \end{bmatrix}$.

Proof (Proof of Theorem 3). Let $\mathcal{A} = (Q, I, E, T)$ be a finite unambiguous automaton accepting the code X such that $I = \{i\}$, $T = \{t\}$, and which has no edge coming in i and no edge going out of t. Such an automaton, called a normalized automaton, can be obtained by standard constructions (see for instance [2]). By merging i and t into a single state denoted by 0, we get an automaton $\mathcal{B} = (Q, 0, E, 0)$ accepting the set X^*. Note that $\mathcal{B}$ is no more unambiguous unless X is UD.

We build the square automaton $\mathcal{B} \times \mathcal{B}$ and replace the state $\begin{bmatrix} 0 \\ 0 \end{bmatrix}$ by two states $\begin{bmatrix} 0 \\ 0 \end{bmatrix}_s$ and $\begin{bmatrix} 0 \\ 0 \end{bmatrix}_t$ such that the edges going out of $\begin{bmatrix} 0 \\ 0 \end{bmatrix}$ go out of $\begin{bmatrix} 0 \\ 0 \end{bmatrix}_s$ and the edges coming in $\begin{bmatrix} 0 \\ 0 \end{bmatrix}$ come in $\begin{bmatrix} 0 \\ 0 \end{bmatrix}_t$. Note that $\begin{bmatrix} 0 \\ 0 \end{bmatrix}_s$ has no incoming edges and $\begin{bmatrix} 0 \\ 0 \end{bmatrix}_t$ has no outgoing edges. We only keep in $\mathcal{B} \times \mathcal{B}$ the states belonging to paths from $\begin{bmatrix} 0 \\ 0 \end{bmatrix}_s$ to $\begin{bmatrix} 0 \\ 0 \end{bmatrix}_t$ and going at least one time through a state $\begin{bmatrix} p \\ q \end{bmatrix}$ with $p = 0, q \neq 0$ or $p \neq 0, q = 0$. By using the state-elimination technique (see for instance [?]), we remove the states $\begin{bmatrix} p \\ q \end{bmatrix}$ with p and q distinct from 0 and get an automaton $\mathcal{C}$ labelled in regular subsets of A^* whose states are $\begin{bmatrix} 0 \\ 0 \end{bmatrix}_s$, $\begin{bmatrix} 0 \\ 0 \end{bmatrix}_t$, and $\begin{bmatrix} p \\ q \end{bmatrix}$ with $p = 0, q \neq 0$ or $p \neq 0, q = 0$. There is at most one edge between two states and each label is a regular non-empty subset of A^*.

States $\begin{bmatrix} p \\ q \end{bmatrix}$ with $p = 0$ are called upper-zero states while states $\begin{bmatrix} p \\ q \end{bmatrix}$ with $q = 0$ are called lower-zero states. Hence $\begin{bmatrix} 0 \\ 0 \end{bmatrix}_s$ and $\begin{bmatrix} 0 \\ 0 \end{bmatrix}_t$ are both upper and lower-zero states.

We denote by $E\begin{bmatrix} p \\ q \end{bmatrix}\begin{bmatrix} p' \\ q' \end{bmatrix}$ the regular set related to the edge $\begin{bmatrix} p \\ q \end{bmatrix} \rightarrow \begin{bmatrix} p' \\ q' \end{bmatrix}$. With a slight abuse of language, we sometimes say that there is an edge labelled by a word w from a state $\begin{bmatrix} p \\ q \end{bmatrix}$ to state $\begin{bmatrix} p' \\ q' \end{bmatrix}$ whenever $w \in E\begin{bmatrix} p \\ q \end{bmatrix}\begin{bmatrix} p' \\ q' \end{bmatrix}$.

Let p_i, q_i, p_j, q_j be states in Q with q_i and q_j distinct from 0. Let e, f be the edges

$$e = \begin{bmatrix} 0 \\ p_i \end{bmatrix} \rightarrow \begin{bmatrix} q_i \\ 0 \end{bmatrix} \text{ and } f = \begin{bmatrix} q_j \\ 0 \end{bmatrix} \rightarrow \begin{bmatrix} 0 \\ p_j \end{bmatrix}$$

(*i.e.* respectively an edge from an upper-zero state to a lower-zero state and an edge from lower-zero state to an upper-zero state).

We denote by

- $L\begin{bmatrix} q_i \\ 0 \end{bmatrix}\begin{bmatrix} q_j \\ 0 \end{bmatrix}$ the regular set of labels of paths from $\begin{bmatrix} q_i \\ 0 \end{bmatrix}$ to $\begin{bmatrix} q_j \\ 0 \end{bmatrix}$ with *all its states being lower-zero states*.
- $S\begin{bmatrix} q_i \\ 0 \end{bmatrix}\begin{bmatrix} q_j \\ 0 \end{bmatrix}$ the union of the labels of all edges contained in a path from $\begin{bmatrix} q_i \\ 0 \end{bmatrix}$ to $\begin{bmatrix} q_j \\ 0 \end{bmatrix}$ with *all its states being lower-zero states*.

Note that we may have $q_i = q_j$. In this case, $L\begin{bmatrix} q_i \\ 0 \end{bmatrix}\begin{bmatrix} q_j \\ 0 \end{bmatrix}$ contains the empty word and $S\begin{bmatrix} q_i \\ 0 \end{bmatrix}\begin{bmatrix} q_j \\ 0 \end{bmatrix}$ may be the empty set.

We define the regular sets

$$Y = E\begin{bmatrix} 0 \\ p_i \end{bmatrix}\begin{bmatrix} q_i \\ 0 \end{bmatrix} \cdot L\begin{bmatrix} q_i \\ 0 \end{bmatrix}\begin{bmatrix} q_j \\ 0 \end{bmatrix} \cdot E\begin{bmatrix} q_j \\ 0 \end{bmatrix}\begin{bmatrix} 0 \\ p_j \end{bmatrix} + S\begin{bmatrix} q_i \\ 0 \end{bmatrix}\begin{bmatrix} q_j \\ 0 \end{bmatrix},$$

$$S_{ef} = \begin{cases} Y & \text{if } p_i \neq 0, p_j \neq 0, \\ Y + E\begin{bmatrix} 0 \\ 0 \end{bmatrix}_s\begin{bmatrix} q_i \\ 0 \end{bmatrix} & \text{if } p_i = 0, p_j \neq 0, \\ Y + E\begin{bmatrix} q_j \\ 0 \end{bmatrix}\begin{bmatrix} 0 \\ 0 \end{bmatrix}_t & \text{if } p_i \neq 0, p_j = 0, \\ Y + E\begin{bmatrix} 0 \\ 0 \end{bmatrix}_s\begin{bmatrix} q_i \\ 0 \end{bmatrix} + E\begin{bmatrix} q_j \\ 0 \end{bmatrix}\begin{bmatrix} 0 \\ 0 \end{bmatrix}_t & \text{if } p_i = p_j = 0, \end{cases}$$

where the symbol $+$ is the union symbol and the dot symbol is the concatenation symbol.

Let $p_i, q_i, p_j, q_j, p_k, q_k$ be states in Q with q_i, q_j, p_j, p_k distinct from 0. Let e, f, g be the edges

$$e = \begin{bmatrix} 0 \\ p_i \end{bmatrix} \to \begin{bmatrix} q_i \\ 0 \end{bmatrix}, \quad f = \begin{bmatrix} q_j \\ 0 \end{bmatrix} \to \begin{bmatrix} 0 \\ p_j \end{bmatrix} \text{ and } g = \begin{bmatrix} 0 \\ p_k \end{bmatrix} \to \begin{bmatrix} q_k \\ 0 \end{bmatrix}.$$

We define the regular set

$$S_{efg} = E\begin{bmatrix} 0 \\ p_i \end{bmatrix}\begin{bmatrix} q_i \\ 0 \end{bmatrix} \cdot L\begin{bmatrix} q_i \\ 0 \end{bmatrix}\begin{bmatrix} q_j \\ 0 \end{bmatrix} \cdot E\begin{bmatrix} q_j \\ 0 \end{bmatrix}\begin{bmatrix} 0 \\ p_j \end{bmatrix} + E\begin{bmatrix} q_j \\ 0 \end{bmatrix}\begin{bmatrix} 0 \\ p_j \end{bmatrix} \cdot L\begin{bmatrix} 0 \\ p_j \end{bmatrix}\begin{bmatrix} 0 \\ p_k \end{bmatrix} \cdot E\begin{bmatrix} 0 \\ p_k \end{bmatrix}\begin{bmatrix} q_k \\ 0 \end{bmatrix}.$$

We define similar sets S_{ef} and S_{efg} when e, g are edges from a lower-zero state to an upper-zero state and f is an edge from an upper-zero state to a lower-zero state, by exchanging the roles played by the upper and lower states.

We get a finite number of regular subsets of X. Some of these states may have a nonempty intersection. We replace two parts having a non-empty intersection by their union. After a finite number of steps we get a finite number of regular subsets of X whose two by two intersections are empty. We denote these sets by $X_1, X_2, \ldots, X_r$. We define the set $X_0 = X - \bigcup_{i=1}^{r} X_i$. We claim that $(X_i)_{0 \le i \le r}$ is the canonical coding partition of X, which proves the proposition.

To prove our claim, we show that any two code words which belong to a same prime relation belong to a same component X_i. Let $z = x_1 x_2 \ldots x_n = y_1 y_2 \ldots y_m$ be a prime relation where x_i, y_j are codewords. The existence of such a factorization is equivalent to the existence of a path in $\mathcal{C}$:

$$\begin{bmatrix} 0 \\ 0 \end{bmatrix}_s \xrightarrow{(e_1)} \begin{bmatrix} q_{01} \\ 0 \end{bmatrix} \ldots \begin{bmatrix} q_{0j_0} \\ 0 \end{bmatrix} \xrightarrow{(e_2)} \begin{bmatrix} 0 \\ p_{11} \end{bmatrix} \ldots \begin{bmatrix} 0 \\ p_{1i_1} \end{bmatrix} \xrightarrow{(e_3)} \begin{bmatrix} q_{11} \\ 0 \end{bmatrix} \ldots \begin{bmatrix} q_{1j_1} \\ 0 \end{bmatrix} \xrightarrow{(e_4)} \begin{bmatrix} 0 \\ p_{21} \end{bmatrix} \ldots$$

$$\xrightarrow{(e_{k-2})} \begin{bmatrix} 0 \\ p_{r1} \end{bmatrix} \ldots \begin{bmatrix} 0 \\ p_{ri_r} \end{bmatrix} \xrightarrow{(e_{k-1})} \begin{bmatrix} q_{r1} \\ 0 \end{bmatrix} \ldots \begin{bmatrix} q_{rj_r} \\ 0 \end{bmatrix} \xrightarrow{(e_k)} \begin{bmatrix} 0 \\ 0 \end{bmatrix}_t.$$

In this path, we denote by e_i the edges going from an upper-zero state to a lower-zero one or the converse. Note that this path encodes two paths in the automaton $\mathcal{A}$. One is read on the upper track, the other one on the lower track. The label of any path read on the upper (or lower track) going from 0 to 0 without going through 0 in between belongs to X. Hence

$$i_1 + \cdots + i_r + 1 = n$$
$$j_0 + j_1 + \cdots + j_r + 1 = m.$$

By renumbering the lower coefficients p_{ij} of the upper-zero states of this path p_1 to p_n, and the upper coefficients q_{ij} of the lower-zero states of this path q_1 to q_m, the label of each part of this path going from a state $\begin{bmatrix} 0 \\ p_{i-1} \end{bmatrix}$ to a state $\begin{bmatrix} 0 \\ p_i \end{bmatrix}$ is labelled by x_i. The label of each part of this path going from a state $\begin{bmatrix} q_{j-1} \\ 0 \end{bmatrix}$ to a state $\begin{bmatrix} q_j \\ 0 \end{bmatrix}$ is labelled by y_j.

By the definition of the sets $S_{e_i e_{i+1}}$ and the sets $S_{e_i e_{i+1} e_{i+2}}$, we get that all x_i and all y_j belong to a same part of the canonical coding partition.

Conversely, we prove that if two words x and y belong to a same component of the partition, then there is a finite chain of words $x = w_0, w_1, \ldots, w_n = y$ such that w_i and w_{i+1} belong to a same prime relation for $0 \le i < n$.

Let q_1, q_2 be two non null states in Q. We first show that if two words $y, y' \in S\begin{bmatrix} q_1 \\ 0 \end{bmatrix}\begin{bmatrix} q_2 \\ 0 \end{bmatrix}$, then there is a finite chain of words $y = w_0, w_1, \ldots, w_n = y'$ such that w_i and w_{i+1} belong to a same prime relation for $0 \le i < n$.

Since $y, y' \in S\begin{bmatrix} q_1 \\ 0 \end{bmatrix}\begin{bmatrix} q_2 \\ 0 \end{bmatrix}$, there are in $\mathcal{C}$ two paths labelled xyz and $x'y'z'$, with $x, x', z,' \in A^*$, containing respectively an edge labelled by y and an edge labelled by y', with the following form:

$$\begin{bmatrix} q_1 \\ 0 \end{bmatrix} \xrightarrow{x} \begin{bmatrix} q_{11} \\ 0 \end{bmatrix} \xrightarrow{y} \begin{bmatrix} q_{12} \\ 0 \end{bmatrix} \xrightarrow{z} \begin{bmatrix} q_2 \\ 0 \end{bmatrix},$$

$$\begin{bmatrix} q_1 \\ 0 \end{bmatrix} \xrightarrow{x'} \begin{bmatrix} q'_{11} \\ 0 \end{bmatrix} \xrightarrow{y'} \begin{bmatrix} q'_{12} \\ 0 \end{bmatrix} \xrightarrow{z'} \begin{bmatrix} q_2 \\ 0 \end{bmatrix}.$$

Since $\begin{bmatrix} q_1 \\ 0 \end{bmatrix}$ is accessible from $\begin{bmatrix} 0 \\ 0 \end{bmatrix}_s$ and $\begin{bmatrix} q_2 \\ 0 \end{bmatrix}$ is co-accessible from $\begin{bmatrix} 0 \\ 0 \end{bmatrix}_t$, these paths can be extended in $\mathcal{C}$ by a shortest path from $\begin{bmatrix} 0 \\ 0 \end{bmatrix}_s$ to $\begin{bmatrix} q_1 \\ 0 \end{bmatrix}$ labelled by a word u, and by a shortest path from $\begin{bmatrix} q_2 \\ 0 \end{bmatrix}$ to $\begin{bmatrix} 0 \\ 0 \end{bmatrix}_t$ labelled by a word w. The resulting paths are

$$\begin{bmatrix} 0 \\ 0 \end{bmatrix}_s \xrightarrow{u} \begin{bmatrix} q_1 \\ 0 \end{bmatrix} \xrightarrow{x} \begin{bmatrix} q_{11} \\ 0 \end{bmatrix} \xrightarrow{y} \begin{bmatrix} q_{12} \\ 0 \end{bmatrix} \xrightarrow{z} \begin{bmatrix} q_2 \\ 0 \end{bmatrix} \xrightarrow{v} \begin{bmatrix} 0 \\ 0 \end{bmatrix}_{t'}$$

$$\begin{bmatrix} 0 \\ 0 \end{bmatrix}_s \xrightarrow{u} \begin{bmatrix} q_1 \\ 0 \end{bmatrix} \xrightarrow{x'} \begin{bmatrix} q'_{11} \\ 0 \end{bmatrix} \xrightarrow{y'} \begin{bmatrix} q'_{12} \\ 0 \end{bmatrix} \xrightarrow{z'} \begin{bmatrix} q_2 \\ 0 \end{bmatrix} \xrightarrow{v} \begin{bmatrix} 0 \\ 0 \end{bmatrix}_t.$$

Let for instance $\begin{bmatrix} 0 \\ 0 \end{bmatrix}_s \xrightarrow{u_1} \begin{bmatrix} q \\ 0 \end{bmatrix}$ be the first edge of the path $\begin{bmatrix} 0 \\ 0 \end{bmatrix}_s \xrightarrow{u} \begin{bmatrix} q_1 \\ 0 \end{bmatrix}$. Hence u_1 and y belong to a same prime relation, and u_1 and y' belong to a same prime relation.

Let now x and y be two words in S_{ef}, where

$$e = \begin{bmatrix} 0 \\ p_i \end{bmatrix} \to \begin{bmatrix} q_i \\ 0 \end{bmatrix} \text{ and } f = \begin{bmatrix} q_j \\ 0 \end{bmatrix} \to \begin{bmatrix} 0 \\ p_j \end{bmatrix}.$$

Let us consider the first case in the definition of S_{ef}. For instance, one can assume that

$$x \in E\begin{bmatrix} 0 \\ p_1 \end{bmatrix}\begin{bmatrix} q_1 \\ 0 \end{bmatrix} \cdot L\begin{bmatrix} q_1 \\ 0 \end{bmatrix}\begin{bmatrix} q_2 \\ 0 \end{bmatrix} \cdot E\begin{bmatrix} q_2 \\ 0 \end{bmatrix}\begin{bmatrix} 0 \\ p_2 \end{bmatrix},$$

$$y \in S\begin{bmatrix} q_1 \\ 0 \end{bmatrix}\begin{bmatrix} q_2 \\ 0 \end{bmatrix}.$$

It follows that there is in $\mathcal{C}$ a path labelled by x containing an edge labelled by $y' \in S\begin{bmatrix} q_1 \\ 0 \end{bmatrix}\begin{bmatrix} q_2 \\ 0 \end{bmatrix}$ which has the following form:

$$\begin{bmatrix} 0 \\ p_1 \end{bmatrix} \to \begin{bmatrix} q_1 \\ 0 \end{bmatrix} \to \cdots \to \begin{bmatrix} q_{1i} \\ 0 \end{bmatrix} \xrightarrow{y'} \begin{bmatrix} q_{1(i+1)} \\ 0 \end{bmatrix} \to \cdots \to \begin{bmatrix} q_2 \\ 0 \end{bmatrix} \to \begin{bmatrix} 0 \\ p_2 \end{bmatrix}.$$

Since $\begin{bmatrix} 0 \\ p_1 \end{bmatrix}$ is accessible from $\begin{bmatrix} 0 \\ 0 \end{bmatrix}_s$ and $\begin{bmatrix} 0 \\ p_2 \end{bmatrix}$ is co-accessible from $\begin{bmatrix} 0 \\ 0 \end{bmatrix}_t$, this path can be extended in $\mathcal{C}$ by a shortest path from $\begin{bmatrix} 0 \\ 0 \end{bmatrix}_s$ to $\begin{bmatrix} 0 \\ p_1 \end{bmatrix}$ labelled by a word u and, by a path from $\begin{bmatrix} 0 \\ p_2 \end{bmatrix}$ to $\begin{bmatrix} 0 \\ 0 \end{bmatrix}_t$ labelled by a word w. The resulting path is

$$\begin{bmatrix} 0 \\ 0 \end{bmatrix}_s \xrightarrow{u} \begin{bmatrix} 0 \\ p_1 \end{bmatrix} \to \begin{bmatrix} q_1 \\ 0 \end{bmatrix} \to \cdots \to \begin{bmatrix} q_{1i} \\ 0 \end{bmatrix} \xrightarrow{y'} \begin{bmatrix} q_{1(i+1)} \\ 0 \end{bmatrix} \to \cdots \to \begin{bmatrix} q_2 \\ 0 \end{bmatrix} \to \begin{bmatrix} 0 \\ p_2 \end{bmatrix} \xrightarrow{v} \begin{bmatrix} 0 \\ 0 \end{bmatrix}_t.$$

This defines a prime relation containing the words x and y'. Furthermore, we know that there is a word w such that y and w belong to a same prime relation, and y' and w belong to a same prime relation.

We consider similarly all cases in the definitions of S_{ef} and S_{efg} to conclude that for any two words x and y in a such a set, there is a finite chain of words $w_0 = x, w_1, \ldots, w_n = y$ such that w_i and w_{i+1} belong to a same prime relation for $0 \le i < n$.   □

Note that, since the definition of the part X_0 is $X_0 = X - \bigcup_{i=1}^{r} X_i$, the computation of the canonical coding partition cannot be achieved in a polynomial time. The computation of the sets S_{ef} and S_{efg} can be performed in polynomial time. Since it is necessary to compute some intersections to get the automata accepting X_i, the computation of the components X_i for $i \ne 0$ also is exponential.

When the code X is not regular, even when context-free, the canonical coding partition may have an infinite number of classes, as shows the following example.

Example 1. Let
$$X = \cup_{n \ge 1} (a^n b + a^n bc^n + c^n a^n b).$$

The code X is context free and its canonical coding partition is $(X_i)_{i \ge 1}$ with $X_i = a^i b + a^i bc^i + c^i a^i b$ for $i \ge 1$ and $X_0 = \emptyset$.

It is also possible to get a finite canonical coding partition with non regular classes.

Example 2. Let X be a code, for instance a uniquely decipherable code. Let Y be the code
$$Y = \{ax, xb \mid x \in X\} + \{a, b\},$$

where a, b are two symbols which do not appear in the words of X. The canonical coding partition of Y is made of a unique class since $axb = ax \cdot b = a \cdot xb$. Such a code is GA.

4   Maximality

In this section we introduce the notion of maximality of a coding partition. Actually two different notions of maximality can be introduced: maximality with respect to one component (Definition1) and maximality with respect to all the components (Definition2). The main result of this section states that the two notions coincide for regular codes.

Definition 1. *Let $P = \{X_1, X_2, \ldots\}$ be a non-trivial coding partition of a code $X \in A^+$. We say that P is maximal with respect to the component X_i if $\forall\, w \in A^+$, the partition $P' = \{X_1, \ldots, X_i \cup \{w\}, \ldots\}$ is a coding partition of $X \cup \{w\}$ iff $w \in X_i^+$.*

Definition 2. *A non-trivial coding partition P is said to be maximal if it is maximal with respect to every component of P.*

Remark 2. It is straightforward that if P is a maximal coding partition of a code X and $P' > P$ then also P' is a maximal coding partition of X.

Theorem 5. *Let X be a code and let $P = \{X_1, X_2, \dots\}$ be a non-trivial coding partition of X. If P is maximal with respect to at least one component, then X is complete.*

Proof. Let X be a code over the alphabet A, with $card(A) \geq 2$ (the case $card(A) < 2$ is trivial). We will first prove that, if X is not complete, then there exists a word $w \in A^* \setminus X$ such that the partition $P_1 = \{\{w\}, X_1, X_2, \dots\}$ is a coding partition of $X \cup \{w\}$. Indeed, if X is not complete, there exists a word $v \in A^*$ such that v does not belong to $F(X^*)$. Let a be the first letter of v and let $b \in A \setminus \{a\}$. Consider the word $w = vb^{|v|-1}$. By construction, w is *unbordered*, i.e. no proper prefix of w is a suffix of w. Since v does not belong to $F(X^*)$, we have that also w does not belong to $F(X^*)$.

Let us first remark that $X^+ \cap \{w\}^+ = \emptyset$. We now prove that every word $t \in (X \cup \{w\})^*$ admits a unique P_1-factorization. Indeed, since w is unbordered, we can uniquely distinguish all occurrences of w in t, i.e. t has a unique factorization of the form

$$t = u_1 w u_2 w \cdots w u_n,$$

with $n \geq 1$ and $u_i \in X^*$, for $i = 1, \dots, n$. From this factorization, since P is a coding partition, we obtain a unique P_1-factorization of t and therefore, by definition, P_1 is a coding partition. From this is trivial that $\forall i \, P' = \{X_1, \dots, X_i \cup \{w\}, \dots\}$ is still a coding partition and so X_i is not maximal. This concludes the proof. $\qquad\square$

The next lemma and its proof is just a little variation of a lemma due to Schutzenberger (see Theorem 7.4 in [5]).

Lemma 1. *Let $X \subseteq A^+$ be a regular and complete code and let $x_1, x_2 \in X^*$. There exist a word $v_1 \in X^+$ and a positive integer m such that for any word $w \in A^*$, $(vwv)^m \in X^+$ where $v = x_1 v_1 x_2$.*

Proof. Since X is a regular set, X^+ is a regular set too. Let

$$\mathcal{A} = (A, Q, \delta, i, F)$$

be a finite state automaton recognizing X^+. For any set of states $S \subseteq Q$ and for any word $u \in A^+$, denote by Su the set $\{\delta(q, u); q \in S\}$ of states reached by paths having label u and starting at any state of S. Let $n = \min\{card(Qu)\}$ with u ranging over A^+, and choose u such that $n = card(Qu)$. Since X is complete, we have $xuy \in X^+$ for some $x, y, \in A^*$ and so $v' := x_1 xuy x_2 \in X^+$. Since $card(Qx_1 xuy x_2) \leq card(Qx_1 xu)$ and $Qx_1 xu \subseteq Qu$, it follows that $card(Qv') \leq card(Qu)$. Thus, by minimality, $card(Qv') = n$. Let $P = Qv'$. Since $Pv' = Qv'v' \subseteq Qv' = P$, it follows from the minimality of n that $Qv'v' = Qv'$ and $Pv' = P$; thus v' defines a permutation of P. Thus, put v a suitable power of v' and wrote $v = x_1 v_1 x_2$ for a certain $v_1 \in X^+$, we may assume that $pv = p$ for all $p \in P$ and $Qv = Qv' = P$. Consider now a word $w \in A^*$ and let $z = vwv$. Again we have $Pz = Qvvwv \subseteq Qv = P$ and thus $Pz = P$. Then for $m = n!$ we have $pz^m = p$ for all $p \in P$. To prove that

$$z^m = (vwv)^m \in X^+,$$

it suffices to show that $qz^m = qv$ for all $q \in Q$. Since $Qv = P$ and $pv = p$ for all $p \in P$, then $qvv = qv$. It follows that $qz = qvwv = qvvwv = qvz$ and therefore that $qz^m = qvz^m$. Since $pz^m = p$ for all $p \in P$, we have that $qvz^m = qv$. Thus $qz^m = qv$ as required. This completes the proof.   □

Theorem 6. *Let X be a regular code and let $P = \{X_1, X_2, \dots\}$ be a non-trivial coding partition of X. If X is complete then P is maximal.*

Proof. Let $w \in A^+$ and $i \geq 1$ such that $P' = \{X_1, \dots, X_i \cup \{w\}, \dots\}$ is a coding partition of $X' = X \cup \{w\}$. Since P is non-trivial, $\exists\, x \in X_j \neq X_i$. By previous lemma there exist $v_1 \in X^+$ and a positive integer m such that $z = (xv_1xwxv_1x)^m \in X^+$. Since $x \notin X_i$ the P' factorization of z is of the form:

$$z = z_1 \cdots z_{s_1} w z_{s_1+1} \cdots z_{s_m} w z_{s_m+1} \cdots z_t$$

where z_h, $1 \leq h \leq t$ are the blocks of the factorization. But $z \in X^+$ so there exists a factorization without w that is again a P' factorization. By the uniqueness of the P' factorization the block corresponding to w must be the same and so $\exists\, y_1, y_2, \dots, y_k \in X_i$ s.t. $w = y_1y_2 \cdots y_k$. This shows that P is maximal.   □

From Theorem 5 and Theorem 6 we get the following corollary.

Corollary 1. *Let $P = \{X_1, X_2, \dots\}$ be a non-trivial coding partition of a regular code X. If P is maximal with respect to a component X_i, then P is maximal.*

5 UD Codes Versus GA Codes

In this section we consider an application of previous results to maximal UD codes. By definition, a UD code X is maximal if any code Y containing strictly X is ambiguous. We here prove that, if a regular UD code X is maximal, then any code Y containing strictly X is globally ambiguous. Moreover, if X is a finite maximal UD code, we prove that for a given word $v \in A^+$, there exists a prime relation involving all the elements of $X \cup \{v\}$.

A generalization of this result to the case of non-UD codes, is given at the end of the paper.

Theorem 7. *Let $X \subseteq A^+$ be a maximal UD code. If X is regular then, for all $v \in A^+$ such that $v \notin X^+$, $X \cup \{v\}$ is GA.*

The proof is an immediate consequence of the next proposition that has an independent interest and that follows from Theorems 5 and 6.

Proposition 1. *Let $X \subseteq A^+$ be a regular code and let $P = \{X_1, X_2, \dots\}$ be a non-trivial coding partition of X. If P is maximal then, for all $v \in A^+$ such that $v \notin X^+$, $X \cup \{v\}$ is GA.*

In the case the code X is *finite* we can derive stronger results.

Recall that a code X is called a *base* if X is a minimal set of generators of X^*.

Theorem 8. *Let $C \subseteq A^+$ be a finite maximal UD code. If $C \neq A$ then there exists a word $v \in A^+$ such that $C' := C \cup \{v\}$ has the following properties:*

- *C' is a base*
- *C' is GA*
- *there exists a prime relation involving all the elements of C', i.e. a relation $x_1 x_2 \cdots x_s = x_{s+1} x_{s+2} \cdots x_t$ such that $\{x_1, x_2, \ldots, x_t\} = C'$.*

Proof. Let us first consider the case that C is a prefix code.

Let $C = \{c_1, c_2, \ldots, c_n\}$ and let $u := c_1 c_2 \cdots c_n$. By hypothesis there is $c_{i_0} \in C$ with $|c_{i_0}| > 1$, and let w be a prefix of c_{i_0} s.t. $|w| = |c_{i_0}| - 1$. Let us put $v := uw$ and $C' := C \cup \{v\}$. We claim that C' is a base. Indeed since $|v| > |c_i|, 1 \leq i \leq n$, it is sufficient to show that $v \notin C^+$. If, by contradiction, $v \in C^+$, being C^+ right unitary (see [2]), we have $w \in C^+$ with w prefix of c_{i_0}, and this is a contradiction because C is a prefix code. Finally since C is maximal C' is not UD so there is a prime relation involving v. This relation by definition of v, being C a prefix code, must have the form $v x_1 \cdots x_s = c_1 \cdots c_n y_1 \cdots y_t$ for some $x_i, y_j \in C'$.

We now consider the case that C is not a prefix set.

We recall that a code $Y \subseteq A^+$ is *right complete* if for all $w \in A^*$ there exists a word $w' \in A^*$ such that $ww' \in Y^*$. Since C is not a prefix set then C is not right complete (see [2]) and let $w \in A^+$ s.t. w is not right completable. Of course also w^m is not right completable and we can choose $m \geq 2$ in such a way that $|w^m| > |c_i| + |w| \ \forall c_i \in C$. Now we put $w_1 := w^m w'$ with $w' \in A^*$ s.t. w_1 is unbordered. Since C is maximal then put $C' := C \cup \{w_1\}$ there exists a prime relation $x_1 x_2 \cdots x_s w_1 x_{s+1} \cdots x_l = y_1 \cdots y_k$, $x_i, y_i \in C'$, $s \geq 1, l \geq s$, $k \geq 2$. Let $p \geq 1$ the first index s.t. $|y_1 \cdots y_p| > |x_1 \cdots x_s w|$: by choice of m and w_1, $|y_1 \cdots y_p| < |x_1 \cdots x_s w^m|$ so $y_1 \cdots y_p = x_1 \cdots x_s w^q u$, with $1 \leq q < m$ and, since w is not right completable, $u \in A^+$. Now we put $v := w^q u c_1 \cdots c_n z w^{m-q-1} w'$ with $z = u^{-1} w$. We have the relation $x_1 x_2 \cdots x_s v x_{s+1} \cdots x_l = y_1 \cdots y_p c_1 \cdots c_n y_{p+1} \cdots y_k$ that is clearly prime. Finally, by definition, $v \notin C^+$ and, by a length argument, one has that $C \cup \{v\}$ is a base and the proof is complete. $\square$

Remark 3. We observe as consequence of Theorem 7 that, if X is a base and it is not GA, then any regular set $Y \subsetneq X$ is not a maximal UD code.

Theorem 9. *Let $X \subseteq A^+$ be a non-GA finite code that is a base. If X is complete then there exists a word $v \in A^+$ such that $X' := X \cup \{v\}$ has the following properties:*

- *X' is a base*
- *X' is GA*
- *there exists a prime relation involving all the elements of X', i.e. a relation $x_1 x_2 \cdots x_s = x_{s+1} x_{s+2} \cdots x_t$ such that $\{x_1, x_2, \ldots, x_t\} = X'$.*

Proof. We recall that a code $Y \subseteq A^+$ is right complete iff $Y \setminus Y A^+$ is a maximal prefix UD code (see [2]). Let P be a non-trivial coding partition of X then, by Theorem 6, P is maximal. Because of Theorem 8 we can suppose that X is not a prefix UD code and then $\emptyset \neq X \setminus X A^+ \subsetneq X$. Moreover, because of previous remark, $X \smallsetminus XA^+$ is not a maximal prefix UD code and so X is not right complete. Then there exists $w \in A^+$ s.t. w is not right completable and we can proceed like in the previous theorem.   □

References

1. Béal, M.-P., Perrin, D.: Codes, unambiguous automata and sofic systems. Theoret. Comput. Sci. 356, 6–13 (2006)
2. Berstel, J., Perrin, D.: Theory of codes, Orlando, FL. Pure and Applied Mathematics, vol. 117. Academic Press Inc, San Diego (1985),
   `http://www-igm.univ-mlv.fr/~berstel/LivreCodes/Codes.html`
3. Burderi, F., Restivo, A.: Coding partitions, Discret. Math. Theor. Comput. Sci (to appear)
4. Dalai, M., Leonardi, R.: Non prefix-free codes for constrained sequences. In: International Symposium on Information Theory, 2005. ISIT 2005, pp. 1534–1538. IEEE, New York (2005)
5. Eilenberg, S.: Automata, Languages and Machines, vol. A. Academic Press, New York (1974)
6. Gönenç, G.: Unique decipherability of codes with constraints with application to syllabification of Turkish words. In: COLING 1973: Computational And Mathematical Linguistics: Proceedings of the International Conference on Computational Linguistics, vol. 1, pp. 183–193 (1973)
7. Guzmán, F.: Decipherability of codes. J. Pure Appl. Algebra 141, 13–35 (1999)
8. Karhumäki, W.P.J., Rytter, W.: Generalized factorizations of words and their algorithmic properties. Theoret. Comput. Sci. 218, 123–133 (1999)
9. Lempel, A.: On multiset decipherable codes. IEEE Trans. Inform. Theory 32, 714–716 (1986)
10. Restivo, A.: A note on multiset decipherable codes. IEEE Trans. Inform. Theory 35, 662–663 (1989)
11. Sakarovitch, J.: Éléments de théorie des automates, Vuibert, Paris. Cambridge University Press, Cambridge (English translation to appear) (2003)

Multi-letter Reversible and Quantum Finite Automata*

Aleksandrs Belovs, Ansis Rosmanis, and Juris Smotrovs

University of Latvia, Raiņa bulv. 29, Rīga, LV-1459, Latvia
stiboh@inbox.lv, sd30056@lanet.lv, juriss@lanet.lv

Abstract. The regular language $(a + b)^*a$ (the words in alphabet $\{a, b\}$ having a as the last letter) is at the moment a classical example of a language not recognizable by a one-way quantum finite automaton (QFA). Up to now, there have been introduced many different models of QFAs, with increasing capabilities, but none of them can cope with this language.

We introduce a new, quite simple modification of the QFA model (actually even a deterministic reversible FA model) which is able to recognize this language. We also completely characterise the set of languages recognizable by the new model FAs, by finding a "forbidden construction" whose presence or absence in the minimal deterministic (not necessarily reversible) finite automaton of the language decides the recognizability.

Thus, the new model still cannot recognize the whole set of regular languages, however it enhances the understanding of what *can* be done in a finite-state real-time quantum process.

1 Introduction

Finite automata (FA) models constitute an important theoretical paradigm for exploring what can be done algorithmically in praxis. After all, no computer has, or — at least in foreseeable future — will have any really infinite resource.

The classical FA models — e.g. deterministic (DFA) or probabilistic (PFA) finite automata — essentially rely on classical physics, particularly on Newtonian mechanics. With the advance of quantum mechanics as the theory describing best the laws governing our physical world on the very basic level, it became natural to investigate what capabilities would have the counterparts of the classical models governed by quantum rules.

First studies on quantum finite automata (QFA) appeared in 1997. From the very start the research did not concentrate on one model, instead new QFA models were introduced regularly. The main reason for such diversity was that the simplest, most "natural" model by Moore and Crutchfield [9], the so-called measure-once one-way read-only QFA (MO-QFA) turned out to have very limited capabilities. These automata work quantumly all the time while reading input, and only at the end of input a measurement of the quantum state is

---

* Supported by the European Social Fund.

T. Harju, J. Karhumäki, and A. Lepistö (Eds.): DLT 2007, LNCS 4588, pp. 60–71, 2007.

performed to determine the outcome (word accepted or rejected). They can recognize only the so-called group languages [4] which is a rather restricted subset of the set of regular languages recognizable by DFAs.

There are QFA models that can recognize all regular languages: Kondacs and Watrous model of two-way QFA [8], Ciamarra's model of one-way read-write QFA [5], Bertoni, Mereghetti and Palano's model of QFA with control language [3], and other. However, these models have been granted such capabilities (two-wayness, ability to modify tape, a regular language controlling the acceptance) which usually are not attributed to the basic finite automata models (e.g. DFA, PFA).

Thus researchers tried to invent different ways to enhance the capabilities of one-way read-only QFAs, mostly by enabling the automaton to perform different kinds of classical actions in addition to the quantum state transition. Kondacs and Watrous [8] introduced the measure-many QFA (MM-QFA) which performs a restricted measurement at each step determining whether to accept or reject the input, or to continue reading input. A. Nayak [10] proposed a further generalisation by allowing the QFA to perform several arbitrary measurements with intermediate unitary transformations at each step. Some other QFA models can be found in [6], [1].

However, all these one-way QFA models can recognize only a proper subset of regular languages, and there is a certain class of languages, best characterised by the presence of a certain forbidden construction in its minimal DFA, which cannot be recognized by any of these QFAs. The regular language $(a + b)^*a$ is a typically mentioned example of this class.

In this paper we shrink this class by introducing a new QFA model which can deal with the characteristic forbidden construction on some occasions, particularly, in the case of language $(a + b)^*a$.

In our model, the automaton is not limited to seeing only one, the just-incoming input letter, but can see several earlier received letters as well. That is, the quantum state transition which the automaton performs at each step depends on the last k letters received, where k is fixed for any automaton.

In fact, as we show, we do not need any of the advantages given by quantum mechanics: already a group FA (which is a DFA having only the restrictions of the quantum mechanics — the reversibility) with the ability to see several letters can recognize as much as its quantum counterpart.

This new model is by far not the most powerful in itself. It cannot recognize extensive language classes which other QFA models can. However, any other QFA model enhanced with this ability to see several input letters should enhance its power by enabling to deal with languages from the class represented by $(a+b)^*a$.

2 Preliminaries

2.1 Notation

An *alphabet* is a finite set of letters. The set of all finite sequences of letters from an alphabet Σ is denoted as Σ^*, and elements of Σ^* are called *words*. We denote the length of a word w by $|w|$, the *empty word* having no letters by ϵ, and the set of all words of length l by Σ^l. Σ^+ is an abbreviation for the set of *non-empty words* $\Sigma^* \setminus \Sigma^0$. The concatenation of words u and v is denoted by uv and a word w repeated m times by w^m.

A subset of Σ^* is called a *language*. A language $L \subseteq \Sigma^*$ is *recognized* by an automaton iff the automaton accepts those and only those words which are in L. The meaning of *'accept'* depends on automata model and is discussed later.

2.2 Deterministic Finite Automata

A *deterministic finite automaton* DFA is a quintuple $(Q, Q_{acc}, q_0, \Sigma, \gamma)$ where Q is a finite set of *states*, $Q_{acc} \subseteq Q$ is the *set of accepting states*, $q_0 \in Q$ is the *initial state*, Σ is a finite *input alphabet*, and γ is a *transition function* that maps $Q \times \Sigma$ to Q.

The computation of a DFA for an input word $w = \sigma_1\sigma_2...\sigma_l$ starts in the initial state and involves $|w|$ steps. At the i-th step the automaton receives a letter σ_i and changes its state from its current state q to $\gamma(q, \sigma_i)$. The DFA *accepts* w iff the *final state* after all $|w|$ steps is in Q_{acc}.

We use $\mathcal{L}_{DFA}$ to denote the class of languages recognized by at least one DFA. It is well known that it is the set of regular languages (see e.g. [12]).

We define an extended transition function $\gamma^* : Q \times \Sigma^* \to Q$ as follows:

$$\begin{cases} \gamma^*(q, \epsilon) = q, \\ \gamma^*(q, w\sigma) = \gamma(\gamma^*(q, w), \sigma), \quad \text{where } w \in \Sigma^*, \sigma \in \Sigma. \end{cases} \tag{1}$$

If the DFA is in a state q and receives a word w, it changes its state to $\gamma^*(q, w)$. Clearly, $\gamma^*(q, uv) = \gamma^*(\gamma^*(q, u), v)$ for all $q \in Q$ and all $u, v \in \Sigma^*$.

We will use a standard state transition diagram notation. '$+$' or '$-$' within a circle denotes accordingly an accepting or non-accepting state. '$\Rightarrow$' points to the initial state and '$\neq$' expresses that two states are distinct.

The DFA having the minimal number of states among all DFA recognizing a language L, is called *the minimal DFA of L*. It is well known that in the minimal DFA all states are reachable and distinguishable (in the sense that, for any two states q, q' there is a word w such that one of the states $\gamma^*(q, w)$ and $\gamma^*(q', w)$ is accepting, while the other is non-accepting).

2.3 Quantum Finite Automata

We will use essentially the Moore and Crutchfield's definition of a *quantum finite automaton* (QFA) [9].

A QFA is a quintuple $(Q, Q_{acc}, |\psi_0\rangle, \Sigma, \mu)$ where $Q = \{q_1, q_2, \ldots, q_n\}$ is a finite set of states which form *a basis* in the Hilbert space $\mathbb{C}^n$, $Q_{acc} \subseteq Q$ is the set of accepting states, $|\psi_0\rangle = (\alpha_1, \alpha_2, \ldots, \alpha_n)^\dagger \in \mathbb{C}^n$ is the *initial state superposition* of basis states with the normalization condition $\langle \psi_0 \mid \psi_0 \rangle = \sum_i |\alpha_i|^2 = 1$, Σ is a finite input alphabet, and μ is a function that maps Σ to the set of n-dimensional *unitary transition matrices*. We denote $\mu(\sigma)$ by U_σ.

At any moment, the state of a QFA can be described by a normalized column vector $|\psi\rangle = (\beta_1, \beta_2, \ldots, \beta_n)^\dagger \in \mathbb{C}^n$ called *superposition* of basis states, β_i being the component corresponding to q_i. After reading a letter σ, the QFA changes its state to $U_\sigma|\psi\rangle$.

After receiving the whole input word $w = \sigma_1\sigma_2\ldots\sigma_l$, the QFA is in a final superposition $|\psi_f\rangle = U_{\sigma_l}\ldots U_{\sigma_2}U_{\sigma_1}|\psi_0\rangle$. Let us denote $|\psi_f\rangle = (\beta_1^f, \beta_2^f, \ldots, \beta_n^f)^\dagger$. The acceptance of a word is decided by a *measurement* according to basis states: for each i, the basis state q_i is obtained as the result of the measurement with probability $p_i = |\beta_i^f|^2$. The word is *accepted* with probability $p = \sum_{i:\, q_i \in Q_{acc}} p_i$. A language L is said to be *recognized by a QFA with bounded error* $(p_1; p_2)$ iff $p_1 < p_2$ and the QFA accepts any word from L with probability at least p_2, and accepts any word not from L with probability at most p_1.

We denote by $\mathcal{L}_{QFA}$ the class of languages recognized by QFA.

2.4 Group Finite Automata

A *group finite automaton (GFA)* is a DFA for which all functions $\gamma_\sigma(q) = \gamma(q, \sigma)$ are bijections from Q to Q (see e.g. [4]). $\mathcal{L}_{GFA}$ denotes the corresponding language class (of *group languages*).

For each $\sigma \in \Sigma$ let us consider the matrix U_σ containing 1 in i-th position of j-th column iff $\gamma(q_j, \sigma) = q_i$; the rest being filled with 0. If we represent each state q_i by a column vector $|q_i\rangle = (0, \ldots, 0, 1, 0, \ldots, 0)^T$ where the unique 1 is in the i-th position, we get that $|q'\rangle = U_\sigma|q\rangle$ iff $\gamma(q, \sigma) = q'$. Since γ_σ is a bijection, U_σ has exactly one 1 in each column and exactly one 1 in each row, therefore U_σ is a permutation matrix, thus also a unitary matrix. Hence it is easy to see that GFA are essentially the intersection of DFA and QFA.

γ_σ being a bijection implies also each transition of GFA being reversible: from the resulting state q' and the input letter σ one can determine the state before receiving the letter.

Group languages can be described by means of forbidden constructions in their minimal DFA (see [2] for similar results for a slightly different reversible FA model). Let us say that a DFA contains an A-construction iff it has two distinct states q_1 and q_2 leading by some word x to the same state q_3 (the latter can coincide with either q_1, or q_2), and a B-construction iff it has two such distinct states q_4 and q_5 and there exists such word y that $\gamma^*(q_4, y) = q_5$ and $\gamma^*(q_5, y) = q_5$ (see Fig. 1). It can be proved that both constructions are equivalent, and the minimal DFA contains an A-construction iff it contains a B-construction iff the regular language is *not* in $\mathcal{L}_{QFA} = \mathcal{L}_{GFA}$.

Fig. 1. Forbidden constructions (A and B) for group languages

3   Multi-letter Automata

For the automata models defined above state changes depend only on the present state and the input letter. We now consider a model of deterministic, quantum and group finite automata which sees several earlier received letters as well. It is essentially a special case of one-way multi-head DFA [7].

3.1   Multi-letter DFA and GFA

Definition 1. A k-letter deterministic finite automaton (DFA$_k$) *is defined by a quintuple* $(Q, Q_{acc}, q_0, \Sigma, \gamma')$, *where* Q *is a finite set of states,* $Q_{acc} \subseteq Q$ *is the set of accepting states,* $q_0 \in Q$ *is the initial state,* Σ *is a finite input alphabet, and* γ' *is a transition function that maps* $Q \times T^k$ *to* Q, *where* $T = \{\Lambda\} \cup \Sigma$ *and letter* $\Lambda \notin \Sigma$ *denotes an* empty input letter.

The computation of a DFA$_k$ starts in the initial state q_0. After receiving a letter, a state transition corresponding to the current state and the last k received letters is applied. If so far only $m < k$ letters are received, the missing letters are replaced with Λ and a corresponding transition applied. When the last letter of the input word is received, the last transition is applied and the computation stops. The input word is *accepted* by the DFA$_k$ iff the computation stops in a state that belongs to Q_{acc}.

We might say that the DFA$_k$ has a tape which contains the letter Λ in its first $k - 1$ positions followed by the input word. The automaton has k reading heads which initially are on the first k positions of the tape. During one computation step each head reads one letter from the tape, the automaton makes the transition corresponding to the word they have read and then all heads move one position forward.

Definition 2. *A DFA$_k$ is called a* k-letter group finite automaton (GFA$_k$) *iff for any word* $w \in T^k$ *the function* $\gamma'_w(q) = \gamma'(q, w)$ *is a bijection from* Q *to* Q.

We use $\mathcal{L}_{DFA_k}$ as a notation for the class of languages recognized by at least one k-letter deterministic finite automaton, and $\mathcal{L}_{DFA_*} = \bigcup_{k=1}^{\infty} \mathcal{L}_{DFA_k}$. Similarly we use the notation $\mathcal{L}_{GFA_k}$ and $\mathcal{L}_{GFA_*}$.

Since for DFA$_1$ and GFA$_1$ state transitions depend only on the last received letter, they are equal accordingly to DFA and GFA. Thus, $\mathcal{L}_{DFA} \subseteq \mathcal{L}_{DFA_*}$ and $\mathcal{L}_{GFA} \subseteq \mathcal{L}_{GFA_*}$.

It is easy to show that one can simulate DFA$_k$ with a DFA encoding the $k - 1$ earlier received letters in its states. However, this transformation does not preserve reversibility, hence is not applicable to the group FA case. Thus $\mathcal{L}_{GFA_*} \subseteq \mathcal{L}_{DFA_*} = \mathcal{L}_{DFA}$. We will show later that $\mathcal{L}_{GFA} \subset \mathcal{L}_{GFA_*} \subset \mathcal{L}_{DFA}$.

3.2   Multi-letter QFA

The same principles of state transitions corresponding to last k received letters can be applied for quantum finite automata.

Definition 3. *A k-letter quantum finite automaton (QFA$_k$) is defined by a quintuple $(Q, Q_{acc}, |\psi_0\rangle, \Sigma, \mu')$ where Q is a set of states, $Q_{acc} \subseteq Q$ is the set of accepting states, $|\psi_0\rangle$ is the initial state superposition obeying normalization condition, Σ is a finite input alphabet, and μ' is a function that assigns an unitary transition matrix U_w on $\mathbb{C}^n$ for each word $w \in (\{\Lambda\} \cup \Sigma)^k$.*

The computation of a QFA$_k$ works in the same way as the computation of a QFA, except that it applies unitary transformations corresponding not only to the last letter received, but the last k letters received (like a DFA$_k$).

We will use $\mathcal{L}_{QFA_k}$ as notation for the class of languages recognized by at least one k-letter quantum finite automaton with bounded error, and $\mathcal{L}_{QFA_*} = \bigcup_{k=1}^{\infty} \mathcal{L}_{QFA_k}$. With a slight modification of the proof for Proposition 6 in [8] we obtain that all languages in $\mathcal{L}_{QFA_*}$ are regular.

GFA$_k$ can be considered as a special case of QFA$_k$ because, since the function $\gamma'_w(q) = \gamma'(q, w)$ is bijection, a state transition corresponding to any word is a unitary transformation. Therefore any language recognized by a GFA$_k$ is recognized by a QFA$_k$ as well.

4   Capabilities

Our multi-letter QFA and GFA can recognize the language $(a + b)^*a$ not recognized by any standard QFA. E.g., consider the following QFA$_2$ / GFA$_2$ (Fig. 2):

$$Q = \left\{ \begin{pmatrix} 1 \\ 0 \end{pmatrix}, \begin{pmatrix} 0 \\ 1 \end{pmatrix} \right\}, \; Q_{acc} = \left\{ \begin{pmatrix} 0 \\ 1 \end{pmatrix} \right\}, \; |\psi_0\rangle = \begin{pmatrix} 1 \\ 0 \end{pmatrix}, \; \Sigma = \{a, b\}, \; U_{\Lambda b} = U_{bb} =$$

$$U_{aa} = \begin{pmatrix} 1 & 0 \\ 0 & 1 \end{pmatrix} \text{ and } U_{\Lambda a} = U_{ba} = U_{ab} = \begin{pmatrix} 0 & 1 \\ 1 & 0 \end{pmatrix}.$$

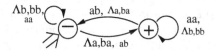

Fig. 2. The GFA$_2$ recognizing the language $(a + b)^*a$

Still there are languages not recognized by any QFA$_k$. We will describe these languages by forbidden constructions in their minimal DFA.

Definition 4. *A DFA contains a C_k-construction iff there are states q_1, q_2, q_3, q_4, q_5 and a word $w = \sigma_1 \sigma_2 \ldots \sigma_k$ of length k such that $q_2 \neq q_5$, $\gamma(q_2, \sigma_k) = \gamma(q_5, \sigma_k) = q_3$, $\gamma^*(q_1, \sigma_1 \ldots \sigma_{k-1}) = q_2$ and $\gamma^*(q_4, \sigma_1 \ldots \sigma_{k-1}) = q_5$ (see Fig. 3).*

Definition 5. *A C_k-construction where exists an $m > 0$ such that $\gamma^*(q_3, w^{m-1}) = q_4$, where q_3, q_4 and w have the same meaning as in Def. 4, we call a D_k-construction.*

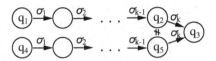

Fig. 3. C_k-construction

Lemma 1. *A C_k-construction implies a D_k-construction.*

Proof. Suppose we have a C_k-construction. If there is an $m' > 0$ such that $\gamma^*(q_3, w^{m'}) = q_3$, then for $q'_4 = \gamma^*(q_3, w^{m'-1})$ and $q'_5 = \gamma^*(q'_4, \sigma_1 \ldots \sigma_{k-1})$ we have $\gamma(q'_5, \sigma_k) = q_3$, and the lemma holds since $q'_5 \neq q_2$ or $q'_5 \neq q_5$ (or both).

If there is no such an m', since Q is finite, there is an $i \geq 2$ and an $m'' > 0$ such that $\gamma^*(q_1, w^i) = \gamma^*(q_1, w^{i+m''})$ and $\gamma^*(q_1, w^{i-1}) \neq \gamma^*(q_1, w^{i+m''-1})$. Hence, there is a $j \in [1; k]$ such that states $q''_2 = \gamma^*(q_1, w^{i-1}\sigma_1 \ldots \sigma_{j-1})$ and $q''_5 = \gamma^*(q_1, w^{i+m''-1}\sigma_1 \ldots \sigma_{j-1})$ are distinct, but $\gamma(q''_2, \sigma_j) = \gamma(q''_5, \sigma_j) = q''_3$. Let $q''_1 = \gamma^*(q_1, w^{i-2}\sigma_1 \ldots \sigma_j)$, $q''_4 = \gamma^*(q_1, w^{i+m''-2}\sigma_1 \ldots \sigma_j)$, $w'' = \sigma_{j+1} \ldots \sigma_k \sigma_1 \ldots \sigma_j$. m'', the word w'' and the states $q''_1, q''_2, q''_3, q''_4, q''_5$ form a D_k-construction. $\square$

4.1   Simulation of DFA by QFA$_k$

At first we will show that, if the minimal DFA of a language L contains at least one C_k-construction, there is no QFA$_k$ with bounded error that recognizes L. Because of Lemma 1 we will use D_k-constructions instead of C_k-construction. We use the following lemma, which is slightly modified Theorem 6 in [9]:

Lemma 2. *Let us consider quantum system with basis states $|q_1\rangle, |q_2\rangle, \ldots, |q_n\rangle$, arbitrary subset of these states Q_{acc} and function $p : \mathbb{C}^n \to [0; 1]$ that for any superposition of the basis states gives the probability that after the measurement according to the basis states, any state from Q_{acc} is obtained. For any superposition of basis states $|\psi\rangle$, any three unitary transformation X, Y, Z on the system and any $\delta > 0$ there is an $h > 0$ such that $|p(ZX|\psi\rangle) - p(ZY^h X|\psi\rangle)| < \delta$.*

Theorem 1. *If the minimal DFA of a language L contains a D_k-construction, then there is no QFA$_k$ recognizing L with bounded error.*

Proof. Suppose we have a D_k-construction: since $q_2 \neq q_5$, there is a word $v \in \Sigma^*$ such that $\gamma^*(q_2, v) \in Q_{acc}$ iff $\gamma^*(q_5, v) \notin Q_{acc}$. Due to the symmetry we assume $\gamma^*(q_2, v) \in Q_{acc}$. Since we consider minimal DFA, where each state is reachable, there is a word $u' \in \Sigma^*$ such that $\gamma^*(q_0, u') = q_1$ (see Fig. 4). Let us denote $u'\sigma_1 \ldots \sigma_{k-1}$ by u and $(\sigma_k \sigma_1 \ldots \sigma_{k-1})^m$ by $s = \sigma'_1 \sigma'_2 \ldots \sigma'_l$, where $l = k \cdot m$. The last $k - 1$ letters of both u and s are $\sigma'_{l-k+2}, \sigma'_{l-k+3}, \ldots, \sigma'_l$.

Among the input words $t_i = us^i v$ for all $i \geq 0$ only t_0 is in the language L recognized by the DFA. Suppose the contrary: there is a QFA$_k$ which recognizes L with bounded error $(p_1; p_2)$.

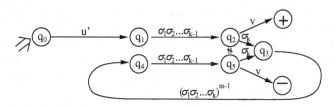

Fig. 4. D_k construction

For all i after receiving the initial fragment u of the word t_i, the QFA$_k$ has applied $|u|$ unitary transformations $X_1, X_2, \ldots, X_{|u|}$ and it is in the superposition $X|\psi_0\rangle$, where $X = X_{|u|} \ldots X_2 X_1$, and the last $k - 1$ letters received are $\sigma'_{l-k+2}, \ldots, \sigma'_l$.

If the last $k - 1$ letters received by the QFA$_k$ are $\sigma'_{l-k+2}, \ldots, \sigma'_l$ and it is in some superposition $|\psi\rangle$, then after receiving the word $s = \sigma'_1 \ldots \sigma'_l$ always the same l unitary transformations $Y_1, Y_2, \ldots, Y_l$ are applied. After these transformations the QFA$_k$ is in the superposition $Y|\psi\rangle$, where $Y = Y_l \ldots Y_2 Y_1$, and the last $k - 1$ letters received still are $\sigma'_{l-k+2}, \ldots, \sigma'_l$.

For all i, when the QFA$_k$ has received the whole input word t_i except its last fragment v, the last $k - 1$ letters received are $\sigma'_{l-k+2}, \ldots, \sigma'_l$ and it is in the superposition $Y^i X|\psi_0\rangle$. When the QFA$_k$ receives the last fragment, it applies $|v|$ unitary transformations $Z_1, Z_2, \ldots, Z_{|v|}$ and the final superposition is $ZY^i X|\psi_0\rangle$, where $Z = Z_{|v|} \ldots Z_2 Z_1$.

By Lemma 2 we get that $\exists h > 0 \left(|p(ZX|\psi_0\rangle) - p(ZY^h X|\psi_0\rangle)| < p_2 - p_1 \right)$. Thus, $t_0 \in L$ iff $t_h \in L$. Contradiction.   □

4.2   Prefix Extension Method

We provide a method we will call *prefix extension method* which for any DFA and any k creates a DFA$_k$ which recognizes the same language as the DFA.

For the created DFA$_k$ we use the same set of states (Q), the same accepting states (Q_{acc}), the same initial state (q_0) and the same alphabet (Σ) as the DFA. A creation of the function γ' consists of three steps:

1. If there is a transition $\gamma(q_2, \sigma) = q_3$ in the DFA, then in the created DFA$_k$ we add transition $\gamma'(q_2, w\sigma) = q_3$ for all words $w \in T^{k-1}$ which satisfies $\exists v \in \Sigma^*(w = \Lambda^* v \wedge \gamma^*(q_0, v) = q_2)$,
2. If there is a transition $\gamma(q_2, \sigma) = q_3$ in the DFA, then in the created DFA$_k$ we add transition $\gamma'(q_2, w\sigma) = q_3$ for all words $w \in T^{k-1}$ which satisfies $\exists q_1 \in Q(\gamma^*(q_1, w) = q_2)$,
3. Clearly, there is no pair $(q', w') \in Q \times T^k$ such that the function γ' maps (q', w') to more than one state, but still there may be a pair (or pairs) $(q'', w'') \in Q \times T^k$ such that $\gamma'(q'', w'')$ is yet undefined. We can choose any state as the value of $\gamma'(q'', w'')$ because after receiving a letter there cannot be a situation when the DFA$_k$ is in the state q'' and the last k letters received form the word w''.

In other words: if there is a transition in the DFA from a state q_2 to a state q_3 by a letter σ, then what we do is adding in front of σ all words of length $k-1$ which could occur as the earlier $k-1$ letters before this transition in this DFA.

An example how a DFA$_k$ is created from a minimal DFA which recognizes the language $a(a+b)^*a$ is given in Fig. 5 (unimportant transitions made during the step 3 are not displayed).

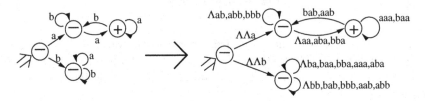

Fig. 5. The DFA$_k$ created from a DFA recognizing $a(a+b)^*a$

Proposition 1. *A DFA$_k$ created from a DFA by the prefix extension method recognizes the same language as the DFA.*

Proof. Let $u = \sigma_1\sigma_2...\sigma_l$ be an input word and let $q_1, q_2, \ldots, q_l$ be a sequence of states such that $\gamma(q_i, \sigma_{i+1}) = q_{i+1}$. Thus, u is accepted by the DFA iff $q_l \in Q_{acc}$. For all $i \in [0, k-1]$, since $\gamma^*(q_0, \sigma_1 \ldots \sigma_i) = q_i$, by the step 1 of the method we get $\gamma'(q_i, \Lambda^{k-i-1}\sigma_1 \ldots \sigma_i\sigma_{i+1}) = q_{i+1}$. But for all $i \in [k-1, l-1]$, since $\gamma^*(q_{i-k+1}, \sigma_{i-k+2} \ldots \sigma_i) = q_i$, by the step 2 we get $\gamma'(q_i, \sigma_{i-k+2} \ldots \sigma_i\sigma_{i+1}) = q_{i+1}$. Thus, both the DFA$_k$ and the DFA accepts the same words. $\square$

During a creation of a DFA$_k$, if after the first two steps there is no such distinct states $q_2, q_5 \in Q$ and a word $w = \sigma_1\sigma_2 \ldots \sigma_k \in T^k$ that $\gamma'(q_2, w) = \gamma'(q_5, w)$ (i.e., so far the function $\gamma'_v(q)$ is injection for all words v), then during the step 3 we can choose such values of the function γ' for arguments where it is still undefined that the created DFA$_k$ is a GFA$_k$ (the function $\gamma'_v(q)$ is bijection for all words v). For example, we can consider the DFA$_3$ in Fig. 5 as a GFA$_3$.

But if there is such w and q_2, q_5, then due to the determinism the values of $\gamma'(q_2, w)$ and $\gamma'(q_5, w)$ cannot be defined during the step 1. So they are defined exactly during the step 2. Thus, there are states $q_1, q_3, q_4 \in Q$ such that $\gamma(q_2, \sigma_k) = \gamma(q_5, \sigma_k) = q_3$, $\gamma^*(q_1, \sigma_1 \ldots \sigma_{k-1}) = q_2$ and $\gamma^*(q_4, \sigma_1 \ldots \sigma_{k-1}) = q_5$. Hence there is a C_k-construction and:

Theorem 2. *For any k, if a DFA does not contain any C_k-construction, then by the prefix extension method from the DFA we can create a GFA$_k$, which recognizes the same language as the DFA.*

Since any GFA$_k$ is a QFA$_k$ as well, from Theorems 1 and 2 we get that the C_k-construction is indeed forbidden construction:

Corollary 1. *A language L is in $\mathcal{L}_{QFA_k}$ iff the minimal DFA of L contains no C_k-construction.*

4.3   Forbidden Constructions for $\mathcal{L}_{QFA_*}$

For any k the C_k-construction decides whether the language recognized by a DFA is in $\mathcal{L}_{QFA_k}$. In order to show that a language L is not recognized by any multi-letter quantum finite automata we have to show that for any k there is a C_k-construction in the minimal DFA of L. In this section we provide two simple constructions that decides whether the language is in $\mathcal{L}_{QFA_*}$.

Let us denote the number of states in a DFA by n.

Definition 6. *A DFA* contains *an E-construction iff there are non-empty words* $x, y \in \Sigma^+$ *and states* $q_6, q_7, q_8 \in Q$ *such that* $q_6 \neq q_7$, $\gamma^*(q_6, y) = \gamma^*(q_7, y) = q_8$, $\gamma^*(q_6, x) = q_6$ *and* $\gamma^*(q_7, x) = q_7$.

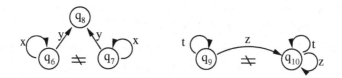

Fig. 6. E-construction and F-construction

Definition 7. *A DFA* contains *an F-construction iff there are non-empty words* $t, z \in \Sigma^+$ *and two distinct states* $q_9, q_{10} \in Q$ *such that* $\gamma^*(q_9, z) = \gamma^*(q_{10}, z) = q_{10}$, $\gamma^*(q_9, t) = q_9$ *and* $\gamma^*(q_{10}, t) = q_{10}$.

Theorem 3. *A DFA contains a C_k-construction for any k iff it contains an E-construction.*

Proof. Suppose for $k = n^2 - n + 1$ we have a C_k-construction. $q_2 \neq q_5$ gives us $\gamma^*(q_1, \sigma_1 \ldots \sigma_l) \neq \gamma^*(q_4, \sigma_1 \ldots \sigma_l)$ for all $l \leq k - 1$. At least two pairs in the set $\{(\gamma^*(q_1, \sigma_1 \ldots \sigma_i), \gamma^*(q_4, \sigma_1 \ldots \sigma_i)) \mid i \in [0; k-1]\}$ are equal since there are only $n^2 - n$ pairs of distinct states. Thus, exists an i and a $j > i$ such that $q_6 = \gamma^*(q_1, \sigma_1 \ldots \sigma_i) = \gamma^*(q_1, \sigma_1 \ldots \sigma_j) \neq \gamma^*(q_4, \sigma_1 \ldots \sigma_i) = \gamma^*(q_4, \sigma_1 \ldots \sigma_j) = q_7$. For $x = \sigma_{i+1} \ldots \sigma_j$ and $y = \sigma_{j+1} \ldots \sigma_k$ we have $\gamma^*(q_6, y) = \gamma^*(q_7, y) = q_8$, $\gamma^*(q_6, x) = q_6$ and $\gamma^*(q_7, x) = q_7$.

Suppose we have an E-construction. For any k let us write $x^k y$ as $\sigma'_1 \sigma'_2 \ldots \sigma'_l \in \Sigma^l$. Since $q_6 \neq q_7$ and $\gamma^*(q_6, x^k y) = \gamma^*(q_7, x^k y)$, exists an $i \in [k|x| + 1; l]$ such that states $q_2 = \gamma^*(q_6, \sigma'_1 \ldots \sigma'_{i-1})$ and $q_5 = \gamma^*(q_7, \sigma'_1 \ldots \sigma'_{i-1})$ are distinct, but $\gamma(q_2, \sigma'_i) = \gamma(q_5, \sigma'_i) = q_3$. For $q_1 = \gamma^*(q_6, \sigma'_1 \ldots \sigma'_{i-k})$, $q_4 = \gamma^*(q_7, \sigma'_1 \ldots \sigma'_{i-k})$ we get $\gamma^*(q_1, \sigma'_{i-k+1} \ldots \sigma'_{i-1}) = q_2$ and $\gamma^*(q_4, \sigma'_{i-k+1} \ldots \sigma'_{i-1}) = q_5$. $\square$

Corollary 2. *A language L can be recognized by multi-letter QFA iff the minimal DFA of L does not contain any E-construction.*

Lemma 3. *For each state $q \in Q$ in a DFA we have $\gamma^*(q, uv^m) = \gamma^*(q, uv^m v^{n!})$ for all words $u, v \in \Sigma^*$ and $m \geq n$.*

Proof. At least two states among $\gamma^*(q, uv^0), \gamma^*(q, uv^1), \ldots, \gamma^*(q, uv^n)$ are equal, so exists an i and a j such that $0 \leq i < j \leq n$ and $\gamma^*(q, uv^i) = \gamma^*(q, uv^j)$. Hence $\gamma^*(q, uv^m) = \gamma^*(q, uv^m v^{j-i})$. Since $j - i$ divides $n!$, the lemma holds.   □

Theorem 4. *An E-construction implies an F-construction.*

Proof. Suppose we have an E-construction. Let us denote $x^{n!}$ by t, $(yx^n)^{n!}$ by z, and z without its last fragment x^n by z'. By Lemma 3 for $u_1 = \epsilon$, $v_1 = yx^n$ and $m_1 = n!$ we get that $\gamma^*(q_6, (yx^n)^{n!}) = \gamma^*(q_6, (yx^n)^{n!}(yx^n)^{n!})$, and for $u_2 = z'$, $v_2 = x$ and $m_2 = n$ we get that $\gamma^*(q_6, (yx^n)^{n!}) = \gamma^*(q_6, (yx^n)^{n!}x^{n!})$. So for the state $q_{10} = \gamma^*(q_6, (yx^n)^{n!})$ we have $\gamma^*(q_6, z) = q_{10} = \gamma^*(q_{10}, z)$, $\gamma^*(q_6, t) = q_6$ and $\gamma^*(q_{10}, t) = q_{10}$. Actually, $\gamma^*(q_7, z) = q_{10}$ and $\gamma^*(q_7, t) = q_7$ as well. One of $q_6 \neq q_{10}$ and $q_7 \neq q_{10}$ is definitely true, thus, there is an F-construction.   □

Substituting $q_6 = q_9$, $q_7 = q_{10}$, $q_8 = q_{10}$, $x = t$ and $y = z$ shows that an F-construction implies an E-construction. Hence:

Corollary 3. *A language L can be recognized by multi-letter QFA iff the minimal DFA of L does not contain any F-construction.*

5   Closure Properties

Theorem 5. *The class $\mathcal{L}_{QFA_*}$ is closed against language intersection, language union and language complement.*

Proof. Suppose $\mathcal{A}$ and $\mathcal{B}$ are two minimal DFA recognizing accordingly languages $L_{\mathcal{A}}$ and $L_{\mathcal{B}}$ and neither contains an F-construction, but the minimal DFA $\mathcal{C}$ of the language $L_{\mathcal{A}} \cap L_{\mathcal{B}}$ contains an F-construction. Easy to prove that any DFA recognizing $L_{\mathcal{A}} \cap L_{\mathcal{B}}$ contains an F-construction as well. Let us consider a DFA $\mathcal{C}'$ such that $Q^{\mathcal{C}'} = Q^{\mathcal{A}} \times Q^{\mathcal{B}}$, $Q^{\mathcal{C}'}_{acc} = Q^{\mathcal{A}}_{acc} \times Q^{\mathcal{B}}_{acc}$, the initial state is $(q_0^{\mathcal{A}}, q_0^{\mathcal{B}})$, and $\gamma^{\mathcal{C}'}((q^{\mathcal{A}}, q^{\mathcal{B}}), \sigma) = (\gamma^{\mathcal{A}}(q^{\mathcal{A}}, \sigma), \gamma^{\mathcal{B}}(q^{\mathcal{B}}, \sigma))$. Easy to see that $\mathcal{C}'$ and $\mathcal{C}$ recognize the same language $L_{\mathcal{A}} \cap L_{\mathcal{B}}$. Thus, there are two words $x, y \in \Sigma^+$ and two distinct states $(q_1^{\mathcal{A}}, q_1^{\mathcal{B}}), (q_2^{\mathcal{A}}, q_2^{\mathcal{B}}) \in Q_{\mathcal{C}'}$ that form an F-construction for $\mathcal{C}'$. At least of one of $q_1^{\mathcal{A}} \neq q_2^{\mathcal{A}}$ and $q_1^{\mathcal{B}} \neq q_2^{\mathcal{B}}$ is true, therefore $\mathcal{A}$ or $\mathcal{B}$ (or both) contains an F-construction. Contradiction, which means $\mathcal{L}_{QFA_*}$ is closed against intersection.

Since for any QFA recognizing a language L substituting Q_{acc} by $Q \setminus Q_{acc}$ gives us a QFA recognizing the complement of L, $\mathcal{L}_{QFA_*}$ is also closed against union.   □

Theorem 6. *The class $\mathcal{L}_{QFA_*}$ is not closed against Kleene star.*

Proof. In the alphabet $\{a, b\}$ the minimal DFA of the language a does not contain any F-construction, but the minimal DFA of the language a^* does.   □

Theorem 7. *The class $\mathcal{L}_{QFA_*}$ is not closed against concatenation.*

Proof. The regular languages $(a + b)^*a$ and $(a + b)^*$ are in $\mathcal{L}_{QFA_*}$, but the minimal DFA of the language $(a + b)^*a(a + b)^*$ contains an F-construction.   □

6   Conclusion

We have shown that for GFA and QFA, contrary to DFA, seeing multiple input letters cannot be simulated by the one-input-letter model (e.g. by encoding into states). Thus, seeing several letters really enhance capabilities of GFA and QFA. Since these enhanced models can deal with the language $(a+b)^*a$ not recognized by any of the up-to-now introduced one-way read-only QFA models, a similar enhancement for those models should improve also their capabilities.

These results also indicate that our notion of what can be done by a finite-state real-time read-only quantum automaton is not yet completely clear, and there could be other enhancements or perhaps modifications of the word acceptance or language recognition notions leading to further improvements.

References

1. Ambainis, A., Beaudry, M., Golovkins, M., Kikusts, A., Mercer, M., Thérien, D.: Algebraic results on quantum automata. In: Diekert, V., Habib, M. (eds.) STACS 2004. LNCS, vol. 2996, pp. 93–104. Springer, Heidelberg (2004)
2. Ambainis, A., Freivalds, R.: 1-way quantum finite automata: strengths, weaknesses and generalizations. In: Proceedings of the 39th Annual Symposium on Foundations of Computer Science. pp. 332–341 (1998)
3. Bertoni, A., Mereghetti, C., Palano, B.: Quantum computing: 1-way quantum automata. In: Ésik, Z., Fülöp, Z. (eds.) DLT 2003. LNCS, vol. 2710, pp. 1–20. Springer, Heidelberg (2003)
4. Brodsky, A., Pippenger, N.: Characterizations of 1-way quantum finite automata. SIAM Journal on Computing 31(5), 1456–1478 (2002) Appeared earlier as Technical Report TR-99-03, University of British Columbia, 1999
5. Ciamarra, M.P.: Quantum reversibility and a new model of quantum automaton. In: Freivalds, R. (ed.) FCT 2001. LNCS, vol. 2138, pp. 376–379. Springer, Heidelberg (2001)
6. Dzelme, I.: Kvantu automāti ar jauktajiem stāvokļiem. Technical Report, University of Latvia (2003)
7. Hromkovič, J.: One-way multihead deterministic finite automata. Acta. Informatica 19, 377–384 (1983)
8. Kondacs, A., Watrous, J.: On the power of quantum finite state automata. In: Kondacs, A., Watrous, J. (eds.) Proceedings of the 38th IEEE Conference on Foundations of Computer Science, pp. 66–75 (1997)
9. Moore, C., Crutchfield, J.: Quantum automata and quantum grammars. Theoretical Computer Science 237, 97–97 (1997) Appeared in preprint form as Santa-Fe Institute Working Paper 97-07-062, 1997
10. Nayak, A.: Optimal lower bounds for quantum automata and random access codes. In: Proceedings of the 40th Annual Symposium on Foundations of Computer Science, pp. 369–377 (1999)
11. Paschen, K.: Quantum finite automata using ancilla qubits. Technical Report, University of Karlsruhe (2000)
12. Sipser, M.: Introduction to the Theory of Computation, pp. 31–90. PWS, Boston (1997)

Approximability and Non-approximability Results in Computing the Mean Speedup of Trace Monoids⋆

Alberto Bertoni and Roberto Radicioni

Università degli Studi di Milano
Dipartimento di Scienze dell'Informazione
Via Comelico 39, 20135 Milano, Italy
{bertoni,radicioni}@dsi.unimi.it

Abstract. The "mean speedup" of a trace monoid can be interpreted as an index of the "intrinsic parallelism". We study the problem of computing the mean speedup under two conditions: (1) uniform distribution on the words of given length and (2) uniform distribution on the traces of given height. In the first case, we give an approximability result showing a probabilistic fully polynomial time approximation scheme, while, in the second case, we prove that the problem is NP-hard to approximate within $n^{1-\epsilon}$ for every $\epsilon > 0$, unless $NP = coR$.

Introduction

Partial commutation and trace monoids have been initially introduced in a combinatorial context [4] and then as an abstract description of concurrent processes [9]. In a concurrent system, traces codify a process as a sequence of events chosen in a finite set Σ, some of which can occur simultaneously in according to a commutation relation C on Σ. In this interpretation, the length and the height of a trace are respectively intended as the sequential and the parallel execution time of a process, so that their ratio is the speedup, i.e. a parameter representing how much faster is parallel execution with respect to sequential one.

The asymptotic behaviour of the mean speedup has been analyzed considering the uniform distribution over the words of a given length [10,2,11] and over the traces of a given length or height [8], for every concurrent alphabet (Σ, C).

In all these cases, for all trace monoids the asymptotic behaviour converges to a real number (with an abuse of language we call it *mean speedup*) representing an index of the "intrinsic parallelism" of the trace monoid. While in the cases of uniform distribution on traces of given length or height the mean speedup is an algebraic number [8], in the case of uniform distribution on words of given length the mean speedup generally is transcendental [8].

In this paper, we study the problem of computing the mean speedup considering the uniform distribution over the words of a given length (MSWL problem)

⋆ This work is partially supported by the MIUR PRIN Project "Automata and Formal languages: mathematical and application driven studies".

T. Harju, J. Karhumäki, and A. Lepistö (Eds.): DLT 2007, LNCS 4588, pp. 72–83, 2007.
© Springer-Verlag Berlin Heidelberg 2007

and the uniform distribution over the traces of a given height (MSTH problem). The MSWL problem is a restriction of the mean spectral radius problem for sets of max-algebra matrices, which has been proved to be NP-hard to approximate [1]. Nevertheless, in this work we show a probabilistic fully polynomial time approximation scheme for the MSWL problem. The algorithm takes as input a parallel alphabet (Σ, C) and error $\epsilon > 0$ and works in time $\Theta((|\Sigma|^5 / \epsilon^3) \log(|\Sigma| / \epsilon))$. It exploits of some subwords in the words of Σ^*, that we called *pivot words*.

On the contrary, in the case of uniform distribution over the traces of given height, we prove that the mean speedup is NP-hard to approximate within a factor $n^{1-\epsilon}$, unless $NP = coR$. The proof is based on a particular relation between the the size of the maximal clique in (Σ, C) and the mean speedup of $(\Sigma^{(3)}, C^{(3)})$, that is, the disjoint union of three isomorphic copies of (Σ, C).

The paper is organized as follows. In Section 1, preliminary notions on trace monoids and approximation algorithms are recalled. In Section 2, we introduce a general notion of the mean speedup of a trace monoid and formally define the problems MSWL and MSTH. Then, in Section 3 and Section 4, we present the two main results of this work: a fully polynomial approximation scheme for MSWL and a strong non approximability result for MSTH.

1 Preliminaries

Given an alphabet Σ and a word $w \in \Sigma^*$, we denote its length by $l(w)$ and the number of occurrences of a symbol σ in w by $l_\sigma(w)$. Moreover, given a function $f : A \longrightarrow B$, we denote the preimage of f for $b \in B$ by $f^{-1}(b)$. For example, $l^{-1}(n) = \Sigma^n$.

A *commutation relation* is a symmetric, irreflexive relation $C \subseteq \Sigma \times \Sigma$ over Σ. If $(a, b) \in C$ then we say that the symbols a and b commute (aCb). We call *dependence relation* the complementary relation $D = (\Sigma \times \Sigma) \setminus C$. The graph (Σ, C) is usually called a *independence graph* or a *parallel alphabet*, as well. Conversely, we call (Σ, D) the *conflict graph* of (Σ, C). The *partially commutative monoid* (or *trace monoid*) generated by a parallel alphabet (Σ, C) is the monoid $\mathcal{M} = \Sigma^* / \equiv$, where $\equiv$ is congruence induced by the equalities

$$\{ab = ba \mid (a, b) \in C\}.$$

A *trace* is an element of $\mathcal{M}$. The notions of length and number of occurrences are directly extended to the traces. Hence, from this point on, $l^{-1}(n) = \{t \in \mathcal{M} \mid l(t) = n\}$. Moreover, we denote by $\mathrm{Alph}(t)$ the set $\{\sigma \in \Sigma \mid l_\sigma(t) > 0\}$. The natural congruence defined by C induces a morphism $\psi : \Sigma^* \longrightarrow \mathcal{M}$ mapping every word Σ^* in a trace in $\mathcal{M}$.

A univocal representation of a trace can be obtained in terms of cliques in the graph (Σ, C). In this view, a clique is a trace whose letters are all mutually independent. A pair of cliques (c_1, c_2) is said *CF-admissible* if for each σ_2 in $\mathrm{Alph}(c_2)$ there exists a σ_1 in $\mathrm{Alph}(c_1)$ such that $(\sigma_1, \sigma_2) \notin C$. Every trace $t \in \mathcal{M}$ univocally identifies a sequence $(c_1, \ldots, c_m)$ of cliques in (Σ, C) such that

(c_i, c_{i+1}) is a CF-admissible pair, for $1 \le i < m$, and $t = c_1 \cdots c_m$. This representation has been introduced in [4] and is called Cartier-Foata decomposition. The integer m is the *height* of t and is denoted by $h(t)$.

In this context, the *Cartier-Foata graph* (or *graph of cliques*) of (Σ, C) is a directed graph $\Gamma = (\Psi, \mathrm{Adm})$ where Ψ is the set of cliques of (Σ, C) and $(c_1, c_2) \in \mathrm{Adm}$ if and only if (c_1, c_2) is a CF-admissible pair. Therefore, each trace t in $\mathcal{M} = \Sigma^* / \equiv$ identifies a walk path of length $h(t)$ in the graph of cliques of (Σ, C).

An alternative definition of $h(t)$ can be given inductively by considering the height of every symbol σ through t. Let $(c_1, \ldots, c_m)$ be the CF-decomposition of a trace t. We denote with $h_\sigma(t)$ the largest i such that c_i contains σ. Then:

$$h_\sigma(t\sigma_i) = \begin{cases} h_\sigma(t) & \text{if } \sigma_i \ne \sigma, \\ \max_{(\hat{\sigma}, \sigma) \notin C} \{h_{\hat{\sigma}}(t)\} + 1 & \text{if } \sigma_i = \sigma, \end{cases} \tag{1}$$

where $h_\sigma(\varepsilon) = 0$ for all $\sigma \in \Sigma$. Then, $h(t) = \max_{\sigma \in \Sigma} \{h_\sigma(t)\}$.

We finally recall some definitions concerning approximation algorithms. Given a problem P, we denote by $sol(I)$ the solution of P on the instance I and by $A(I)$ the output of an algorithm A on input I. We are interested in problems whose solutions are nonnegative numbers. In this context, given a problem P and a function $r(n) : \mathbf{N} \longrightarrow \mathbf{N}$, a probabilistic algorithm A working in polynomial time is a *$r(n)$-approximation algorithm* for P if, for every $n \ge 0$ and every instance I of P of size n, with probability at least $1/2$ it holds that

$$\max\left\{ \frac{A(I)}{sol(I)}, \frac{sol(I)}{A(I)} \right\} \le r(n)$$

The value $\max\{A(I)/sol(I), sol(I)/A(I)\}$ is called the *performance ratio* of A on input I. This parameter is clearly always greater than or equal to 1.

A problem P admits a *probabilistic fully polynomial time approximation scheme* (pFPTAS) if there exists a probabilistic algorithm A that, having as input an instance I of P and a real number $\epsilon > 0$, works in polynomial time in the size of I and $1/\epsilon$ and returns a solution with performance ratio $1 + \epsilon$ with probability at least $1/2$.

2   The Speedup

In parallel computing, the speedup is the ratio between the execution time of a sequential algorithm and the execution time of its parallel version. The length $l(t)$ of a trace t can be viewed as its sequential execution time, while the height $h(t)$ represents the parallel execution time. Hence, the ratio $l(t)/h(l)$ can be interpreted as the speedup of an algorithm codified by a trace t.

Given a trace monoid $\mathcal{M}$ and a probability measure on its traces, we study the mean speedup of $\mathcal{M}$. Formally, a *weight function* on $\mathcal{M}$ is a function $\xi : \mathcal{M} \longrightarrow \mathbf{N}$ such that $|\xi^{-1}(n)| < \infty$ for every $n > 0$. For example, the length and the height of a trace are weight functions.

Definition 1. *Consider a trace monoid $\mathcal{M}$, a weight function ξ over $\mathcal{M}$ and, for every $n > 0$, a probability distribution P_n over the set $\xi^{-1}(n)$. Let $\rho_{(\xi,P)}(n)$ be the mean value of the speedup of the traces in $\xi^{-1}(n)$*

$$\rho_{(\mathcal{M},\xi,P)}(n) \;=\; \sum_{t \in \xi^{-1}(n)} P_n(t) \frac{l(t)}{h(t)}.$$

If the limit $\rho_{(\mathcal{M},\xi,P)} = \lim_{n\to\infty} \rho_{(\mathcal{M},\xi,P)}(n)$ exists, we call $\rho_{(\mathcal{M},\xi,P)}$ the mean speedup *of the triple $(\mathcal{M},\xi,P)$.*

The three most studied cases consider the mean speedup computed over the uniform distribution over the words of a given length [10,2,3,11], over the traces of a given length [5,6,8] and over the traces of a given height [8]. In this work, we analyze the first and the third case.

Uniform Distribution over the Words of a Given Length. By setting $\xi(t) = l(t)$ and $P_n(t) = |\psi^{-1}(t)| / |\Sigma|^n$, the probability of t is proportional to the number of words of Σ^* mapped in t by ψ. Denoting the mean speedup by λ_*, we have

$$\lambda_*(\Sigma, C) \;=\; \lim_{n\to\infty} \sum_{t \in l^{-1}(n)} \frac{|\psi^{-1}(t)|}{|\Sigma|^n} \frac{n}{h(t)} \;=\; \lim_{n\to\infty} \frac{n}{|\Sigma|^n} \sum_{w \in \Sigma^n} \frac{1}{h(\psi(w))}.$$

The existence of this limit has been proved in [10]. Formally, the problem of computing λ_* is defined as

PROBLEM: *Mean Speedup on Word Length (MSWL)*
INPUT: a graph (Σ, C);
OUTPUT: $\lambda_*(\Sigma, C)$.

Our aim is to develop an efficient approximation algorithm for MSWL.

Uniform Distribution over the Traces of a Given Height. By setting $\xi(t) = h(t)$ and $P_n(t) = 1/|h^{-1}(n)|$, the probability of a trace t is uniform over all the traces of height n. Denoting the mean speedup by $\eta_\mathcal{M}$, we have

$$\eta_\mathcal{M}(\Sigma, C) \;=\; \lim_{n\to\infty} \sum_{t \in h^{-1}(n)} \frac{1}{|h^{-1}(n)|} \frac{l(t)}{n} \;=\; \lim_{n\to\infty} \frac{1}{n\,|h^{-1}(n)|} \sum_{t \in h^{-1}(n)} l(t).$$

This limit has been proved to exist and to be an algebraic number in [8]. Formally, the problem of computing $\eta_\mathcal{M}$ can be defined as

PROBLEM: *Mean Speedup on Trace Height (MSTH)*
INPUT: a graph (Σ, C);
OUTPUT: $\eta_\mathcal{M}(\Sigma, C)$.

Our aim is to classify MSTH from a computational complexity point of view.

3   MSWL Is Approximable

The mean speedup $\lambda_*(\Sigma, C)$ has been widely studied [10,2,3,11]. In particular, in [10] it has been proved the existence of $\lambda_*(\Sigma, C)$ for any parallel alphabet (Σ, C) by means of Klingman's Ergodic Theorem; exact computation for simple monoids are proposed in [11,2], while it is known that generally $\lambda_*(\Sigma, C)$ is not algebraic [8]. As observed in [8], MSWL is a restriction of the max-algebra spectral radius estimation problem, whose approximation is NP-hard [1]. On the contrary, the main result of this section is the approximability of MSWL by means of a pFPTAS.

Let $(x_n)_{n \in \mathbf{N}^+}$ be a sequence of independent uniform random variables with values in Σ, so that the sequence $(x_1, \ldots, x_n)$ is generated with uniform distribution over Σ^n. In [10] has been proved that,

$$\text{Prob}\left\{ \lim_{n \to \infty} \frac{n}{h(\psi(x_1 \cdots x_n))} = \lambda_*(\Sigma, C) \right\} = 1. \tag{2}$$

A first consequence is the following

Lemma 1. *Let (Σ, C) be a parallel alphabet. Then*

$$\lim_{n \to \infty} \frac{1}{n |\Sigma|^n} \sum_{w \in \Sigma^n} h(\psi(w)) = \lambda_*(\Sigma, C)^{-1}.$$

Proof. Equation (2) implies that $h(\psi(x_1 \cdots x_n))/n \longrightarrow \lambda_*(\Sigma, C)^{-1}$ with probability 1, as n goes to the infinity. The thesis immediately follows (see also [11, Theorem 3.1]).

As a second consequence, let (Σ, D) be the conflict graph of (Σ, C) and let $(\Sigma_s, D_s)_{s \in S}$ be the set of the connected components of (Σ, D). For all $s \in S$, denote by $C_s = (\Sigma_s \times \Sigma_s) \setminus D_s$ the commutation relation restricted to Σ_s.

Lemma 2. *The mean speedup of (Σ, C) can be expressed as a function of the mean speedups of the subgraphs $(\Sigma_s, C_s)_{s \in S}$:*

$$\lambda_*(\Sigma, C) = \min_{s \in S}\left\{ \frac{|\Sigma|}{|\Sigma_s|} \lambda_*(\Sigma_s, C_s) \right\}.$$

Proof. The lemma is proved in [10, Theorem 5.7] by using Strong Law of Large Numbers and Equation (2).

Lemma 2 allows to extend a pFPTAS $\hat{A}$ for MSWL restricted to the case of parallel alphabets with connected conflicts graph to a pFPTAS A for MSWL.

Lemma 3. *Let $\hat{A}$ be a pFPTAS for MSWL restricted to the case of (Σ, C) is a parallel alphabet with connected conflict graph, then*

$$A(\Sigma, C, \epsilon) = \min_{s \in S}\left\{ \frac{|\Sigma|}{|\Sigma_s|} \hat{A}(\Sigma_s, C_s, \epsilon) \right\}$$

is a pFPTAS for MSWL.

Proof. Let $\hat{A}(s) = \hat{A}(\Sigma_s, C_s, \epsilon)$. By hypothesis, for every $s \in S$, we have that $(1+\epsilon)^{-1}\hat{A}(s) \le \lambda_*(\Sigma_s, C_s) \le (1+\epsilon)\hat{A}(s)$, which implies

$$\frac{1}{1+\epsilon} \min_{s \in S} \left\{ \frac{|\Sigma|}{|\Sigma_s|} \hat{A}(s) \right\} \le \min_{s \in S} \left\{ \frac{|\Sigma|}{|\Sigma_s|} \lambda_*(\Sigma_s, C_s) \right\} \le (1+\epsilon) \min_{s \in S} \left\{ \frac{|\Sigma|}{|\Sigma_s|} \hat{A}(s) \right\}.$$

Then, by Lemma 2, $(1+\epsilon)^{-1}A(\Sigma, C, \epsilon) \le \lambda_*(\Sigma, C) \le (1+\epsilon)A(\Sigma, C, \epsilon)$. Since A works in polynomial time, the thesis follows.

In light of these results, we prove the following

Theorem 1. *The* MSWL *problem admits a pFPTAS.*

Proof. From Lemma 1 and Lemma 3, it is sufficient to prove the existence of a pFPTAS for approximating $\lambda_*(\Sigma, C)^{-1}$, having as input a parallel alphabet (Σ, C) whose conflict graph is connected. The algorithm

Input: (Σ, C, ϵ)
1: $j \leftarrow \lceil 12(6n^2/\epsilon) \log{(6n/\epsilon)} \rceil$;
2: $t \leftarrow \lceil (4\log 2)(n/\epsilon)^2 \rceil$;
3: generate uniformly at random t words $x_1, x_2, \ldots, x_t \in \Sigma^j$.
4: compute the sample mean $S_j = \frac{1}{t} \sum_{i=1}^{t} h(x_i)/j$.
Output: S_j

approximates $\lambda_*(\Sigma, C)^{-1}$ with performance ratio at most $1+\epsilon$, for every (Σ, C) having a connected conflict graph (Σ, D). It clearly works in polynomial time with respect of $|\Sigma|$ and $1/\epsilon$. Indeed, the computation of the height of a word $w \in \Sigma^k$ can be made in time $O(|\Sigma| \cdot k)$ by exploiting Equation (1). Hence, the complexity of the algorithm is $O(n^5 \epsilon^{-3} \log(n/\epsilon))$.

Its correctness is proved in the next subsection.

3.1   Correctness of the Algorithm

Let X be the random variable $X(w) = h(w)/l(w)$, where $w \in \Sigma^j$. We denote by $\lambda_j = E[X(w) \mid w \in \Sigma^j]$ the expected value of X over the whole set Σ^j, and write $\lambda = \lim_{j \to \infty} \lambda_j$. It is straightforward to observe that, if $j < j'$, then $\lambda_j \ge \lambda_{j'}$. Henceforward, we assume that the conflict graph (Σ, D) is connected, and let $n = |\Sigma|$. Let T be a spanning tree for (Σ, D) and fix a root σ_n, then give an order $\sigma_1, \ldots, \sigma_n$ to the vertices in Σ, such that the sons of a vertex σ_k precede σ_k in the word $\sigma_1 \sigma_2 \cdots \sigma_n$. The word $p = \sigma_1 \cdots \sigma_{n-1} \sigma_n \sigma_{n-1} \cdots \sigma_1$ is called a *pivot word* for (Σ, D) and satisfies the following property:

Lemma 4. $h(w_1 p w_2) > h(w_1) + h(w_2)$.

Proof. Let c_π be the clique containing the symbol σ_n of p in the CF-decomposition of $w_1 p w_2$ and let x_i be a symbol in $w_1 \sigma_1 \cdots \sigma_{n-1}$. If $x_i = \sigma_n$, then $x_i D \sigma_n$ and it does not belong to c_π. Otherwise, by the particular form of the pivot word, there exists a path of dependences $x_i D \alpha_1, \alpha_1 D \alpha_2, \ldots, \alpha_s D \sigma_n$ in T such that $x_i \alpha_1 \alpha_2, \ldots, \alpha_s \sigma_n$ is a subword of $w_1 \sigma_1 \cdots \sigma_n$. Then, x_i does not belong to

c_π. It can be proved symmetrically that none of the symbols in $\sigma_{n-1}\cdots\sigma_1 w_2$ are in c_π and, then, the clique c_π is exactly $\{\sigma_n\}$.

Now, let $(d_1,\ldots,d_k)$ and $(e_1,\ldots,e_h)$ be the CF-decompositions of the words $w_1\sigma_1\cdots\sigma_{n-1}$ and $\sigma_{n-1}\cdots\sigma_1 w_2$, respectively. The CF-decomposition of $w_1 p w_2$ is then $(d_1,\ldots,d_k,\{\sigma_n\},e_1,\ldots,e_h)$ and its height is

$$h(w_1 p w_2) = h(w_1\sigma_1\cdots\sigma_{n-1}) + 1 + h(\sigma_{n-1}\cdots\sigma_1 w_2) > h(w_1) + h(w_2),$$

which proves the lemma.

Corollary 1. *If p is a subword of y, then $h(w_1 y w_2) > h(w_1) + h(w2)$.*

The probability that a word of length H contains p as a subword goes rapidly to 1 as H goes to infinity. Indeed:

Lemma 5. *Let P be the probability that $x \in \Sigma^H$ contains p as a subword. Then,*

$$P \geq 1 - 2ne^{-\frac{H}{2n^2}}.$$

Proof. Split x in $2n-1$ factors $x_1,\ldots,x_{2n-1}$ of length $\left\lceil\frac{H}{2n}\right\rceil$ and a remaining factor x_{2n} of smaller or equal length. P is at least the probability that x_1 contains σ_1, ..., x_n contains σ_n, x_{n+1} contains σ_{n-1}, ..., x_{2n-1} contains σ_1. In formula:

$$P \geq 1 - \mathrm{Prob}\{\sigma_1 \notin \mathrm{Alph}(x_1) \vee \sigma_2 \notin \mathrm{Alph}(x_2) \vee \cdots \vee \sigma_1 \notin \mathrm{Alph}(x_{2n-1})\}$$

$$\geq 1 - (2n-1)\left(\frac{n-1}{n}\right)^{\left\lceil\frac{H}{2n}\right\rceil} \geq 1 - 2ne^{-\frac{H}{2n^2}}.$$

Lemma 5 guarantees that the property of Corollary 1 holds for a fraction of strings which tends rapidly to 1 and provides a further property for λ_j:

Lemma 6. *For every $n, H, j \in \mathbf{N}$ such that $0 < 2n-1 \leq H \leq 2j$, it holds*

$$\lambda_j \geq \lambda_{2j} \geq \lambda_j\left(1 - \frac{H}{2j} - 2ne^{-\frac{H}{2n^2}}\right). \tag{3}$$

Proof. Let $w_1 y w_2 \in \Sigma^{2j+H}$, where $l(w_1) = l(w_2) = j$ and $l(y) = H$. From Corollary 1 and Lemma 5 we have that $h(w_1 y w_2) \geq h(w_1) + h(w_2)$ with probability $1 - 2ne^{-\frac{H}{2n^2}}$ or, equivalently,

$$\frac{h(w_1 y w_2)}{2j+H} \geq \frac{h(w_1)+h(w_2)}{2j+H} \geq \left(1 - \frac{H}{2j}\right)\frac{h(w_1)+h(w_2)}{2j}.$$

That is, with probability $1 - 2ne^{-\frac{H}{2n^2}}$,

$$X(w_1 y w_2) \geq \left(1 - \frac{H}{2j}\right)\frac{1}{2}(X(w_1) + X(w_2)). \tag{4}$$

Let us denote by A the set of all the words in Σ^H having the pivot word as a subword. Then, by (4),

$$\lambda_{2j+H} = \sum_{\substack{w_1, w_2 \in \Sigma^j \\ y \in \Sigma^H}} X(w_1 y w_2) n^{-(2j+H)} \geq \sum_{\substack{w_1, w_2 \in \Sigma^j \\ y \in A}} X(w_1 y w_2) n^{-(2j+H)}$$

$$= \left(1 - \frac{H}{2j}\right) \cdot \lambda_j \cdot \sum_{y \in A} n^{-H}.$$

Recalling Lemma 5, we have

$$\lambda_{2j} \geq \lambda_{2j+H} \geq \lambda_j \left(1 - \frac{H}{2j}\right)\left(1 - 2ne^{-\frac{H}{2n^2}}\right) \geq \lambda_j \left(1 - \frac{H}{2j} - 2ne^{-\frac{H}{2n^2}}\right).$$

Lemma 6 allows to prove the following

Lemma 7. *Let* $\lambda = \lim_{n \to \infty} \lambda_j$. *Then, for every* $j > n$,

$$1 \leq \frac{\lambda_j}{\lambda} \leq \exp\left(6n^2 \frac{\log j}{j}\right).$$

Proof. Consider Lemma 6 and write H as a function of j such that, for j going to the infinity, the quantity at the last member of Equation (3) tends to λ_j. Hence, let $H = H(j) = 2n^2 \log j$. Then, for $j > n$, we have

$$\lambda_{2j} \geq \lambda_j \left(1 - n^2 \frac{\log j}{j} - \frac{2n}{j}\right) \geq \lambda_j \left(1 - 2n^2 \frac{\log j}{j}\right).$$

Extending such inequality to $2^k j$, we can write it as

$$\lambda_{2^k j} \geq \lambda_j \prod_{s=0}^{k-1} \left(1 - 2n^2 \frac{\log(2^s j)}{2^s j}\right) \geq \lambda_j \exp\left[\sum_{s=0}^{\infty} \log\left(1 - 2n^2 \frac{\log(2^s j)}{2^s j}\right)\right].$$

Since $\log(1 - 2x) > -3x$ for $0 < x < \bar{x}$, a lower bound for the sum is

$$\sum_{s=0}^{\infty} \log\left(1 - 2n^2 \frac{\log(2^s j)}{2^s j}\right) \geq \sum_{s=0}^{\infty} \left(-3n^2 \frac{\log j + s \log 2}{2^s j}\right) \geq -6n^2 \frac{\log j}{j}$$

and we obtain, for every $k > 0$, the following inequality

$$\lambda_j \geq \lambda_{2^k j} \geq \lambda_j \exp\left(-6n^2 \frac{\log j}{j}\right).$$

Hence, for k going to the infinity, we have $1 \leq \frac{\lambda_j}{\lambda} \leq \exp\left(6n^2 \frac{\log j}{j}\right)$.

Now, we are ready to prove the correctness of the algorithm. First of all, by Lemma 7 we have $(\lambda_j - \lambda) \leq \lambda(\exp(6n^2 \log j/j) - 1)$.

Observe that $\exp(6n^2 \log \bar{j}/\underline{j}) - 1 \leq \epsilon/2$ whereas

$$\bar{j} = \left\lceil 12 \cdot \frac{6n^2}{\epsilon} \log\left(\frac{6n}{\epsilon}\right) \right\rceil \qquad \text{(Line 1 of the algorithm)}.$$

Therefore, $\lambda_{\bar{j}} - \lambda \leq \lambda \cdot \epsilon/2$.

Since $\lambda \geq 1/n$, applying Hoeffding's Inequality, we obtain

$$\text{Prob}\left\{|S_{\bar{j}} - \lambda_{\bar{j}}| \geq \lambda\frac{\epsilon}{2}\right\} \leq \text{Prob}\left\{\left|\frac{1}{t}\sum_{i=1}^{t} X_i - \lambda_{\bar{j}}\right| \geq \frac{1}{n}\frac{\epsilon}{2}\right\} \leq 2e^{-\frac{t\epsilon^2}{2n^2}}.$$

Then, by setting $\bar{t} = \lceil(4\log 2)(n/\epsilon)^2\rceil$ (Line 2 of the algorithm):

$$\text{Prob}\left\{|S_{\bar{j}} - \lambda_{\bar{j}}| \geq \lambda\frac{\epsilon}{2}\right\} \leq \frac{1}{2}.$$

Hence, with probability at least $1/2$, we have $|S_{\bar{j}} - \lambda| \leq |S_{\bar{j}} - \lambda_{\bar{j}}| + |\lambda_{\bar{j}} - \lambda| < \lambda \cdot \epsilon$ or, equivalently, $\max\left\{\frac{S_{\bar{j}}}{\lambda}, \frac{\lambda}{S_{\bar{j}}}\right\} < 1 + \epsilon$.

4   MSTH Is Not Approximable (Unless $NP = coR$)

The mean speedup over traces of the same height has been studied by Krob et al. in [8]. They proved that, given a parallel alphabet (Σ, C), the bivariate generating function counting the traces having a given length and height is rational and that $\eta_{\mathcal{M}}(\Sigma, C)$ is an algebraic number. In this section we prove that, unless $NP = coR$, any approximation algorithm for MSTH must have performance ratio $\Omega(n^{1-\epsilon})$, for any $\epsilon > 0$.

The idea is that of reducing to MSTH the MAX-CLIQUE problem, which is difficult to approximate [7]. The reduction is based on the following construction. Let $G = (\Sigma, C)$ be a parallel alphabet. We denote by $G^{(k)} = (\Sigma^{(k)}, C^{(k)})$ the disjoint union of k graphs $G_i = (\Sigma_i, C_i)$, $1 \leq i \leq k$, isomorphic to G, where

$$\Sigma^{(n)} = \bigsqcup_1^k \Sigma_i \qquad \text{and} \qquad C^{(k)} = \bigsqcup_1^k C_i.$$

Notice that, for every clique c of $G^{(k)}$, there exists a unique $l \in \{1, 2, \ldots, k\}$ such that c is a clique in G_l. We denote this situation by $T(c) = l$. Let $\Gamma(k)$ be the graph of cliques of $G^{(k)}$ and let c' and c'' be two cliques in $G^{(k)}$ such that $T(c') \neq T(c'')$. Then, (c', c'') is an arc in $\Gamma(k)$.

Given a path $t = x_1 \cdots x_n$ in $\Gamma(k)$, we set the weight function $\xi(t) = |c_1| + \cdots + |c_n|$ (if t is intended as a trace, $\xi(t) = l(t)$). Then, considering the uniform distribution over the paths of length n in $\Gamma(k)$, we define the expected value

$$\xi_n(k) = E[\xi(x_1 \cdots x_n) \mid x_1 \cdots x_n \text{ is a path in } \Gamma(k)].$$

Let $G^{(k)} = (\Sigma^{(k)}, C^{(k)})$. By definition, we have

$$\eta_{\mathcal{M}}(\Sigma^{(k)}, C^{(k)}) = \lim_{n \to \infty} \frac{\xi_k(n)}{n}.$$

From this point on, we write $\eta_{\mathcal{M}}(G^{(k)})$ for $\eta_{\mathcal{M}}(\Sigma^{(k)}, C^{(k)})$ for simplicity of notation. Moreover, we denote by M_G the maximal size of a clique of G, and by m_G the mean size of a clique in G, under uniform distribution over the cliques.

The following result states that, for large k, m_G can be considered a lower bound for $\eta_{\mathcal{M}}(G^{(k)})$.

Lemma 8. *For every* $k > 0$, $M_G \geq \eta_{\mathcal{M}}(G^{(k)}) \geq m_G(1 - 2/k)$.

Proof. The first inequality is trivial, since $\xi(t) \leq n \cdot M_G$ for every path t of length n in $\Gamma^{(k)}$ by definition. Hence, we focus our attention on the second inequality.

We say that a path $x_1 \cdots x_l \cdots x_n$ in $\Gamma(k)$ has a "hit" in position l if and only if $T(x_{l-1}) \neq T(x_l) \neq T(x_{l+1})$. If $x_1 \cdots x_l \cdots x_n$ has a hit in position l, then every sequence $x_1 \cdots x_{l-1} c \, x_{l+1} \cdots x_n$ with $T(c) = T(x_l)$ is a path in $\Gamma(k)$. Therefore,

$$E[\xi(x_l) \mid x_1 \cdots x_l \cdots x_n \text{ has a hit in position } l] = m_G. \tag{5}$$

The probability P_l that a path $x_1 \cdots x_l \cdots x_n$ has a hit in position l is

$$P_l \geq \frac{k-2}{k}. \tag{6}$$

Indeed, fixed a path $x_1 \cdots x_{l-1} x_l x_{l+1} \cdots x_n$, the probability that $T(x_{l-1}) \neq T(x_l) \neq T(x_{l+1})$ is at least $1 - 1/k$ if $T(x_{l-1}) = T(x_{l+1})$ and at least $1 - 2/k$ otherwise. Then, it follows that

$$\xi_n(k) = E[\xi(x_1 \cdots x_n) \mid x_1 \cdots x_n \text{ is a path in } \Gamma(k)] =$$
$$= E[\xi(x_1) + \cdots + \xi(x_n) \mid x_1 \cdots x_n \text{ is a path in } \Gamma(k)] =$$
$$= \sum_{l=1}^{n} E[\xi(x_l) \mid x_1 \cdots x_n \text{ is a path in } \Gamma(k)] \geq$$
$$\geq \sum_{l=1}^{n} E[\xi(x_l) \mid x_1 \cdots x_n \text{ is a path in } \Gamma(k) \text{ with a hit in pos. } l] \cdot P_l.$$

From Equations (5) and (6), we have $\xi_n(k) \geq n \cdot m_G \cdot (1 - 2/k)$. Hence,

$$\lim_{n \to \infty} \frac{\xi_n(k)}{n} \geq m_G \left(1 - \frac{2}{k}\right)$$

The next lemma states that m_G is close to M_G.

Lemma 9. *Let* $G(\Sigma, C)$ *be a graph and let* $|\Sigma| = n$. *Then* $m_G \geq M_G/(4 \log_2 n)$.

Proof. Let $\text{Prob}\{|c| \leq s\}$ be the probability that the size of a random clique, selected with uniform probability over the cliques of G, is smaller than s. If $M_G < 5$, then the thesis is immediately proved, since $M_G/(4 \log_2 n) \leq 1$.

Now, suppose that $M_G \geq 5$. Recalling that G contains at least 2^{M_G} cliques and at most $\binom{n}{l}$ cliques of size l, we have

$$\text{Prob}\{|c| \leq s\} \leq \frac{\sum_{l=1}^{s} \binom{n}{l}}{2^{M_G}} \leq \frac{e \cdot n^s}{2^{M_G}}.$$

Observe that $e \cdot n^s / 2^{M_G} \leq 1/2$ if and only if $s \leq (M_G - 1 - \log_2 e)/\log_2 n$. Since

$$\frac{M_G}{2 \log_2 n} \leq \frac{M_G - 1 - \log_2 e}{\log_2 n}$$

for every $M_G \geq 5$, for $\bar{s} = M_G/(2 \log_2 n)$ it holds that $\text{Prob}\{|c| \geq \bar{s}\} \leq 1/2$. By applying Markov's Inequality, we finally obtain

$$\frac{1}{2} \leq \text{Prob}\{|c| \geq \bar{s}\} \leq \frac{m_G}{\bar{s}} = m_G \frac{2 \log_2 n}{M_G},$$

which proves the lemma.

The reduction of MAX-CLIQUE to MSTH is given in the following

Lemma 10. *Let A be an approximation algorithm with performance ratio $r(n)$ for MSTH. Then, the algorithm $B(\Sigma, C) = A(\Sigma^{(3)}, C^{(3)})$ is an approximation algorithm with performance ratio $12 \cdot r(3n) \log(n)$ for MAX-CLIQUE.*

Proof. Given the oracle A, the algorithm clearly works in polynomial time with respect to $n = |\Sigma|$.

For the sake of simplicity, we denote $\eta_\mathcal{M}(\Sigma, C)$ by $\eta(G)$. From Lemma 8, we know that $M_G \geq \eta(G^{(3)}) \geq m_G/3$. By Lemma 9, $m_G \geq M_G/(4 \log_2 n)$. Hence,

$$M_G \geq \eta(G^{(3)}) \geq \frac{M_G}{12 \log_2 n}.$$

Since A has performance ratio $r(n)$ and works with an input of size $3n$, then we have

$$r(3n) \geq \max\left\{\frac{A(G^{(3)})}{\eta(G^{(3)})}, \frac{\eta(G^{(3)})}{A(G^{(3)})}\right\}$$

$$\geq \max\left\{\frac{A(G^{(3)})}{M_G}, \frac{M_G}{12 \log_2 n A(G^{(3)})}\right\}$$

$$\geq \frac{1}{12 \log_2 n} \max\left\{\frac{A(G^{(3)})}{M_G}, \frac{M_G}{A(G^{(3)})}\right\},$$

that proves the lemma.

Now, the main result of this section is immediately proved:

Theorem 2. *If $NP \neq CoR$, then any approximation algorithm for MSTH has performance ratio $r(n) = \Omega(n^{1-\epsilon})$, for every $\epsilon > 0$.*

Proof. By Håstad [7], given $\epsilon > 0$, any approximation algorithm for MAX-CLIQUE has performance ratio $\Omega(n^{1-\epsilon})$, unless $NP = coR$. Hence, by Lemma 10, $12r(3n) \log_2 n > \Omega(n^{1-\epsilon})$ for every $\epsilon > 0$, which is equivalent to the thesis.

5 Conclusions and Open Problems

In this paper we proved that the MSWL problem admits a probabilistic FPTAS, obtained by exploiting concentration results [10] and the properties of some words that we called *pivot words*. In this direction, it seems to be possible to improve the performance of the randomized algorithm by exploiting the ergodic theorem, as in [11]. The existence of a deterministic FPTAS for MSWL is a still open problem.

In addition, we proved that the MSTH problem is NP-hard to approximate within $|\Sigma|^{1-\epsilon}$, unless $NP = coR$. The approximability of the mean speedup considering the uniform distribution over the traces of a given length remains an open problem.

References

1. Blondel, V.D., Gaubert, S., Tsitsiklis, J.N.: Approximating the spectral radius of sets of matrices in the max-algebra is NP-hard. IEEE Trans. Automat. Control 45(9), 1762–1765 (2000)
2. Brilman, M., Vincent, J.M.: Dynamics of synchronized parallel systems. Comm. Statist. Stochastic Models 13(3), 605–617 (1997)
3. Brilman, M., Vincent, J.-M.: On the estimation of the throughput for a class of stochastic resources sharing systems. Math. Oper. Res. 23(2), 305–321 (1998)
4. Cartier, P., Foata, D.: Problèmes combinatoires de commutation et réarrangements. Lecture Notes in Mathematics, vol. 85. Springer-Verlag, Berlin Heidelberg (1969)
5. Gaubert, S.: Performance evaluation of (max, +) automata. IEEE Trans. Automat. Control 40(12), 2014–2025 (1995)
6. Gaubert, S., Mairesse, J.: Task resource models and (max, +) automata. In: Idempotency (Bristol, 1994). Publ. Newton Inst, vol. 11, pp. 133–144. Cambridge Univ. Press, Cambridge (1998)
7. Håstad, J.: Clique is hard to approximate within $n^{1-\epsilon}$. Acta Math. 182(1), 105–142 (1999)
8. Krob, D., Mairesse, J., Ioannis, M.: Computing the average parallelism in trace monoids. Discrete Math. 273(1-3), 131–162 (2003) (EuroComb'01 Barcelona)
9. Mazurkiewicz, A.: Concurrent program schemes and their interpretation. Technical Report DAIMI PB-78, Aarhus University, Comp. Science Depart. (July 1977)
10. Saheb, N.: Concurrency measure in commutation monoids. Discrete Appl. Math. 24(1-3), 223–236 (1989) First Montreal Conference on Combinatorics and Computer Science, 1987
11. Saheb, N., Zemmari, A.: Methods for computing the concurrency degree of commutation monoids. In: Formal power series and algebraic combinatorics (Moscow, 2000), pp. 731–742. Springer, Berlin Heidelberg (2000)

The Dynamics of Cellular Automata in Shift-Invariant Topologies

Laurent Bienvenu[1] and Mathieu Sablik[2]

[1] Laboratoire d'Informatique Fondamentale, CNRS & Université de Provence,
Marseille, France
laurent.bienvenu@lif.univ-mrs.fr
[2] ENS Lyon, Unité de Mathématiques Pures et Appliquées, Lyon, France
msablik@umpa.ens-lyon.fr

Abstract. We study the dynamics of cellular automata, and more specifically their transitivity and expansivity, when the set of configurations is endowed with a shift-invariant (pseudo-)distance. We first give an original proof of the non-transitivity of cellular automata when the set of configurations is endowed with the Besicovitch pseudo-distance. We then show that the Besicovitch pseudo-distance induces a distance on the set of shift-invariant measures and on the whole space of measures, and we prove that in these spaces also, cellular automata cannot be expansive nor transitive.

1 Introduction

Cellular automata were introduced by J. von Neumann as a simple formal model for cellular growth and replication. They consist in a discrete lattice of finite-state machines, called *cells* which evolve sequentially and synchronously according to a local rule. This local rule is the same for all cells and determines how a cell will evolve given the states of a finite number of neighboring cells. A snapshot of the states of the cells is called a *configuration*, and a cellular automaton can be seen as a map from the set of configurations to itself. Despite the apparent simplicity of their definition, cellular automata, seen as discrete dynamical systems, can have very complex behaviors, some of which not even being fully understood yet. This behavior is typically studied by endowing the set of configurations with the Cantor distance. For this distance the so-called shift maps, which spacially shift the states of cells according to a fixed vector, can have highly chaotic behaviors.

Other distances can also be defined on the space of configurations for which the shift maps are non-chaotic. An example of such a distance is the Besicovitch distance (in fact, pseudo-distance), introduced by Cattaneo et al. [CFMM97]. It was proven by Blanchard et al. [BCF03] that no cellular automaton can be transitive for this pseudo-distance. Their proof uses Kolmogorov complexity, which is an algorithmic measure of information content. We first provide new simple proof of this fact, also based on Kolmogorov complexity, and we show that our proof can be turned into a purely analytic one, based on Hausdorff dimension.

T. Harju, J. Karhumäki, and A. Lepistö (Eds.): DLT 2007, LNCS 4588, pp. 84–95, 2007.

Suppose now that a measure μ is defined on the set of configurations. A cellular automaton acts on the set of configurations and canonically transforms μ into another measure. Hence, instead of its action on the set of configurations, a cellular automaton can be studied via its action on the set of measures. If μ is shift-invariant, then its image by any cellular automaton is also shift-invariant. Hence, cellular automata also have a natural action on the set of shift-invariant measures. In [Sab07], it is shown that any pseudo-distance on the set of configurations induces a pseudo-distance on the set of shift-invariant measures. Thus, both the Cantor and the Besicovitch distances induce a distance on the set of shift-invariant measures. We show that in this framework also, no cellular automaton is transitive nor expansive on the set of shift-invariant measures endowed with the distance induced by the Besicovitch distance.

The last section of the paper unifies the two proofs of non-transitivity, in the space of configurations and in the space of shift-invariant measures respectively, by embedding these two spaces in the (much) bigger one containing all measures (non-necessarily shift-invariant). Here again, Kolmogorov complexity and effective Hausdorff dimension turn out to be the cornerstone of the proof.

Before moving on to our discussion, we recall the formal definition of the main concepts of the paper, namely transitivity and expansivity. Let (X, d) be a metric space, and $f : X \to X$. The map f is said to be *transitive* if for any $x, y \in X$ and any $\varepsilon > 0$, there exists $x', y' \in X$ and $n \in \mathbb{N}$ such that $d(x, x') < \varepsilon$, $d(y, y') < \varepsilon$ and $f^n(x') = y'$. It is said to be *expansive* if there exists $\varepsilon > 0$ such that for all x, y with $x \neq y$, there exists an $n \in \mathbb{N}$ such that $d(f^n(x), f^n(y)) > \varepsilon$.

Informally, transitivity is a mixing property, while expansivity is a sign of sensitivity to initial conditions. Hence, both these conditions are often seen as symptomatic of chaotic dynamical systems.

2   Action of Cellular Automata on $\mathcal{A}^{\mathbb{M}}$

Formally speaking, a *cellular automaton* is a tuple $\langle \mathcal{A}, \mathbb{M}, \mathbb{U}, \delta \rangle$ where $\mathcal{A}$ is a finite alphabet (the *set of states*), $\mathbb{M}$ is a semi-group (the set of indices of cells), $\mathbb{U}$ is a finite subset of $\mathbb{M}$ (the *neighborhood*), and δ is a function from $\mathcal{A}^{\mathbb{U}}$ into $\mathcal{A}$ (*the local rule*). In this setting, the set of configurations is the set $\mathcal{A}^{\mathbb{M}}$. The cellular automaton acts on it via its *global rule*, defined as follows: for all $x \in \mathcal{A}^{\mathbb{M}}$, and all $i \in \mathbb{M}$, the i-th coordinate of $F(x)$ is given by the rule $F(x)_i = \delta((x_{i+k} : k \in \mathbb{U}))$. In the sequel, when this create no confusion, we will make no distinction between a CA and its global rule.

In this paper, the semi-group $\mathbb{M}$ will be of the form $\mathbb{M} = \mathbb{Z}^{d'} \times \mathbb{N}^{d''}$, but most of the results we will present can be generalized to a larger class of semi-groups. Let $\mathbb{M} = \mathbb{Z}^{d'} \times \mathbb{N}^{d''}$. For all $m \in \mathbb{M}$, we denote by $|m|$ the distance of m to the origin point. This allows us to define the radius of the cellular automaton: $r(F) = \max\{|m| : m \in \mathbb{U}\}$ where $\mathbb{U}$ is the neighborhood of F.

Cantor Topology. One can define a topology on $\mathcal{A}^{\mathbb{M}}$ by endowing $\mathcal{A}$ with the discrete topology, and considering the product topology (or Cantor topology) on $\mathcal{A}^{\mathbb{M}}$. For this topology, $\mathcal{A}^{\mathbb{M}}$ is compact, perfect and totally disconnected. Moreover one can define a metric (which we call the *Cantor distance*) on $\mathcal{A}^{\mathbb{M}}$ which is compatible with the Cantor topology:

$$\forall x, y \in \mathcal{A}^{\mathbb{M}}, \quad d_C(x, y) = 2^{-\min\{|i|:x_i \neq y_i\ i \in \mathbb{M}\}}.$$

Let $\mathbb{U} \subset \mathbb{M}$. For $x \in \mathcal{A}^{\mathbb{M}}$, we denote by $x_{\mathbb{U}} \in \mathcal{A}^{\mathbb{U}}$ the restriction of x to $\mathbb{U}$. For a pattern $w \in \mathcal{A}^{\mathbb{U}}$, one defines the cylinder centered on w by $[w]_{\mathbb{U}} = \{x \in \mathcal{A}^{\mathbb{Z}} : x_{\mathbb{U}} = w\}$.

The action of $\mathbb{M}$ on itself allows to define an action on $\mathcal{A}^{\mathbb{M}}$ by *shift*. For all $m \in \mathbb{M}$ this action is defined by:

$$\sigma^m : \mathcal{A}^{\mathbb{M}} \longrightarrow \mathcal{A}^{\mathbb{M}}$$
$$(x_i)_{i \in \mathbb{M}} \longmapsto (x_{i+m})_{i \in \mathbb{M}}$$

Cellular automata commute with the shift maps: for every cellular automaton $F : \mathcal{A}^{\mathbb{M}} \to \mathcal{A}^{\mathbb{M}}$ and all $m \in \mathbb{M}$, $F \circ \sigma^m = \sigma^m \circ F$. In fact, this a fundamental characteristic of CA. Indeed, Hedlund's theorem [Hed69] states that the cellular automata on $(\mathcal{A}^{\mathbb{M}}, d_C)$ are exactly the continuous functions which commute with the shift maps. It is easy to remark that any cellular automaton F is Lipschitz for the distance d_C. More precisely, for all $x, y \in \mathcal{A}^{\mathbb{M}}$, one has:

$$d_C(F(x), F(y)) \leq 2^{-r(F)} d_C(x, y).$$

It is well-known and easy to see that the action of any shift σ^m on $(\mathcal{A}^{\mathbb{M}}, d_C)$ is transitive. More generally, for all surjective cellular automaton $F : \mathcal{A}^{\mathbb{M}} \to \mathcal{A}^{\mathbb{M}}$ of neighborhood $\mathbb{U}$ one can easily check that the action of $F \circ \sigma^m$ on $(\mathcal{A}^{\mathbb{M}}, d_C)$ is transitive for all $m \in \mathbb{M} \setminus \mathbb{U}$. The reason for this is that the distance d_C is non-homogeneous, hence a simple transport of information is enough to obtain transitivity. This can seem counter-intuitive, and a natural way to overcome this problem is to look at the action of cellular automata on spaces where the distance is shift-invariant or even where the points of the space are themselves shift-invariant. In such spaces, transitivity will not come from transport of information, but rather from *creation* of information.

Besicovitch Topology. Thus, it seems that a shift-invariant distance on $\mathcal{A}^{\mathbb{M}}$ would be very appropriate to study the dynamics of cellular automata. Following this idea, Cattaneo et al. introduced the *Besicovitch pseudo-distance*:

Definition 1 ([CFMM97]). *The Besicovitch pseudo-distance d_B is defined on $\mathcal{A}^{\mathbb{M}}$ by*

$$d_B(x, y) = \limsup_{n \to +\infty} \frac{\operatorname{Card}(\{i \in \mathbb{U}_n : x_i \neq y_i\})}{\operatorname{Card}(\mathbb{U}_n)}.$$

Informally speaking, it measures the asymptotic density of the cells on which x and y differ. It is clearly a pseudo-distance, i.e. it satisfies both the symetry property and triangular inequality. However, $d_B(x, y)$ does not imply $x = y$: if x and y coincide everywhere except on a very sparse set of cells, their Besicovitch pseudo-distance is zero, and yet they are different configurations. Hence, the topology induced on $\mathcal{A}^{\mathbb{M}}$ by d_B is not separated. Notice also that d_B is shift-invariant.

It was proven by Blanchard et al. that CA cannot be expansive with respect to d_B:

Theorem 1 ([BFK97]). *There is no expansive CA on $(\mathcal{A}^{\mathbb{M}}, d_B)$.*

Cattaneo et al. asked whether there exist transitive CA for the Besicovitch pseudo-distance. It remained a recurrent open question (see [BFK97], [Man98], [DFM00]) until it was negatively answered by Blanchard et al. [BCF03]. The original proof of this theorem uses the notion of Kolmogorov complexity, but is quite involved. We present here a simpler proof also based on Kolmogorov complexity, which we will extend later to a much more general framework. We assume that the reader is familiar with Kolmogorov complexity (see [LV97] for an extensive survey, see also [Cal02] for Kolmogorov complexity of strings over a non-binary alphabet).

Theorem 2. *There is no transitive CA on $(\mathcal{A}^{\mathbb{M}}, d_B)$.*

Proof. For all $x \in \mathcal{A}^{\mathbb{M}}$, we set

$$\dim_1(x) = \liminf_{n \to +\infty} \frac{K(x_{\mathbb{U}_n})}{\mathrm{Card}(\mathbb{U}_n)}$$

where K denotes Kolmogorov complexity (what version of Kolmogorov complexity we use does not matter, since all versions coincide up to a logarithmic term). Notice that the quantity $\dim_1(x)$ lies in $[0, \log |\mathcal{A}|]$ (here and in the rest of the paper, log is the logarithm of base 2). The notation $\dim_1$ is justified by a result of Mayordomo [May00] who (elaborating on the work of Staiger and others) showed that this quantity is an effectivization of Hausdorff dimension. We start with two easy lemmas, which we will need again later on:

Lemma 1. *For every $x \in \mathcal{A}^{\mathbb{M}}$ and every CA F, one has $\dim_1(F(x)) \leq \dim_1(x)$*

Indeed, to compute $F(x)_{\mathbb{U}_n}$, one only needs to know $x_{\mathbb{U}_{n+r(F)}}$, by definition of a CA. Hence $K(F(x)_{\mathbb{U}_n}) \leq K(x_{\mathbb{U}_{n+r(F)}})$. But as $\mathbb{M} = \mathbb{N}^{d'} \times \mathbb{Z}^{d''}$, there are at most $O(n^{d'+d''-1})$ cells in $\mathbb{U}_{n+r(F)} \setminus \mathbb{U}_n$. Hence,

$$K(F(x)_{\mathbb{U}_n}) \leq K(x_{\mathbb{U}_{n+r(F)}}) \leq K(x_{\mathbb{U}_n}) + O(n^{d'+d''-1}).$$

Since the quantity $O(n^{d'+d''-1})$ is a $o(\mathrm{Card}(\mathbb{U}_n))$ (because $\mathrm{Card}(\mathbb{U}_n) = O(n^{d'+d''})$), the lemma is proved.

Lemma 2. *For all $x, y \in \mathcal{A}^{\mathbb{M}}$:*

$$|\dim_1(x) - \dim_1(y)| \leq \hbar(d_B(x, y))$$

with $\hbar(x) = -(1-x)\log(1-x) - x\log(x) + x\log|\mathcal{A}|$ (notice that $\hbar(x)$ is concave, and tends towards 0 as x tends towards 0, which proves that $\dim_1$ is uniformly continuous w.r.t d_B).

Let $k = |\mathcal{A}|$. We identify $\mathcal{A}$ with $(\mathbb{Z}/k\mathbb{Z}) = \{\bar{0}...\overline{k-1}\}$, and hence $\mathcal{A}^{\mathbb{M}}$ with $(\mathbb{Z}/k\mathbb{Z})^{\mathbb{M}}$, which is a group (and we denote its addition by $\oplus$). If $d_B(x, y) \leq \varepsilon$ then by definition of d_B, one can write $x = y \oplus z$, where z is a configuration such that for all n, $\frac{\mathrm{Card}\left(\{i \in \mathbb{U}_n : z_i \neq \bar{0}\}\right)}{\mathrm{Card}(\mathbb{U}_n)} \leq \varepsilon + o(1)$. For a given n, setting $N = \mathrm{Card}(\mathbb{U}_n)$, the number of patterns consisting of N cells, with at least $(1-\varepsilon)N$ cells labeled by $\bar{0}$ is bounded by

$$\varepsilon N \binom{N}{\varepsilon N} |\mathcal{A}|^{\varepsilon N}$$

Hence, the Kolmogorov complexity of $z_{\mathbb{U}_n}$ is not greater than the logarithm of this quantity, which, by Stirling's formula, is equal to $\hbar(\varepsilon)N + o(N)$. Since $x_{\mathbb{U}_n}$ can be computed from $y_{\mathbb{U}_n}$ and $z_{\mathbb{U}_n}$, it follows that for all n,

$$K(x_{\mathbb{U}_n}) \leq K(y_{\mathbb{U}_n}) + K(z_{\mathbb{U}_n}) + o(\mathrm{Card}(\mathbb{U}_n))$$
$$\leq K(y_{\mathbb{U}_n}) + \hbar(\varepsilon)\mathrm{Card}(\mathbb{U}_n) + o(\mathrm{Card}(\mathbb{U}_n))$$

By definition of $\dim_1$, the lemma follows. We are now ready to prove Theorem 2. Let F be a CA on $\mathcal{A}^{\mathbb{M}}$. Let x be a configuration such that $\dim_1(x) = 0$ and y such that $\dim_1(y) = \log|\mathcal{A}|$ (such sequences exist, see for example [Lut00]). Let $\varepsilon > 0$. If F were transitive, then there would exist $x', y' \in \mathcal{A}^{\mathbb{M}}$ and $n \in \mathbb{N}$ such that $d_B(x, x') \leq \varepsilon$, $d_B(y, y') \leq \varepsilon$ and $F^n(x') = y'$. By Lemma 2, we would then have $\dim_1(x') \leq \hbar(\varepsilon)$, and $\dim_1(y') \geq 1 - \hbar(\varepsilon)$. But also, applying inductively Lemma 1 on x', we would have $\dim_1(F^n(x')) \leq \dim_1(x') \leq \hbar(\varepsilon)$, i.e, $\dim_1(y') \leq \hbar(\varepsilon)$. For ε small enough, this contradicts $\dim_1(y') \geq 1 - \hbar(\varepsilon)$. □

3   Action of Cellular Automata on $\mathcal{M}_\sigma(\mathcal{A}^{\mathbb{M}})$

Measures on $\mathcal{A}^{\mathbb{M}}$. Let $\mathfrak{B}$ be the Borel sigma-algebra of $\mathcal{A}^{\mathbb{M}}$. We denote by $\mathcal{M}(\mathcal{A}^{\mathbb{M}})$ the set of probability measures on $\mathcal{A}^{\mathbb{M}}$ defined on the sigma-algebra $\mathfrak{B}$. Usually $\mathcal{M}(\mathcal{A}^{\mathbb{M}})$ is endowed with weak* topology: a sequence $(\mu_n)_{n \in \mathbb{N}}$ of $\mathcal{M}(\mathcal{A}^{\mathbb{M}})$ converges to $\mu \in \mathcal{M}(\mathcal{A}^{\mathbb{M}})$ if and only if for all finite subset $\mathbb{U} \subset \mathbb{M}$ and for all pattern $u \in \mathcal{A}^{\mathbb{U}}$, one has $\lim_{n \to \infty} \mu_n([u]_{\mathbb{U}}) = \mu([u]_{\mathbb{U}})$.

In the weak* topology, the set $\mathcal{M}(\mathcal{A}^{\mathbb{M}})$ is compact and metrizable. One defines a distance compatible with the weak* topology by for all $\mu, \nu \in \mathcal{M}(\mathcal{A}^{\mathbb{M}})$:

$$d_*^{\mathcal{M}}(\mu, \nu) = \sum_{n \in \mathbb{N}} \frac{1}{\mathrm{Card}(\mathbb{U}_n)} \sum_{u \in \mathcal{A}^{\mathbb{U}_n}} \left|\mu([u]_{\mathbb{U}_n}) - \nu([u]_{\mathbb{U}_n})\right|,$$

where $\mathbb{U}_n = \{m \in \mathbb{M} : |m| \leq n\}$.

Let $F : X \to Y$ be a mesurable function between the measurable spaces X and Y and let $\mu \in \mathcal{M}(X)$. It is possible to consider the mesure $F_*\mu$ on Y defined by $F_*\mu(B) = \mu(F^{-1}(B))$ for all measurable set $B \subset Y$. Thus, the $\mathbb{M}$-action σ acts naturally on $\mathcal{M}(\mathcal{A}^{\mathbb{M}})$ by:

$$\sigma_*^m(\mu(B)) = \mu(\sigma^{-m}(B)), \text{ for all } m \in \mathbb{M}, \ \mu \in \mathcal{M}(\mathcal{A}^{\mathbb{M}}) \text{ and } B \in \mathfrak{B}.$$

A measure $\mu \in \mathcal{M}(\mathcal{A}^{\mathbb{M}})$ is said σ-*invariant* if $\sigma_*^m \mu = \mu$ for all $m \in \mathbb{M}$; denote $\mathcal{M}_\sigma(\mathcal{A}^{\mathbb{M}})$ the set of σ-invariant probability measure.

The Distance $d_B^{\mathcal{M}}$. In [Sab07], a general framework to define a distance on $\mathcal{M}_\sigma(\mathcal{A}^{\mathbb{M}})$ is given: let d be a pseudo-distance on $\mathcal{A}^{\mathbb{M}}$, we want to introduce a pseudo-distance on $\mathcal{M}_\sigma(\mathcal{A}^{\mathbb{M}})$ induced by the pseudo-distance d. Let $\mu, \nu \in \mathcal{M}_\sigma(\mathcal{A}^{\mathbb{M}})$, the intuitive idea is to calculate the mean of $d(x, y)$ when x is chosen according to the probability measure μ and y according to the probability measure ν. If we just take (x, y) according to the probability $\mu \times \nu$, when $\nu = \mu$, one obtains $\int d(x, y) \mathrm{d}(\mu \times \mu)$ which is in general positive. Hence it is important to allow some kind of correlation in the choice of x and y. This is why we introduce the notion of joint measure.

Let μ and ν be two σ-invariant probability measures on $\mathcal{A}^{\mathbb{M}}$. A probability measure λ on $\mathcal{A}^{\mathbb{M}} \times \mathcal{A}^{\mathbb{M}}$ is a *joint measure* according to μ and ν if λ is $\sigma \times \sigma$-invariant and $\pi_*^1 \lambda = \mu$ and $\pi_*^2 \lambda = \nu$, where π^1 and π^2 are respectively the projections according the first and second coordinate. Denote $\mathcal{J}(\mu, \nu)$ the set of joint measures according μ and ν. Of course, one has $\mathcal{J}(\mu, \nu) \subset \mathcal{M}_{\sigma \times \sigma}(\mathcal{A}^{\mathbb{M}} \times \mathcal{A}^{\mathbb{M}})$. Moreover $\mathcal{J}(\mu, \nu)$ is convex and compact for the weak topology.

Definition 2. *Let d be a pseudo-distance on $\mathcal{A}^{\mathbb{M}}$ such that $(x, y) \mapsto d(x, y)$ is Borel-measurable (this is the case for d_C and d_B). One defines a function $d^{\mathcal{M}}$ from $\mathcal{M}_\sigma(\mathcal{A}^{\mathbb{M}}) \times \mathcal{M}_\sigma(\mathcal{A}^{\mathbb{M}})$ on $\mathbb{R}^+$ by:*

$$d^{\mathcal{M}}(\mu, \nu) = \inf_{\lambda \in \mathcal{J}(\mu, \nu)} \int d(x, y) \mathrm{d}\lambda(x, y) \quad \text{for all } \mu, \nu \in \mathcal{M}_\sigma(\mathcal{A}^{\mathbb{M}}).$$

In [Sab07], we prove that $d_C^{\mathcal{M}}$ is equivalent to $d_*^{\mathcal{M}}$ and that $d_B^{\mathcal{M}}$ defines a distance on $\mathcal{M}_\sigma(\mathcal{A}^{\mathbb{M}})$, which is not equivalent to $d_*^{\mathcal{M}}$. Moreover we give general properties about this type of measure. In particular we have the following lemma:

Lemma 3. *Let $\mu, \nu \in \mathcal{M}_\sigma(\mathcal{A}^{\mathbb{M}})$ and let $\mathbb{U} \subset \mathbb{M}$ be a finite subset. One has:*

$$d_B^{\mathcal{M}}(\mu, \nu) \geq \frac{1}{\mathrm{Card}(\mathbb{U})} \inf_{\lambda \in \mathcal{J}(\mu, \nu)} \lambda([u]_{\mathbb{U}} \times [v]_{\mathbb{U}} : u, v \in \mathcal{A}^{\mathbb{U}}, u \neq v).$$

Proof. Let $\mu, \nu \in \mathcal{M}_\sigma(\mathcal{A}^{\mathbb{M}})$ and let $\lambda \in \mathcal{J}(\mu, \nu)$. Let $u, v \in \mathcal{A}^{\mathbb{U}}$, one has:

$$\bigcup_{u, v \in \mathcal{A}^{\mathbb{U}}, u \neq v} [u]_{\mathbb{U}} \times [v]_{\mathbb{U}} \subset \bigcup_{m \in \mathbb{U}} \left(\bigcup_{a, b \in \mathcal{A}, a \neq b} [a]_m \times [b]_m \right).$$

One deduces the following inequality:

$$\lambda([u]_{\mathbb{U}} \times [v]_{\mathbb{U}} : u, v \in \mathcal{A}^{\mathbb{U}}, u \neq v) \leq \sum_{m \in \mathbb{U}} \lambda([a]_m \times [b]_m : a, b \in \mathcal{A}, a \neq b)$$

$$\underset{(\star)}{=} \mathrm{Card}(\mathbb{U}) \, \lambda([a]_0 \times [b]_0 : a, b \in \mathcal{A}, a \neq b),$$

where $(\star)$ follows from the $\sigma \times \sigma$-invariance of λ.

This lemma allows in particular to prove that $d_B^{\mathcal{M}}$ is a distance.

Action of a Cellular Automaton on $\mathcal{M}_\sigma(\mathcal{A}^\mathbb{M})$**.** Let $(\mathcal{A}^\mathbb{M}, F)$ be a CA and $\mu \in \mathcal{M}_\sigma(\mathcal{A}^\mathbb{M})$. Since F commutes with the shift, if $\mu \in \mathcal{M}_\sigma(\mathcal{A}^\mathbb{M})$ then $F_*\mu \in \mathcal{M}_\sigma(\mathcal{A}^\mathbb{M})$. Let d be a pseudo-distance on $\mathcal{A}^\mathbb{M}$. To study the $\mathbb{N}$-action of F_* on $(\mathcal{M}_\sigma(\mathcal{A}^\mathbb{M}), d^\mathcal{M})$ as a dynamical system, we are going to prove the continuity of the function F_* on $(\mathcal{M}_\sigma(\mathcal{A}^\mathbb{M}), d^\mathcal{M})$.

Proposition 1. *Let d be a pseudo-distance on $\mathcal{A}^\mathbb{M}$ and let $F : \mathcal{A}^\mathbb{M} \to \mathcal{A}^\mathbb{M}$ be a function d-Lipschitz of constant K on $\mathcal{A}^\mathbb{M}$. For all $\mu, \nu \in \mathcal{M}_\sigma(\mathcal{A}^\mathbb{M})$, one has:*

$$d^\mathcal{M}(F_*\mu, F_*\nu) \leq K d^\mathcal{M}(\mu, \nu).$$

In particular F_* is continuous on $(\mathcal{M}_\sigma(\mathcal{A}^\mathbb{M}), d^\mathcal{M})$.

Proof. Let $\lambda \in \mathcal{J}(\mu, \nu)$, one has $(F_* \times F_*)\lambda \in \mathcal{J}(F_*\mu, F_*\nu)$, thus:

$$\int d(x, y) \mathrm{d}(F_* \times F_*)\lambda = \int d(F(x), F(y)) \, \mathrm{d}\lambda \leq \int K d(x, y) \, \mathrm{d}\lambda.$$

One deduces that $d(F_*\mu, F_*\nu) \leq K d(\mu, \nu)$.

Since all CA are Lipschitz for d_C and d_B, this proposition holds for all CA. Thus one can study the dynamical system $F_* : \mathcal{M}_\sigma(\mathcal{A}^\mathbb{M}) \to \mathcal{M}_\sigma(\mathcal{A}^\mathbb{M})$ according to the distance $d_*^\mathcal{M}$ or $d_B^\mathcal{M}$.

Non-expansivity of CA on $(\mathcal{M}_\sigma(\mathcal{A}^\mathbb{M}), d_B^\mathcal{M})$**.** In the space of measures $(\mathcal{M}_\sigma(\mathcal{A}^\mathbb{M}), d_B^\mathcal{M})$, we have the following counterpart to Theorem 1:

Proposition 2. *Let $(\mathcal{A}^\mathbb{M}, F)$ be a CA. F_* does not act expansively on $(\mathcal{M}_\sigma(\mathcal{A}^\mathbb{M}), d_B^\mathcal{M})$.*

Proof. Let $\mu, \nu \in (\mathcal{M}_\sigma(\mathcal{A}^\mathbb{M}))$ and $\varepsilon > 0$. Consider $\mu' = (1 - \varepsilon)\mu + \varepsilon\nu$. Let $\lambda' \in \mathcal{J}(\mu, \mu)$ and $\lambda'' \in \mathcal{J}(\mu, \nu)$, such that $\lambda = (1 - \varepsilon)\lambda' + \varepsilon\lambda'' \in \mathcal{J}(\mu, \mu')$. One then has:

$$(1 - \varepsilon) \int d(x, y) \mathrm{d}\lambda' + \varepsilon \int d(x, y) \mathrm{d}\lambda'' = \int d(x, y) \mathrm{d}\lambda \geq d^\mathcal{M}(\mu, \mu')$$

Thus,

$$\varepsilon d^\mathcal{M}(\mu, \nu) = (1 - \varepsilon) d^\mathcal{M}(\mu, \mu) + \varepsilon d^\mathcal{M}(\mu, \nu) \geq d^\mathcal{M}(\mu, (1 - \varepsilon)\mu + \varepsilon\nu)$$

Since F_* preserves convex combinations, one has $F_*^n \mu' = (1 - \varepsilon)F_*^n \mu + \varepsilon F_*^n \nu$ for all $n \in \mathbb{N}$, so $d^\mathcal{M}(F_*^n \mu, F_*^n \mu') \leq \varepsilon d^\mathcal{M}(F_*^n \mu, F_*^n \nu)$. Hence, F_* is not enxpansive in $(\mathcal{M}_\sigma(\mathcal{A}^\mathbb{M}), d_B^\mathcal{M})$.

Continuity of the Entropy of σ**.** The information contained in a a generic configuration can be expressed by the entropy of the shift. A comparative study of the entropy of the shift and Kolmogorov complexity was carried out by Brudno [Bru82]. As we will see, the entropy of the shift is continuous with respect to the underlying measure.

Definition 3. *Let $\mu \in \mathcal{M}_\sigma(\mathcal{A}^\mathbb{M})$, the entropy of the shift $\mathbb{M}$-action can be defined as:*

$$h_\mu(\sigma) = \lim_{n\to\infty} \frac{H_\mu(\mathcal{P}_{\mathbb{U}_n})}{\mathrm{Card}(\mathbb{U}_n)},$$

where $\mathcal{P}_{\mathbb{U}_n}$ is the partition of cylinders centered on $\mathbb{U}_n$ and $H_\mu(\mathcal{P}_{\mathbb{U}_n})$ is the entropy of the partition $\mathcal{P}_{\mathbb{U}_n}$ according to the measure μ, defined by:

$$H_\mu(\mathcal{P}_{\mathbb{U}_n}) = -\sum_{u\in\mathcal{A}^{\mathbb{U}_n}} \mu([u]_{\mathbb{U}_n}) \log(\mu([u]_{\mathbb{U}_n})).$$

One recalls that $\mathbb{U}_n = \{m \in \mathbb{M} : |m| \le n\}$.

Let $\mathcal{P}_1$ and $\mathcal{P}_2$ be two partitions of $\mathcal{A}^{\mathbb{M}}$. We define the refinement of $\mathcal{P}_1$ and $\mathcal{P}_2$ by

$$\mathcal{P}_1 \vee \mathcal{P}_2 = \{A \cap B : A \in \mathcal{P}_1 \text{ and } B \in \mathcal{P}_2\}.$$

Moreover it is possible to define the conditional entropy of $\mathcal{P}_1$ given $\mathcal{P}_2$:

$$H_\mu(\mathcal{P}_1|\mathcal{P}_2) = -\sum_{B\in\mathcal{P}_2} \mu(B) \sum_{A\in\mathcal{P}_1} \frac{\mu(A\cap B)}{\mu(B)} \log(\mu(A)).$$

Thanks conditional entropy, it is possible to decompose the entropy of a refinement:

$$H_\mu(\mathcal{P}_1 \vee \mathcal{P}_2) = H_\mu(\mathcal{P}_2) + H_\mu(\mathcal{P}_1|\mathcal{P}_2).$$

It is well known that the function $\mu \mapsto h_\mu(\sigma)$ is upper semi-continuous in $(\mathcal{M}_\sigma(\mathcal{A}^{\mathbb{M}}), d_*^{\mathcal{M}})$, see [DGS76] for more detail.

Theorem 3. *The function $\mu \mapsto h_\mu(\sigma)$ is uniformly continuous in $(\mathcal{M}_\sigma(\mathcal{A}^{\mathbb{M}}), d_B^{\mathcal{M}})$.*

Proof. Let μ and ν in $\mathcal{M}_\sigma(\mathcal{A}^{\mathbb{M}})$. By definition of the entropy of σ, one has

$$h_\mu(\sigma) = \lim_{n\to\infty} \frac{H_\mu(\mathcal{P}_{\mathbb{U}_n})}{\mathrm{Card}(\mathbb{U}_n)} \text{ and } h_\nu(\sigma) = \lim_{n\to\infty} \frac{H_\nu(\mathcal{P}_{\mathbb{U}_n})}{\mathrm{Card}(\mathbb{U}_n)}.$$

However, for all $\lambda \in \mathcal{J}(\mu,\nu)$ one has:

$$\begin{aligned}
|H_\mu(\mathcal{P}_{\mathbb{U}_n}) - H_\nu(\mathcal{P}_{\mathbb{U}_n})| &= |H_\lambda(\mathcal{P}_{\mathbb{U}_n} \times \mathcal{A}^{\mathbb{M}}) - H_\lambda(\mathcal{A}^{\mathbb{M}} \times \mathcal{P}_{\mathbb{U}_n})| \\
&= |(H_\lambda(\mathcal{P}_{\mathbb{U}_n} \times \mathcal{A}^{\mathbb{M}}) - H_\lambda(\mathcal{P}_{\mathbb{U}_n} \times \mathcal{A}^{\mathbb{M}} \vee \mathcal{A}^{\mathbb{M}} \times \mathcal{P}_{\mathbb{U}_n})) \\
&\quad - (H_\lambda(\mathcal{A}^{\mathbb{M}} \times \mathcal{P}_{\mathbb{U}_n}) - H_\lambda(\mathcal{P}_{\mathbb{U}_n} \times \mathcal{A}^{\mathbb{M}} \vee \mathcal{A}^{\mathbb{M}} \times \mathcal{P}_{\mathbb{U}_n}))| \\
&\le H_\lambda(\mathcal{P}_{\mathbb{U}_n} \times \mathcal{A}^{\mathbb{M}}|\mathcal{A}^{\mathbb{M}} \times \mathcal{P}_{\mathbb{U}_n}) + H_\lambda(\mathcal{A}^{\mathbb{M}} \times \mathcal{P}_{\mathbb{U}_n}|\mathcal{P}_{\mathbb{U}_n} \times \mathcal{A}^{\mathbb{M}}).
\end{aligned}$$

Moreover, one has:

$$\begin{aligned}
H_\lambda(\mathcal{P}_{\mathbb{U}_n} \times \mathcal{A}^{\mathbb{M}}|\mathcal{A}^{\mathbb{M}} \times \mathcal{P}_{\mathbb{U}_n}) &\le \sum_{i\in\mathbb{U}_n} H_\lambda(\mathcal{P}_i \times \mathcal{A}^{\mathbb{M}}|\mathcal{A}^{\mathbb{M}} \times \mathcal{P}_{\mathbb{U}_n}) \\
&\le \mathrm{Card}(\mathbb{U}_n) H_\lambda(\mathcal{P}_0 \times \mathcal{A}^{\mathbb{M}}|\mathcal{A}^{\mathbb{M}} \times \mathcal{P}_{\mathbb{U}_n}) \\
&\le \mathrm{Card}(\mathbb{U}_n) H_\lambda(\mathcal{P}_0 \times \mathcal{A}^{\mathbb{M}}|\mathcal{A}^{\mathbb{M}} \times \mathcal{P}_0),
\end{aligned}$$

where $\mathcal{P}_0 = \mathcal{P}_{\mathbb{U}_0}$. Symmetrically one obtains

$$H_\lambda(\mathcal{A}^{\mathbb{M}} \times \mathcal{P}_{\mathbb{U}_n}|\mathcal{P}_{\mathbb{U}_n} \times \mathcal{A}^{\mathbb{M}}) \le \mathrm{Card}(\mathbb{U}_n) H_\lambda(\mathcal{A}^{\mathbb{M}} \times \mathcal{P}_0|\mathcal{P}_0 \times \mathcal{A}^{\mathbb{M}}).$$

Thus, by summation one has:

$$|h_\mu(\sigma) - h_\nu(\sigma)| \leq H_\lambda(\mathcal{P}_0 \times \mathcal{A}^{\mathbb{M}}|\mathcal{A}^{\mathbb{M}} \times \mathcal{P}_0) + H_\lambda(\mathcal{A}^{\mathbb{M}} \times \mathcal{P}_0|\mathcal{P}_0 \times \mathcal{A}^{\mathbb{M}}).$$

Consider $\alpha = (\cup_{a,b\in\mathcal{A},a\neq b}[a]_0 \times [b]_0; \cup_{a\in\mathcal{A}}[a]_0 \times [a]_0)$, the partition of $\mathcal{A}^{\mathbb{M}} \times \mathcal{A}^{\mathbb{M}}$ formed of two elements. Set $\delta = \lambda(\cup_{a,b\in\mathcal{A},a\neq b}[a]_0 \times [b]_0)$. One has:

$$H_\lambda(\mathcal{P}_0 \times \mathcal{A}^{\mathbb{M}}|\mathcal{A}^{\mathbb{M}} \times \mathcal{P}_0) \leq H_\lambda(\alpha) \leq -(\delta\log(\delta) + (1-\delta)\log(1-\delta)).$$

Let $\varepsilon > 0$. The function $\delta \rightarrow \delta\log(\delta)+(1-\delta)\log(1-\delta)$ tends towards 0 when δ tends towards 0. Thus, there exists $\delta_0 > 0$ such that $\delta\log(\delta)+(1-\delta)\log(1-\delta) \leq \frac{\varepsilon}{2}$ for all $\delta < \delta_0$. Let $\mu,\nu \in \mathcal{M}_\sigma(\mathcal{A}^{\mathbb{M}})$ such that $d_B^{\mathcal{M}}(\mu,\nu) < \delta_0$. According to Lemma 3, there exists $\lambda \in \mathcal{J}(\mu,\nu)$ such that

$$\lambda([a]_0 \times [b]_0 : a, b \in \mathcal{A}, a \neq b) \leq d_B^{\mathcal{M}}(\mu,\nu) < \delta_0.$$

In this case, one has $H_\lambda(\mathcal{P}_0 \times \mathcal{A}^{\mathbb{M}}|\mathcal{A}^{\mathbb{M}} \times \mathcal{P}_0) \leq \frac{\varepsilon}{2}$, and symmetrically $H_\lambda(\mathcal{A}^{\mathbb{M}} \times \mathcal{P}_0, \mathcal{P}_0 \times \mathcal{A}^{\mathbb{M}}) \leq \frac{\varepsilon}{2}$. We deduce that for all $\varepsilon > 0$, there exists δ_0 such that if $d_B^{\mathcal{M}}(\mu,\nu) \leq \delta_0$ then

$$|h_\mu(\sigma) - h_\nu(\sigma)| \leq H_\lambda(\mathcal{P}_0 \times \mathcal{A}^{\mathbb{M}}|\mathcal{A}^{\mathbb{M}} \times \mathcal{P}_0) + H_\lambda(\mathcal{A}^{\mathbb{M}} \times \mathcal{P}_0|\mathcal{P}_0 \times \mathcal{A}^{\mathbb{M}}) \leq \varepsilon.$$

This proves the uniform continuity of $\mu \rightarrow h_\mu(\sigma)$ in $(\mathcal{M}_\sigma(\mathcal{A}^{\mathbb{M}}), d_B^{\mathcal{M}})$.

Application to Transitivity

Theorem 4. *Let $(\mathcal{A}^{\mathbb{M}}, F)$ be a CA. F_* cannot be transitive in $(\mathcal{M}_\sigma(\mathcal{A}^{\mathbb{M}}), d_B^{\mathcal{M}})$.*

Proof. Let

$$\mathcal{U} = \{\mu \in \mathcal{M}_\sigma(\mathcal{A}^{\mathbb{M}}) : h_\mu(\sigma) < 1/3\} \text{ and } \mathcal{V} = \{\mu \in \mathcal{M}_\sigma(\mathcal{A}^{\mathbb{M}}) : h_\mu(\sigma) > 2/3\}.$$

By Theorem 3, $\mathcal{U}$ and $\mathcal{V}$ are open sets of $(\mathcal{M}_\sigma(\mathcal{A}^{\mathbb{M}}), d_B^{\mathcal{M}})$. Since F commutes with σ, it can be view as a factor map from $(\mathcal{A}^{\mathbb{M}}, \mu, \sigma)$ to $(\mathcal{A}^{\mathbb{M}}, F_*\mu, \sigma)$, so one has $h_\mu(\sigma) \geq h_{F_*\mu}(\sigma)$. Thus $F_*(\mathcal{U}) \subset \mathcal{U}$. One deduces that $\mathcal{V} \cap F_*^n(\mathcal{U}) = \emptyset$ for all $n \in \mathbb{N}$, thus F_* can not be transitive in $(\mathcal{M}_\sigma(\mathcal{A}^{\mathbb{M}}), d_B^{\mathcal{M}})$.

In $(\mathcal{M}_\sigma(\mathcal{A}^{\mathbb{M}}), d_*^{\mathcal{M}})$, the function $\mu \rightarrow h_\mu(\sigma)$ is just upper semi-continuous, so $\mathcal{V}$ is not open and the previous proof does not hold. In the space $(\mathcal{M}_\sigma(\mathcal{A}^{\mathbb{M}}), d_*^{\mathcal{M}})$, the existence of transitive CA is open.

4   Action of Cellular Automata on $\mathcal{M}(\mathcal{A}^{\mathbb{M}})$

In this section, we do not restrict ourselves to the space of shift-invariant measures: we instead consider the whole space $\mathcal{M}(\mathcal{A}^{\mathbb{M}})$. The distance $d_B^{\mathcal{M}}$ defined in the previous section can be extended to arbitrary measures, hence endowing $\mathcal{M}(\mathcal{A}^{\mathbb{M}})$ with a Besicovitch-like topology. On the space $\mathcal{M}(\mathcal{A}^{\mathbb{M}})$, $d_B^{\mathcal{M}}$ is only a pseudo-distance, as for example two measures which are equal up to a shift are at distance 0 from each other. Similarly to $(\mathcal{A}^{\mathbb{M}}, d_B)$, the space $(\mathcal{M}(\mathcal{A}^{\mathbb{M}}), d_B^{\mathcal{M}})$ is not separated.

The space $(\mathcal{M}_\sigma(\mathcal{A}^{\mathbb{M}}), d_B^{\mathcal{M}})$ can clearly be viewed as a subspace of $(\mathcal{M}(\mathcal{A}^{\mathbb{M}}), d_B^{\mathcal{M}})$. Moreover, $(\mathcal{A}^{\mathbb{M}}, d_B)$ can also be viewed as a subspace of $(\mathcal{M}(\mathcal{A}^{\mathbb{M}}), d_B^{\mathcal{M}})$ via the isometric embedding

$$\mathcal{A}^{\mathbb{M}} \longrightarrow \mathcal{M}(\mathcal{A}^{\mathbb{M}})$$
$$x \longmapsto \delta_x$$

where δ_x is the measure concentrated on x (i.e. $\delta_x(A) = 1$ if $x \in A$, $\delta_x(A) = 0$ otherwise).

The proof of non-expansivity of CA which was proven in the previous section naturally extends to the whole space $(\mathcal{M}(\mathcal{A}^{\mathbb{M}}), d_B^{\mathcal{M}})$. On the other hand, the proofs of non-transitivity we presented respectively for $(\mathcal{A}^{\mathbb{M}}, d_B)$ and $(\mathcal{M}_\sigma(\mathcal{A}^{\mathbb{M}}), d_B^{\mathcal{M}})$ cannot be extended in a completely straightforward way to $(\mathcal{M}(\mathcal{A}^{\mathbb{M}}), d_B)$. It is true however that no CA is transitive in this space. In fact, non-transitivity happens in the larger class of Lipschitz funtions:

Theorem 5. *Let $F : \mathcal{A}^{\mathbb{M}} \to \mathcal{A}^{\mathbb{M}}$ be a function that is Lipschitz w.r.t. the distance d_C. The action of F_* on $(\mathcal{M}(\mathcal{A}^{\mathbb{M}}), d_B^{\mathcal{M}})$ is not transitive.*

Proof. We adapt the proof of Theorem 2. First notice that for a function F that is Lipschitz in d_C with constant 2^r one only needs to know $F\,x_{\mathbb{U}_{n+r}}$ to compute $F(x)_{\mathbb{U}_n}$, hence Lemma 1 remains true if one takes F to be a Lipschitz function w.r.t. d_C and one replaces Kolmogorov complexity K by $K^{(F)}$, i.e. Kolmogorov complexity relativized to oracle F (F being a Lipschitz function, it can be given as an oracle), and $\dim_1$ by $\dim_1^{(F)}$.

For a measure μ, we set

$$\mathbb{E}\dim_1(\mu) = \int \dim_1(x)\,\mathrm{d}\mu(x)$$

We will need the following analogue of Lemma 2 (which can be relativized to any given oracle):

Lemma 4. *There exists a constant c such that for all μ, ν:*

$$d_B^{\mathcal{M}}(\mu, \nu) < c \Rightarrow \left|\mathbb{E}\dim_1(\mu) - \mathbb{E}\dim_1(\nu)\right| \leq \hbar(d_B^{\mathcal{M}}(\mu, \nu))$$

and thus, $\mathbb{E}\dim_1$ is uniformly continuous w.r.t. $d_B^{\mathcal{M}}$.

Let c be the constant such that $\hbar$ is increasing on $[0, c]$, and let μ, ν be such that $d_B^{\mathcal{M}}(\mu, \nu) < c$. Let $\varepsilon \in (d_B^{\mathcal{M}}(\mu, \nu), c)$. By definition of $d_B^{\mathcal{M}}$, there exists a measure $\lambda \in \mathcal{J}(\mu, \nu)$ such that

$$\int d_B(x, y)\,\mathrm{d}\lambda \leq \varepsilon$$

Since $\hbar$ is increasing on $[0, c]$ and concave:

$$\int \hbar(d_B(x, y)) \leq \hbar\left(\int d_B(x, y)\,\mathrm{d}\lambda\right) \leq \hbar(\varepsilon)$$

which by Lemma 1 implies:

$$\int |\dim_1(x) - \dim_1(y)|\,\mathrm{d}\lambda \leq \hbar(\varepsilon)$$

and thus
$$\left|\mathbb{E}\dim_1(\mu) - \mathbb{E}\dim_1(\nu)\right| \le \hbar(\varepsilon)$$
which implies the desired result, as ε can be chosen arbitrarily close to $d_B^{\mathcal{M}}(\mu, \nu)$.

We are now ready to prove Theorem 5. Let F be a Lipschitz function w.r.t. d_C. Let δ_0 be the measure concentrated on the configuration where all cells have state 0, and ν be Lebesgue measure. Let μ' be a measure such that $d_B^{\mathcal{M}}(\delta_0, \mu') \le \varepsilon$ and μ'' be a measure such that $d_B^{\mathcal{M}}(\nu, \mu'') \le \varepsilon$ with ε small enough. Since $\mathbb{E}\dim_1(\delta_0) = 0$ and $\mathbb{E}\dim_1(\nu) = \log|\mathcal{A}|$, by Lemma 4, one has $\mathbb{E}\dim_1(\mu') \le \hbar(\varepsilon)$ and $\mathbb{E}\dim_1(\mu'') \ge 1 - \hbar(\varepsilon)$.

By Lemma 1, for all $x \in \mathcal{A}^{\mathbb{M}}$, one has $\dim_1^{(F)}(F(x)) \le \dim_1^{(F)}(x)$, hence for every $\mu \in \mathcal{M}(\mathcal{A}^{\mathbb{M}})$, $\mathbb{E}\dim_1^{(F)}(F_*(\mu)) \le \mathbb{E}\dim_1^{(F)}(\mu)$. Hence, by the above discussion, if $d_B^{\mathcal{M}}(\delta_0, \mu) \le \varepsilon$, for all $n \in \mathbb{N}$, $\mathbb{E}\dim_1^{(F)}(F_*^n(\mu)) \le \hbar(\varepsilon)$, which (still by the above discussion) means that $F_*^n(\mu)$ will never be $d_B^{\mathcal{M}}$-close to Lebesgue measure. This finishes the proof of the theorem.

The non-tranisitivity of CA in $(\mathcal{M}_\sigma(\mathcal{A}^{\mathbb{M}}), d_B^{\mathcal{M}})$ (as stated in Theorem 4) immediately follows from the above proof, as δ_0 and Lebesgue measure are shift-invariant measures. On can also modifiy the above proof to get Theorem 2: instead of Lebesgue measure, take ν equal to δ_z for some $z \in \mathcal{A}^{\mathbb{M}}$ such that $\dim_1^{(F)}(z) = \log|\mathcal{A}|$, the rest of the proof remaining the same.

5   Conclusion

It appears that in the shift-invariant topologies we considered, cellular automata cannot be expansive nor transitive. This is mainly due to the unability of cellular automata to create information. Indeed, for the non-transitivity of CA, the three proofs we gave all have the same scheme. First, we define on a (pseudo-)metric space (E, d) where d is a shift-invariant distance (in this paper, resp. $(\mathcal{A}^{\mathbb{M}}, d_B)$, $(\mathcal{M}_\sigma(\mathcal{A}^{\mathbb{M}}), d_B^{\mathcal{M}})$ and $(\mathcal{M}(\mathcal{A}^{\mathbb{M}}), d_B^{\mathcal{M}})$) a quantity which in some sense measures the amount of information, $\mathcal{I} : E \to \mathbb{R}_+$, (resp. $\dim_1$, $h_\mu(\sigma)$ and $\mathbb{E}\dim_1$), which we prove to be uniformly continuous w.r.t. the distance d. This amount of information is non-increasing under the action of a cellular automaton (or even Lipschitz functions), i.e. $\mathcal{I}(F(x)) \le \mathcal{I}(x)$ for all $x \in E$. Since in all cases there are elements of the space which contain little information (i.e. $\mathcal{I}(x) = 0$) and some which contain a lot of information (i.e. in our case $\mathcal{I}(x) = \log|\mathcal{A}|$). Hence, the two open sets $\mathcal{U} = \{x \in E : \mathcal{I}(x) < \varepsilon\}$ and $\mathcal{V} = \{x \in E : \mathcal{I}(x) > \log|\mathcal{A}| - \varepsilon\}$ witness, for ε small enough, the non-transitivity of cellular automata.

References

[BCF03]   Blanchard, F., Cervelle, J., Formenti, E.: Periodicity and transitivity for cellular automata in Besicovitch topologies. In: Rovan, B., Vojtáš, P. (eds.) MFCS 2003. LNCS, vol. 2747, pp. 228–238. Springer, Heidelberg (2003)

[BFK97]   Cellular automata in the Cantor, Besicovitch, and Weyl topological spaces. Complex Systems 11(2), 107–123 (1997)

[Bru82]   Brudno, A.A.: Entropy and the complexity of the trajectories of a dynamic system. Trudy Moskov. Mat. Obshch. 44, 124–149 (1982)

[Cal02]   Calude, C.: Information theory and randomness: an algorithmic perspective, 2nd edn. Spinger, Berlin Heidelberg (2002)

[CFMM97]   Cattaneo, G., Formenti, E., Margara, L., Mazoyer, J.: A shift-invariant metric on S^Z inducing a non-trivial topology. In: Privara, I., Ružička, P. (eds.) MFCS 1997. LNCS, vol. 1295, pp. 179–188. Springer, Heidelberg (1997)

[DFM00]   Delorme, M., Formenti, E., Mazoyer, J.: Open problems on cellular automata. Technical report RR-2000-25, Ecole Normale Supérieure de Lyon (2000)

[DGS76]   Denker, M., Grillenberger, C., Sigmund, K.: Ergodic theory on compact spaces. Lecture Notes in Mathematics, vol. 527. Springer-Verlag, Berlin Heidelberg (1976)

[Egg49]   Eggleston, H.G.: The fractional dimension of a set defined by decimal properties. Quarterly Journal of Mathematics, Oxford Series 20, 31–36 (1949)

[Fal85]   Falconer, K.: The geometry of fractal sets. Cambridge University Press, Cambridge (1985)

[Hed69]   Hedlund, G.A.: Endomorphisms and automorphisms of the shift dynamical system. Math. Systems Theory 3, 320–375 (1969)

[Lut00]   Lutz, J.: Dimension in complexity classes. In: Conference on Computational Complexity, CCC'00, pp. 158–169. IEEE Computer Society, Los Alamitos (2000)

[LV97]   Li, M., Vitanyi, P.: An introduction to Kolmogorov complexity and its applications, 2nd edn. Spinger, Berlin Heidelberg (1997)

[Man98]   Manzini, G.: Characterization of sensitive linear automata with respect to the couting distance. In: Mathematical Foundations of Computer Science 1998 (1998)

[May00]   Mayordomo, E.: A Kolmogorov complexity characterization of constructive Hausdorff dimension. Information Processing Letters 84, 1–3 (2000)

[Sab07]   M. Sablik. Action of Cellular automata on σ-invariant probability measure. Prépublication de l'UMPA (September 2006),
             www.umpa.ens-lyon.fr/~msablik

Two Element Unavoidable Sets of Partial Words[*]

F. Blanchet-Sadri[1], N.C. Brownstein[2], and Justin Palumbo[3]

[1] Department of Computer Science, University of North Carolina,
P.O. Box 26170, Greensboro, NC 27402–6170, USA
[2] Department of Mathematics, University of Central Florida,
P.O. Box 161364, Orlando, FL 32816–1364, USA
[3] Department of Mathematics,
Rutgers The State University of New Jersey,
110 Frelinghuysen Road, Piscataway, NJ 08854–8019, USA

Abstract. The notion of an unavoidable set of words appears frequently in the fields of mathematics and theoretical computer science, in particular with its connection to the study of combinatorics on words. The theory of unavoidable sets has seen extensive study over the past twenty years. In this paper we extend the definition of unavoidable sets of words to unavoidable sets of partial words. Partial words, or finite sequences that may contain a number of "do not know" symbols or holes, appear in natural ways in several areas of current interest such as molecular biology, data communication, DNA computing, etc. We demonstrate the utility of the notion of unavoidability on partial words by making use of it to identify several new classes of unavoidable sets of full words. Along the way we begin work on classifying the unavoidable sets of partial words of small cardinality. We pose a conjecture, and show that affirmative proof of this conjecture gives a sufficient condition for classifying all the unavoidable sets of partial words of size two. Lastly we give a result which makes the conjecture easy to verify for a significant number of cases.

1 Introduction

An *unavoidable* set of words X over an alphabet A is a set for which any sufficiently long word over A will have a factor in X. It is clear from the definition that from each unavoidable set we can extract a finite unavoidable subset, so the study can be reduced to finite unavoidable sets. This concept was explicitly introduced in 1983 in connection with an attempt to characterize the rational languages among the context-free ones [8]. Since then it has been consistently studied by researchers in both mathematics and theoretical computer science. Testing the unavoidability of X can be done in different ways [7]: Check whether

[*] This material is based upon work supported by the National Science Foundation under Grant No. DMS–0452020. A World Wide Web server interface at www.uncg.edu/mat/research/unavoidablesets has been established for automated use of the program. We thank the referees of preliminary versions of this paper for their very valuable comments and suggestions.

T. Harju, J. Karhumäki, and A. Lepistö (Eds.): DLT 2007, LNCS 4588, pp. 96–107, 2007.
© Springer-Verlag Berlin Heidelberg 2007

there is a loop in the finite automaton of Aho and Corasick [1] recognizing $A^* \setminus A^* X A^*$, or simplify X as much as possible. There is a large literature on unavoidable sets of words and we refer the reader to [6,14] for more information.

Another concept relevant to this paper is that of a *partial word*, or a finite sequence of symbols from a finite alphabet that may contain a number of "do not know" symbols or "holes". Partial words appear in natural ways in several areas of current interest such as molecular biology, data communication, and DNA computing. In this paper, we introduce unavoidable sets of partial words. In terms of unavoidability, sets of partial words serve as efficient representations of sets of full words. This is strongly analogous to the study of unavoidable patterns, in which sets of patterns are used to represent infinite sets of full words [13]. The main goal here is to demonstrate that the study of unavoidable sets of partial words leads to new insights both on the theory of unavoidable sets and on the combinatorial structure of the set of words A^* as a whole. In accomplishing this we mainly focus on the problem of classifying the unavoidable sets of size two.

The contents of our paper are summarized as follows. In Section 2, we review some basic definitions related to words and partial words. In Section 3, we recall the definition of unavoidable sets of words and some useful elementary properties. There, we present our definition of unavoidable sets of partial words and we introduce the problem of classifying such sets of small cardinality and in particular those with two elements, x_1, x_2, with respect to the regular constraints: x_1 matches the pattern $(a\diamond^*)^* a$ and x_2 the pattern $(b\diamond^*)^* b$ where $\diamond$ denotes the "do not know" symbol and a, b denote distinct letters of the alphabet. In Section 4, we give an elegant characterization of the particular case of this problem when x_1 matches $a\diamond^* a$ and x_2 matches $b\diamond^* b$, propose a conjecture characterizing the case where x_1 matches $a\diamond^* a$ and x_2 matches $b\diamond^* b\diamond^* b$, and prove that verifying this conjecture is sufficient for solving the problem in general. There, we also prove one direction of our conjecture. In Section 5, we give partial results towards the other direction of our conjecture and in particular prove that it is easy to verify in a large number of cases. Finally in Section 6, we pose several natural and interesting questions related to unavoidable sets of partial words.

2 Preliminaries

We begin this section with the following basic terms and definitions.

Throughout this paper A is a fixed finite set called the *alphabet* whose elements we call *letters*. We use A^* to denote the set of words over A, that is the set of finite sequences of letters in the alphabet. For $u \in A^*$ we write $|u|$ for the length of u. Under the concatenation operation of words, A^* forms a free monoid whose identity is the empty word which we denote by ε. If there exist $x, y \in A^*$ such that $u = xvy$ then we say that v is a *factor* of u.

A *two-sided infinite word* w is a function $w : \mathbb{Z} \to A$. A finite word u is a factor of the two-sided infinite word w if u is a finite subsequence of w, that is if there exists some $i \in \mathbb{Z}$ so that $u = w(i+1)\ldots w(i+|u|)$. For a positive integer p, we say that w has *period* p if $w(i) = w(j)$ for all $i, j \in \mathbb{Z}$ satisfying $i \equiv j \bmod p$. If

w has period p for some p then we call w *periodic*. We can now define infinite powers of a word: if v is a nonempty finite word, then we denote by $v^{\mathbb{Z}}$ the unique two-sided infinite word w such that w has period $|v|$ and $w(0) \ldots w(|v| - 1) = v$.

A word of finite length n over an alphabet A can be defined as a total function $w : \{0, \ldots, n - 1\} \rightarrow A$. Analogously a *partial word* (or, *pword*) of length n over A is a partial function $u : \{0, \ldots, n - 1\} \rightarrow A$. For $0 \leq i < n$, if $u(i)$ is defined, then we say that i belongs to the domain of u (denoted by $i \in D(u)$). Otherwise we say that i belongs to the *set of holes* of u (denoted by $i \in H(u)$). In cases where $H(u)$ is empty we say that u is a *full word*.

Let $A_{\diamond} = A \cup \{\diamond\}$. If u is a partial word of length n over A, then the *companion* of u is the total function $u_{\diamond} : \{0, \ldots, n - 1\} \rightarrow A_{\diamond}$ defined by

$$
u_{\diamond} = \begin{cases} u(i) & \text{if } i \in D(u) \\ \diamond & \text{otherwise} \end{cases}
$$

Throughout this paper we identify a partial word with its companion. We reserve the term *letter* for members of A. We will refer to an occurrence of the symbol $\diamond$ in a partial word as a *hole*.

Two partial words u and v of equal length are said to be *compatible*, denoted by $u \uparrow v$, if $u(i) = v(i)$ for every $i \in D(u) \cap D(v)$. If X is a set of partial words, we use $\hat{X}$ to denote the set of all full words compatible with a member of X.

3   Unavoidable Sets

We first recall the definition of an unavoidable set of full words and some relevant properties. Let $X \subseteq A^*$. A two-sided infinite word w *avoids* X if no factor of w is a member of X. We say that X is *unavoidable* if no two-sided infinite word avoids X. In other words X is unavoidable if every two-sided infinite word has a factor in X.

Following are two useful facts giving alternative characterizations of unavoidable sets: (1) The set $X \subseteq A^*$ is unavoidable if and only if there are only finitely many words in A^* with no member of X as a factor; and (2) If the set $X \subseteq A^*$ is finite, then X is unavoidable if and only if no periodic two-sided infinite word avoids it. Proofs can be found in [13].

We now give our extension of the definition of unavoidable sets of words to unavoidable sets of partial words.

Definition 1. *Let $X \subseteq A_{\diamond}^*$. A two-sided infinite word w avoids X if no factor of w is a member of $\hat{X}$. We say that X is unavoidable if no two-sided infinite word avoids X. In other words X is unavoidable if every two-sided infinite word has a factor compatible with a member of X.*

There is a simple connection between sets of partial words and sets of full words that is worth noting. By the definition of $\hat{X}$, w has a factor in $\hat{X}$ if and only if that same factor is compatible with a member of X. Thus the two-sided infinite words which avoid X are exactly those which avoid $\hat{X}$, and X is unavoidable if and only if $\hat{X}$ is unavoidable.

With regards to unavoidability X is then essentially a representation of a set of full words. This representation makes possible new approaches to unavoidable sets of full words. It is easier to consider the two-sided infinite words avoiding $X = \{aa, b \diamond^3 b\}$ as those without an occurrence of aa and no two occurrences of b separated by three letters rather as the words avoiding

$$\hat{X} = \{aa, baaab, baabb, babab, babbb, bbaab, bbabb, bbbab, bbbbb\}.$$

It is most natural to look first for the unavoidable sets of partial words that have small cardinality. Insight into the structure of A^* can be gained by identifying an unavoidable set, especially if that set contains few elements.

Any set of partial words containing the empty word or $\diamond^n$ for some $n \in \mathbb{N}$ will be called a trivial unavoidable set. To find nontrivial unavoidable sets of size 2 we may assume that $A = \{a, b\}$. Classifying the unavoidable sets of size 2 is a daunting task and is the focus of this paper.

Say $X = \{x_1, x_2\}$ is unavoidable. As mentioned before if X is nontrivial it must be that one member of X is compatible with a power of a and the other is compatible with a power of b, as that is the only way to guarantee that both $a^{\mathbb{Z}}$ and $b^{\mathbb{Z}}$ will not avoid X. So in order to classify the unavoidable sets of size 2, it is sufficient to determine for which $m_1, m_2, \ldots, m_k$ and $n_1, n_2, \ldots, n_l$ the set

$$X_{m_1,\ldots,m_k \mid n_1,\ldots,n_l} = \{a \diamond^{m_1} a \ldots a \diamond^{m_k} a, b \diamond^{n_1} b \ldots b \diamond^{n_l} b\}$$

is unavoidable. We can in fact simplify the situation a little further. The following lemma tells us that it is enough to solve the problem for cases where $m_1 + 1, m_2 + 1, \ldots, m_k + 1$ and $n_1 + 1, n_2 + 1, \ldots, n_l + 1$ are relatively prime.

Lemma 1. *Let $p \in \mathbb{N}$. The set $X_{m_1,\ldots,m_k \mid n_1,\ldots,n_l}$ is unavoidable if and only if the set*

$$Y = \{a \diamond^{p(m_1+1)-1} a \ldots a \diamond^{p(m_k+1)-1} a, b \diamond^{p(n_1+1)-1} b \ldots b \diamond^{p(n_l+1)-1} b\}$$

is unavoidable.

Proof. In terms of notation it will be helpful to define

$$M_j = \sum_{i=1}^{j} m_i + 1$$

Now suppose the two-sided infinite word w avoids $X_{m_1,\ldots,m_k \mid n_1,\ldots,n_l}$, and let

$$v = \ldots w(-1)^p w(0)^p w(1)^p \ldots$$

We claim that v avoids Y. Suppose otherwise. Then v has a factor compatible with some $x \in Y$. Without loss of generality say that

$$x = a \diamond^{p(m_1+1)-1} a \ldots \diamond^{p(m_k+1)-1} a$$

Then to say that v has a factor compatible with x is equivalent to saying that there exists $i \in \mathbb{Z}$ for which

$$v(i) = v(i + pM_1) = \cdots = v(i + pM_k) = a$$

But if we set $h = \lfloor \frac{i}{p} \rfloor$ then this implies that

$$w(h) = w(h + M_1) = \cdots = w(h + M_k) = a$$

contradicting the fact that w avoids $X_{m_1,\ldots,m_k \mid n_1,\ldots,n_l}$.

We prove the other direction analogously. Suppose now that the two-sided infinite word w avoids Y, and set $v = \ldots w(-p)w(0)w(p)\ldots$. We claim that v avoids $X_{m_1,\ldots,m_k \mid n_1,\ldots,n_l}$. Otherwise v has a factor compatible with some $x \in p$ which we may suppose without loss of generality is $a \diamond^{m_1} a \ldots \diamond^{m_k} a$. Then there exists $i \in \mathbb{Z}$ for which

$$v(i) = v(i + M_1) = \cdots = v(i + M_k)$$

but this implies that

$$w(pi) = w(pi + pM_1) = \cdots = w(pi + pM_k)$$

which contradicts the fact that w avoids Y.

In order to solve the problem of identifying when $X_{m_1,\ldots,m_k \mid n_1,\ldots,n_l}$ is unavoidable we start with small values of k and l. The set $\{a, b \diamond^{n_1} b \ldots b \diamond^{n_l} b\}$ is unavoidable for if w is a two-sided infinite word which lacks a factor compatible with a it must be $b^{\mathbb{Z}}$. This handles the case where $k = 0$ (and symmetrically $l = 0$).

4   Special Cases

We first consider the case where $k = 1$ and $l = 1$, that is, we consider the set $X_{m \mid n} = \{a \diamond^m a, b \diamond^n b\}$. In this case, we can give a nice characterization of which integers m, n make this set avoidable.

Theorem 1. *Write $m + 1 = 2^s r_0, n + 1 = 2^t r_1$ where r_0, r_1 are odd. Then $X_{m \mid n} = \{a \diamond^m a, b \diamond^n b\}$ is avoidable if and only if $s = t$.*

Proof. Let w be a two-sided infinite word avoiding $X_{m \mid n}$. Then w also avoids $b \diamond^m b$. Otherwise for some $i \in \mathbb{Z}$, $w(i) = b$ and $w(i + m + 1) = b$. Since w avoids $b \diamond^n b$ we must have that $w(i + n + 1) = a$ and $w(i + m + 1 + n + 1) = a$, which contradicts the fact that w avoids $a \diamond^m a$. A symmetrical argument shows that w avoids $a \diamond^n a$.

For ease of notation, write $\bar{a} = b$ and $\bar{b} = a$. Let $p \in \mathbb{N}$. We will say that a two-sided infinite word is p-alternating if for all $i \in \mathbb{Z}$, $w(i) = \overline{w(i + p)}$. By our previous observation it is easy to see that w avoids $X_{m \mid n}$ if and only if w is $m + 1$-alternating and $n + 1$-alternating. Thus to prove the theorem it is sufficient to show that a two-sided infinite word exists which is p-alternating and q-alternating if and only if $s = t$ where $p = 2^s r_0$ and $q = 2^t r_1$ with r_0 and r_1 odd.

Notice that if w is p-alternating then it has period $2p$: for $i \in \mathbb{Z}$,

$$w(i) = \overline{w(i+p)} = \overline{\overline{w(i+2p)}} = w(i+2p)$$

Now suppose $s \neq t$. Without loss of generality say $s < t$. Then $s+1 \leq t$. Let l be the least common multiple of p and q. The prime factorization of l must have no greater power of 2 than the prime factorization of q. Thus there exists an odd number k such that $kq \equiv 0 \bmod 2p$. If there were a two-sided infinite word w which was p-alternating and q-alternating we would have $w(0) = w(2p) = w(kq)$ since w has period $2p$. But since k is odd and w is q-alternating we also have $w(0) = \overline{w(kq)}$. This is a contradiction. We have half of the necessary implication.

Now suppose $s = t$. Then $p = 2^s r_0$, $q = 2^s r_1$. We only need to prove that there exists some w which is p-alternating and q-alternating and we do this by induction on s. If $s = 0$, then p and q are odd. Then the word $\ldots ababab \ldots$ is p-alternating and q-alternating. This handles our base case. Now say w is $2^s r_0$ and $2^s r_1$-alternating. Then $v = \ldots w(-1)w(-1)w(0)w(0)w(1)w(1) \ldots$ is $2^{s+1} r_0$ and $2^{s+1} r_1$-alternating. This finishes the induction and our proof.

We next consider the case where $k = 1$ and $l = 2$, that is, sets of the form $X_{m|n_1,n_2} = \{a\diamond^m a, b\diamond^{n_1} b\diamond^{n_2} b\}$. We believe, based on extensive experimental evidence, that we have identified the cases for which $X_{m|n_1,n_2}$ is unavoidable which we state in this section (Conjecture 1). As a result of this conjecture, $X_{m_1,\ldots,m_k|n_1,\ldots,n_l}$ is avoidable for all larger k, l. Here we prove one direction of our conjecture, and in Section 5, we give partial results towards the other direction which turns out to be easy for even values of m.

There is a delicate tension in the change of difficulty of the problem as we increase k and l. On the one hand, we have identified a large number of avoidable sets of the form $\{a\diamond^m a, b\diamond^n b\}$. For $X_{m|n_1,n_2}$ to be avoidable it is sufficient that $\{a\diamond^m a, b\diamond^{n_1} b\}$, $\{a\diamond^m a, b\diamond^{n_2} b\}$ or $\{a\diamond^m a, b\diamond^{n_1+n_2+1} b\}$ be avoidable. Thus by first identifying the avoidable sets for smaller values of k and l our job has gotten a little easier. On the other hand the structure of words avoiding $\{a\diamond^m a, b\diamond^{n_1} b\diamond^{n_2} b\}$ is not nearly as nice as those avoiding $\{a\diamond^m a, b\diamond^n b\}$. There is no simple characterization akin to p-alternation.

In proving that a set of the form $X_{m|n_1,n_2}$ is unavoidable our strategy is to derive a contradiction using structural properties that any potential two-sided infinite word w avoiding X would have. These properties take the form of certain rules involving the occurrences of letters in w. For example, whenever $w(i) = w(i+n_1+1) = b$ in w, we must have that $w(i+n_1+n_2+2) = a$. The presence of an a also has implications: if $w(i) = a$ then $w(i-m-1) = b$ and $w(i+m+1) = b$. Often particular values of m, n_1 and n_2 have a relationship that cause these patterns to reoccur and perpetuate themselves, making a contradiction easy to find. In order for this to happen we also need a starting point for the perpetuation the ground. For this Theorem 1 is a very handy tool.

We give an example of this in action. The set $\{a\diamond^7 a, b\diamond b\diamond^3 b\}$ is unavoidable. Suppose instead that there exists a two-sided infinite word w which avoids it.

We know from Theorem 1 that $\{a \diamond^7 a, b \diamond b\}$ is unavoidable, thus w must have a factor compatible with $b \diamond b$. Say without loss of generality that $w(0) = w(2) = b$. This implies that $w(6) = a$, which in turn implies that $w(-2) = b$. Then we have that $w(-2) = w(0) = b$, forcing $w(4) = a$. This propagation continues: $w(-4) = w(-2) = b$ and so $w(2) = a$, which makes $w(-6) = b$ giving $w(0) = a$, a contradiction. This example is part of a more general phenomenon. Notice how in this example as the patterns reoccur, we have a sequence of a's traveling to the left toward the b at $w(0)$. There is a symmetric situation in which the b's travel to the right towards the a at $w(n_1 + 1)$. Both scenarios are covered by the following proposition.

Proposition 1. *Suppose either* $m = 2n_1 + n_2 + 2$ *or* $m = n_2 - n_1 - 1$, *and* $n_1 + 1$ *divides* $n_2 + 1$. *Then* $X_{m|n_1,n_2}$ *is unavoidable if and only if* $\{a \diamond^m a, b \diamond^{n_1} b\}$ *is unavoidable.*

One notable consequence of Proposition 1 is that if m is odd, then both $\{a \diamond^m a, b b \diamond^{m+1} b\}$ and $\{a \diamond^m a, b b \diamond^{m-2} b\}$ are unavoidable.

The next theorem takes advantage of the perpetuating pattern phenomenon in a more complicated context. Proposition 1 held because each a forced a b into the next position of an occurence of $w(i) = w(i + n_1 + 1) = b$, which in turn forced a new a in w. This created a single traveling sequence of a's and b's, causing an a overlap with the b at $w(0)$, yielding a contradiction. In the next argument, we take notice of the fact that each a occurring in w may contribute to two occurrences of $w(i) = w(i + n_1 + 1) = b$ simultaneously so that a contradiction will occur after many traveling sequences of letters appear and overlap.

Theorem 2. *Say that* $m = n_2 - n_1 - 1$ *or* $m = 2n_1 + n_2 + 2$, *and that the highest power of 2 dividing* $n_1 + 1$ *is less than the highest power of 2 dividing* $m + 1$. *Then* $X_{m|n_1,n_2}$ *is unavoidable.*

Proof. Since the highest power of 2 dividing $n_1 + 1$ is different than the highest power of 2 dividing $m+1$, we have that the set $Y = \{a \diamond^m a, b \diamond^{n_1} b\}$ is unavoidable. Consider first the case where $m = n_2 - n_1 - 1$ and suppose for contradiction that there exists a two-sided infinite word w that avoids X. Then w has no factor compatible with $\{a \diamond^m a\}$, and so since Y is unavoidable it must have a factor compatible with $\{b \diamond^{n_1} b\}$. Assume without loss of generality that $w(0) = b$ and $w(n_1 + 1) = b$.

We now generate an infinite table of facts about w. Two horizontally adjacent entries in the table will represent positions in w which are $n_1 + 1$ letters apart. Two vertically adjacent entries in the table will represent positions in w which are $m + 1 = n_2 - n_1$ letters apart. The two upper left entries of our table are $w(0) = b$ and $w(n_1 + 1) = b$, two facts we have already assumed. Since w avoids $X_{m|n_1,n_2}$ we have more information relevant to the table: two horizontally adjacent b entries force an a entry diagonally down and to the right from them, and an a entry forces a b entry in the vertically adjacent positions. From these rules we can build the following table, labeling the columns $C_0, C_1, \ldots$:

$$C_0 \qquad\qquad C_1 \qquad\qquad\qquad C_2 \qquad\qquad\qquad\qquad C_3 \qquad\qquad \dots$$

$$w(0) = b \; w(n_1 + 1) = b \quad w(2n_1 + 2) = b \qquad w(3n_1 + 3) = b$$
$$w(n_1 + n_2 + 2) = a \; w(2n_1 + n_2 + 3) = a$$
$$w(2n_2 + 2) = b \qquad w(n_1 + 2n_2 + 3) = b$$

For $i \in \mathbb{N}$, we shall define v_i to be the factor of w represented by C_i. If i is odd then C_i has i entries, and if i is even then C_i has $i + 1$ entries. Thus we define

$$v_i = \begin{cases} w(in_1 + i)w(in_1 + i + 1)\dots w(in_2 + i) & \text{if } i \text{ even} \\ w((in_1 + i)w(in_1 + i + 1)\dots w(n_1 + (i - 1)n_2 + i) & \text{if } i \text{ odd} \end{cases}$$

Two adjacent entries in C_i represent a distance of $m+1$ positions between letters in v_i. Thus for i even we have that $|v_i| = im + 1$ and for i odd we have that $|v_i| = (i - 1)m + 1$. We can also use the table to get some partial information about the positions of a's and b's in v_i. For $j \in \mathbb{N}$, $v_i(j) = b$ if $j \equiv 0 \bmod 2m + 2$, and $v_i(j) = a$ if $j \equiv m + 1 \bmod 2m + 2$.

Because the highest power of 2 dividing $n_1 + 1$ is no greater than the highest power of 2 dividing $m_1 + 1$, there exists some k for which $k(n_1 + 1) \equiv m + 1 \bmod 2m + 2$. Take i sufficiently large so that $|v_i| > kn_1 + k$. Because of how k was chosen, we have that $v_i(kn_1 + k) = a$. However examining the table we see that

$$w((i + k)n_1 + i + k) = v_i(kn_1 + k) = v_{i+k}(0) = b$$

a contradiction. This handles the situation where $m = n_2 - n_1 - 1$. The proof for the case where $m = 2n_1 + n_2 + 2$ is similar, the only difference is that the table will represent increasingly negative positions of w, rather than increasingly positive ones.

As an application of Theorem 2, take $m = 1$. Let us see for which $n_1 \in \mathbb{N}$ the hypotheses of the theorem hold to make $X_{m|n_1,n_2}$ unavoidable. The highest power of 2 dividing $n_1 + 1$ should be less than the highest power of 2 dividing $m + 1 = 2$. Thus $n_1 + 1$ must be odd, n_1 is even. Since $m = 1$ we cannot have $m = 2n_1 + n_2 + 2$. Say we have $m = n_2 - n_1 - 1$. Then $n_2 = n_1 + 2$. So we have that for any even n_1, the set $\{a \diamond a, b \diamond^{n_1} b \diamond^{n_1 + 2} b\}$ is unavoidable. We will prove in Section 5 that this is a complete characterization of unavoidability of $X_{m|n_1,n_2}$ for $m = 1$.

The next proposition identifies another large class of unavoidable sets using a modification of the strategies discussed so far.

Proposition 2. *Suppose $n_1 < n_2$, $2m = n_1 + n_2$ and $|m - n_1|$ divides $m + 1$. Then $X_{m|n_1,n_2}$ is unavoidable.*

We believe that together Lemma 1, Proposition 1, Proposition 2, and Theorem 2 nearly give a complete characterization of when $X_{m|n_1,n_2}$ is unavoidable. Following is what we believe to be the only exception.

Proposition 3. *The set $X_{6|1,3} = \{a \diamond^6 a, b \diamond b \diamond^3 b\}$ is unavoidable.*

We now state our conjecture.

Conjecture 1. The set $X_{m|n_1,n_2}$ is unavoidable precisely when the hypotheses of at least one of Lemma 1, Proposition 1, Proposition 2, Proposition 3 or Theorem

2 hold. Restated, $X_{m|n_1,n_2}$ is unavoidable for relatively prime $m+1$, n_1+1 and n_2+1 with $n_1 \leq n_2$ if and only if one of the following conditions (or their symmetric equivalents) hold:

- Proposition 1: The case where the set $\{a\diamond^m a, b\diamond^{n_1} b\}$ is unavoidable, $m = 2n_1 + n_2 + 2$ or $m = n_2 - n_1 - 1$, and $n_1 + 1$ divides $n_2 + 1$.
- Theorem 2: The case where $m = n_2 - n_1 - 1$ or $m = 2n_1 + n_2 + 2$, and the highest power of 2 dividing $n_1 + 1$ is less than the highest power of 2 dividing $m + 1$.
- Proposition 2: The case where $n_1 < n_2$, $2m = n_1 + n_2$ and $|m - n_1|$ divides $m + 1$.
- Proposition 3: The case where $m = 6$, $n_1 = 1$ and $n_2 = 3$.

The reader may verify that for any fixed m the only one of the above conditions that contributes infinitely many unavoidable sets to $X_{m|n_1,n_2}$ is Theorem 2, and that this theorem never applies to even m. Thus the conjecture states that there are only finitely many values of m, n_1, n_2 with m fixed and even and $X_{m|n_1,n_2}$ unavoidable. We will prove in Section 5 that this is indeed the case.

Using Lemma 1 we may assume without loss of generality that $m + 1, n_1 + 1, n_2 + 1$ are relatively prime. An important consequence of the conjecture is that in order for $X_{m|n_1,n_2}$ to be unavoidable it is necessary that either $m = 6$ and $n_1, n_2 = 1, 3$, or that one of the following equations hold:

$$m = 2n_1 + n_2 + 2 \tag{1}$$

$$m = 2n_2 + n_1 + 2 \tag{2}$$

$$m = n_1 - n_2 - 1 \tag{3}$$

$$m = n_2 - n_1 - 1 \tag{4}$$

$$2m = n_1 + n_2 \tag{5}$$

Using this fact we can show that an affirmative proof of the conjecture has a powerful consequence.

We end this section with the following proposition which implies that if Conjecture 1 is true then we have completely classified the unavoidable sets of size two.

Proposition 4. *If Conjecture 1 holds, then* $X_{m_1,\ldots,m_k|n_1,\ldots,n_l}$ *is avoidable for all* $k \geq 2$ *and* $l \geq 3$.

Proof. Assuming Conjecture 1 holds, it is enough to prove that both $X_{m_1,m_2|n_1,n_2}$ and $X_{m|n_1,n_2,n_3}$ are avoidable for all m_1, m_2, n_1, n_2. We handle the case of $X_{m_1,m_2|n_1,n_2}$. Assume without loss of generality that m_1, m_2, n_1, n_2 are relatively prime. In order for this set to be unavoidable it is necessary that the sets $\{a\diamond^{m_1} a, b\diamond^{n_1} b\diamond^{n_2} b\}$, $\{a\diamond^{m_2} a, b\diamond^{n_2} b\diamond^{n_2} b\}$, $\{a\diamond^{m_1} a\diamond^{m_2} a, b\diamond^{n_1} b\}$ and the set $\{a\diamond^{m_1} a\diamond^{m_2} a, b\diamond^{n_2} b\}$ are unavoidable as well. For each of these sets, Conjecture 1 gives a necessary condition: either $m = 6$ and $n_1 = 1, n_2 = 3$ (or symmetrically $n_1 = 3, n_2 = 1$) or one of Equations 1, 2, 3, 4 or 5 must hold. Consider the following tables:

$m_1 = 2n_1 + n_2 + 2$	$m_2 = 2n_1 + n_2 + 2$
$m_1 = 2n_2 + n_1 + 2$	$m_2 = 2n_2 + n_1 + 2$
$m_1 = n_1 - n_2 - 1$	$m_2 = n_1 - n_2 - 1$
$m_1 = n_2 - n_1 - 1$	$m_2 = n_2 - n_1 - 1$
$m_1 = 6, n_1 = 1, n_2 = 3$	$m_2 = 6, n_1 = 1, n_2 = 3$
$m_1 = 6, n_2 = 1, n_1 = 3$	$m_2 = 6, n_2 = 1, n_1 = 3$
$2m_1 = n_1 + n_2$	$2m_2 = n_1 + n_2$

$n_1 = 2m_1 + m_2 + 2$	$n_2 = 2m_1 + m_2 + 2$
$n_1 = 2m_2 + m_1 + 2$	$n_2 = 2m_2 + m_1 + 2$
$n_1 = m_1 - m_2 - 1$	$n_2 = m_1 - m_2 - 1$
$n_1 = m_2 - m_1 - 1$	$n_2 = m_2 - m_1 - 1$
$n_1 = 6, m_1 = 1, m_2 = 3$	$n_2 = 6, m_1 = 1, m_2 = 3$
$n_1 = 6, m_2 = 1, m_1 = 3$	$n_2 = 6, m_2 = 1, m_1 = 3$
$2n_1 = m_1 + m_2$	$2n_2 = m_1 + m_2$

In order for $X_{m_1,m_2|n_1,n_2}$ to be unavoidable it is necessary that at least one equation from each column be satisfied. It is easy to verify using a computer algebra system that this is impossible except in the case where the last equation in each column is satisfied. However in this case $m_1 = m_2 = n_1 = n_2$ and so by Theorem 1 the set is avoidable.

5   Avoidability Results for $k = 1$ and $l = 2$

In order to prove the conjecture, only one direction remains. We must show that if none of the hypotheses of Lemma 1, Proposition 1, Proposition 2, Proposition 3 or Theorem 2 hold then $X_{m|n_1,n_2}$ is avoidable. In this section we give partial results towards this goal.

We have found that in general identifying sets of the form $X_{m|n_1,n_2}$ as avoidable tends to be a more difficult task than identifying them as unavoidable. In the case of unavoidability we needed only consider a single word then derive a contradiction from its necessary structural properties. To find a class of avoidable sets we must invent some general procedure for producing a two-sided infinite word which avoids each such set. This is precisely what we move towards in the following propositions in which we verify that the conjecture holds for certain values of m and n_1.

It is easy to see that none of Equations 1, 2, 3, 4 or 5 are satisfied when $n_1, n_2 < m \leq n_1 + n_2 + 2$. Thus the conjecture for such values is that $X_{m|n_1,n_2}$ is avoidable. The following fact verifies that this is indeed the case.

Proposition 5. *If $n_1, n_2 < m < n_1 + n_2 + 2$ then $X_{m|n_1,n_2}$ is avoidable.*

The next proposition makes the conjecture easy to verify for even values of m.

Proposition 6. *Assume m is even and that $2m \leq n_1, n_2$. Then $X_{m|n_1,n_2}$ is avoidable.*

For any fixed even m there are then only finitely many values of n_1, n_2 which might be unavoidable. The reader may verify that this is consistent with the conjecture. The reader may also verify that the conjecture for $m = 0$ is that $X_{0|n_1,n_2}$ is always avoidable, and indeed this is given by Proposition 6. Similarly the conjecture for $m = 2$ is that $X_{2|n_1,n_2}$ is avoidable except for $n_1 = 1, n_2 = 3$ or $n_2 = 3, n_1 = 1$. It is easy to find avoiding two-sided infinite words for other values of n_1 and n_2 less than 5 when $m = 2$. By Proposition 6 this is all that is necessary to confirm the conjecture for $m = 2$. In this way we have been able to verify the conjecture for all even m up to very large values. The odd values of m seem to be much more difficult and will most likely require more sophisticated techniques. The following proposition gives our confirmation of the conjecture for $m = 1$.

Proposition 7. *The conjecture holds for $m = 1$; that is $X_{1|n_1,n_2}$ is unavoidable if and only if n_1 and n_2 are even numbers with $|n_1 - n_2| = 2$.*

The following and final proposition says that if m and n_1 are close enough in value then $X_{m|n_1,n_2}$ is avoidable for large enough n_2.

Proposition 8. *Let $s \in \mathbb{N}$ with $s < m - 2$. Then for $n > 2(m + 1)^2 + m - 1$, $X_{m|m+s,n} = \{a\diamond^m a, b\diamond^{m+s} b\diamond^n b\}$ is avoidable.*

6   Open Questions

Conjecture 1, although tested in numerous cases via computer, and verified for $m = 1$ and a large number of even values of m, still remains to be proven. As was shown in Section 4, an affirmative answer to this question would imply that $X_{m_1,\dots,m_k|n_1,\dots,n_l}$ is avoidable for all $k, l \geq 3$. Given that avoidable sets of the form $X_{m_1,\dots,m_k|n_1,\dots,n_l}$ for small k and l translate directly to avoidable sets for larger k and l, it might seem intuitive that for some sufficiently large fixed k and l there exists an easy proof that $X_{m_1,\dots,m_k|n_1,\dots,n_l}$ is always avoidable, and thus all larger values are. This is a deceptively difficult question. There is an interesting tension occurring between the increase in avoidability of $X_{m_1,\dots,m_k|n_1,\dots,n_l}$ and the structural complication of $X_{m_1,\dots,m_k|n_1,\dots,n_l}$ as k and l increase.

We pose two open questions that propose direction for further research.

Open question 1. *Can one find some sufficiently large values of k and l for which it is easy to prove that $X_{m_1,\dots,m_k|n_1,\dots,n_l}$ is always avoidable?*

Efficient algorithms to determine if a finite set of full words is unavoidable are well known, see for example [6]. These same algorithms can be used to decide if a finite set of partial words X is unavoidable by determining the unavoidability of $\hat{X}$. However this incurs a dramatic loss in efficiency, as each pword u in X can contribute as many as $\|A\|^{\|H(u)\|}$ elements to $\hat{X}$. There are algorithms for finding repetitions with gaps that could be useful for answering Open question 2, for instance [9,10,11,12,15].

Open question 2. *Is there an efficient procedure to determine if a finite set of partial words is unavoidable?*

References

1. Aho, A.V., Corasick, M.J.: Efficient string machines, an aid to bibliographic research. Comm. ACM 18, 333–340 (1975)
2. Berstel, J., Boasson, L.: Partial words and a theorem of Fine and Wilf. Theoret. Comput. Sci. 218, 135–141 (1999)
3. Blanchet-Sadri, F.: Codes, orderings, and partial words. Theoret. Comput. Sci. 329, 177–202 (2004)
4. Blanchet-Sadri, F.: Primitive partial words. Discrete Appl. Math. 148, 195–213 (2005)
5. Blanchet-Sadri, F., Duncan, S.: Partial Words and the Critical Factorization Theorem. J. Combin. Theory Ser. A 109, 221–245 (2005),
   http://www.uncg.edu/mat/cft/
6. Choffrut, C., Culik II, K.: On extendibility of unavoidable sets. Discrete Appl. Math. 9, 125–137 (1984)
7. Choffrut, C., Karhumäki, J.: Combinatorics of Words. In: Rozenberg, G., Salomaa, A. (eds.) Handbook of Formal Languages, vol. 1, pp. 329–438. Springer, Heidelberg (1997)
8. Ehrenfeucht, A., Haussler, D., Rozenberg, G.: On regularity of context-free languages. Theoret. Comput. Sci. 27, 311–322 (1983)
9. Kolpakov, R., Kucherov, G.: Finding Approximate Repetitions Under Hamming Distance. In: Meyer auf der Heide, F. (ed.) ESA 2001. LNCS, vol. 2161, pp. 170–181. Springer-Verlag, Heidelberg (2001)
10. Kolpakov, R., Kucherov, G.: Finding Approximate Repetitions Under Hamming Distance. Theoret. Comput. Sci. 33, 135–156 (2003)
11. Landau, G., Schmidt, J.: An Algorithm for Approximate Tandem Repeats. In: Apostolico, A., Crochemore, M., Galil, Z., Manber, U. (eds.) Combinatorial Pattern Matching. LNCS, vol. 684, pp. 120–133. Springer-Verlag, Heidelberg (1993)
12. Landau, G.M., Schmidt, J.P., Sokol, D.: An Algorithm for Approximate Tandem Repeats. J. Comput. Biology 8, 1–18 (2001)
13. Lothaire, M.: Algebraic Combinatorics on Words. Cambridge University Press, Cambridge (2002)
14. Rosaz, L.: Inventories of unavoidable languages and the word-extension conjecture. Theoret. Comput. Sci. 201, 151–170 (1998)
15. Schmidt, J.P.: All Highest Scoring Paths in Weighted Grid Graphs and Their Application to Finding All Approximate Repeats in Strings. SIAM J. Comput. 27, 972–992 (1998)

Hairpin Finite Automata

Henning Bordihn[1], Markus Holzer[2], and Martin Kutrib[3]

[1] Institut für Informatik, Universität Potsdam,
August-Bebel-Straße 89, D-14482 Potsdam, Germany
henning@cs.uni-potsdam.de
[2] Institut für Informatik, Technische Universität München,
Boltzmannstraße 3, D-85748 Garching bei München, Germany
holzer@informatik.tu-muenchen.de
[3] Institut für Informatik, Universität Giessen,
Arndtstraße 2, D-35392 Giessen, Germany
kutrib@informatik.uni-giessen.de

Abstract. We introduce and investigate nondeterministic finite automata with the additional ability to apply the hairpin inversion operation to the remaining part of the input. We consider three different modes of hairpin operations, namely left-most hairpin, general hairpin, and right-most hairpin. We show that these operations do not increase the computation power, when the number of operations is bounded by a constant. An unbounded number of these operations leads to language families that are properly contained in the family of context-sensitive languages and are supersets of the family of regular languages. Moreover, we show that in most cases we obtain incomparability results for the language families under consideration. Finally, we summarize closure properties of language families accepted by variants of hairpin finite automata.

1 Introduction

Since the origin of life it evolves by replication (and mutation) of DNA or RNA, but it was not until one and a half decade ago that DNA computing, or more generally computing with molecules, was discovered for solving problems in computer science. For instance, in [1] it was shown how to solve the NP-complete Hamiltonian Path Problem with tools from molecular biology. Also the gene assembly performed during the replication of certain single-cell organisms, namely ciliates, has inspired several computational models, e.g., see [9,11]. For further information about the biological process underlying ciliate genetics we refer to, e.g., [12,13]. Although the models proposed in [9] and [11] are different, both are based on simple operations on the DNA molecule guided by pointers. In the latter model the used operations are inspired by the way in which a DNA molecule can fold, namely hi (hairpin loop with inverted pointers) which reverses a substring between a pointer p and the reversal of p, ld (loop with direct repeat of pointers) which deletes a substring between two occurrences of a pointer, and $dlad$ (double loop with alternating direct repeat of pointers) which swaps two substrings marked by pointer-pairs. The hairpin inverse of a word w in Σ^+ is defined as

T. Harju, J. Karhumäki, and A. Lepistö (Eds.): DLT 2007, LNCS 4588, pp. 108–119, 2007.

$hi(w) = \{ xpy^R p^R z \mid w = xpyp^R z \text{ and } x, y, z \in \Sigma^*, p \in \Sigma^+ \}$, where p is called pointer. By using appropriate morphisms it is enough to consider pointers of length 1 only, namely $hi(w) = \{ xay^R az \mid w = xayaz \text{ and } x, y, z \in \Sigma^*, a \in \Sigma \}$.

In order to understand the very nature of the generative power of the aforementioned operations several authors studied them from a purely language theoretical perspective, see e.g., [4,5,6,7,8]. It is worth mentioning that the application of the operations is only guided by the pointers in the word, and thus no additional (other) control on the application of the operation is present. In this paper we develop the novel approach to combine bio-inspired operations with an additional control mechanism, namely a (finite) automaton. This leads to the recently introduced extended finite state automata, which are finite state machines equipped with an additional operation on the unread part of the input. These machines have been investigated in several papers and led to the devices of flip-pushdown automata [14], the "flip-pushdown input-reversal" theorem [10], input-reversal automata [2], and revolving-input automata [3]. Here we restrict ourself to extended finite automata with the additional operation of hairpin inversion and distinguish three different modes of hairpin inversion, namely left-most hairpin (the involved pointers are as close as possible), general hairpin (no restriction on the pointers), and right-most hairpin (the pointers are as far away as possible).

Obviously, if the number of operations applied is zero, the family of regular languages is characterized. We show that this remains true as long as the number of hairpin operations is arbitrarily constant, regardless of the interpretation of the hairpin operation. In most cases the induced language families are incomparable (or properly included) with each other. To this end, we develop a pumping argument for languages accepted by right-most and general hairpin finite automata. Moreover, we investigate the relation of the hairpin finite automata language families to the context-free languages and their most important sub-families. There it turns out that whenever the letter structure of a left-most hairpin, general hairpin, or right-most hairpin language is "simple" in a certain sense, then the language is regular. Concerning the closure properties of the language families under consideration, we show quite negative results. For all investigated formal language operations, except the union, both the right-most hairpin and general hairpin finite automata language families are not closed. This is quite surprising, since we are dealing with language families defined by automata. Nevertheless, this nicely resembles some non-closure properties recently obtained for revolving finite automata languages [3].

2   Preliminaries

We denote the powerset of a set S by 2^S. The empty word is denoted by λ, the reversal of a word w by w^R, and for the length of w we write $|w|$. For the number of occurrences of a symbol a in w we use the notation $|w|_a$. Set inclusion is denoted by $\subseteq$, and strict set inclusion by $\subset$.

In the following we consider finite automata with the ability to apply the hairpin operation to the unread part of the input. We may start with a uniform definition.

Definition 1. *A* (nondeterministic) extended finite state automaton *is a 6-tuple* $A = (Q, \Sigma, \delta, \Delta, q_0, F)$, *where Q is a finite set of states, Σ is the input alphabet, δ and Δ are mappings from $Q \times (\Sigma \cup \{\lambda\})$ to 2^Q, where δ is called the transition function, $q_0 \in Q$ is the initial state, and $F \subseteq Q$ is the set of accepting states. Furthermore, A is said to be λ-free, if both δ and Δ are restricted to $Q \times \Sigma$.*

The different modes are formally distinguished by different interpretations of the mapping Δ. To this end, we consider *configurations* of extended finite state automata to be tuples (q, w), where $q \in Q$ is the current state, and $w \in \Sigma^*$ is the still unread part of the input. If a is in $\Sigma \cup \{\lambda\}$ and w in Σ^*, then we write $(q, aw) \vdash_A (p, w)$, if p is in $\delta(q, a)$. Those transitions will be referred to as ordinary transitions.

A hairpin operation is performed by applying the mapping Δ. For $a \in \Sigma$ and $v, w \in \Sigma^*$,

1. a *left-most hairpin transition* is defined by $(q, avaw) \vdash_A (p, av^Raw)$, for p in $\Delta(q, a)$ or $\Delta(q, \lambda)$, and $v \in (\Sigma \setminus \{a\})^*$,
2. a *general hairpin transition* is defined by $(q, avaw) \vdash_A (p, av^Raw)$, for p in $\Delta(q, a)$ or $\Delta(q, \lambda)$, and
3. a *right-most hairpin transition* is defined by $(q, avaw) \vdash_A (p, av^Raw)$, for p in $\Delta(q, a)$ or $\Delta(q, \lambda)$, and $w \in (\Sigma \setminus \{a\})^*$.

The corresponding transitions will be referred to as non-ordinary transitions.

An extended finite state automaton $A = (Q, \Sigma, \delta, \Delta, q_0, F)$ with left-most hairpin, general hairpin, or right-most hairpin transitions is called a *left-most hairpin finite automaton (lh-FA), general hairpin finite automaton (h-FA),* or *right-most hairpin finite automaton (rh-FA),* respectively.

For any type of hairpin automata, whenever there is a choice between an ordinary transition or a hairpin operation, the automaton nondeterministically chooses the next move. As usual, the reflexive transitive closure of $\vdash_A$ is denoted by $\vdash_A^*$. The subscript A will be dropped from $\vdash_A$ and $\vdash_A^*$ whenever the meaning remains clear.

Let k be a non-negative integer. We define $T_k(A)$, the language *accepted with at most k non-ordinary steps* to be $T_k(A) = \{ w \in \Sigma^* \mid (q_0, w) \vdash_A^* (q, \lambda)$ with at most k non-ordinary steps and $q \in F \}$. If the number of non-ordinary steps is not bounded, the language accepted is analogously defined as above and denoted by $T(A)$.

In order to clarify our notation we give an example. In the sequel we often deal with languages where the symbols are embedded in marker symbols #. To this end, we define an homomorphism that maps any symbol a to a#, and define L# = #$h(L)$, for any language L over some alphabet not containing #. That is,

$$a_1 a_2 \cdots a_n \in L \iff \#a_1 \# a_2 \# \cdots \# a_n \# \in L\# \quad \text{and} \quad \lambda \in L \iff \# \in L\#.$$

In what follows, when specifying an automaton we will list only those transitions which do not map to the empty set.

Example 2. The non-regular context-free language

$$L\#, \text{ where } L = \{ w \in \{a, b\}^* \mid |w|_a = |w|_b \}$$

is accepted by the general hairpin automaton $A = (Q, \{a, b, \#\}, \delta, \Delta, q_0, \{q_\lambda\})$ with state set $Q = \{q_0, q_\lambda, q_a, q_b, q_{a?}, q_{b?}, q_{a\#}, q_{b\#}, q_{a\#b}, q_{b\#a}\}$ and

1. $\delta(q_0, \#) = \{q_\lambda\}$
2. $\delta(q_\lambda, a) = \{q_a\}$
3. $\Delta(q_a, \#) = \{q_{a?}\}$
4. $\delta(q_{a?}, \#) = \{q_{a\#}\}$

5. $\delta(q_{a\#}, b) = \{q_{a\#b}\}$
6. $\delta(q_{a\#b}, \#) = \{q_\lambda\}$
7. $\delta(q_\lambda, b) = \{q_b\}$
8. $\Delta(q_b, \#) = \{q_{b?}\}$

9. $\delta(q_{b?}, \#) = \{q_{b\#}\}$
10. $\delta(q_{b\#}, a) = \{q_{b\#a}\}$
11. $\delta(q_{b\#a}, \#) = \{q_\lambda\}$

Automaton A accepts $L\#$ as follows: When starting in state q_0 the first transition reads the leading letter # and changes to state q_λ. Next, the transitions (2)–(6) read the letters a, #, b, and # in sequence while doing a hairpin transition on the first # letter. If this is possible the automaton will be in state q_λ again. More formally, for $v \in \{a\#, b\#\}^*\{a, b\}$ and $w \in \{a\#, b\#\}^*$ we find $(q_\lambda, a\#v\#b\#w) \vdash_A (q_a, \#v\#b\#w) \vdash_A (q_{a?}, \#b\#v^R\#w) \vdash_A (q_{a\#}, b\#v^R\#w) \vdash_A (q_{a\#b}, \#v^R\#w) \vdash_A (q_\lambda, v^R\#w)$, and $(q_\lambda, a\#b\#w) \vdash_A (q_a, \#b\#w) \vdash_A (q_{a?}, \#b\#w) \vdash_A (q_{a\#}, b\#w) \vdash_A (q_{a\#b}, \#w) \vdash_A (q_\lambda, w)$. A similar reasoning applies for the transitions (7)–(11).

Example 3. By a straightforward modification of the automaton in Example 2, we obtain (1) a general hairpin automaton that accepts the non-context-free language $L\#$, where $L = \{ w \in \{a, b, c\}^* \mid |w|_a = |w|_b = |w|_c \}$, and (2) a rightmost hairpin automaton that accepts the non-regular context-free language $L\#$, where $L = \{ a^n b^n \mid n \geq 0 \}$.

By standard techniques one can prove that λ-moves do not increase the computational power of hairpin automata. So, in the sequel we may consider λ-free automata for convenience.

Theorem 4. *Let k be a non-negative integer. For a hairpin automaton A of any type, one can construct a λ-free hairpin automaton B of the same type, such that $T_k(A) = T_k(B)$. The statements remain true if an unbounded number of hairpin steps is allowed.*

3   Hairpin Finite Automata

Now we turn to investigate the computational capacities of the devices in question. Our first results concern the weakest automata. The following theorem shows that providing finite automata with a bounded number of hairpin transitions does not increase their computational capacity. To this end, we need the following result shown in [5]. Recall that a language family is called a *trio*, if it is closed under λ-free homomorphism, inverse homomorphism, and intersection with regular languages.

Lemma 5. *Let $\mathscr{L}$ be a trio closed under hairpin inversion. Then for each $L \in \mathscr{L}$ over Σ, and for each $a \in \Sigma$, the a-projected hairpin inverse of L is in $\mathscr{L}$. Here the a-projected hairpin inverse of a language $L \subseteq \Sigma^*$ is defined to be $hi_a(L) = \{ hi_a(w) \mid w \in L \}$, where $hi_a(w) = \{ xay^Raz \mid w = xayaz \text{ and } x, y, z \in \Sigma^* \}$, for $w \in \Sigma^*$ and $a \in \Sigma$.*

Theorem 6. *Let k be a non-negative integer. A language L is accepted by any type of hairpin automaton A with at most k hairpin steps, i.e., $T_k(A) = L$, if and only if L is regular.*

Proof. Since any given nondeterministic finite automaton is a hairpin automaton whose transition Δ is not defined, clearly, any regular language is accepted by any type of hairpin automaton.

For the converse implication we argue as follows: Let $L \subseteq \Sigma^*$ be accepted by a hairpin automaton $A = (Q, \Sigma, \delta, \Delta, q_0, F)$ with at most k hairpin steps. Then L can be written as $L = \cup_{i=0}^k T_{=i}(A)$, where $T_{=i}(A)$ is the language accepted by A with *exactly* i hairpin steps. It suffices to show that each $T_{=i}(A)$ is regular. The proof is by induction on k. We start with the left- and right-most hairpin operations.

If $k = 0$, the statement is obviously true. Now consider an accepting computation on input w such that A performs exactly $k + 1$ left-most (right-most, respectively) hairpin operations. Let $A_{p,q} = (Q, \Sigma, \delta, \Delta, p, \{q\})$ be defined from A. Moreover, let $u, v, x \in \Sigma^*$ and $a \in \Sigma$. We find a decomposition $w = uavax$ obeying the properties for left-most (right-most, respectively) hairpin such that $(q_0, w) = (q_0, uavax) \vdash^* (q, avax) \vdash (p, av^Rax) \vdash^* (q_f, \lambda)$, where $(q_0, uavax) \vdash^* (q, avax)$ is a computation without any hairpin step, $p \in \Delta(q, a)$, $(p, av^Rax) \vdash^* (q_f, \lambda)$ is a computation with exactly k hairpin moves, and $q_f \in F$. The decomposition gives rise to the languages L_{lh} and L_{rh} defined as

$$L_{lh} = \bigcup_{\substack{q \in Q \\ a \in \Sigma, p \in \Delta(q,a)}} T_{=0}(A_{q_0,q}) \cdot \{ avax \in \Sigma^* \mid av^Rax \in T_{=k}(A_{p,q_f}), \\ v \in (\Sigma \setminus \{a\})^*, \text{ and } q_f \in F \}$$

and

$$L_{rh} = \bigcup_{\substack{q \in Q \\ a \in \Sigma, p \in \Delta(q,a)}} T_{=0}(A_{q_0,q}) \cdot \{ avax \in \Sigma^* \mid av^Rax \in T_{=k}(A_{p,q_f}), \\ x \in (\Sigma \setminus \{a\})^*, \text{ and } q_f \in F \}.$$

By induction hypothesis, the languages $T_{=0}$ and $T_{=k}$ are regular. Thus, language L_{lh} can be rewritten as

$$\bigcup_{\substack{q \in Q \\ a \in \Sigma, p \in \Delta(q,a)}} T_{=0}(A_{q_0,q}) \cdot g_a(hi_\#(g_a^{-1}(T_{=k}(A_{p,q_f})) \cap R_a)),$$

where $g_a : (\Sigma \cup \{\#\})^* \to \Sigma^*$ is the homomorphism defined as $g_a(b) = b$, if $b \in \Sigma$, and $g_a(\#) = a$, otherwise, and $R_a = \#(\Sigma \setminus \{a\})^* \# \Sigma^*$. A similar construction can be given for L_{rh}. By Lemma 5 and since the family of regular languages is a concatenation closed trio, the languages L_{lh} and L_{rh} are regular, too. Hence we conclude that $T_{=(k+1)}(A)$ is regular and the stated claim follows for the hairpin operations under consideration.

For the general hairpin the idea of the above given proof is not directly applicable, because the pointers for the general hairpin operation are not uniquely determined as in the case of left- and right-most hairpin. Nevertheless, by another construction we succeed showing that at most k general hairpin operations remain regular. We sketch the construction for $k = 2$, but it is easy to see that it generalizes to arbitrary k. Performing exactly two general hairpin operations results in three different situation schemes, namely, the pointers are (i) non-overlapping, (ii) nested overlapping, and (iii) cross dependent. In case (i) the input to the automaton is of the form $u_1 a u_2 a u_3 b u_4 b u_5$, and the two general hairpins are done on the pointers a and b, which are not necessarily different. Similarly, for (ii) and (iii) the input is of the form $u_1 a u_2 b u_3 b u_4 a u_5$ and $u_1 a u_2 b u_3 a u_4 b u_5$, respectively. To construct a finite automaton for the hairpin language under consideration, the machine guesses the situation scheme, e.g., cross dependent, and then the input is assumed to be of the form $u_1 a u_2 b u_3 a u_4 b u_5$ and will be processed as follows: During the computation from left to right of the simulating finite state automaton, $2 \cdot 5$ mappings of the form $Q \times 2^Q$ induced by the transition function δ of the hairpin automaton for the words u_i and u_i^R, for $1 \leq i \leq 5$, are computed online. Finally these mappings are appropriately combined together with the transition function Δ of the hairpin automaton such that a run on the word $u_1 a u_3^R b u_4^R a u_2 b u_5$ is simulated. Since the number of situation schemes is a constant depending on k we are done. The tedious details are omitted due to the space limitation. □

In the sequel we consider automata whose number of hairpin transitions is not restricted to be constant. We call a hairpin transition or a series of consecutive hairpin transitions *void*, if they change the state of the automaton only, but do not change the remaining input word. For example, if in between the pointers there is a palindrome, or if two consecutive hairpin transitions on the same pointer pair are performed, then the transitions are void. Observe that the latter case is not useless. It can be used to test whether there is a matching symbol in the remaining input. If not, the computation gets stuck, otherwise it continues.

We turn to compare the computational capacities of the devices under consideration. To this end, next we present a tool for showing that certain languages are not accepted by right-most hairpin automata.

Lemma 7. *Let $L \subseteq \Sigma^*$ be a right-most hairpin automaton language. Then there exist constants $n > 0$ and ℓ, r, where $\ell + r > 0$, such that any word of L that admits a factorization $v^x u (v^R)^y w$, where $x, y \geq n$, $u, v \in \Sigma^+$, all symbols of v are different, and $w \in \Sigma^*$ contains no symbols of v, implies $v^x v^{i \cdot \ell} u (v^R)^y v^{i \cdot r} w \in L$, for any $i \geq 0$.*

Proof. Let A be a λ-free, s-state right-most hairpin automaton accepting language L. We set $n = 2s + 1$, let $v = a_1 \cdots a_k \in \Sigma^+$, and consider an input $v^x u (v^R)^y w$, where $x, y \geq n$, $u \in \Sigma^+$.

While A consumes subwords v, hairpin transitions have the following form:

$$(q, a_i a_{i+1} \cdots a_k v^{x'} u(v^R)^{y'} a_k \cdots a_{i+1} a_i \cdots a_1 w)$$
$$\vdash (p, a_i a_{i+1} \cdots a_k v^{y'} u^R (v^R)^{x'} a_k \cdots a_{i+1} a_i \cdots a_1 w).$$

So, the state may change, the subword u is reversed, and the numbers x' and y' of still unread subwords v are interchanged.

Whenever automaton A reads the first symbol a_1 of some subword v, it does it in some state q having performed either an even or an odd number of hairpin transitions. For this there are at most $n - 1 = 2s$ possibilities. Since x and y are at least n, there are at least two configurations which are identical except for the number of consumed subwords v, say $(q, v^{x_1} u(v^R)^{y_1} w)$ and $(q, v^{x_2} u(v^R)^{y_2} w)$, for an even number of hairpin transitions performed, and $(q, v^{y_1} u^R (v^R)^{x_1} w)$ and $(q, v^{y_2} u^R (v^R)^{x_2} w)$, for an odd number of hairpin transitions performed, where $1 \le x_2 \le x_1 \le x$, $1 \le y_2 \le y_1 \le y$, and $x_2 + y_2 \le x_1 + y_1$.

Now let $\ell = x_1 - x_2$ and $r = y_1 - y_2$ in the case of an even number of hairpin operations, and $\ell = y_1 - y_2$ and $r = x_1 - x_2$ in the other case. We obtain $\ell + r = x_1 - x_2 + y_1 - y_2 > 1$. Moreover, there are computations

$$(q, v^{x_1} v^{x_1 - x_2} u(v^R)^{y_1} (v^R)^{y_1 - y_2} w) \vdash^* (q, v^{x_1} u(v^R)^{y_1} w) \quad \text{or}$$
$$(q, v^{y_1} v^{y_1 - y_2} u^R (v^R)^{x_1} (v^R)^{x_1 - x_2} w) \vdash^* (q, v^{y_1} u^R (v^R)^{x_1} w).$$

Repeating the computations completes the proof.   □

Theorem 8. *There is a general hairpin automaton language, which is neither a left-most, nor a right-most hairpin automaton language.*

Proof. By Example 2, the language $L\#$, where $L = \{ w \in \{a, b\}^* \mid |w|_a = |w|_b \}$ is a general hairpin automaton language. In order to disprove that $L\#$ is a right-most hairpin automaton language we apply Lemma 7. Contrarily we assume $L\#$ is such a language, set $v = \#a$, $w = \lambda$, and consider words $v^n \#(b\#)^{2n}(v^R)^n$, where n is the constant of Lemma 7. Since these words belong to $L\#$, there are constants $\ell + r > 0$ such that $v^{n+\ell}\#(b\#)^{2n}(v^R)^{n+r}$ belongs to $L\#$, too, which is a contradiction.

In order to disprove that $L\#$ is a left-most hairpin automaton language we consider words $(\#a)^m (\#b)^m \#$, $m \ge 1$. While reading the first half of the input, any left-most hairpin transition is a void one. Therefore, if m is large enough, a corresponding acceptor would run into cycles, and words with more symbols a than symbols b would be accepted.   □

At first glance, one could expect that general hairpin automata are more powerful than right-most hairpin automata. In fact, both automata classes are incomparable with respect to their language acceptance capacities. In order to prove this result, we present a tool for showing that certain languages are not accepted by general hairpin automata.

Lemma 9. *Let $L \subseteq \Sigma^*$ be a language accepted by some general hairpin automaton A, and let $uavawaz$ be a word in L, where $a \in \Sigma$, and $u, v, w, z \in \Sigma^*$.*

1. *If there is an accepting computation such that* $(q, avawaz) \vdash (p, aw^R av^R az)$ *is the first non-void hairpin transition, then* $uawav^R az$ *belongs to* L, *too.*
2. *If there is an accepting computation such that* $(q, avawaz) \vdash (p, av^R awaz)$ *is the first non-void hairpin transition, then* $uaw^R avaz$ *belongs to* L, *too.*

Proof. We consider an accepting computation of general hairpin automaton A on input $uavawaz$. We have $(q_0, uavawaz) \vdash^* (q, avawaz)$, since there are only void hairpin transitions on the prefix u, $(q, avawaz) \vdash (p, aw^R av^R az)$ (respectively $(q, avawaz) \vdash (p, av^R awaz)$) is the first non-void hairpin transition, and finally $(p, aw^R av^R az) \vdash^* (q_f, \lambda)$ (respectively $(p, av^R awaz) \vdash^* (q_f, \lambda)$) is the rest of the computation. Due to the nondeterministic choice of the matching pointer a, we obtain $(q_0, uavav^R az) \vdash^* (q, awav^R az) \vdash (p, aw^R av^R az) \vdash^* (q_f, \lambda)$ (respectively $(q_0, uaw^R avaz) \vdash^* (q, aw^R avaz) \vdash (p, av^R awaz) \vdash^* (q_f, \lambda)$), and the assertion follows. □

Theorem 10. *There is a right-most hairpin automaton language, which is neither a left-most nor a general hairpin automaton language.*

Proof. By Example 3, language $L\#$, where $L = \{ a^n b^n \mid n \geq 0 \}$ is a right-most hairpin automaton language.

In order to disprove that $L\#$ is a general hairpin automaton language we apply Lemma 9. Contrarily, we assume $L\#$ is accepted by some general hairpin automaton A. We observe that A has to perform a non-void hairpin transition while reading the first half $\#a\#a \cdots \#a$ of the input. Otherwise it would run into a loop, if n is large enough. Whenever A performs a hairpin transition on pointers a, it is a void transition since in between the pointers there is a palindrome. Therefore, the non-void transition appears on pointers $\#$. Moreover, in between both pointers there are both symbols a and b. Otherwise it would be a void transition. We conclude that the input can be factorized $u\#x\#z$, where the first non-void transition appears on the pointers $\#$, and x contains at least one symbol a and at least one symbol b. So, we obtain $u\#v\#w\#z$, where v contains no b and w contains no a. By Lemma 9 (1), the input $u\#w\#v^R\#z$ belongs to $L\#$, too. This is a contradiction since some symbol a in v^R follows some symbol b in w.

In order to disprove that $L\#$ is a left-most hairpin automaton language, we observe that L is bounded but non-regular, and apply Corollary 14. □

Now we turn to compare the devices under consideration with some standard language families. Trivially, all types of hairpin finite state automata accept the regular languages. Obviously, *unary* languages accepted by hairpin automata are regular since a hairpin transition does not change the remaining part of the input. Therefore, it can be omitted.

Theorem 11. *A unary language L is accepted by a hairpin automaton of any type if and only if L is regular.*

Example 3 shows that general hairpin and right-most hairpin automata accept non-regular languages. On the other hand, the context-sensitive languages form a proper upper bound.

Theorem 12. *The families of languages accepted by hairpin finite state automata of any type are properly included in the family of context-sensitive languages. For left- and right-most hairpin automata, in addition, they are properly included in NP.*

Proof. The first statement follows by a straightforward simulation of a hairpin automaton A by a nondeterministic linear space-bounded Turing machine. The properness of the inclusion follows by unary languages and Theorem 12. For the second statement, in case of left- and right-most hairpin automata the Turing machine obeys a polynomial time bound. We consider an accepting computation of a right- or left-most hairpin automaton A on an input of length n. Let s be the number of states of A. Since A is a one-way device, the input head may stay at most $s \cdot n$ time steps on one input tape cell, otherwise the computation would run into cycles. Therefore, the computation obeys a time-bound of $O(n^2)$. The details are left to the reader. The properness follows by Theorem 8, since there is a non-accepted context-free language, and the context-free languages belong to NP.   □

Observe that a straightforward simulation of a general hairpin automaton by a Turing machine does not lead to a polynomial time bound in general, since for instance on a word of the form $(\#01)^n\#$ the hairpin inverse operation may be used to produce all possible words w with $w \in (\#\{01, 10\})^n\#$. Since the number of words of this form is exponential in n, the running time is not necessarily bounded by a polynomial anymore.

Next we focus our interest on context-free languages and some of their most important sub-families. In particular, we consider linear, deterministic, and bounded context-free languages. We recall that a language over some alphabet $\{a_1, \ldots, a_k\}$ is bounded, if it is a subset of $a_1^* a_2^* \cdots a_k^*$.

Theorem 13. *Let L be a bounded language. If L is accepted by a hairpin automaton of any type, then it is regular.*

Proof. Let A be an automaton of a type under consideration, and let A accept a bounded language L. We construct an equivalent nondeterministic finite automaton A'. Basically, the construction is based on the observation that any hairpin transition, regardless of left-most hairpin, right-most hairpin, or general hairpin, is a void one. Automaton A' may simulate ordinary transitions of A directly. Whenever A applies a hairpin transition, automaton A' simulates the state change directly and, in addition, remembers that it has to check whether the next input symbol is a matching pointer, i.e., whether there is a matching pointer at all. If the check fails, automaton A' rejects.   □

For left-most hairpin automata the following generalization follows immediately.

Corollary 14. *Let L be a bounded language. If $L\#$ is accepted by a left-most hairpin automaton, then L is regular.*

Now we can collect the first parts of the comparisons.

Theorem 15. *There is a bounded, linear, deterministic context-free language, which is neither a general hairpin, nor a left-most hairpin, nor a right-most hairpin automaton language.*

Proof. The language $\{\, a^n b^n \mid n \geq 0 \,\}$ is bounded, linear, deterministic context free, but not regular. Hence, the statement holds due to Theorem 13. □

For the converse comparison we derive the following corollary from Example 3.

Corollary 16. *There is a non-context-free general hairpin automaton language.*

The non-context-free language of Example 3 is not accepted by right-most hairpin automata. But nevertheless, the next theorem says that there is a suitable language.

Theorem 17. *There is a non-context-free right-most hairpin automaton language.*

Proof. The witness language is $L = \{\, (a\#\$)^n (b\#)^n (c\$)^n \mid n \geq 1 \,\}$. Clearly, language L is not context free.

A (deterministic) right-most hairpin automaton A which accepts L works in cycles. A cycle starts in the initial state q_0. By reading a symbol a, state q_1 is reached. Next, a right-most hairpin transition on $\#$ leads to state q_2. Reading $\#$ and b consecutively leads to state q_3. If automaton A finds an input symbol $\$$ in state q_3, it tries to read the sequence $\$\#c\$$. If this sequence is the rest of the input, automaton A accepts. If A finds an input symbol $\#$ in state q_3, it performs a right-most hairpin transition and changes to state q_4. Subsequently, it reads the $\#$, which leads to state q_5, and tries to perform a right-most hairpin transition on $\$$, which leads to state q_6. The computation continues by reading $\$$ and c consecutively. Now A is in state q_7. Next a right-most hairpin transition on $\$$ leads to state q_8. Finally, reading the symbol $\$$ completes the cycle, and A changes back to state q_0. So, automaton A computes cycles as follows:

$$(q_0, (a\#\$)^n (b\#)^n (c\$)^n) \vdash (q_1, \#\$(a\#\$)^{n-1}(b\#)^{n-1}b\#(c\$)^n)$$
$$\vdash (q_2, \#b(\#b)^{n-1}(\$\#a)^{n-1}\$\#(c\$)^n) \vdash^2 (q_3, (\#b)^{n-1}(\$\#a)^{n-1}\$\#(c\$)^n)$$
$$\vdash (q_4, \#\$(a\#\$)^{n-1}(b\#)^{n-2}b\#(c\$)^n) \vdash (q_5, \$(a\#\$)^{n-1}(b\#)^{n-1}(c\$)^{n-1}c\$)$$
$$\vdash (q_6, \$c(\$c)^{n-1}(\#b)^{n-1}(\$\#a)^{n-1}\$) \vdash^2 (q_7, \$c(\$c)^{n-2}(\#b)^{n-1}(\$\#a)^{n-1}\$)$$
$$\vdash (q_8, \$(a\#\$)^{n-1}(b\#)^{n-1}(c\$)^{n-2}c\$) \vdash (q_0, (a\#\$)^{n-1}(b\#)^{n-1}(c\$)^{n-1}).$$

The acceptance is completed by:

$$(q_0, a\#\$b\#c\$) \vdash (q_1, \#\$b\#c\$) \vdash (q_2, \#b\$\#c\$) \vdash^2 (q_3, \$\#c\$) \vdash^4 (q_f, \lambda).$$

It remains to be shown that every accepted word belongs to L. To this end, we observe that for all states there is just one unique transition to get in. Therefore, we can reconstruct accepted words simply by reversing the above computations. □

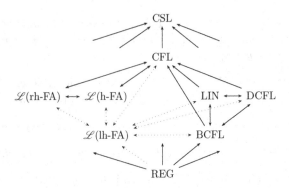

Fig. 1. Inclusion structure, where a solid line with one arrow indicates a proper inclusion, a solid line with two arrows links incomparable families, a dotted line with one arrow indicates either proper inclusion or equality, and a dotted line with two arrows links families which are either incomparable or related by a proper inclusion. In addition, language families that are not linked by a path are pairwise incomparable. $\mathscr{L}$(lh-FA), $\mathscr{L}$(rh-FA), and $\mathscr{L}$(h-FA) denote the families of languages which are accepted by left-most hairpin, right-most hairpin, and general hairpin automata, respectively. Moreover, BCFL refers to the family of bounded context-free languages. The context-sensitive languages are a proper superset of all depicted families (indicated by the arrows pointing to CSL). The regular languages are properly contained in all families other than $\mathscr{L}$(lh-FA) (indicated by the arrows starting at REG).

4   Conclusions

We have studied the power of hairpin finite automata, which are finite machines equipped with the additional ability to apply a hairpin inversion operation on the unread part of the input. The proven inclusion relations are summarized in Figure 1. Moreover, we also considered closure and non-closure properties, as well as basic computational complexity problems of these language families. The closure properties are summarized in the following table.

	$\cup$	$\cap$	$\sim$	$\cdot$	$*$	h^{-1}	h_λ	$.^R$
h-FA	yes	no	no	no	no	no	no	no
rh-FA	yes	no	no	no	no	no	no	?

Nevertheless, several questions for hairpin finite automata remain unanswered. We mention a few of them: (1) What is the computational power of left-most hairpin automata? We were not even able to determine the relationship between the automata and ordinary finite automata. So, is there a non-regular language accepted by left-most hairpin automata? Or are these hairpin automata another characterization of regular languages? In the latter case, the closure properties would be trivial, but descriptional complexity issues would be of natural interest. In the former case, negative closure properties would be of interest. (2) The family of general hairpin automata languages is not even closed under intersection with regular sets. This fact sheds some light on the relations between nondeterminism (with respect to the choice of matching pointers) and the structure of words. Is the family of right-most hairpin languages closed under intersection with regular sets? (3) The hairpin inversion operation deals with the reversal of subwords, but the family of general hairpin automata languages is not closed under reversal. What about the family of right-most hairpin automata languages?

References

1. Adleman, L.: Molecular computation of solutions to combinatorial problems. Science 266, 1021–1024 (1994)
2. Bordihn, H., Holzer, M., Kutrib, M.: Input reversals and iterated pushdown automata—a new characterization of Khabbaz geometric hierarchy of languages. In: Calude, C.S., Calude, E., Dinneen, M.J. (eds.) DLT 2004. LNCS, vol. 3340, pp. 102–113. Springer, Heidelberg (2004)
3. Bordihn, H., Holzer, M., Kutrib, M.: Revolving-input finite automata. In: De Felice, C., Restivo, A. (eds.) DLT 2005. LNCS, vol. 3572, pp. 168–179. Springer, Heidelberg (2005)
4. Daley, M., Ibarra, O., Kari, L.: Closure properties and decision questions of some language classes under ciliate bio-operations. Theoretical Computer Science 306, 19–38 (2003)
5. Daley, M., Kari, L., McQuillan, I.: Families of languages defined by ciliate bio-operations. Theoretical Computer Science 320, 51–69 (2004)
6. Dassow, J., Holzer, M.: Language families defined by a ciliate bio-operation: Hierarchies and decision problems. International Journal of Foundations of Computer Science 16, 645–662 (2005)
7. Dassow, J., Mitrana, V., Salomaa, A.: Operations and language generating devices suggested by genome evolution. Theoretical Computer Science 270, 701–738 (2002)
8. Dassow, J., Păun, Gh.: Remarks on operations suggested by mutations in genomes. Fundamenta Informaticae 36, 183–200 (1998)
9. Ehrenfeucht, A., Prescott, D.M., Rozenberg, G.: Computational aspects of gene (un)scrambling in ciliates. In: Landweber, L.F., Winfree, E. (eds.) Evolution as Computation, pp. 45–86. Springer, Heidelberg (2001)
10. Holzer, M., Kutrib, M.: Flip-pushdown automata: $k + 1$ pushdown reversals are better than k. In: Baeten, J.C.M., Lenstra, J.K., Parrow, J., Woeginger, G.J. (eds.) ICALP 2003. LNCS, vol. 2719, pp. 490–501. Springer, Heidelberg (2003)
11. Kari, L., Landweber, L.F.: Computational power of gene rearrangement. In: Winfree, E., Gifford, D. (eds.) DNA Based Computers V, DIMACS 54, pp. 207–216 (2000)
12. Prescott, D.M.: Cutting, slicing, reordering, and elimination of DNA sequences in hypotrichous ciliates. BioEssays 14, 317–324 (1992)
13. Prescott, D.M.: Genome gymnastics: Unique modes of DNA evolution and processing in ciliates. Nature Review Genetics 1, 191–198 (2000)
14. Sarkar, P.: Pushdown automaton with the ability to flip its stack. Report TR01-081, Electronic Colloquium on Computational Complexity (ECCC) (November 2001)

Characterizing Reduction Graphs for Gene Assembly in Ciliates*

Robert Brijder and Hendrik Jan Hoogeboom

Leiden Institute of Advanced Computer Science, Universiteit Leiden,
Niels Bohrweg 1, 2333 CA Leiden, The Netherlands
rbrijder@liacs.nl

Abstract. The biological process of gene assembly has been modeled based on three types of string rewriting rules, called string pointer rules, defined on so-called legal strings. It has been shown that reduction graphs, graphs that are based on the notion of breakpoint graph in the theory of sorting by reversal, for legal strings provide valuable insights into the gene assembly process. We characterize which legal strings obtain the same reduction graph (up to isomorphism), and moreover we characterize which graphs are (isomorphic to) reduction graphs.

1 Introduction

Ciliates form a large group of one-cellular organisms that are able to transform one nucleus, called the micronucleus, into an astonishing different one, called the macronucleus. This intricate DNA transformation process is called *gene assembly*. Each gene in the micronucleus, called micronuclear gene, is transformed to a gene in the macronucleus, called macronuclear gene. The string pointer reduction system models gene assembly based on three types of string rewriting rules, called string pointer rules, defined on so-called legal strings [1]. In this model, a micronuclear gene is represented by a legal string u, while its macronuclear gene (with its waste products) is represented by the reduction graph of u [2,3]. The *reduction graph* is based on the notion of breakpoint graph in the theory of sorting by reversal [4,5,6].

In this paper we characterize which graphs are (isomorphic to) reduction graphs (cf. Theorem 13). This characterization corresponds to an efficient algorithm. In this way we obtain a restriction on the form of the macronuclear structures that can possibly occur. We also provide a characterization that determines, given two legal strings, whether or not they have the same reduction graph (cf. Theorem 15). This may allow one to determine which micronuclear genes obtain the same macronuclear structure. It turns out that two legal strings obtain the same reduction graph (up to isomorphism) exactly when they can be transformed into each other by two types of string rewriting rules, which surprisingly are in a sense dual to the string positive rules and the string double rules (two of the three types of string pointer rules).

---

* This research was supported by the Netherlands Organization for Scientific Research (NWO) project 635.100.006 "VIEWS".

T. Harju, J. Karhumäki, and A. Lepistö (Eds.): DLT 2007, LNCS 4588, pp. 120–131, 2007.

The latter characterization has other uses as well. In a sense, the reduction graph allows for a complete characterization of applicability of *string negative rules*, the other type of string pointer rule, during the transformation process [2,7,8,3]. Moreover, it has been shown that the reduction graph does not retain much information about the applicability of the other two types of rules [7]. Therefore, the legal strings that obtain the same reduction graph are exactly the legal strings that have similar characteristics concerning the string negative rule. From a biological point of view, this may allow for a way to determine whether or not the strategies regarding the string negative rule are different among the different kinds of (genes in) ciliates. Due to space constraints, proofs of the results are omitted, but can be found in an extended version [9].

2   String Pointer Reduction System

The string pointer reduction system is the model of gene assembly that is used in this paper. In this section we give a concise description of this system, omitting examples and motivation. We refer to [10] for an in-depth description of this model including motivation and examples.

We fix $\kappa \geq 2$, and define the alphabet $\Delta = \{2, 3, \ldots, \kappa\}$. For $D \subseteq \Delta$, we define $\bar{D} = \{\bar{a} \mid a \in D\}$ and $\Pi = \Delta \cup \bar{\Delta}$. The elements of Π will be called *pointers*. We use the "bar operator" to move from Δ to $\bar{\Delta}$ and back from $\bar{\Delta}$ to Δ. Hence, for $p \in \Pi$, $\bar{\bar{p}} = p$. For a string $u = x_1 x_2 \cdots x_n$ with $x_i \in \Pi$, the *inverse* of u is the string $\bar{u} = \bar{x}_n \bar{x}_{n-1} \cdots \bar{x}_1$. For $p \in \Pi$, we define $\mathbf{p}$ to be p if $p \in \Delta$, and $\bar{p}$ if $p \in \bar{\Delta}$, i.e., $\mathbf{p}$ is the "unbarred" variant of p. The *domain* of a string $v \in \Pi^*$ is $\mathrm{dom}(v) = \{\mathbf{p} \mid p \text{ occurs in } v\}$. A *legal string* is a string $u \in \Pi^*$ such that for each $p \in \Pi$ that occurs in u, u contains exactly two occurrences from $\{p, \bar{p}\}$. For a pointer p and a legal string u, if both p and $\bar{p}$ occur in u then we say that both p and $\bar{p}$ are *positive* in u; if on the other hand only p or only $\bar{p}$ occurs in u, then both p and $\bar{p}$ are *negative* in u. We say that legal strings u and v are *equivalent*, denoted by $u \approx v$, if there is homomorphism $\varphi : \Pi^* \to \Pi^*$ with $\varphi(p) \in \{p, \bar{p}\}$ and $\varphi(\bar{p}) = \overline{\varphi(p)}$ for all $p \in \Pi$ such that $\varphi(u) = v$. E.g., legal strings $2\bar{2}33$ and $\bar{2}233$ are equivalent, while $2\bar{2}33$ are $2\bar{2}\bar{3}3$ are not. Note that $\approx$ is an equivalence relation. Equivalent legal strings are characterized by their "unbarred version" and their set of positive pointers.

The string pointer reduction system consists of three types of reduction rules, called *string pointer rules*, operating on legal strings. Here we will not consider these rules directly, but rather study the reduction graph (which is recalled in the next section) that captures essential properties of the rewriting system.

3   Reduction Graph

First we give some general notions w.r.t. graphs. A *coloured base* B is a 4-tuple (V, f, s, t) such that V is a finite set of *vertices*, $s, t \in V$ are the *source and target vertices* respectively, and $f : V \backslash \{s, t\} \to \Gamma$ for some set of *vertex labels* Γ. The elements of $\{\{x, y\} \mid x, y \in V, x \neq y\}$ are called *edges* for B. A *n-edge coloured graph*, $n \geq 1$, is a tuple $G = (V, E_1, E_2, \cdots, E_n, f, s, t)$ where $B = (V, f, s, t)$ is a coloured base and, for $i \in \{1, \ldots, n\}$, E_i is a set of edges for B. We also denote G by $B(E_1, E_2, \cdots, E_n)$. We define $\text{dom}(G) = \text{rng}(f)$, where $\text{rng}(f)$ is the range $f(V)$ of f. As usual, graphs G and G' are considered *isomorphic*, denoted $G \approx G'$, when they are equal modulo the identity of the vertices. However, the source (target, resp.) vertex of G needs to correspond to the source (target, resp.) vertex of G'.

We now recall the definition of reduction graph. This definition is equal to the one in [7], and is in slightly less general form compared to the one in [2]. We refer to [2], where it was introduced, for a motivation and for more examples and results. The notion of reduction graph uses the intuition from the notion of breakpoint graph (or reality-and-desire diagram) known from another branch of DNA processing theory called sorting by reversal, see e.g. [6] and [11].

Definition 1. Let $u = p_1 p_2 \cdots p_n$ with $p_1, \ldots, p_n \in \Pi$ be a legal string. The *reduction graph of u*, denoted by $\mathcal{R}_u$, is a 2-edge coloured graph (V, E_1, E_2, f, s, t), where

$$V = \{I_1, I_2, \ldots, I_n\} \cup \{I'_1, I'_2, \ldots, I'_n\} \cup \{s, t\},$$

$$E_1 = \{e_0, e_1, \ldots, e_n\} \text{ with } e_i = \{I'_i, I_{i+1}\} \text{ for } 1 < i < n, e_0 = \{s, I_1\}, e_n = \{I'_n, t\},$$

$$E_2 = \{\{I'_i, I_j\}, \{I_i, I'_j\} \mid i, j \in \{1, 2, \ldots, n\} \text{ with } i \neq j \text{ and } p_i = p_j\} \cup$$
$$\{\{I_i, I_j\}, \{I'_i, I'_j\} \mid i, j \in \{1, 2, \ldots, n\} \text{ and } p_i = \bar{p}_j\}, \text{ and}$$

$$f(I_i) = f(I'_i) = \mathbf{p}_i \text{ for } 1 \leq i \leq n.$$

The edges of E_1 are called the *reality edges*, and the edges of E_2 are called the *desire edges*. Notice that for each $p \in \text{dom}(u)$, the reduction graph of u has exactly two desire edges containing vertices labelled by p. It follows from the definition that, given legal strings u and v, $u \approx v$ iff $\mathcal{R}_u = \mathcal{R}_v$. Reality edges follow the linear order of the legal string, whereas desire edges connect positions in the string that will be joined when performing reduction rules, see [2].

In depictions of reduction graphs, we will represent the vertices (except for s and t) by their labels, because the exact identity of the vertices is not essential for the problems considered in this paper. We will also depict reality edges as "double edges" to distinguish them from the desire edges.

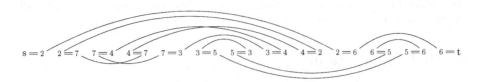

Fig. 1. The reduction graph $\mathcal{R}_u$ of u in Example 1

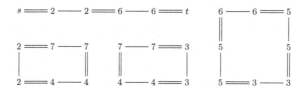

Fig. 2. The reduction graph of Figure 1 obtained by rearranging the vertices

Example 1. The reduction graph of $u = 2\bar{7}47353\bar{4}2656$ is depicted in Figure 1. Note how positive pointers (here 4 and 7) are connected by crossing desire edges, while those for negative pointers are parallel. By rearranging the vertices we can depict the graph as shown in Figure 2.

4   Abstract Reduction Graphs and Extensions

We now generalize the notion of reduction graph as a starting point to consider which graphs are (isomorphic to) reduction graphs. Moreover, we extend the reduction graphs by a set of edges, called *merge edges*, such that, along with the reality edges, the linear structure of the legal string is preserved in the graph. We will first define a set of edges for a given coloured base which has features in common with desire edges of a reduction graph.

Definition 2. Let $B = (V, f, s, t)$ be a coloured base. We say that a set of edges E for B is *desirable* if *(i)* for all $\{v_1, v_2\} \in E$, $f(v_1) = f(v_2)$, and *(ii)* for each $v \in V \backslash \{s, t\}$ there is exactly one $e \in E$ such that $v \in e$.

We now generalize the concept of reduction graph. We define for $f : X \to Y$ and $y \in Y$, $f^{-1}(y) = \{x \in X \mid f(x) = y\}$.

Definition 3. A 2-edge coloured graph $B(E_1, E_2)$ with $B = (V, f, s, t)$ is called an *abstract reduction graph* if

1. $\text{rng}(f) \subseteq \Delta$, and for each $p \in \text{rng}(f)$, $|f^{-1}(p)| = 4$,
2. for each $v \in V$ there is exactly one $e \in E_1$ such that $v \in e$,
3. E_2 is desirable for B.

The set of all abstract reduction graphs is denoted by $\mathcal{G}$. Clearly, if $G \approx \mathcal{R}_u$ for some u, then $G \in \mathcal{G}$. Therefore, for abstract reduction graphs $G = B(E_1, E_2)$, the edges in E_1 are called *reality edges* and the edges in E_2 are called *desire edges*. For graphical depictions of abstract reduction graphs we will use the same conventions as we have for reduction graphs. Thus, edges in E_1 will be depicted as "double edges", vertices are represented by their label, etc.

The definition of abstract reduction graph captures the "look and feel" of reduction graphs (see the next example). Each vertex label occurs four times, etc. The first goal of this paper is to set additional properties that characterize reduction graphs. It will turn out that the next example will not pass the test.

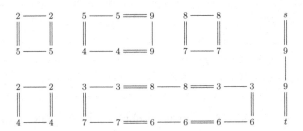

Fig. 3. An abstract reduction graph

Example 2. The 2-edge coloured graph in Figure 3 is an abstract reduction graph.

Next we introduce an extension to reduction graphs such that the "generic" linear order of the vertices $s, I_1, I_1', \ldots, I_n, I_n', t$ is retained, even when we consider the graphs up to isomorphism.

Definition 4. Let u be a legal string. The *extended reduction graph of* u, denoted by $\mathcal{E}_u$, is a 3-edge coloured graph $B(E_1, E_2, E_3)$, where $\mathcal{R}_u = B(E_1, E_2)$ and $E_3 = \{\{I_i, I_i'\} \mid 1 \leq i \leq n\}$ with $n = |u|$.

The edges in E_3 are called the *merge edges of* u, denoted by M_u. In this way, the reality edges and the merge edges form a unique path which passes through the vertices in the generic linear order. This is illustrated in the next example. In figures merge edges will be depicted by "dashed edges".

Fig. 4. The extended reduction graph $\mathcal{E}_u$ of u given in Example 1

Example 3. The extended reduction graph $\mathcal{E}_u$ of u given in Example 1 is shown in Figure 4, cf. Figure 1.

Merge edges can be generalized for abstract reduction graphs as follows.

Definition 5. Let $G = B(E_1, E_2) \in \mathcal{G}$, and let E be a set of edges for B. We say that E is *merge-legal* for G if E is desirable for B, and $E_2 \cap E = \varnothing$. We denote the set $\{E \mid E$ merge-legal for $G\}$ by ω_G. The set of all $E \in \omega_G$ where $B(E_1, E)$ is a connected graph is denoted by θ_G.

For legal string u, we also denote $\omega_{\mathcal{R}_u}$ and $\theta_{\mathcal{R}_u}$ by ω_u and θ_u, respectively. Notice that $M_u \in \theta_u \subseteq \omega_u$. Therefore, merge-legal edges will also be depicted by "dashed edges".

Fig. 5. An abstract reduction graph with two different sets of merge-legal edges

Example 4. Figure 5 depicts an abstract reduction graph twice. On the left-hand side it is augmented with a merge-legal set that is not in θ_G, and on the right-hand side is augmented with a merge-legal set that is in θ_G.

We say that $G = B(E_1, E_2, E)$ is an *extended abstract reduction graph* if $G' = B(E_1, E_2) \in \mathcal{G}$ and $E \in \theta_{G'}$. Since $M_u \in \theta_u$ for each legal string u, this notion is a natural abstraction of the notion of extended reduction graph, and hence the edges in E will be called *merge edges (of G)*. Moreover, for each graph $G' = B(E_1, E_2) \in \mathcal{G}$ isomorphic to a reduction graph we must have $\theta_{G'} \neq \varnothing$. In the next section we show that this condition is sufficient.

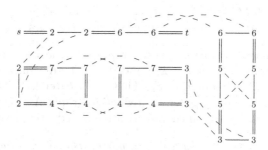

Fig. 6. An extended abstract reduction graph obtained by augmenting the reduction graph of Figure 2 with merge edges

Example 5. If we consider the reduction graph $\mathcal{R}_u = B(E_1, E_2)$ of Example 1 shown in Figures 1 and 2, then, of course, $B(E_1, E_2, M_u) = \mathcal{E}_u$ shown in Figure 4 is a extended abstract reduction graph. In Figure 6 another extended reduction graph is shown – it is $\mathcal{R}_u$ augmented with a set of merge edges E in θ_u. It is easy to see that indeed $E \in \theta_u$: simply notice that the path from s to t induced by the reality and merge edges will go through every vertex of the graph.

5   Back to Legal Strings

We now show that for extended abstract reduction graphs G we can "go back" in the sense that there are legal strings u such that G is isomorphic to $\mathcal{E}_u$. Due to space constraints, we do this by example. The idea is that the reality and merge edges together form the linear structure of the legal string, while the desire edges determine whether the pointers are positive or negative.

Fig. 7. The extended abstract reduction graph G given in Example 6

Example 6. Let us consider the extended abstract reduction graph G of Figure 6. By rearranging the vertices we obtain Figure 7. From this figure and the definition of extended reduction graph, it is clear that $v = 274265374356$ is a legal string that corresponds to G. Moreover, every legal string equivalent to v also corresponds to G.

Let L_G be the set of all legal strings that corresponds to extended abstract reduction graph G. It turns out that L_G is an non-empty equivalence class w.r.t. to the $\approx$ relation (for legal strings).

Theorem 6. (i) *Let G and G' be extended abstract reduction graphs. Then $G \approx G'$ iff $L_G = L_{G'}$. (ii) Let u and v be legal strings. Then $u \approx v$ iff $\mathcal{E}_u \approx \mathcal{E}_v$.*

Let G be an extended abstract reduction graph, and take $u \in L_G$ (such a u exists since L_G is nonempty). Since $u \in L_{\mathcal{E}_u}$ and the L_G's are equivalence classes, we have $L_{\mathcal{E}_u} = L_G$ and therefore $G \approx \mathcal{E}_u$. Thus every extended abstract reduction graph G is isomorphic to an extended reduction graph. As a corollary we have the following graph theoretical characterization of reduction graphs.

Theorem 7. *Let G be a 2-edge coloured graph. Then G is isomorphic to a reduction graph iff $G \in \mathcal{G}$ and $\theta_G \neq \varnothing$.*

It is easy to verify that $\theta_G = \varnothing$ for graph G in Figure 8. Therefore this graph is not isomorphic to a reduction graph. From an algorithmic point of view, Theorem 7 is not very useful since it requires one to check for each $E \in \omega_G$, whether or not $E \in \theta_G$ (this is e.g. not trivial for the graph in Figure 3).

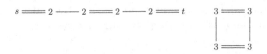

Fig. 8. An abstract reduction graph G for which $\theta_G = \varnothing$

6   Flip Edges

In this section and the next two we provide a characterization of the statement $\theta_G \neq \varnothing$. This allows, using Theorem 7, for an efficient algorithm that determines whether or not a given $G \in \mathcal{G}$ is isomorphic to a reduction graph. Moreover, it allows for an efficient algorithm that determines a legal string u for which $G \approx \mathcal{R}_u$.

Fig. 9. Flip operation for p. All vertices are labelled by p.

Let $G \in \mathcal{G}$. Then a merge-legal set for G is easily obtained as follows. Let $p \in \text{dom}(G)$ and let $\{v_1, v_2\}$ and $\{v_3, v_4\}$ be the two desire edges with vertices labelled by p. A merge-legal set for G must have either the edges $\{v_1, v_3\}$ and $\{v_2, v_4\}$ or the edges $\{v_1, v_4\}$ and $\{v_2, v_3\}$, see both sides in Figure 9. By assigning such edges for each $p \in \text{dom}(G)$ we obtain a merge-legal set for G. Thus, $\omega_G \neq \varnothing$ for each $G \in \mathcal{G}$. Note that in particular, if $\text{dom}(G) = \varnothing$, then $\omega_G = \{\varnothing\}$.

We now formally define a type of operation that in Figure 9 transforms the situation on the left-hand side to the situation on the right-hand side, and the other way around. Informally speaking it "flips" edges of merge-legal sets.

Definition 8. Let $G = B(E_1, E_2) \in \mathcal{G}$, let f be the vertex labeling function of G, and let $p \in \text{dom}(G)$. The *flip operation for p (w.r.t. G)*, denoted by $\text{flip}_{G,p}$, is the function $\omega_G \to \omega_G$ defined by:

$$\text{flip}_{G,p}(E) = \{\{v_1, v_2\} \in E \mid f(v_1) \neq p \neq f(v_2)\} \cup \{e_1, e_2\},$$

where e_1 and e_2 are the two edges with vertices labelled by p such that $e_1, e_2 \notin E_2 \cup E$.

When G is clear from the context, we also denote $\text{flip}_{G,p}$ by flip_p. By Figure 9, the flip operations are self-inverse, i.e. $\text{flip}_{G,p}^2$ is the identity function on ω_G.

Example 7. Consider Figure 5, and let G be the abstract reduction graph (ignoring the merge-legal edges) of this figure. If we apply $\text{flip}_{G,2}$ to the set of merge-legal edges depicted on the left-hand side of the figure, then we obtain the set of merge-legal edges depicted in on the right-hand side of the figure.

Let $G \in \mathcal{G}$, and let $D = \{p_1, \ldots, p_l\} \subseteq \text{dom}(G)$. Then we define $\text{flip}_D = \text{flip}_{p_l} \cdots \text{flip}_{p_1}$. Note that flip_D is well defined. Also, if $D_1, D_2 \subseteq \text{dom}(G)$ and $D_1 \neq D_2$, then $\text{flip}_{D_1}(E) \neq \text{flip}_{D_2}(E)$. Moreover, for each $E \in \omega_G$, $\omega_G = \{\text{flip}_D(E) \mid D \subseteq \text{dom}(G)\}$.

7   Merging and Splitting Connected Components

We now recall the definition of pointer-component graph, introduced in [7]. We generalize it here (trivially) for abstract reduction graphs in general. Surprisingly however, this graph has different uses in this paper compared to its original uses in [7], where it is used to characterize which string negative rules are used in sequences of string pointer rules that transforms u into the empty string.

First, we recall the notion of multigraph. A *multigraph* is a (undirected) graph $G = (V, E, \epsilon)$, where parallel edges are possible. Therefore, E is a finite set of edges and $\epsilon : E \rightarrow \{\{x, y\} \mid x, y \in V\}$ is the *endpoint mapping*. We allow $x = y$, and therefore edges can be of the form $\{x, x\} = \{x\}$ — an edge of this form should be seen as a "loop" for x. The *order* $|V|$ of G is denoted by $o(G)$. Again, multigraphs are considered *isomorphic* when they are equal modulo the identity of the vertices: multigraphs $G = (V, E, \epsilon)$ and $G' = (V', E, \epsilon')$ are *isomorphic*, denoted $G \approx G'$, if there is a bijection $\alpha : V \rightarrow V'$ such that $\alpha\epsilon = \epsilon'$, or more precisely, for $e \in E$, $\epsilon(e) = \{v_1, v_2\}$ implies $\epsilon'(e) = \{\alpha(v_1), \alpha(v_2)\}$.

Definition 9. Let $G \in \mathcal{G}$. The *pointer-component graph of* G, denoted by $\mathcal{PC}_G$, is a multigraph (ζ, E, ϵ), where ζ is the set of connected components of G, $E = \mathrm{dom}(G)$, and ϵ is, for $e \in E$, defined by $\epsilon(e) = \{C \in \zeta \mid C$ contains vertices labelled by $e\}$.

If $G = \mathcal{R}_u$ for some legal string u, then we also write $\mathcal{PC}_u = \mathcal{PC}_G$ and we say that $\mathcal{PC}_u$ is the *pointer-component graph of* u.

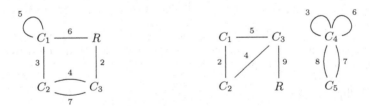

Fig. 10. The pointer-component graphs of the (abstract) reduction graphs from Figure 2 (left-hand side) and from Figure 3 (right-hand side)

Example 8. The pointer-component graphs of the (abstract) reduction graphs from Figure 2 and from Figure 3 are shown in Figure 10.

Let $G = B(E_1, E_2)$ be an abstract reduction graph and let $E \in \omega_G$. We consider the effect of the flip operation on the pointer-component graph defined on the abstract reduction graph $H = B(E_1, E)$. Note that when $G = B(E_1, E_2) \in \mathcal{G}$ and $E \in \omega_G$, then E is desirable for B, and thus $H = B(E_1, E) \in \mathcal{G}$. Therefore, e.g., $\mathcal{PC}_H$ is defined. We distinguish the pointers that form loops in the pointer-component graph: for $G \in \mathcal{G}$, $\mathrm{bridge}(G) = \{e \in E \mid |\epsilon(e)| = 2\}$ where $\mathcal{PC}_G = (V, E, \epsilon)$. E.g., by the right-hand side of Figure 10 we have $\mathrm{bridge}(G) = \mathrm{dom}(G) \backslash \{3, 6\}$ for G depicted in Figure 3.

Definition 10. For each edge p, the *p-merge rule*, denoted by merge_p, is a rule applicable to (defined on) multigraphs $G = (V, E, \epsilon)$ with $p \in \text{bridge}(G)$. It is defined by $\text{merge}_p(G) = (V', E, \epsilon')$, where $V' = (V \backslash \epsilon(p)) \cup \{v'\}$ with $v' \notin V$, and $\epsilon'(e) = \{h(v_1), h(v_2)\}$ iff $\epsilon(e) = \{v_1, v_2\}$ where $h(v) = v'$ if $v \in \epsilon(p)$, otherwise it is the identity.

It is easy to see that merge rules commute. We are now ready to state the following result. The (proof of the) result shows that depending on whether or not p is a loop, the flip operation on p can merge connected components or modify/split a connected component.

Theorem 11. *Let $G = B(E_1, E_2) \in \mathcal{G}$, let $E \in \omega_G$, let $H = B(E_1, E)$, and let, for $p \in \text{dom}(G)$, $H_p = B(E_1, \text{flip}_p(E))$.*

- *If $p \in \text{bridge}(H)$, then $\mathcal{PC}_{H_p} \approx \text{merge}_p(\mathcal{PC}_H)$ (and therefore $o(\mathcal{PC}_{H_p}) = o(\mathcal{PC}_H) - 1$).*
- *If $p \in \text{dom}(H) \backslash \text{bridge}(H)$, then $o(\mathcal{PC}_H) \leq o(\mathcal{PC}_{H_p}) \leq o(\mathcal{PC}_H) + 1$.*

Example 9. Consider again Figure 5, and let $G = B(E_1, E_2)$ be the abstract reduction graph (ignoring the merge-legal edges) of this figure. If we let $E \in \omega_G$ be the set of merge-legal edges depicted on the left-hand side of the figure, then $2 \in \text{bridge}(H)$ with $H = B(E_1, E)$. Therefore, by Theorem 11 and the fact that G has exactly two connected components, $H_2 = B(E_1, \text{flip}_2(E))$ is a connected graph. Indeed, this is clear from the right-hand side of the figure (by ignoring the edges from E_2).

8   Connectedness of Pointer-Component Graphs

We now characterize the requirement $\theta_G \neq \varnothing$ found in Theorem 7. The proof of this result depends heavily on Theorem 11.

Theorem 12. *Let $G \in \mathcal{G}$. Then $\mathcal{PC}_G$ is a connected graph iff $\theta_G \neq \varnothing$.*

Example 10. By Theorem 12 and the left-hand side of Figure 10, for the (abstract) reduction graph G_1 in Figure 2 we have $\theta_{G_1} \neq \varnothing$. By the right-hand side of Figure 10, for the abstract reduction graph G_2 in Figure 3 we have $\theta_{G_2} = \varnothing$.

By Theorem 12 and Theorem 7 we obtain the first main result of this paper. It shows that one needs to check only a few computationally easy conditions to determine whether or not a 2-edge coloured graph is (isomorphic to) a reduction graph. Surprisingly, the "high-level" notion of pointer-component graph is crucial in this characterization.

Theorem 13. *Let G be a 2-edge coloured graph. Then G isomorphic to a reduction graph iff $G \in \mathcal{G}$ and $\mathcal{PC}_G$ is a connected graph.*

Not only is it computationally efficient to determine whether or not a 2-edge coloured graph G is isomorphic to a reduction graph, but, when this is the case, then it is also computationally easy to determine a legal string u for which $G \approx \mathcal{R}_u$. Indeed, we can determine such a u from $G = B(E_1, E_2)$ as follows:

1. Determine an $E \in \omega_G$. As we have seen, such an E is easily obtained.
2. Compute $\mathcal{PC}_H$ with $H = B(E_1, E)$, and determine a set of edges D such that $\mathcal{PC}_H|_D$ is a tree, where $\mathcal{PC}_H|_D$ is $\mathcal{PC}_H$ restricted to the set of edges D,
3. Compute $G' = B(E_1, E_2, \mathrm{flip}_D(E))$, and determine a $u \in L_{G'}$.

As a consequence, every connected multigraph $G = (V, E, \epsilon)$ with $E \subseteq \Delta$ is isomorphic to a pointer-component graph of a legal string. Thus pointer-component graphs of legal strings can, surprisingly, take all imaginable forms.

9   Dual String Rules

If we let $\mathcal{R}$ be the function that assigns to each legal string u its reduction graph $\mathcal{R}_u$, then Theorem 13 characterizes the range of $\mathcal{R}$. Here we characterize the fibers $\mathcal{R}^{-1}(\mathcal{R}_u)$ modulo graph isomorphism, i.e., we characterize when two legal strings have isomorphic reduction graphs.

First we define the dual string rules. These rules turn out to characterize the effect of flip operations on the underlying legal string, cf. Theorem 14 below. For all $p, q \in \Pi$ with $\mathbf{p} \neq \mathbf{q}$ we define

- the *dual string positive rule* for p is defined by $\mathbf{dspr}_p(u_1 p u_2 p u_3) = u_1 p \bar{u}_2 p u_3$,
- the *dual string double rule* for p, q is defined by $\mathbf{dsdr}_{p,q}(u_1 p u_2 q u_3 p u_4 \bar{q} u_5) = u_1 p u_4 q u_3 \bar{p} u_2 \bar{q} u_5$,

where $u_1, u_2, \ldots, u_5$ are arbitrary (possibly empty) strings over Π. Notice that the dual string rules are self-inverse. Also notice the strong similarities between these rules and the string pointer rules, given, e.g., in [10,7].

Let u and v be legal strings. We say that u and v are *dual*, denoted by $\approx_d$ if there is a (possibly empty) sequence φ of dual string rules applicable to u such that $\varphi(u) \approx v$. Notice that $\approx_d$ is an equivalence relation.

We define $\mathrm{dom}(\mathbf{dspr}_p) = \{\mathbf{p}\}$ and $\mathrm{dom}(\mathbf{dsdr}_{p,q}) = \{\mathbf{p}, \mathbf{q}\}$ for $p, q \in \Pi$. For a composition $\varphi = \rho_n \cdots \rho_2 \rho_1$ of dual string rules we define $\mathrm{dom}(\varphi) = \cup_i \mathrm{dom}(\rho_i)$. We call φ *reduced* if $\mathrm{dom}(\rho_i) \cap \mathrm{dom}(\rho_j) = \varnothing$ for all $1 \leq i < j \leq n$.

Let $G = B(E_1, E_2, E_3)$ be an extended abstract reduction graph, and let $D \subseteq \mathrm{dom}(G)$. Then we define $\mathrm{flip}_D(G) = B(E_1, E_2, \mathrm{flip}_{G',D}(E_3))$, where $G' = B(E_1, E_2)$.

Theorem 14. *Let u be a legal string, and let $D \subseteq \mathrm{dom}(u)$. There is a reduced sequence φ of dual string rules applicable to u such that $\mathrm{dom}(\varphi) = D$ iff $\mathrm{flip}_D(M_u) \in \theta_u$. In this case, $\mathcal{E}_{\varphi(u)} \approx \mathrm{flip}_D(\mathcal{E}_u)$, and consequently $\mathcal{R}_{\varphi(u)} \approx \mathcal{R}_u$.*

Using the previous result, we are ready to show the second (and final) main result of this paper. It shows that $\mathcal{R}^{-1}(\mathcal{R}_u)$ (modulo graph isomorphism) for legal string u is the "orbit" of u under the dual string rules.

Theorem 15. *Let u and v be legal strings. Then $u \approx_d v$ iff $\mathcal{R}_u \approx \mathcal{R}_v$.*

10 Discussion

This paper characterizes the range of $\mathcal{R}$ (Theorem 13) and each fiber $\mathcal{R}^{-1}(\mathcal{R}_u)$ modulo graph isomorphism (Theorem 15).

The first characterization corresponds to a computationally efficient algorithm that determines whether or not a graph G is isomorphic to a reduction graph. It turns out that once G satisfies the "look and feel" of reduction graphs, e.g., each vertex label should occur exactly four times, then reduction graphs are characterized as having a connected pointer-component graph. Moreover, if G is isomorphic to a reduction graph, then the algorithm given below Theorem 13 allows for an efficient determination of a legal string u such that $G \approx \mathcal{R}_u$.

The second characterization determines, given u, the whole set $\mathcal{R}^{-1}(\mathcal{R}_u)$ modulo graph isomorphism. From a biological point of view, the fibers characterize which micronuclear genes obtain the same macronuclear structure. It turns out that $\mathcal{R}^{-1}(\mathcal{R}_u)$ is the orbit of u under the dual string rules. Surprisingly, these two types of string rewriting rules are very similar to the string positive rules and the string double rules that are used to define the model.

References

1. Cavalcanti, A., Clarke, T., Landweber, L.: MDS_IES_DB: a database of macronuclear and micronuclear genes in spirotrichous ciliates. Nucleic Acids Res. 33, D396–D398 (2005)
2. Brijder, R., Hoogeboom, H., Rozenberg, G.: Reducibility of gene patterns in ciliates using the breakpoint graph. Theor. Comput. Sci. 356, 26–45 (2006)
3. Brijder, R., Hoogeboom, H., Rozenberg, G.: The breakpoint graph in ciliates. In: Berthold, M.R., Glen, R.C., Diederichs, K., Kohlbacher, O., Fischer, I. (eds.) CompLife 2005. LNCS (LNBI), vol. 3695, pp. 128–139. Springer, Heidelberg (2005)
4. Hannenhalli, S., Pevzner, P.: Transforming cabbage into turnip: Polynomial algorithm for sorting signed permutations by reversals. J. ACM 46, 1–27 (1999)
5. Bergeron, A., Mixtacki, J., Stoye, J.: On sorting by translocations. In: Miyano, S., Mesirov, J., Kasif, S., Istrail, S., Pevzner, P., Waterman, M. (eds.) Research in Computational Molecular Biology. LNCS (LNBI), vol. 3500, pp. 615–629. Springer, Heidelberg (2005)
6. Setubal, J., Meidanis, J.: Introduction to Computional Molecular Biology. PWS Publishing Company, London (1997)
7. Brijder, R., Hoogeboom, H., Muskulus, M.: Strategies of loop recombination in ciliates. LIACS Technical Report 2006-01, [arXiv:cs.LO/0601135] (2006)
8. Brijder, R., Hoogeboom, H., Muskulus, M.: Applicability of loop recombination in ciliates using the breakpoint graph. In: Berthold, M.R., Glen, R.C., Fischer, I. (eds.) CompLife 2006. LNCS (LNBI), vol. 4216, pp. 97–106. Springer, Heidelberg (2006)
9. Brijder, R., Hoogeboom, H.: The fibers and range of reduction graphs in ciliates. LIACS Technical Report 2007-01, [arXiv:cs.LO/0702041] (2007)
10. Ehrenfeucht, A., Harju, T., Petre, I., Prescott, D., Rozenberg, G.: Computation in Living Cells – Gene Assembly in Ciliates. Springer Verlag, Heidelberg (2004)
11. Pevzner, P.: Computational Molecular Biology: An Algorithmic Approach. MIT Press, Cambridge (2000)

2-Visibly Pushdown Automata*

Dario Carotenuto, Aniello Murano, and Adriano Peron

Università degli Studi di Napoli "Federico II", Via Cinthia, I-80126 Napoli, Italy

Abstract. Visibly Pushdown Automata (*VPA*) are a special case of
pushdown machines where the stack operations are driven by the input. In this paper, we consider *VPA with two stacks*, namely *2-VPA*.
These automata introduce a useful model to effectively describe concurrent pushdown systems using a simple communication mechanism
between stacks. We show that 2-VPA are strictly more expressive than
VPA. Indeed, 2-VPA accept some context-sensitive languages that are
not context-free and some context-free languages that are not accepted
by any VPA. Nevertheless, the class of languages accepted by 2-VPA is
closed under all boolean operations and determinizable in ExpTime, but
does not preserve decidability of emptiness problem. By adding an ordering constraint on stacks (2-OVPA), decidability of emptiness can be
recovered (preserving desirable closure properties) and solved in PTime.
Using these properties along with the automata-theoretic approach, we
prove that the model checking problem over 2-OVPA models against
2-OVPA specifications is ExpTime-complete.

1 Introduction

In the area of formal design verification, one of the most significant developments
has been the discovery of the *model checking* technique, that automatically allows
to verify on-going behaviors of reactive systems ([4, 9, 12]). In this verification
method (for a survey see [5]), one checks the correctness of a system with respect
to a desired behavior by checking whether a mathematical model of the system
satisfies a formal specification of this behavior.

Traditionally, model checking is applied to finite-state systems, typically modeled by labeled state-transition graphs. Recently, model checking has been extended to infinite-state sequential systems (e.g., see [13, 2]). These are systems in
which each state carries a finite, but unbounded, amount of information, e.g., a
pushdown store. *Pushdown automata* (PDA) naturally model the control flow of
sequential programs with nested and recursive procedure calls. Therefore, PDA
are the proper model to tackle with program analysis, compiler optimization,
and model checking questions that can be formulated as decision problems for
PDA. While many analysis problems, such as identifying dead code and accesses
to uninitialized variables, can be captured as regular requirements, many others
require inspection of the stack or matching of calls and returns, and are non-
regular context-free. More examples of useful non-regular properties are given

---

* Work partially supported by MIUR FIRB Project no. RBAU1P5SS.

T. Harju, J. Karhumäki, and A. Lepistö (Eds.): DLT 2007, LNCS 4588, pp. 132–144, 2007.
© Springer-Verlag Berlin Heidelberg 2007

in [10], where the specification of unbounded message buffers is considered. Since checking context-free properties on PDA is proved in general to be undecidable [7], weaker models have been proposed to decide different kinds of non-regular properties. One of the most promising approaches is that of *Visibly Pushdown Automata* (VPA) [1]. These are PDA where the push or pop actions on the stack are controlled externally by the input alphabet. Such a restriction on the use of the stack allows to enjoy all desirable closure properties and tractable decision problems, though retaining an expressiveness adequate to formulate program analysis questions (as summarized in Figure 1). Therefore, checking pushdown properties of pushdown models is feasible as long as the calls and returns are made visible. This visibility requirement seems quite natural while writing requirements about pre/post conditions or for inter-procedural flow properties. In particular, requirements that can be verified in this manner include all regular properties, and non-regular properties such as: partial correctness (if P holds when a procedure is invoked, then, if the procedure returns, P' holds upon return), total correctness (if P holds when a procedure is invoked, then the procedure must return and P' must hold at the return state), local properties (the computation within a procedure by skipping over calls to other procedures satisfies a regular property, for instance, every request is followed by a response), access control (a procedure A can be invoked only if another procedure B is in the current stack), and stack limits (whenever the stack size is bounded by a given constant, a property A holds). Unfortunately, some natural context-free properties like "the number of calls to procedures A and B is the same" cannot be captured by any VPA [1]. Moreover, VPA cannot explicitly represent concurrency: for instance, properties of two threads running in parallel, each one exploiting its own pushdown store.

  In this paper, we propose an extension of VPA in order to enrich with further expressiveness the model though maintaining some desirable closure properties and decidability results. We first consider VPA with an additional, input driven, pushdown store and we call the proposed model *2-Visibly Pushdown Automaton* (2-VPA). As in the VPA case, 2-VPA input symbols are partitioned in subclasses, each one triggering a transition belonging to a specific class, i.e., push/pop/local transition, which also selects the operating stack, i.e., the first or the second or both. Moreover, visibility in 2-VPA affects the transfer of information from one stack to the other. 2-VPA turn out to be strictly more expressive than VPA and they also accept some context-sensitive languages that are not context-free. Unfortunately, this extension does not preserve decidability of the emptiness problem as we prove by a reduction from the halting problem over Minsky Machines. In the automata-theoretic approach, to gain with a decidable model checking procedure, decidability of the emptiness problem is crucial. For this reason, we add to 2-VPA a suitable restriction on stack operations, namely we consider 2-VPA in which pop operations on the second stack are allowed only if the first stack is empty. We call such a variant *ordered* 2-VPA (2-OVPA). The ordering constraint is inspired from the class of *multi-pushdown automata* (MPDA), defined in [3]. These are pushdown automata exploiting an ordered collection of arbitrary number of pushdown stores in which a pop action on the i-th stack can occur only if all previous stacks are empty. In [3], it has been shown that the class of languages accepted by MPDA is strictly included into context-sensitive languages, it has the emptiness problem decidable, it is closed under union, but not under intersection and complement.

From an expressive point of view, 2-OVPA are a proper subclass of MPDA with two stacks (PD^2). Differently from PD^2, exploiting visibility allows to recover in 2-OVPA closure under intersection and complement thus allowing to face the model checking problem following the automata-theoretic approach. In such an approach, to verify whether a system, modeled as a 2-OVPA S, satisfies a correctness requirement expressed by a 2-OVPA P, we check for emptiness the intersection between the language accepted by S and the complement of the language accepted by P (i.e., $L(S) \cap \overline{L(P)} = \emptyset$). Since we prove for 2-OVPA that intersection and emptiness can be performed in polynomial time while complementation in exponential time, and since inclusion for VPA is ExpTime-complete [1], we get that model checking an 2-OVPA model against an 2-OVPA specification is ExpTime-complete. This is notable since checking context-free properties on PDA is proved to be undecidable [7], as well as model checking multi-pushdown properties over MPDA.

The extension we propose for VPA does not only affect expressiveness, but also gives us a way to naturally describe distributed pushdown systems behavior. In fact, we show that 2-OVPA capture the behavior of systems built on pairs of VPA running in a suitable synchronous way according to a distributed computing paradigm. To this purpose, we introduce a composition operator on VPA parameterized on a communication interface. Given a pair of VPA, this operator allows to build a *Synchronized System* of VPA (S-VPA), which behaves synchronously and in parallel. A communication between two synchronous VPA consists in a transfer of information from the top of the stack of one VPA to the top of stack of the other. If we interpret each one of the involved VPA as a process with its pushdown store (containing activation records of procedure calls, for instance), the enforced communication form can be seen as a *Remote Procedure Call* [11], widely exploited in the client-server paradigm of distributed computing. In our case, ordering of VPA modules can be interpreted as follows: we can see the former one acting as a client and the latter as a server. The client can always demand to the server the execution of a task and the server can return a result to the client whenever this is available (its stack is empty). The properties of languages accepted by 2-VPA and 2-OVPA we obtain along the paper are summarized in Figure 1. Due to page limitations, proofs are omitted and reported in the extended version[1].

2   Preliminaries

Let Σ be a finite alphabet partitioned into three pairwise disjoint sets Σ_c, Σ_r, and Σ_l standing respectively for *call*, *return*, and *local* alphabets. We denote

---

[1] http://people.na.infn.it/∼carotenuto/research/2vpaTechRep.pdf.

Languages	Closure Properties			Decision problems	
	$\cup$	$\cap$	Complement	Emptiness	Inclusion
Regular	Yes	Yes	Yes	NLOGSPACE	PSPACE
CFL	Yes	No	No	PTIME	Undecidable
VPL	Yes	Yes	Yes	PTIME	EXPTIME
$\mathcal{L}_{PD^2}$	Yes	No	No	PTIME	Undecidable
2-VPL	**Yes**	**Yes**	**Yes**	**Undecidable**	**Undecidable**
2-OVPL	**Yes**	**Yes**	**Yes**	**Ptime**	**ExpTime**

Fig. 1. A comparison between closure properties and decision problems

the tuple $\widetilde{\Sigma} = \langle \Sigma_c, \Sigma_r, \Sigma_l \rangle$ a *visibly pushdown alphabet*. A *(nondeterministic) visibly pushdown automaton* (*VPA*) on finite words over $\widetilde{\Sigma}$ [1] is a tuple $M = (Q, Q_{in}, \Gamma, \bot, \delta, Q_F)$, where Q, Q_{in}, Q_F, and Γ are respectively finite sets of *states, initial states, final states,* and *stack symbols*; $\bot \notin \Gamma$ is the stack *bottom* symbol and we use $\Gamma_\bot$ to denote $\Gamma \cup \{\bot\}$; and $\delta \subseteq \delta_c \cup \delta_r \cup \delta_l$, is the *transition relation* where $\delta_c = Q \times \Sigma_c \times Q \times \Gamma$, $\delta_r = Q \times \Sigma_r \times \Gamma_\bot \times Q$, and $\delta_l = Q \times \Sigma_l \times Q$. We call $(q, a, q', \gamma) \in \delta_c$ a *push transition*, where on reading a the symbol γ is pushed onto the stack and the control state changes to q'; $(q, a, \gamma, q') \in \delta_r$ a *pop transition*, where γ is popped from the stack leading to the control state q'; and $(q, a, q') \in \delta_l$ a *local transition*, where the automaton on reading a only changes its control to q'. A *configuration* for a VPA M is a pair $(q, \sigma) \in Q \times (\Gamma^*.\bot)$ where σ is the stack content. A *run* $\rho = (q_0, \sigma_0) \ldots (q_k, \sigma_k)$ of M on a word $w = a_1 \ldots a_k$ is a sequence of configurations such that $q_0 \in Q_{in}$, $\sigma_0 = \bot$, and for every $i \in \{0, \ldots, k\}$, one of the following holds: [**Push**]: $(q_i, a_i, q_{i+1}, \gamma) \in \delta_c$, and $\sigma_{i+1} = \gamma.\sigma_i$; [**Pop**]: $(q_i, a_i, \gamma, q_{i+1}) \in \delta_r$, and either $\gamma \in \Gamma$ and $\sigma_i = \gamma.\sigma_{i+1}$, or $\gamma = \sigma_i = \sigma_{i+1} = \bot$; or [**Local**]: $(q_i, a_i, q_{i+1}) \in \delta_l$ and $\sigma_{i+1} = \sigma_i$.

A run is *accepting* if its last configuration contains a final state. The language accepted by a VPA M is the set of all words w with an accepting run of M on w, say it *L(M)*. A language of finite words $L \subseteq \Sigma^*$ is a *visibly pushdown language* (*VPL*) with respect to a pushdown alphabet $\widetilde{\Sigma}$, if there is a VPA M such that $L = L(M)$. VPLs are a subclass of deterministic context-free languages, a superclass of regular languages, and are closed under intersection, union, complementation, concatenation, and Kleene-*. Furthermore, the emptiness problem for a VPA M, i.e., deciding whether $L(M) \neq \emptyset$, is decidable with time complexity $O(n^3)$, where n is the number of states in M.

In the literature, different extensions of classical pushdown automata with multiple stacks have been considered. Here, we recall *multiple-pushdown automata* as they were introduced in [3]. These machines are pushdown automata endowed with an ordered set of an arbitrary number of stacks and the constraint that pop operations occur sequentially and only operate on the first non-empty stack. Thus, push operations are never constrained and they can be performed independently on every stack. The formal definition follows.

A multi-pushdown automaton with $n \geq 1$ stacks (PD^n, for short) is a tuple $M = (\Sigma, Q, Q_{in}, \Gamma, Z_0, \delta, Q_F)$, where Σ, Q, Q_{in}, Γ, and Q_F are respectively finite sets of input symbols, states, initial states, stack symbols, and final states, $Z_0 \notin \Gamma$ is the *bottom stack symbol* and used to identify the initial non-empty stack, and δ is the transition relation defined as a partial function from $Q \times (\Sigma \cup \{\varepsilon\}) \times \Gamma$ to $2^{Q \times (\Gamma^*)^n}$. If $(q', \alpha_1, \ldots, \alpha_n) \in \delta(q, a, \gamma)$, on reading a the automaton changes its control state from q to q', the stack symbol $\gamma \in \Gamma$ is popped from the first non-empty stack, and for each i in $\{1, \ldots, n\}$, and $\alpha_i \in \Gamma^*$ is pushed on the i-th stack. A configuration of M is a $n+2$-tuple $\langle q, x; \sigma^0, \ldots, \sigma^n \rangle$, where $q \in Q$, $x \in \Sigma^*$, $\sigma^0, \ldots, \sigma^n \in \Gamma^*$, and σ^i is the content of the i-th stack. The above configuration is initial if $q = q_0$, $\sigma^0 = Z_0$, and all other stacks are empty, and it is final if $q \in F$. The transition relation $\vdash_M$ over configurations is defined in the following way: $\langle q, ax; \varepsilon, \ldots, \varepsilon, \gamma.\gamma_i, \ldots, \gamma_n \rangle \vdash_M \langle q', x; \alpha_1, \ldots, \alpha_{i-1}, \alpha_i \gamma_i, \ldots, \alpha_n \gamma_n \rangle$ if $(q', \alpha_1, \ldots, \alpha_n) \in \delta(q, a, \gamma)$. A word w is accepted by a PD^n M iff $\langle q, w; Z_0, \varepsilon \ldots, \varepsilon \rangle \vdash_M^* \langle q_F, \varepsilon; \gamma_1, \ldots, \gamma_n \rangle$, where $\vdash_M^*$ is the Kleene-closure of $\vdash_M$ and $q_F \in Q_F$. The language of a PD^n M is the set of words accepted by M. We denote the class of languages accepted by PD^n as $\mathcal{L}_{PD^n}$. The following theorem summarizes the main results about PD^n.

Theorem 1 ([3]). *For every $n \geq 1$, we have that $\mathcal{L}_{PD^n}$ subsumes CFLs, it is strictly included in CSLs as well as in $\mathcal{L}_{PD^{n+1}}$. It is closed under union, concatenation and Kleene-*. Moreover, it has a decidable emptiness problem and solvable in $O(|Q|^3)$, where $|Q|$ is the number of states of the automaton.*

3   Visibly Pushdown Automata with Two Stacks

A *2-pushdown alphabet* is a pair of pushdown alphabets $\widetilde{\Sigma} = \langle \widetilde{\Sigma}^0, \widetilde{\Sigma}^1 \rangle$, where $\widetilde{\Sigma}^0 = \langle \Sigma_c^0, \Sigma_r^0, \Sigma_l^0 \rangle$ and $\widetilde{\Sigma}^1 = \langle \Sigma_c^1, \Sigma_r^1, \Sigma_l^1 \rangle$ are a possibly different partitioning of the same input alphabet Σ. The intuition is that the $\widetilde{\Sigma}^0$ drives the operations over the first stack and $\widetilde{\Sigma}^1$ those over the second. Symbols in $\widetilde{\Sigma}$ belonging to call, return or local partitions of both $\widetilde{\Sigma}^0$ and $\widetilde{\Sigma}^1$ are simply denoted by $\Sigma_c, \Sigma_r, \Sigma_l$, respectively. Furthermore, input symbols that drive a call operation on the first (resp., second) stack and a return on the second (resp., first) stack are called *synchronized communication* symbols and formally denoted as $\Sigma_{s_1} = \Sigma_c^0 \cap \Sigma_r^1$ (resp., $\Sigma_{s_0} = \Sigma_r^0 \cap \Sigma_c^1$). Finally, we denote with Σ_{c_i} (resp., Σ_{r_i}) the set of call (resp., return) symbols for the stack i and local for the other, with $i = 0, 1$. In the following, we use $\widetilde{\Sigma}$ to denote both a 2-pushdown alphabet and a (1-)pushdown alphabet, when the meaning is clear from the context.

Definition 1 (2-Visibly Pushdown Automaton). *A (nondeterministic) 2-Visibly Pushdown Automaton (2-VPA) on finite words over a 2-pushdown alphabet $\widetilde{\Sigma}$ is a tuple $M = (Q, Q_{in}, \Gamma, \bot, \delta, Q_F)$, where Q, Q_{in}, Q_F, and Γ are respectively finite sets of states, initial states, final states and stack symbols, $\bot \notin \Gamma$ is the stack bottom symbol (with $\Gamma_\bot$ used to denote $\Gamma \cup \{\bot\}$), and δ is the transition relation defined as the union of the following sets, for $i \in \{0, 1\}$:*

- $\delta_{c_i} \subseteq (Q \times \Sigma_{c_i} \times Q \times \Gamma)$,
- $\delta_c \subseteq (Q \times \Sigma_c \times Q \times \Gamma \times \Gamma)$,
- $\delta_{s_i} \subseteq (Q \times \Sigma_{s_i} \times \Gamma_\bot \times Q \times \Gamma)$,
- $\delta_{r_i} \subseteq (Q \times \Sigma_{r_i} \times \Gamma_\bot \times Q)$,
- $\delta_r \subseteq (Q \times \Sigma_r \times \Gamma_\bot \times \Gamma_\bot \times Q)$,
- $\delta_l \subseteq Q \times \Sigma_l \times Q$.

We say that M is deterministic if Q_{in} is a singleton, and for every $q \in Q$, $a \in \Sigma$, and $\gamma_r, \gamma_r' \in \Gamma_\perp$, there is at most one transition of the form (q, a, q'), (q, a, q', γ), $(q, a, q', \gamma, \gamma')$, (q, a, γ_r, q'), $(q, a, \gamma_r, \gamma_r', q')$, or $(q, a, \gamma_r, q', \gamma')$ belonging to δ.

Transitions in δ_l, δ_{c_i}, and δ_{r_i} extend VPA's local, call, and return transitions to deal with two stacks, in a natural way. We call $(q, a, q', \gamma, \gamma') \in \delta_c$ a *double-call* transition where on reading a the automaton changes its control state from q to q', and the symbols γ and γ' are pushed on the first and second stack, respectively; we call $(q, a, \gamma, \gamma', q') \in \delta_r$ a *double-pop* transition where on reading a the automaton changes its control state from q to q', and the symbols γ and γ' are popped from the first and second stack, respectively; finally, we call $(q, a, \gamma, q', \gamma') \in \delta_{s_i}$, with $i \in \{0, 1\}$, a *synchronous (communication)* transition between stacks, where on reading a the automaton changes its control state from q to q' and the symbol γ is popped from the stack i and γ' pushed on the other.

A *configuration* of a 2-VPA M is a triple (q, σ^0, σ^1) where $q \in Q$ and $\sigma^0, \sigma^1 \in \Gamma^*.\perp$. For an input word $w = a_1 \ldots a_k \in \Sigma^*$, a run of M on w is a sequence $\rho = (q_0, \sigma_0^0, \sigma_0^1) \ldots (q_k, \sigma_k^0, \sigma_k^1)$ where $q_0 \in Q_{in}$, $\sigma_0^0 = \sigma_0^1 = \perp$, and for all $i \in \{0, \ldots, k-1\}$, there are $j, j' \in \{0, 1\}$, $j \neq j'$, such that one of the following holds:

Push: $(q_i, a_i, q_{i+1}, \gamma) \in \delta_{c_j}$, then $\sigma_{i+1}^j = \gamma.\sigma_i^j$ and $\sigma_{i+1}^{j'} = \sigma_i^{j'}$;

2Push: $(q_i, a_i, q_{i+1}, \gamma, \gamma') \in \delta_c$ then $\sigma_{i+1}^j = \gamma.\sigma_i^j$ and $\sigma_{i+1}^{j'} = \gamma'.\sigma_i^{j'}$;

Pop: $(q_i, a_i, \gamma, q_{i+1}) \in \delta_{r_j}$, then either $\gamma = \sigma_i^j = \sigma_{i+1}^j = \perp$, or $\gamma \neq \perp$ and $\sigma_i^j = \gamma.\sigma_{i+1}^j$. In both cases $\sigma_{i+1}^{j'} = \sigma_i^{j'}$;

2Pop: $(q_i, a_i, \gamma_0, \gamma_1, q_{i+1}) \in \delta_r$ then, for $k \in \{0, 1\}$, either $\gamma_k = \sigma_i^k = \sigma_{i+1}^k = \perp$, or $\gamma_k \neq \perp$ and $\sigma_i^k = \gamma.\sigma_{i+1}^k$;

Local: $(q_i, a_i, q_{i+1}) \in \delta_l$ then $\sigma_{i+1}^0 = \sigma_i^0$ and $\sigma_{i+1}^1 = \sigma_i^1$;

Synch: $(q_i, a_i, \gamma, q_{i+1}, \hat{\gamma}) \in \delta_{s_j}$ then either $\gamma = \sigma_i^j = \sigma_{i+1}^j = \perp$, or $\gamma \neq \perp$ and $\sigma_i^j = \gamma.\sigma_{i+1}^j$. In both cases $\sigma_{i+1}^{j'} = \hat{\gamma}.\sigma_i^{j'}$.

From the above definition, we notice that communication between stacks is only allowed by applying a *synch.* transition. For a configuration c, we write $c \vdash_M c'$ meaning that c' is obtained from c by applying one of the rules above. We omit M when it is clear from the context. A run ρ is *accepting* when it ends with a configuration containing a final state. A word w is *accepted* if there is an accepting run ρ of M on w. The language accepted by M, denoted by $L(M)$, is the set of all words accepted by M. A language $L \subseteq \Sigma^*$ is a 2-VPL with respect to $\widetilde{\Sigma}$ if there is a 2-VPA M over $\widetilde{\Sigma}$ such that $L(M) = L$.

Theorem 2. *The emptiness problem for 2-VPA is undecidable.*

Proof. [sketch] We prove the result by showing a reduction from the halting problem of two counters Minsky machines. A Minsky machine with two counters C_0 and C_1 is a finite sequence $M = (L_1 : I_1; L_2 : I_2; \ldots; L_n : \text{halt})$ where $n \geq 1$, $L_1, \ldots, L_n$ are pairwise different instruction labels, and $I_1, \ldots, I_n$ are instructions of type *increment*, i.e., $C_m := C_m + 1$; goto L_j, or of type *test and decrement*, i.e., if $C_m = 0$ then goto L_j else $C_m := C_m - 1$; goto L_k, where $0 \leq m \leq 1$ and $1 \leq j, k \leq n$. A configuration of M is a triple (L_i, v_0, v_1) where L_i is an instruction label, and $v_0, v_1 \in \mathbb{N}$ represent the values of the counters C_0 and C_1, respectively. Let *Conf* be the set of all configurations of M, the transition relation $\hookrightarrow \subseteq Conf \times Conf$ between configurations is defined in an obvious way, and $\hookrightarrow^*$ is the transitive and reflexive closure of $\hookrightarrow$. If $(L_1, 0, 0)$ $\hookrightarrow \ldots \hookrightarrow (L_j, v_j^0, v_j^1)$ holds for a Minsky machine M, we say that $(L_1, 0, 0) \ldots$ (L_j, v_j^0, v_j^1) is an execution trace for M. The halting problem for M is to decide whether there exist $v_0, v_1 \in \mathbb{N}$ such that $(L_1, 0, 0) \hookrightarrow^* (L_n, v_0, v_1)$. This problem is known to be undecidable [8].

We now prove that given a two counters Minsky machine M there exists a 2-VPA M' over $\widetilde{\Sigma}$ such that $L(M') \neq \emptyset$ iff M eventually halts. Let $M = (L_1 : I_1; L_2 : I_2; \ldots; L_n : \texttt{halt})$, we define $M' = (Q, Q_{in}, \Gamma, \bot, \delta, Q_F)$ such that $Q = \{L_1, \ldots, L_n\}$, $Q_{in} = \{L_1\}$, $\Gamma = \{A\}$, where A does not appear in M, $Q_F = \{L_n\}$, and $\widetilde{\Sigma}$ is the partitioned set of all instructions I_i, with $i = 1, \ldots, n$, such that $I_i \in \Sigma_{c_0}$ (resp., $I_i \in \Sigma_{c_1}$) if I_i is an increment instruction of the counter C_0 (resp., C_1), or $I_i \in \Sigma_{r_0}$ (resp., $I_i \in \Sigma_{r_1}$) if I_i is a test and decrement instruction over the counter C_0 (resp., C_1). Finally, δ is defined as follows: if I_i is an increment instruction such as $C_m := C_m + 1$; goto L_j, with $m \in \{0, 1\}$, then $(L_i, I_i, L_j, A) \in \delta_{c_m}$; otherwise, if I_i is a test and decrement instruction such as if $C_m = 0$ then goto L_j else $C_m := C_m - 1$; goto L_k, with $m \in \{0, 1\}$ then $(L_i, I_i, \bot, L_j), (L_i, I_i, A, L_k) \in \delta_{r_m}$. It remains to prove that M halts iff M' accepts a word. It is easy to show by induction the following assertion.

Given a sequence of numbers $s = s_1 s_2 \ldots s_k$, with $s_i \in \{1, \ldots, n\}$ for all $i \in \{1, \ldots, k\}$, the sequence $(L_{s_1}, v_{s_1}^0, v_{s_1}^1) \ldots (L_{s_k}, v_{s_k}^0, v_{s_k}^1)$ of elements from $\{L_1, \ldots L_n\} \times \mathbb{N} \times \mathbb{N}$ is an execution trace of M if and only if the sequence $(L_{s_1}, \sigma_{s_1}^0, \sigma_{s_1}^1) \ldots (L_{s_k}, \sigma_{s_k}^0, \sigma_{s_k}^1)$ of elements from $Q \times \Gamma^*.\bot \times \Gamma^*.\bot$ is a run of M', with $|\sigma_{s_i}^j| = v_{s_i}^j + 1$ for each $i \in \{1, \ldots, k\}$ and $j \in 0, 1$.

The above assertion implies that M halts iff M' accepts a word. $\qquad\square$

It is interesting to notice that the reduction we consider in the proof of Theorem 2 also applies to the restricted model of VPA with 2 stacks where operations acting simultaneously on both stacks are avoided. This follows from the fact that two counters Minsky machine instructions only involves one counter at a time, and the sets Σ_c, Σ_r and Σ_{s_i}, with $i \in \{0, 1\}$, are empty.

4   Ordered Visibly Pushdown Automata with Two Stacks

In this section, we consider the subclass of 2-VPA which enforces the ordering constraints on using pushdown stores as defined for MPDA. In more detail, we consider a class of *ordered 2-VPA* (*2-OVPA*) as the class of 2-VPA in which a pop operation on the second stack can occur only if the first stack is empty. Thus, in such a model simultaneous pop operations are not allowed. The formal definition of 2-OVPA follows.

Definition 2. *A 2-OVPA M over $\widetilde{\Sigma}$ is a 2-VPA such that Σ_r is empty and for all input word $w = a_1 \ldots a_k \in \Sigma^*$ and run $\rho = (q_0, \sigma_0^0, \sigma_0^1) \ldots (q, \sigma_k^0, \sigma_k^1)$ of M over w, for all $i \in \{1, \ldots, n\}$, the following hold:*

Pop: $(q_i, a_i, \gamma, q_{i+1}) \in \delta_{r_1}$ *then* $\sigma_i^0 = \sigma_{i+1}^0 = \perp$ *and* $\sigma_{i+1}^1 = \gamma.\sigma_i^1$
Synch: $(q_i, a_i, \gamma, q_{i+1}, \hat{\gamma}) \in \delta_{s_1}$ *then* $\sigma_i^0 = \perp$ *and* $\sigma_{i+1}^0 = \hat{\gamma}.\perp$ *and* $\sigma_{i+1}^1 = \gamma.\sigma_i^1$.

Directly from the fact that 2-OVPA are a subclass of MPDA and the fact that for MPDA the emptiness is solvable in cubic time, we get the following.

Corollary 1. *Given a 2-OVPA M, deciding whether $L(M) \neq \emptyset$ is solvable in $O(n^3)$, where n is the number of states in M.*

While dealing with automata, one interesting question is whether the acceptance power increases while using ε-moves, i.e., transitions that allow to change the state without consuming any input. Here we investigate 2-VPA with the ability of performing a restricted form of ε-moves: we only enable ε-moves on reading the top of the stack symbols on a local action. More formally, the variant 2-VPA$_\varepsilon$ of 2-VPA we consider is obtained by replacing δ_l in Definition 1 with a subset of $Q \times (\Sigma \cup \{\varepsilon\}) \times \Gamma \times \Gamma \times Q$ and by substituting the **Local** rule in the definition of a run for 2-VPA with the following:
Local$_\varepsilon$: $a_i \in \Sigma_l \cup \{\varepsilon\}$ and there exists $(q_i, a_i, \gamma^0, \gamma^1, q_{i+1}) \in \delta$ such that $\sigma_i^j = \sigma_{i+1}^j = \gamma^j.\sigma_i^j$, for all $j \in \{0, 1\}$.

Since at each step, a 2-VPA$_\varepsilon$ can now choose whether to consume an input symbol or take an ε-move, we consider the run definition modified accordingly. In the following theorem, we show that 2-VPA and 2-VPA$_\varepsilon$, as well as 2-OVPA and 2-OVPA$_\varepsilon$, are expressively equivalent.

Theorem 3. $L \in$ *2-VPL iff* $L \in$ *2-VPL$_\varepsilon$ and* $L \in$ *2-OVPL iff* $L \in$ *2-OVPA$_\varepsilon$.*

We conclude the section with an example of a language accepted by a 2-OVPA$_\varepsilon$.

Example 1. Let $L_1 = \{a^n b^n c^n \mid \exists n \in \mathbb{N}\}$. We show a 2-OVPA$_\varepsilon$ M accepting L_1. The alphabet $\widetilde{\Sigma}$ we use for M is partitioned in $\Sigma_{c_0} = \{a\}$, $\Sigma_{s_0} = \{b\}$, and $\Sigma_{r_1} = \{c\}$ (i.e., all the other partition elements are empty). The automaton is the following $M = (Q, Q_{in}, \Gamma, \perp, \delta, Q_F)$, with $Q = \{q_0, q_1, q_2, q_3, q_F\}$, $Q_{in} = \{q_0\}$, $Q_F = \{q_0, q_F\}$, $\Gamma = \{A, B\}$ and $\delta = \{(q_0, a, q_1, A), (q_1, a, q_1, A), (q_1, b, A, q_2, B), (q_2, b, A, q_2, B), (q_2, \varepsilon, \perp, B, q_3), (q_3, c, B, q_3), (q_3, \varepsilon, \perp, \perp, q_F)\}$. The 2-OVPA$_\varepsilon$ M is depicted in Figure 2, where we adopt the following conventions to represent arcs: for a local transition such as (q_i, a, A, B, q_j) we label the arc between q_i

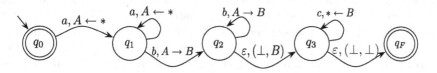

Fig. 2. A 2-OVPA$_\varepsilon$ accepting $L_1 = \{a^n b^n c^n | \exists n \in \mathbb{N}\}$

and q_j as $a, (A, B)$; for a synch transition such as (q_i, a, A, q_j, B) we label the arc as $s, A \rightarrow B$, if $a \in \Sigma_{s_0}$, and as $s, B \leftarrow A$, otherwise; moreover a push or pop transition is labeled like a synch transition but with one part missing. For example, a pop from the second stack (q_i, a, B, q_j) is labeled as $a, * \leftarrow B$.

5   Expressiveness and Closure Properties

In this section, we compare 2-VPLs and 2-OVPLs with VPLs [1], deterministic and (nondeterministic) context-free languages (resp., DCFLs and CFLs) [6], and multi-pushdown languages [3] ($\mathcal{L}_{PD^n}$). Recall that the following chain of inclusions holds: VPLs $\subset$ DCFLs $\subset$ CFLs $\subset \mathcal{L}_{PD^2} \subset$ CSLs.

Theorem 4. *The following assertions hold:*
a) 2-OVPLs $\subset$ 2-VPLs; b) VPLs $\subset$ 2-OVPLs; c) VPLs $\subset$ 2-VPLs;
d) DCFLs $\setminus$ 2-VPLs $\neq \emptyset$; e) DCFLs $\setminus$ 2-OVPLs $\neq \emptyset$;
f) (2-VPLs $\cap$ CFLs)$\setminus$ VPLs $\neq \emptyset$; g) 2-OVPLs $\subset \mathcal{L}_{PD^2}$; h) 2-OVPLs $\subset$ CSLs.

Although 2-VPLs and 2-OVPLs are strictly more expressive than VPLs, we show they preserve union, intersection, complementation (and thus inclusion). These properties, along with the emptiness problem for 2-OVPA being solvable in PTIME, make 2-OVPA a powerful engine for system verification using the automata-theoretic approach. We recall that 2-VPA and MPDA do not support such an approach since MPDA does not enjoy closure under intersection and complementation, and for 2-VPA the emptiness problem is undecidable.

Theorem 5 (Closure Properties). *Let L_1 and L_2 be two 2-VPLs (resp., 2-OVPLs) with respect to the same $\widetilde{\Sigma}$. Then, $L_1 \cap L_2$, $L_1 \cup L_2$ are 2-VPLs (resp., 2-OVPLs) over $\widetilde{\Sigma}$. Also, $L_1 \cdot L_2$, and L_1^* are 2-VPLs over $\widetilde{\Sigma}$. Furthermore, all the mentioned operations can be performed in polynomial-time.*

The closure of 2-VPA and 2-OVPA under complementation can be proved as an immediate consequence of determinization.

Theorem 6 (Determinization). *Given a 2-VPA (resp., 2-OVPA) M over $\widetilde{\Sigma}$, there is a deterministic 2-VPA (resp., deterministic 2-OVPA) M' over $\widetilde{\Sigma}$ such that $L(M) = L(M')$. Moreover, if M has n states, we can construct M' with $O(2^{2n^2})$ states and $O(2^{n^2} \cdot |\Sigma|)$ stack symbols.*

Proof. [**sketch**] The proof we present is inspired from that given in [1] for VPA. There, the main idea is to do a subset construction, postponing handling push transitions. The push transitions are stored into the stack and simulated later, namely at the time of the matching pop transitions. The construction has two components: a set of *summary edges* S, that keeps track of what state transitions are possible from a push transition to the corresponding pop transition, and a set of *path edges* R, that keeps track of all possible state reached from an initial state. In our case, we have to handle two stacks and the communication mechanism.

Therefore, we have to use two summary edges sets S_0 and S_1, and, in order to manage the communication transitions, we augment the structure of states adding information about the top of the stacks. Let M be a 2-VPA (resp., 2-OVPA) over $\widetilde{\Sigma}$. We define a deterministic 2-VPA (resp., 2-OVPA) M' over $\widetilde{\Sigma}$ such that $L(M) = L(M')$ behaving as sketched in the following example. We refer to the extended version for the detailed definition. Let $w = w_1 c_1^0 w_2 c_1^1 w_3$ be an input word, where in w_1 each push, either into the first or into the second stack, is matched by a pop, but there may be unmatched pop transitions; w_2 and w_3 are words in which all push and pop transitions are matched for both stacks; c_1^0 and c_1^1 are push, the former for the first stack and the latter for the second. In M', after reading w, the first stack is $(S_0, R_0, c_1^0).\bot$, the second stack is $(S_1, R_1, c_1^1).\bot$, and the control state is (S_0'', S_1'', R''). S_0 contains all the pair of states (q, q') such that the 2-VPA (resp., 2-OVPA) M can go from q with first stack empty to q' with first stack empty on reading w_1. Analogously, S_1 contains all the pairs (q, q') such that M can go from q with second stack empty to q' with second stack empty on reading $w_1 c_1^0 w_2$. R_0 and R_1 are the sets of all states reachable by M from an initial state on reading w_1 and $w_1 c_1^0 w_2$, respectively. S_0'' and S_1'' are the current summaries for the first and second stack, respectively, and R'' is the set of all states reachable by M on reading w.   □

Corollary 2 (Closure under complementation). *Let $L \in$ 2-VPLs (resp., 2-OVPLs) over $\widetilde{\Sigma}$, then $\Sigma^* \backslash L \in$ 2-VPLs (resp., 2-OVPLs) over $\widetilde{\Sigma}$.*

6   Model Checking and Synchronized Systems of VPA

A model checking procedure verifies the correctness of a system with respect to a desired behavior by checking whether a mathematical model of the system satisfies a formal specification of this behavior. Here, we consider the case whether both the model of the system and the formal specification of the required behavior are given by VPA with two stacks, say them M and P, respectively. The automata-theoretic approach to model checking exploits the combination of closure properties and emptiness decidability: checking whether M *satisfies* P is reduced to check whether $L(M) \cap \overline{L(P)} = \emptyset$ (all the runs of the model M satisfy the behavioral property represented by P).

Recall that the emptiness problem for 2-OVPA is solvable in cubic time (Corollary 1). Since determinization for 2-OVPA is in ExpTime (Theorem 6), and intersection can be done in polynomial-time (Theorem 5), we get an ExpTime algorithm to solve the model checking problem. The completeness follows from the fact that VPA model checking is ExpTime-complete [1].

Theorem 7. *The model checking problem for 2-OVPA is ExpTime-complete.*

In the remaining part of this section we show that 2-OVPA gives a natural way to describe distributed pushdown systems. In fact, we show that 2-OVPA capture the behavior of systems built on pairs of VPA working in a suitable synchronous way according to distributed computing paradigm. To this purpose, we introduce an operator of synchronous composition on VPA that allows to build a Synchronized System of VPA from a pair of VPA M_0 and M_1. The automata M_0 and M_1 run independently on the same input so that each input symbol can drive different transitions on the two, that is a local transition for the former and a push transition for the latter. Only communications between M_0 and M_1 have to be synchronized in accordance with a relation λ (a parameter of the synchronous composition operator) that contains all the transitions that are push transitions for the one and pop transitions for the other. The idea is that λ contains all the pairs of transitions on which the two VPA are allowed to communicate. The only constraint on the pushdown alphabets is that an input symbol can not trigger a pop transition on both VPA. Moreover, we have to prevent that M_1 can pop whenever M_0 has a non-empty stack, and thus every pop transition of M_1 is synchronized with M_0. Two VPA M_0 and M_1 over $\widetilde{\Sigma}^0$ and $\widetilde{\Sigma}^1$, respectively, are *synchronizable* if $\Sigma^0 = \Sigma^1$ and $\Sigma_r^0 \cap \Sigma_r^1$ is empty.

Definition 3 (Synchronized Systems of VPA). *A Synchronized System of VPA (S-VPA) $M_0||_\lambda M_1$ is a pair of synchronizable VPA M_0 and M_1 over $\widetilde{\Sigma}^0$ and $\widetilde{\Sigma}^1$, respectively, together with a communication relation $\lambda \subseteq \delta_c^0 \times \delta_r^1 \cup \delta_r^0 \times \delta_c^1$, where δ^0 and δ^1 are the transition relations of M_0 and M_1, respectively.*

A run ρ on $w = a_1 \ldots a_n \in (\Sigma^0 \cup \Sigma^1)^*$ for $M_0||_\lambda M_1$ is a pair of VPA runs on w, $\pi^0 = (q_0^0, \bot)(q_1^0, \sigma_1^0) \ldots (q_n^0, \sigma_n^0)$ for M_0 and $\pi^1 = (q_0^1, \bot)(q_1^1, \sigma_1^1) \ldots (q_n^1, \sigma_n^1)$ for M_1 such that, for all $k \in \{0, \ldots, n-1\}$, where t_k^0 is the transition applied from (q_k^0, σ_k^0) to $(q_{k+1}^0, \sigma_{k+1}^0)$ in M_0, and t_k^1 is the transition applied from (q_k^1, σ_k^1) to $(q_{k+1}^1, \sigma_{k+1}^1)$ in M_1, such that if t_k^1 is a pop transition then σ_k^0 is empty and if $(t_k^0, t_k^1) \in \delta_c^0 \times \delta_r^1 \cup \delta_r^0 \times \delta_c^1$ then $(t_k^0, t_k^1) \in \lambda$. A run ρ is accepting if both π^0 and π^1 are accepting and thus w is accepted. $L(M_0||_\lambda M_1)$ is the set of words accepted by $M_0||_\lambda M_1$. From Definition 3, it follows that $L(M_0||_\lambda M_1) \subseteq L(M_0) \cap L(M_1)$. Next theorem states that 2-OVPA are more expressive than S-VPA.

Theorem 8. *Let $M_0||_\lambda M_1$ be a S-VPA over $\widetilde{\Sigma}^0$, $\widetilde{\Sigma}^1$, then $L(M_0||_\lambda M_1)$ is a 2-OVPL with respect to $\widetilde{\Sigma} = \langle \widetilde{\Sigma}^0, \widetilde{\Sigma}^1 \rangle$.*

We give an evidence of the power of the introduced S-VPA by means of an example of a system behaving in a context-sensitive way. Consider a client-server system of pushdown processes described by a pair of synchronized VPA (see Figure 3) behaving in the following way: first, the client collects in its pushdown store an ordered pool of jobs on reading a sequence of input $job_i \in JobSet$; after that, the client transfers ($rcall$) the whole ordered sequence of jobs to the

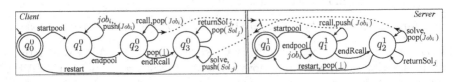

Fig. 3. An example of an S-VPA

server; then the server dispatches to the client a solution for each job (*solve*) in the same order the client has collected the jobs; moreover, the server waits a special commitment from the client ($returnSol_j$) after each dispatching, which is necessary to process next job; when the server runs out of pending jobs, the whole system can restart the computation (*restart*). Notice that the communication interface λ relates each Job_i that the server has to pop, with its solution Sol_j that the client has to push, determining the computation.

7 Conclusions

In this paper, we have investigated *ordered visibly pushdown automata with two stacks* (2-OVPA), obtained by merging the definitions of visibly pushdown automata [1] and multi-pushdown automata with two stacks [3]. We have shown that 2-OVPA are determinizable, closed under intersection and complementation, and have the emptiness problem decidable and solvable in polynomial time. Thus, we get that the inclusion problem is also decidable for 2-OVPA, and in particular, it is EXPTIME-complete. It is worth noticing that dropping visibility or the ordering constraint from 2-OVPA makes inclusion undecidable. The properties satisfied by 2-OVPA, along with the fact that they accept some context-free languages that are not regular as well as some context-sensitive languages that are not context-free, make 2-OVPA a powerful model in system verification while using the automata-theoretic approach. Finally, the model we propose can be also extended to deal with an arbitrary number n of stacks (n-OVPA). We argue (it is left to further investigation) that n-OVPA still retain decidability and closure properties of 2-OVPA and that, from an expressivity viewpoint, n-OVPA define a strict hierarchy based on the number of pushdown stores.

References

[1] Alur, R., Madhusudan, P.: Visibly pushdown languages. In: STOC'04, pp. 202–211. ACM Press, New York (2004)

[2] Bozzelli, L., Murano, A., Peron, A.: Pushdown module checking. In: Sutcliffe, G., Voronkov, A. (eds.) LPAR 2005. LNCS (LNAI), vol. 3835, pp. 504–518. Springer, Heidelberg (2005)

[3] Breveglieri, L., Cherubini, A., Citrini, C., Crespi-Reghizzi, S.: Multi-push-down languages and grammars. Int. J. Found. Comput. Sci. 7(3), 253–292 (1996)

[4] Clarke, E.M., Emerson, E.A.: Design and verification of synchronization skeletons using branching time temporal logic. In: Kozen, D. (ed.) Logics of Programs. LNCS, vol. 131, pp. 52–71. Springer, Heidelberg (1982)

[5] Clarke, E.M., Grumberg, O., Peled, D.A.: Model Checking. MIT Press, Cambridge (1999)

[6] Hopcroft, J.E., Ullman, J.D.: Introduction to Automata Theory, Languages, and Computation. Addison-Wesley, Reading (1979)

[7] Kupferman, O., Piterman, N., Vardi, M.: Pushdown specifications. In: Baaz, M., Voronkov, A. (eds.) LPAR 2002. LNCS (LNAI), vol. 2514, pp. 262–277. Springer, Heidelberg (2002)

[8] Minsky, M.L.: Computation: Finite and Infinite Machines. Prentice-Hall, Englewood Cliffs (1967)

[9] Queille, J.P., Sifakis, J.: Specification and verification of concurrent programs in Cesar. In: Dezani-Ciancaglini, M., Montanari, U. (eds.) International Symposium on Programming. LNCS, vol. 137, pp. 337–351. Springer, Heidelberg (1981)

[10] Sistla, A., Clarke, E.M., Francez, N., Gurevich, Y.: Can message buffers be axiomatized in linear temporal logic. Information and Control 63(1-2), 88–112 (1984)

[11] Van Steen, M., Tanenbaum, A.S.: Tanenbaum. Distributed Systems: Principles and Paradigms. Prentice Hall, Englewood Cliffs (2002)

[12] Vardi, M.Y., Wolper, P.: Automata-theoretic techniques for modal logics of programs. J. of Computer and System Sciences 32(2), 182–221 (1986)

[13] Walukiewicz, I.: Pushdown processes: Games and Model Checking. In: Alur, R., Henzinger, T.A. (eds.) CAV 1996. LNCS, vol. 1102, pp. 62–74. Springer, Heidelberg (1996)

An Efficient Computation of the Equation $\mathbb{K}$-Automaton of a Regular $\mathbb{K}$-Expression

Jean-Marc Champarnaud, Faissal Ouardi, and Djelloul Ziadi

LITIS, University of Rouen, France, 76821 Mont-Saint-Aignan – France
jean-marc.champarnaud@univ-rouen.fr
faissal.ouardi@univ-rouen.fr
djelloul.ziadi@univ-rouen.fr

Abstract. The aim of this paper is to describe a quadratic algorithm to compute the equation $\mathbb{K}$-automaton of a regular $\mathbb{K}$-expression as defined by Lombardy and Sakarovitch. Our construction is based on an extension to regular $\mathbb{K}$-expressions of the notion of c-continuation that we introduced to compute the equation automaton of a regular expression as a quotient of its position automaton.

1 Introduction

The conversion of a regular expression into an automaton is a rather old problem. The first algorithms appeared in the sixties, respectively based on the notion of position [21,15], prebase [22] and ε-transition [25]. In the nineties, different techniques were developed to design quadratic[1] algorithms for the construction of the position automaton: the star normal form [3], the compressed normal NFA [13] and the ZPC-structure [26]. Moreover the notion of partial derivative of a regular expression introduced by Antimirov [2] raised several challenging problems that boosted the research in this topic.

First, what is the relation between these different constructions? It was proved in [10] that the notions of prebase and of partial derivative lead to an identical automaton, the equation automaton[2]. Based on the notion of c-continuation it was shown in [11] that the equation automaton is a quotient of the c-continuation automaton that is itself isomorphic to the position automaton. A new construction [17] based on the follow relation was introduced in [17] and the follow automaton was proved to be a quotient of the position automaton. It was shown in [7] that the follow automaton can be constructed from the ZPC-structure. Finally, as mentioned in [27], the deep relation that exists between position, equation and follow automata can be easily understood through the ZPC-structure and the c-continuation computation.

Second, how to compare the performance of the algorithms that yield a quotient of the position automaton? The equation automaton [11] and the follow automaton [17,7] are both computed in quadratic time and space. Comparing the number of states of these automata is a more intricated issue. A new approach for this problem appears in [8,9]:

---

[1] In the following, complexity depends on the size of the expression.

[2] This name refers to the systems of expression equations used by Mirkin [22]. Other names are partial derivative automaton and Antimirov automaton.

T. Harju, J. Karhumäki, and A. Lepistö (Eds.): DLT 2007, LNCS 4588, pp. 145–156, 2007.
© Springer-Verlag Berlin Heidelberg 2007

a class of normalized expressions is defined such that any expression can be turned into an equivalent normalized one in linear time, and it is proved that the equation automaton of a normalized expression is always smaller than its follow automaton.

Last, how to extend these constructions to the case of $\mathbb{K}$-expressions? Concerning the position $\mathbb{K}$-automaton, the first algorithm, based on an inductive construction, is described in [5], and the first quadratic one, based on a generalization of the notion of ZPC-structure appears in [27,6]. As for the extension of the equation automaton, the notion of $\mathbb{K}$-derivative is used in [19,20,23] to show that the automaton of derived terms of a regular $\mathbb{K}$-expression is the weighted equivalent of the equation automaton. More recently, a unified frame is presented in [1] for the construction of position, equation, and follow automata and $\mathbb{K}$-automata, based on Thompson ε-automaton construction, epsilon-removal and minimization.

This paper addresses the construction of the equation $\mathbb{K}$-automaton. Our algorithm is based on the computation of a $\mathbb{K}$-covering [20] from the c-continuation $\mathbb{K}$-automaton onto the equation $\mathbb{K}$-automaton. It therefore provides a good understanding of this automaton. The structure of the automaton is derived from a set of c-continuations that is computed straightforwardly via the boolean case algorithm, while the set of weights is computed apart, leading to an overall quadratic complexity, i.e. as efficient as in the boolean case. Let us mention that this algorithm has been implemented inside VAU-CANSON platform [14], using the data structure proposed in [12] for the computation of the c-continuations.

The next section contains useful preliminaries and Section 3 is a reminder of the fundamental results presented in [20]. Section 4 generalizes the computation of c-derivatives to regular $\mathbb{K}$-expressions, describes the construction of the equation $\mathbb{K}$-automaton from the c-continuation one and gives an analysis of its complexity.

2   Preliminaries

Let A be a finite alphabet, and $(\mathbb{K}, \oplus, \otimes, \overline{0}, \overline{1})$ be a semiring (commutative or not). The star operator $\circledast$ can be partially defined over $\mathbb{K}$ as follows [16,18]: the scalar $y^{\circledast} \in \mathbb{K}$ is the unique solution (if it exists) of the equations $y \otimes x \oplus \overline{1} = x$ and $x \otimes y \oplus \overline{1} = x$, with $\overline{0}^{\circledast} = \overline{1}$. In this paper, examples come from the semiring $(\mathbb{Q}, +, \times)$.

Definition 1. *A (non-commutative) formal series S with coefficients in $\mathbb{K}$ and variables in A is a mapping from the free monoid A^* to $\mathbb{K}$ that associates with the word $w \in A^*$ a coefficient $\langle S, w \rangle \in \mathbb{K}$.*

A formal series is usually written as an infinite sum: $S = \sum_{u \in A^*} \langle S, u \rangle u$. The *support* of the formal series S is the language $\text{supp}(S) = \{u \in A^* \mid \langle S, u \rangle \neq \overline{0}\}$. The set of formal series over A with coefficients in $\mathbb{K}$ is denoted by $\mathbb{K}\langle\langle A \rangle\rangle$. A structure of semiring is defined on $\mathbb{K}\langle\langle A \rangle\rangle$ as follows [18]:

- $\langle S \oplus T, u \rangle = \langle S, u \rangle \oplus \langle T, u \rangle$,
- $\langle S \otimes T, u \rangle = \bigoplus_{u_1 u_2 = u} \langle S, u_1 \rangle \otimes \langle T, u_2 \rangle$, with $S, T \in \mathbb{K}\langle\langle A \rangle\rangle$.

The star of a series S is defined by: $S^* = \bigoplus_{n \geq 0} S^n$ with $S^0 = \varepsilon$, $S^n = S^{n-1} \otimes S$ if $n > 0$. For clarity the symbol S^* is used for series, whereas the symbol $\circledast$ is kept for the semiring of coefficients $\mathbb{K}$. The star of a formal series does not always exist. The *proper series* S_p associated with a formal series S is defined by $\langle S_p, \varepsilon \rangle = 0$ and $\langle S_p, u \rangle = \langle S, u \rangle$ for any word $u \in A^+$. The star of a proper series always exists. We will use the following construction for the star of a formal series.

Proposition 1. [18] *The star of a formal series $S \in \mathbb{K}\langle\langle A \rangle\rangle$ is defined if and only if $\langle S, \varepsilon \rangle^{\circledast}$ is defined in $\mathbb{K}$. In this case: $S^* = \langle S, \varepsilon \rangle^{\circledast}(S_p \langle S, \varepsilon \rangle^{\circledast})^*$.*

A polynomial is a formal series with finite support. The set of polynomials is denoted by $\mathbb{K}\langle A \rangle$ that is a subsemiring of $\mathbb{K}\langle\langle A \rangle\rangle$.

Definition 2. *The semiring of regular series $\mathbb{K}\,\mathrm{Rat}(A^*) \subset \mathbb{K}\langle\langle A \rangle\rangle$ is the smallest set of $\mathbb{K}\langle\langle A \rangle\rangle$ that contains the semiring $\mathbb{K}\langle A \rangle$ of polynomials, and that is stable under the operations of addition, product and star (when it is defined).*

Definition 3. *A regular $\mathbb{K}$-expression over an alphabet A is inductively defined by:*

 - *$a \in A, k \in \mathbb{K}$ are regular $\mathbb{K}$-expressions that respectively denote the regular series $S_a = a$ and $S_k = \overline{k}$,*
 - *if F, G and H are regular $\mathbb{K}$-expressions that respectively denote the regular series S_{F}, S_{G} and S_{H} (such that $S_{\mathrm{H}}{}^*$ exists), then $(\mathrm{F} + \mathrm{G})$, $(\mathrm{F} \cdot \mathrm{G})$ and (H^*) are regular $\mathbb{K}$-expressions that respectively denote the regular series $S_{\mathrm{F}} \oplus S_{\mathrm{G}}$, $S_{\mathrm{F}} \otimes S_{\mathrm{G}}$, and $S_{\mathrm{H}}{}^*$.*

Let E be a $\mathbb{K}$-expression. We will denote by A_{E} the alphabet of E, and by $|\,\mathrm{E}\,|$ its size that is equal to the size of the syntax tree of E. The linearized version $\overline{\mathrm{E}}$ of E is the $\mathbb{K}$-expression deduced from E by associating with every occurrence of a symbol a of A_{E} its rank i in E. Subscripted symbols a_i are called positions. Let h be the mapping that associates with a position $a_i \in A_{\overline{\mathrm{E}}}$ the symbol $a \in A_{\mathrm{E}}$. Given an expression F over a set of positions, we denote by $h(\mathrm{F})$ the expression obtained by replacing every position x in F by $h(x)$. We write $\mathrm{E} \equiv \mathrm{F}$ if E and F graphically coincid.

An element of $\mathbb{K}$ can be seen as a $\mathbb{K}$-expression (written k) or as a scalar (written $\overline{k}$). It induces a morphism from the semiring of $\mathbb{K}$-expressions with no occurrence of symbol of A_{E} to the semiring $\mathbb{K}$. We denote by $\overline{K}$ the scalar[3] associated to such an expression K. Following [6], the null term $\lambda(\mathrm{E})$ of a $\mathbb{K}$-expression E is the $\mathbb{K}$-expression induced from E by replacing each occurrence of a symbol of A_{E} by the symbol 0 (associated to the scalar $\overline{0}$ of $\mathbb{K}$). For example, if $\mathrm{E} = (\frac{1}{2} \cdot a^* + \frac{1}{3} \cdot b^*)^* \cdot a^*$, we get $A_{\mathrm{E}} = \{a, b\}$, $\overline{\mathrm{E}} = (\frac{1}{2} \cdot a_1^* + \frac{1}{3} \cdot b_2^*)^* \cdot a_3^*$, $A_{\overline{\mathrm{E}}} = \{a_1, b_2, a_3\}$ and $|\overline{\mathrm{E}}| = 13$. It comes $\lambda(\mathrm{E}) = (\frac{1}{2} \cdot 0^* + \frac{1}{3} \cdot 0^*)^* \cdot 0^*$ and $\overline{\lambda(\mathrm{E})} = \overline{6}$.

Definition 4. *A $\mathbb{K}$-automaton $\mathcal{A} = \langle Q, A, q_0, \delta, \gamma, \mu \rangle$ is defined as follows[4]:*

 - *Q is a finite set of states; A is the alphabet; q_0 is the initial state,*
 - *$\delta \subseteq Q \times A \times Q$ is the set of transitions,*
 - *$\gamma : \delta \to \mathbb{K}$ (resp. $\mu : Q \to \mathbb{K}$) is the transition (resp. output) weight function.*

---

[3] Actually there is no conflict with the notation $\overline{\mathrm{E}}$ used for the linearized version of E.

[4] A more general definition is given in [4].

A path p from a state q_0 to a state q_n is a sequence of transitions $p = (p_1, p_2, \cdots, p_n)$ with $p_i = (q_{i-1}, a_i, q_i)$ for $1 \leq i \leq n$. Its label is the word $w(p) = a_1 a_2 \cdots a_n$. We denote by $\text{coef}(p)$ the weight of the path p in $\mathcal{A}$, with $\text{coef}(p) = \gamma(p_1) \otimes \gamma(p_2) \otimes \cdots \otimes \gamma(p_n) \otimes \mu(q_n)$. Let $\mathcal{C}_\mathcal{A}$ be the set of all paths in $\mathcal{A}$ starting from q_0. The $\mathbb{K}$-automaton $\mathcal{A}$ *realizes* the series $S_\mathcal{A}$ defined by:

$$S_\mathcal{A} = \sum_{u \in A^*} \langle S_\mathcal{A}, u \rangle u, \text{ where } \langle S_\mathcal{A}, u \rangle = \bigoplus_{p \in \mathcal{C}_\mathcal{A}, w(p) = u} \text{coef}(p)$$

A series $S \in \mathbb{K}\langle\langle A \rangle\rangle$ is recognizable if there exists a $\mathbb{K}$-automaton that realizes it.

Theorem 1. (Schützenberger [24]) *A formal series is recognizable if and only if it is regular.*

Definition 5. [20] *Let* $\mathcal{A}$ *and* $\mathcal{B}$ *be two* $\mathbb{K}$*-automata. Let* $\Delta(q, q') = \bigoplus_{(q,a,q') \in \delta} \gamma(q, a, q') a$. *A surjective mapping* φ *from the set of states of* $\mathcal{A}$ *onto the set of states of* $\mathcal{B}$ *induces a* $\mathbb{K}$*-covering from* $\mathcal{A}$ *onto* $\mathcal{B}$ *if* $\mathcal{A}$ *is such that:*

$$\forall p, q \in Q_\mathcal{A}, \ \varphi(p) = \varphi(q) \Rightarrow \begin{cases} 1) \ \mu(p) = \mu(q) \\ \\ 2) \ \forall r \in Q_\mathcal{A}, \ \bigoplus_{s \in \varphi^{-1}\varphi(r)} \Delta_\mathcal{A}(p, s) = \bigoplus_{s \in \varphi^{-1}\varphi(r)} \Delta_\mathcal{A}(q, s) \end{cases}$$

and if $\mathcal{B}$ *satisfies the following conditions:*

$$3) \ \forall r \in Q_\mathcal{B}, \quad \mu(r) = \mu(p) \qquad \qquad \text{for any } p \in \varphi^{-1}(r)$$
$$4) \ \forall (r, s) \in Q_\mathcal{B}^2, \Delta_\mathcal{B}(r, s) = \bigoplus_{q \in \varphi^{-1}(s)} \Delta_\mathcal{A}(p, q) \text{ for any } p \in \varphi^{-1}(r)$$

Proposition 2. [20] *Let* $\mathcal{A}$ *and* $\mathcal{B}$ *be two* $\mathbb{K}$*-automata. If* $\varphi : \mathcal{A} \longrightarrow \mathcal{B}$ *is a* $\mathbb{K}$*-covering, then* $S_\mathcal{A} = S_\mathcal{B}$.

Let E be a regular $\mathbb{K}$-expression over an alphabet A and consider the language $L(\overline{E})$ associated to the linearized version of E [6]. The polynomials First(E), Last(E) and Follow($\cdot$, E) in $\mathbb{K}\langle A_{\overline{E}} \rangle$ can be computed as follows (where x is a position of E):

$$\text{First}(k) = \overline{0} \text{ for all } k \in \mathbb{K}$$
$$\text{First}(a) = \overline{1} a_i \ (a_i \text{ is the position associated to } a \text{ in } A_{\overline{E}})$$
$$\text{First}(F + G) = \text{First}(F) \oplus \text{First}(G)$$
$$\text{First}(F \cdot G) = \text{First}(F) \oplus \overline{\lambda(F)} \, \text{First}(G)$$
$$\text{First}(F^*) = \overline{\lambda(F)}^{\circledast} \, \text{First}(F)$$

Similar rules hold for Last except for $\text{Last}(F \cdot G) = \text{Last}(G) \oplus \overline{\lambda(G)} \, \text{Last}(F)$.

$$\text{Follow}(x, k) = \overline{0} \text{ for all } k \in \mathbb{K}$$
$$\text{Follow}(x, a) = \overline{0} \text{ for all } a \in A$$
$$\text{Follow}(x, F + G) = \text{Follow}(x, F) \oplus \text{Follow}(x, G)$$
$$\text{Follow}(x, F \cdot G) = \text{Follow}(x, F) \oplus \langle \text{Last}(F), x \rangle \, \text{First}(G) \oplus \text{Follow}(x, G)$$
$$\text{Follow}(x, F^*) = \text{Follow}(x, F) \oplus \langle \text{Last}(F^*), x \rangle \, \text{First}(F)$$

These polynomials lead to the definition of the position $\mathbb{K}$-automaton of E, that realizes the series denoted by E.

Definition 6. *The position $\mathbb{K}$-automaton $\mathcal{P}_E = \langle Q, A_E, q_0, \delta, \gamma, \mu \rangle$ is defined by:*

- $Q = \{q_0\} \cup A_{\overline{E}}$, *with* $q_0 \notin A_{\overline{E}}$,
- $(q, a, p) \in \delta \Leftrightarrow h(p) = a$ *and* $\begin{cases} \langle \text{First}(E), p \rangle \neq \overline{0} & \text{if } q = q_0, \\ \langle \text{Follow}(q, E), p \rangle \neq \overline{0} & \text{otherwise.} \end{cases}$
- *For all* $(p, a, q) \in \delta$, *it holds:*
  $$\gamma(q, a, p) = \begin{cases} \langle \text{First}(E), p \rangle & \text{if } q = q_0, \\ \langle \text{Follow}(q, E), p \rangle & \text{otherwise.} \end{cases}$$
- $\mu(q) = \begin{cases} \overline{\lambda}(E) & \text{if } q = q_0, \\ \langle \text{Last}(E), q \rangle & \text{otherwise.} \end{cases}$

3 From $\mathbb{K}$-Derivatives to the Equation $\mathbb{K}$-Automaton

In this section, we recall the main results reported in [20] about the computation of the set of $\mathbb{K}$-derivatives of a regular $\mathbb{K}$-expression. Such a $\mathbb{K}$-derivative is defined as a polynomial, which leads to a generalization of the notion of partial derivative [2].

Definition 7. *Let* E *be a regular $\mathbb{K}$-expression[5] and* $a \in A$. *The $\mathbb{K}$-derivative of* E *w.r.t.* a *is the polynomial* $\partial_a(E)$ *inductively defined as follows:*

$$\partial_a(k) = \overline{0} \qquad\qquad \partial_a(E + F) = \partial_a(E) \oplus \partial_a(F)$$
$$\partial_a(b) = \begin{cases} \overline{1} \text{ if } b = a \\ \overline{0} \text{ otherwise} \end{cases} \qquad \partial_a(E \cdot F) = \partial_a(E) \cdot F \oplus \overline{\lambda}(E) \partial_a(F)$$
$$\partial_a(E^*) = \overline{\lambda(E)}^{\circledast} (\partial_a(E) \cdot E^*)$$

By linearity, the $\mathbb{K}$-derivative of a polynomial is given by: $\partial_a(\bigoplus_{i \in I} \overline{k_i} E_i) = \bigoplus_{i \in I} \overline{k_i} \partial_a(E_i)$.

The $\mathbb{K}$-derivative of a regular $\mathbb{K}$-expression E w.r.t. a word $u \in A^+$ is defined by: $\forall u \in A^+, \forall a \in A, \; \partial_{ua}(E) = \partial_a(\partial_u(E))$.

Example 1. *The $\mathbb{K}$-derivatives w.r.t.* a *and* b *of the regular $\mathbb{K}$-expression* $E = \frac{1}{2}a^*(\frac{1}{3}b^* + \frac{1}{6}b^*)^*$ *are the polynomials* $\partial_a(E) = \overline{\frac{1}{2}}a^*(\frac{1}{3}b^* + \frac{1}{6}b^*)^*$ *and* $\partial_b(E) = \overline{\frac{1}{2}}b^*(\frac{1}{3}b^* + \frac{1}{6}b^*)^*$.

Proposition 3. [20] *Let* E *and* F *be regular $\mathbb{K}$-expressions. Let* $u \in A^+$ *and* $k \in \mathbb{K}$. *Then it holds:*

1) $\partial_u(E + F) = \partial_u(E) \oplus \partial_u(F)$,
2) $\partial_u(E \cdot F) = \partial_a(E) \cdot F \oplus \bigoplus_{u = vw} \langle \partial_v(E), 1 \rangle \partial_w(F)$,
3) $\partial_u(E^*) = \bigoplus_{u = u_1 \cdots u_n} \overline{\lambda(E)}^{\circledast} \langle \partial_{u_1}(E), 1 \rangle \overline{\lambda(E)}^{\circledast} \cdots \langle \partial_{u_{n-1}}(E), 1 \rangle \overline{\lambda(E)}^{\circledast} (\partial_{u_n}(E) \cdot E^*)$.

---

[5] In [20] a different definition of a $\mathbb{K}$-regular expression is used, where k is not an expression. Hence a slightly different formulation of the $\mathbb{K}$-derivative.

There exists a set of regular $\mathbb{K}$-expressions that plays a specific role in the computation of the $\mathbb{K}$-derivatives of E: the set of derived terms of E.

Definition 8. *The set* $\mathrm{dt}(E)$ *of the derived terms of a regular $\mathbb{K}$-expression E is inductively defined as follows:*

$$\mathrm{dt}(k) = \emptyset$$
$$\mathrm{dt}(a) = \{1\}$$
$$\mathrm{dt}(F + G) = \mathrm{dt}(F) \cup \mathrm{dt}(G)$$

$$\mathrm{dt}(F \cdot G) = \bigcup_{F_i \in \mathrm{dt}(F)} (F_i \cdot G) \cup \mathrm{dt}(G)$$

$$\mathrm{dt}(F^*) = \bigcup_{F_i \in \mathrm{dt}(F)} (F_i \cdot F^*)$$

The set of derived terms of a regular $\mathbb{K}$-expression E is the equivalent of the prebase of a regular expression, as introduced by Mirkin [22].

Example 2. *For the regular $\mathbb{K}$-expression* $E = \frac{1}{2}a^*(\frac{1}{3}b^* + \frac{1}{6}b^*)^*$*, we have:*

$$\mathrm{dt}(E) = \{\frac{1}{2}a^*(\frac{1}{3}b^* + \frac{1}{6}b^*)^*, a^*(\frac{1}{3}b^* + \frac{1}{6}b^*)^*, b^*(\frac{1}{3}b^* + \frac{1}{6}b^*)^*\}.$$

The computation of the $\mathbb{K}$-derivatives of a regular $\mathbb{K}$-expression E leads to the construction of the equation $\mathbb{K}$-automaton[6] that realizes the series denoted by E.

Definition 9. *The equation $\mathbb{K}$-automaton* $\mathcal{E}_E = \langle Q, A_E, q_0, \delta, \gamma, \mu \rangle$ *of a regular $\mathbb{K}$-expression E is defined as follows:*

- $Q = \mathrm{dt}(E) \cup \{E\}$*;* $q_0 = \{E\}$*,*
- $(E_i, a, E_j) \in \delta \Leftrightarrow \langle \partial_a(E_i), E_j \rangle \neq \overline{0}$ *for all* $E_i, E_j \in Q$*,*
- $\gamma(E_i, a, E_j) = \langle \partial_a(E_i), E_j \rangle$ *for all* $E_i, E_j \in Q$*;* $\mu(E_i) = \overline{\lambda(E_i)}$*.*

Example 3. *(Ex. 1 continued)*

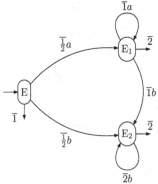

Fig. 1. The equation $\mathbb{K}$-automaton $\mathcal{E}_E$ associated with $E = \frac{1}{2}a^*(\frac{1}{3}b^* + \frac{1}{6}b^*)^*$

Theorem 2. [20] *There exists a $\mathbb{K}$-covering from the position $\mathbb{K}$-automaton onto the equation $\mathbb{K}$-automaton of a regular $\mathbb{K}$-expression.*

---

[6] Also called the automaton of derived terms in [20].

4   From c-Derivatives to the Equation $\mathbb{K}$-Automaton

We now define the notions of c-derivative, c-continuation and c-continuation automaton for a regular $\mathbb{K}$-expression. The main point here is that, following Definition 3, a regular $\mathbb{K}$-expression E can be seen as a regular expression over the alphabet $A_E \cup B_E$, where B_E is the set of elements of $\mathbb{K}$ occurring in E. Therefore, for any symbol $x \in A_{\overline{E}}$, the c-continuation w.r.t. x of the regular $\mathbb{K}$-expression E graphically coincids with the c-continuation w.r.t. x of the associated regular expression. As a consequence, the set of states and the set of transitions of the c-continuation $\mathbb{K}$ automaton of E are computed straightforwardly via the boolean case algorithm described in [12].

We also define the notion of coefficient of a regular $\mathbb{K}$-expression w.r.t. a symbol and show how the weights of the transitions are deduced from the coefficients of the c-continuations. We assume that, for E a regular $\mathbb{K}$-expression, the following identities are satisfied: $0 \cdot E = E \cdot 0 = 0$, $0 + E = E + 0 = E$, $1 \cdot E = E \cdot 1 = E$.

4.1   From c-Derivatives to the c-Continuation $\mathbb{K}$-Automaton

Definition 10. *The c-derivative of a regular $\mathbb{K}$-expression E w.r.t. a symbol a, written $d_a(E)$, is defined by:*

$$d_a(k) = 0$$

$$d_a(x) = \begin{cases} 1 & \text{if } a = x \\ 0 & \text{otherwise} \end{cases} \qquad d_a(F \cdot G) = \begin{cases} d_a(F) \cdot G & \text{if } d_a(F) \neq 0 \\ d_a(G) & \text{if } d_a(F) = 0 \text{ and } \lambda(F) \neq 0 \\ 0 & \text{otherwise} \end{cases}$$

$$d_a(F + G) = \begin{cases} d_a(F) & \text{if } d_a(F) \neq 0 \\ d_a(G) & \text{otherwise} \end{cases} \qquad d_a(F^*) = d_a(F) \cdot F^*$$

The *c*-derivative of E w.r.t. a word u is defined by: $d_\varepsilon(E) = E$, and $d_{u_1 \ldots u_n}(E) = d_{u_2 \ldots u_n}(d_{u_1}(E))$.

The main property of c-derivatives still holds for regular $\mathbb{K}$-expressions, leading to the notion of c-continuation.

Theorem 3. *If E is linear, for every symbol $a \in A_{\overline{E}}$ there exists a $\mathbb{K}$-expression $c_a(E)$, called the c-continuation of E w.r.t. a, such that for every word $u \in A_{E'}^*$ the c-derivative $d_{ua}(E)$ is either 0 or $c_a(E)$.*

Example 4. *(Ex. 1 continued)*
We have $d_{b_2 b_2}(\overline{E}) = b_2^*(\frac{1}{3}b_2^* + \frac{1}{6}b_3^*)^*$ and $d_{b_3 b_2}(\overline{E}) = b_2^*(\frac{1}{3}b_2^* + \frac{1}{6}b_3^*)^*$.

Proposition 4. *For every symbol a of a linear expression E, the c-continuation $c_a(E)$ can be computed as follows:*

$$c_a(a) = 1 \qquad\qquad c_a(F \cdot G) = \begin{cases} c_a(F) \cdot G & \text{if } c_a(F) \text{ exists} \\ c_a(G) & \text{otherwise} \end{cases}$$

$$c_a(F + G) = \begin{cases} c_a(F) & \text{if } c_a(F) \text{ exists} \\ c_a(G) & \text{otherwise} \end{cases} \qquad c_a(F^*) = c_a(F) \cdot F^*$$

In the following we will write c_x instead of $c_x(E)$ when there is no ambiguity.

Definition 11. *The coefficient of a linear $\mathbb{K}$-expression E w.r.t. a symbol a is the scalar $k_a(E)$ inductively defined as follows:*

$$k_a(k) = \overline{0}$$

$$k_a(b) = \begin{cases} \overline{1} \text{ if } b = a \\ \overline{0} \text{ otherwise} \end{cases}$$

$$k_a(F + G) = k_a(F) \oplus k_a(G)$$

$$k_a(F \cdot G) = k_a(F) \oplus \overline{\lambda(F)} \otimes k_a(G)$$

$$k_a(F^*) = \overline{\lambda(F)}^{\circledast} \otimes k_a(F)$$

The coefficient of E w.r.t. a word u is defined by: $k_\varepsilon(E) = \overline{\lambda(E)}$, and $k_{u_1 \ldots u_n}(E) = k_{u_1}(E) \otimes k_{u_2 \ldots u_n}(d_{u_1}(E))$.

Definition 12. *The c-continuation automaton $\mathcal{C}_E = \langle Q_C, A_E, q_C, \delta_C, \gamma_C, \mu_C \rangle$ of a regular $\mathbb{K}$-expression E is defined by:*

- $Q_C = \{(x, c_x) \mid x \in A_{\overline{E}} \cup \{0\}\}; q_C = (0, c_0)$,
- $((x, c_x), a, (y, c_y)) \in \delta_C \Leftrightarrow h(y) = a \text{ and } d_y(c_x) \equiv c_y$,
- $\gamma_C((x, c_x), a, (y, c_y)) = k_y(c_x); \mu_C(c_x) = \overline{\lambda(c_x)}$.

Notice that if $k_y(c_x) = \overline{0}$ then the transition is not considered.

Example 5. *(Ex. 1 continued)*

$c_0(E) = \overline{E}$

$c_{a_1}(E) = a_1^*(\frac{1}{3}b_2^* + \frac{1}{6}b_3^*)^*$

$c_{b_2}(E) = b_2^*(\frac{1}{3}b_2^* + \frac{1}{6}b_3^*)^*$

$c_{b_3}(E) = b_3^*(\frac{1}{3}b_2^* + \frac{1}{6}b_3^*)^*$

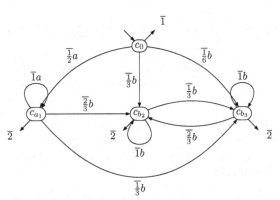

Fig. 2. The c-continuation $\mathbb{K}$-automaton associated with $E = \frac{1}{2}a^*(\frac{1}{3}b^* + \frac{1}{6}b^*)^*$

4.2 From the c-Continuation $\mathbb{K}$-Automaton to the Equation $\mathbb{K}$-Automaton

As for the boolean case, we can relate the c-continuation $\mathbb{K}$-automaton to both the position and the equation $\mathbb{K}$-automata.

Proposition 5. *Let E be a regular $\mathbb{K}$-expression. Then the following equalities hold:*
$\text{First}(E) = \bigoplus_{x \in A_{\overline{E}}} k_x(\overline{E})x, \text{ Last}(E) = \bigoplus_{x \in A_{\overline{E}}} \overline{\lambda(c_x)}x, \text{ Follow}(x, E) = \bigoplus_{y \in A_{\overline{E}}} k_y(c_x)y.$

Theorem 4. *The c-continuation and position $\mathbb{K}$-automata of a regular $\mathbb{K}$-expression are isomorphic.*

As a corollary and according to Proposition 2, the $\mathbb{K}$-automaton $\mathcal{C}_E$ realizes the series denoted by E.

Following the boolean track [11], we now consider the equivalence $\sim$ defined by:

$$\forall (x, c_x), (z, c_z) \in Q_\mathcal{C}, (x, c_x) \sim (z, c_z) \Leftrightarrow h(c_x) \equiv h(c_z)$$

We will denote by $[x]$ the class of the state (x, c_x) and we will write z instead of (z, c_z) whenever there is no ambiguity (for instance $z \in [x]$). Let z be an arbitrary element of $[x]$. We denote by $C_{[x]}$ the expression $h(c_z)$ that is identical for all $z \in [x]$. Notice that $[x]$ is characterized by the expression $C_{[x]}$.

We now define the $\mathbb{K}$-automaton $\mathcal{C}_E/_\sim = \langle Q_\sim, A_E, I_\sim, \delta_\sim, \gamma_\sim, \mu_\sim \rangle$ whose states are the $\sim$-classes and that will be proved to be equivalent to $\mathcal{C}_E$. The relation $\sim$ being right-invariant [11], the function $\delta_\sim$ is well-defined. The soundness of the weight functions $\gamma_\sim$ and $\mu_\sim$ will be proved by exhibiting a $\mathbb{K}$-covering from $\mathcal{C}_E$ onto $\mathcal{C}_E/_\sim$.

Definition 13. *The $\mathbb{K}$-automaton $\mathcal{C}_E/_\sim = \langle Q_\sim, A_E, q_\sim, \delta_\sim, \gamma_\sim, \mu_\sim \rangle$ is defined by:*

- $Q_\sim = \{C_{[x]} \mid x \in A_{\overline{E}} \cup \{0\}\}$; $q_\sim = C_{[0]}$,
- $(C_{[x]}, a, C_{[y]}) \in \delta_\sim \Leftrightarrow$ *for any $z \in [x]$, $\exists t \in [y] \mid h(t) = a$ and $(c_z, a, c_t) \in \delta_\mathcal{C}$,*
- $\gamma_\sim(C_{[x]}, a, C_{[y]}) = \displaystyle\bigoplus_{t \in [y], h(t) = a} \gamma_\mathcal{C}((z, c_z), a, (t, c_t))$, *for any $z \in [x]$,*
- $\mu_\sim(C_{[x]}) = \overline{\lambda(c_z)}$, *for any $z \in [x]$.*

Theorem 5. *Let E be a regular $\mathbb{K}$-expression. Then the mapping h defines a $\mathbb{K}$-covering from $\mathcal{C}_E$ onto $\mathcal{C}_E/_\sim$.*

Proof. Let $h : Q_\mathcal{C} \longrightarrow Q_\sim$ be the surjective mapping defined by: $h(x, c_x) = h(c_x)$, for all $x \in A_{\overline{E}} \cup \{0\}$. We have $h(x, c_x) = h(z, c_z) \Leftrightarrow (x, c_x) \sim (z, c_z) \Leftrightarrow h(c_x) \equiv h(c_z)$.

Condition 1 of the Definition 5 says that two equivalent states in $\mathcal{C}_E$ should have the same output weight. It can be rewritten: $\forall (x, c_x), (z, c_z) \in Q_\mathcal{C}, (x, c_x) \sim (z, c_z) \Rightarrow \mu_\mathcal{C}(x, c_x) = \mu_\mathcal{C}(z, c_z)$. This condition is satisfied since $\mu_\mathcal{C}(x, c_x) = \overline{\lambda(c_x)}$ and $(x, c_x) \sim (z, c_z) \Rightarrow h(c_x) \equiv h(c_z) \Rightarrow \lambda(c_x) = \lambda(c_z)$. Moreover, since $(x, c_x) \sim (z, c_z) \Rightarrow \lambda(c_x) = \lambda(c_z)$, $\mu_\sim(C_{[x]})$ can be computed as $\overline{\lambda(c_z)}$, for any $z \in [x]$. Hence the Condition 3 is also satisfied.

Condition 2 ensures that the transition weights in $\mathcal{C}_E/_\sim$ can be computed from the weights of the transitions outgoing from any state in the origin class. Let us set $S_x = \displaystyle\bigoplus_{t \in [y], h(t) = a} \gamma_\mathcal{C}((x, c_x), a, (t, c_t))$. The transition weight function $\gamma_\mathcal{C}$ must be such that: $\forall (x, c_x), (z, c_z) \in Q_\mathcal{C}, (x, c_x) \sim (z, c_z) \Rightarrow \forall (y, c_y) \in Q_\mathcal{C}, \forall a \in A_E, S_x = S_z$. We have $\gamma_\mathcal{C}((x, c_x), a, (t, c_t)) = k_t(c_x)$. Let $t = a_i$ (resp. $t = a_j$) be the kth symbol occurring in c_x (resp. c_z). Since $h(c_x) \equiv h(c_z)$, we have $h(a_i) = h(a_j)$. Moreover, since $h(d_{a_i}(c_x)) = h(d_{a_j}(c_z))$, we get $h(c_{a_i}) = h(c_{a_j})$, and thus $(a_i, c_{a_i}) \sim (a_j, c_{a_j})$. Hence for all position $t = a_i$ occurring in the left sum there is a corresponding position $t = a_j$ occurring in the right sum. Finally, by a simple induction we get that $k_{a_i}(c_x) = k_{a_j}(c_z)$. Consequently, the two sums are equal and Condition 2 is satisfied. Moreover, since the sum S_z is independent from the choice of z in $[x]$, $\gamma_\sim(C_{[x]}, a, C_{[y]})$ can be computed as S_z for any $z \in [x]$. Hence the Condition 4 is also satisfied. ∎

As a corollary and according to Proposition 2, the $\mathbb{K}$-automaton $\mathcal{C}_{\mathrm{E}}/_\sim$ realizes the series denoted by E.

We now show that the $\mathbb{K}$-automaton $\mathcal{C}_{\mathrm{E}}/_\sim$ and the equation $\mathbb{K}$-automaton $\mathcal{E}_{\mathrm{E}}$ are isomorphic.

Lemma 1. *Let* E *be a regular* $\mathbb{K}$-*expression. Then it holds:* $\mathrm{dt}(\mathrm{E}) = \bigcup_{x \in A_{\overline{\mathrm{E}}}} h(c_x(\mathrm{E}))$.

Lemma 2. *The following equality holds for all positions* x *in* $A_{\overline{\mathrm{E}}} \cup \{0\}$:

$$\partial_a(h(c_x(\mathrm{E}))) = \bigoplus_{a_i \in A_{\overline{\mathrm{E}}}, h(a_i)=a} k_{a_i}(c_x(\mathrm{E}))h(d_{a_i}(c_x(\mathrm{E})))$$

Corollary 1. *The following equality holds for all positions* x, y *in* $A_{\overline{\mathrm{E}}} \cup \{0\}$:

$$\langle \partial_a(h(c_x(\mathrm{E}))), h(c_y(\mathrm{E})) \rangle = \bigoplus_{a_i \in [y], h(a_i)=a} k_{a_i}(c_x(\mathrm{E}))$$

Theorem 6. *Let* E *be a regular* $\mathbb{K}$-*expression. The* $\mathbb{K}$-*automaton* $\mathcal{C}_{\mathrm{E}}/_\sim$ *and the equation* $\mathbb{K}$-*automaton* $\mathcal{E}_{\mathrm{E}}$ *are isomorphic.*

Proof. By Lemma 1, the $\mathbb{K}$-automata $\mathcal{C}_{\mathrm{E}}/_\sim$ and $\mathcal{E}_{\mathrm{E}}$ have identical sets of states and identical output weight functions. By Lemma 2, they have identical sets of transitions, and by Corollary 1, identical transition weight functions.   ∎

Finally, let us notice that combining Theorem 4, Theorem 5 and Theorem 6 provides a proof of Theorem 2.

4.3   The Algorithm for Converting E into $\mathcal{E}_{\mathrm{E}}$

We now give the sketch of the Algorithm *AlgoKExptoEq* for converting a regular $\mathbb{K}$-expression into its equation $\mathbb{K}$-automaton.

The complexity is as follows. Let n (resp. $\tilde{n}$) be the number of states in $\mathcal{C}_E$ (resp. $\mathcal{C}_{\mathrm{E}}/_\sim$). We assume that $O(n) = O(|E|)$. Although $O(\tilde{n}) = O(n)$, time complexity will be expressed as far as possible w.r.t. $\tilde{n}$ for more accuracy.

On the one hand, the computation of the set of states of $\mathcal{C}_{\mathrm{E}}/_\sim$ (Step 1: Lines 3–4) and the computation of its set of transitions (Step 2: Lines 6,7,9,10) are carried out via the boolean case algorithm *AlgoCtoE* [12]. As a straightforward consequence, Step 1 can be implemented in $O(n^2)$ time over the ZPC-structure [26] of the regular expression associated with E, and Step 2 in $O(\tilde{n}n)$ time. Notice that any optimisation of Step 1 or of Step 2, would lead to an improvement of both boolean and weighted algorithms.

On the other hand, the computation of the transition weight function (Step 4: Lines 6,7,9,11) can be implemented in $O(\tilde{n}n)$ time. Indeed, the weight $\gamma_\sim(C_{[x]}, a, C_{[y]})$ only depends on the weights of the transitions outgoing from one arbitrarily chosen element in $[x]$. Moreover, each weight $\gamma_C((z, c_z), h(y), (y, c_y))$ is involved in the computing of at most one weight $\gamma_\sim(C_{[x]}, a, C_{[y]})$. Notice that the output weight function (Step 3: Lines 6–8) is computed in $O(\tilde{n})$ time since $\mu_\sim(C_{[x]}) = \mu_C((z, c_z))$ for any $z \in [x]$.

Finally it comes an $O(n^2 + \tilde{n}n)$ complexity. Let us put emphasis on the fact that it is actually the use of the ZPC-structure that allows us to get this quadratic complexity.

Algorithm 1. AlgoKExptoEq(E)

1: Input: a regular $\mathbb{K}$-expression E
2: Output: the $\mathbb{K}$-automaton $\mathcal{C}_E/\sim$
3: Compute the set of c-continuations of E and the $\mathbb{K}$-automaton $\mathcal{C}_E$
4: Compute $Q_\sim$ as the quotient $Q_\mathcal{C}/\sim$
5: Set $\delta_\sim$ to $\emptyset$ and $\gamma_\sim$ to $\overline{0}$
6: **for all** $C_{[x]} \in Q_\sim$ **do**
7:     Choose an arbitrary element (z, c_z) in $C_{[x]}$.
8:     Set $\mu_\sim(C_{[x]}) = \mu_\mathcal{C}((z, c_z))$.
9:     **for all** $((z, c_z), h(y), (y, c_y)) \in \delta_\mathcal{C}$ **do**
10:         $\delta_\sim = \delta_\sim \cup (C_{[x]}, h(y), C_{[y]})$
11:         $\gamma_\sim(C_{[x]}, h(y), C_{[y]}) = \gamma_\sim(C_{[x]}, h(y), C_{[y]}) \oplus \gamma_\mathcal{C}((z, c_z), h(y), (y, c_y))$
12:     **end for**
13: **end for**

Theorem 7. *Let* E *be a regular $\mathbb{K}$-expression. The Algorithm AlgoKExptoEq computes the equation $\mathbb{K}$-automaton of* E *with a time complexity $O(|Q_\mathcal{C}^2| + |Q_\sim||Q_\mathcal{C}|)$, that is a quadratic time complexity w.r.t. the size of* E.

5 Conclusion

The algorithm we described for converting a regular $\mathbb{K}$-expression into its equation $\mathbb{K}$-automaton makes complete the general approach based on the notion of ZPC-structure and of c-continuation computation that we already used to handle boolean constructions as well as the one of the position $\mathbb{K}$-automaton. Its main advantadge is its robustness: it is straightforwardly deduced from our algorithm for constructing the equation automaton. The role of the set of c-continuations is easy to understand, its computation and its partitionning are well-studied procedures, leading to a quadratic time complexity in both boolean and weighted cases.

References

1. Allauzen, C., Mohri, M.: A Unified Construction of the Glushkov, Follow, and Antimirov Automata. In: Královič, R., Urzyczyn, P. (eds.) MFCS 2006. LNCS, vol. 4162, pp. 110–121. Springer, Heidelberg (2006)
2. Antimirov, V.: Partial derivatives of regular expressions and finite automaton constructions. Theoret. Comput. Sci. 155, 291–319 (1996)
3. Brüggemann-Klein, A.: Regular Expressions into Finite Automata. Theoret. Comput. Sci. 120, 197–213 (1993)
4. Berstel, J., Reutenauer, C.: Les séries rationnelles et leurs langages. In: Études et recherches en informatique, Masson, Paris, 1984, Springer, Berlin Heidelberg English version: Rational series and their languages (1988)
5. Caron, P., Flouret, M.: Glushkov construction for series: The non commutative case. Intern. Journ. Comput. Maths 80(4), 457–472 (2003)
6. Champarnaud, J.-M., Laugerotte, E., Ouardi, F., Ziadi, D.: From Regular Weighted Expressions to Finite Automata. Intern. Journ. of Found. Comput. Sci. 5(15), 687–700 (2004)

7. Champarnaud, J.-M., Nicart, F., Ziadi, D.: From the ZPC-structure of a regular expression to its follow automaton. Intern. Journ. of Alg. and Comp. 16(1), 17–34 (2006)
8. Champarnaud, J.-M., Ouardi, F., Ziadi, D.: Follow automaton versus equation automaton. In: Ilie, L., Wotschke, D. (eds.) DCFS'2004, Descriptional Complexity of Formal Systems Workshop, Proceedings, pp. 145–153 (2004)
9. Champarnaud, J.-M., Ouardi, F., Ziadi, D.: Normalized expressions and finite automata, Intern. Journ. of Alg. and Comp. (to appear)
10. Champarnaud, J.-M., Ziadi, D.: From Mirkin's Prebases to Antimirov's Word Partial Derivatives. Informatica Fundamentae 45(3), 195–205 (2001)
11. Champarnaud, J.-M., Ziadi, D.: Canonical derivatives and finite automaton constructions. Theoret. Compt. Sci. 289, 137–163 (2002)
12. Champarnaud, J.-M., Ziadi, D.: From c-continuations to new quadratic algorithms for automaton synthesis. Intern. Journ. of Alg. and Comp. 11(6), 707–735 (2001)
13. Chang, C.-H., Paige, R.: From Regular Expressions to DFA's Using Compressed NFA's. Theoret. Comput. Sci. 178, 1–36 (1997)
14. Claveirole, T., Lombardy, S., O'Connor, S., Pouchet, L.-N., Sakarovitch, J.: Inside Vaucanson. In: Farré, J., Litovsky, I., Schmitz, S. (eds.) CIAA 2005. LNCS, vol. 3845, pp. 116–128. Springer, Heidelberg (2006)
15. Glushkov, V.-M.: The abstract theory of automata. Russian Mathematical Surveys 16, 1–53 (1961)
16. Hebisch, U., Weinert, H.J.: Semirings: algebraic theory and applications in computer science. World Scientific, Singapore (1993)
17. Ilie, L., Yu, S.: Follow automata. Information and computation 186, 140–162 (2003)
18. Kuich, W., Salomaa, A.: Semirings, automata, languages. In: EATCS Monographs on Theoretical Computer Science, vol. 5, Springer-Verlag, Berlin Heidelberg (1986) Princeton U. Press
19. Lombardy, S., Sakarovitch, J.: Derivations of Rational Expressions with Multiplicity. In: Diks, K., Ritter, W. (eds.) MFCS 2002. LNCS, vol. 2420, pp. 471–482. Springer, Heidelberg (2002)
20. Lombardy, S., Sakarovitch, J.: Derivatives of Rational Expressions with Multiplicity. Theoret. Comput. Sci. 332, 141–177 (2005)
21. McNaughton, R.F., Yamada, H.: Regular expressions and state graphs for automata. IEEE Trans. Electronic Comput. 9, 39–47 (1960)
22. Mirkin, B.G.: An algorithm for constructing a base in a language of regular expressions. Engineering Cybernetics 5, 110–116 (1966)
23. Sakarovitch, J.: Éléments de la théorie des automates. Les classiques de l'informatique, Vuibert Paris (2003)
24. Schützenberger, M.P.: On the definition of a family of automata. Information and control 6, 245–270 (1961)
25. Thompson, K.: Regular expression search algorithm. Comm. ACM 11(6), 419–422 (1968)
26. Ziadi, D., Ponty, J.-L., Champarnaud, J.-M.: Passage d'une expression rationnelle à un automate fini non-déterministe. Bull. Belg. Math. Soc. 4, 177–203 (1997)
27. Ziadi, D.: Quelques aspects théoriques et algorithmiques des automates. Thèse d'habilitation à diriger des recherches, Université de Rouen (2002)

An Extension of Newton's Method to ω-Continuous Semirings*

Javier Esparza, Stefan Kiefer, and Michael Luttenberger

Institute for Formal Methods in Computer Science
Universität Stuttgart, Germany
{esparza,kiefersn,luttenml}@informatik.uni-stuttgart.de

Abstract. Fixed point equations $x = F(x)$ over ω-continuous semi-rings are a natural mathematical foundation of interprocedural program analysis. Equations over the semiring of the real numbers can be solved numerically using Newton's method. We generalize the method to any ω-continuous semiring and show that it converges faster to the least fixed point than the Kleene sequence $0, F(0), F(F(0)), \ldots$ We prove that the Newton approximants in the semiring of languages coincide with finite-index approximations studied by several authors in the 1960s. Finally, we apply our results to the analysis of stochastic context-free grammars.

1 Introduction

In [2] we have argued that fixed point equations over ω-continuous semirings are a natural mathematical foundation of interprocedural program analysis. In this approach a program is mapped (in a syntax-driven way) to a system of fixed point equations over an abstract semiring. The carrier and the operations of the semiring are instantiated depending on the information about the program one wishes to compute. The information is the least solution of the system.

On ω-continuous semirings one can apply Kleene's fixed point theorem, and so the least solution of a system of equations $x = F(x)$ is the supremum of the sequence $0, F(0), F^2(0), \ldots$, where 0 is the vector whose components are all equal to the neutral element of $+$. If the carrier of the semiring is finite, this yields a procedure to compute the solution. However, if the carrier is infinite, the procedure rarely terminates, and its convergence can be very slow. So it is natural to look for "accelerations". Loosely speaking, an acceleration is a function G having the same least fixed point μF as F, but such that $(G^i(0))_{i \geq 0}$ converges faster to μF than $(F^i(0))_{i \geq 0}$.

In [2] we presented a generic acceleration scheme for *commutative* ω-continuous semirings, which we call the *Newton scheme*. We showed that the Newton scheme generalizes two well-known but apparently disconnected acceleration schemes from the literature: Newton's method for approximating a zero of a differentiable function (this is the reason for the name of our scheme) (see

---

* This work was partially supported by the DFG project *Algorithms for Software Model Checking*.

T. Harju, J. Karhumäki, and A. Lepistö (Eds.): DLT 2007, LNCS 4588, pp. 157–168, 2007.

for instance [12]), and the Hopkins-Kozen iteration scheme for Kleene algebras (which are very close to idempotent commutative semirings) [9].

In this paper we further generalize the Newton scheme of [2] to *arbitrary* ω-continuous semirings, commutative or not. In particular, this allows us to solve systems of fixed point equations over the language semiring having the set of languages over a given alphabet as carrier and union and concatenation of languages as sum and product, respectively. For instance, if we consider this semiring for the alphabet $\{(,)\}$, then the least solution of the equation $X = (X) + XX + 1$ is the Dyck language of well-parenthesized expressions. Clearly, the least solution of a system is a context-free language, and every context-free language is the solution of a system.

The Newton acceleration scheme approximates the least solution from below. In the case of languages, it computes a chain $L_0 \subseteq L_1 \subseteq L_2 \ldots$ of approximations of the least solution $L = \bigcup_{i \geq 0} L_i$. Our main theorem characterizes these approximations, and shows that, once again, a well-known concept from the literature "is nothing but" Newton's approximation technique.

The i-th approximation of the Newton scheme turns out to be the *index-$(i+1)$* approximation $L_{i+1}(G)$ of $L(G)$. Recall that a terminal word w is in $L_i(G)$ if there is a derivation $S \Rightarrow \alpha_1 \Rightarrow \cdots \Rightarrow \alpha_r = w$ and every α_i, $0 \leq i \leq r$ contains at most i occurrences of variables [13,8,14,7].

Our result allows to transfer results from language theory to numerical analysis and vice versa. We develop a way of applying finite-index approximations to stochastic context-free grammars and computing the approximation quality.

It is well-known that Newton's method for approximating the zero of a function is based on the notion of differential. Our results require to give a definition of derivative of a polynomial expressions for arbitrary ω-continuous semirings. This can be seen as a generalization of the Brzozowski's definition of derivative for regular languages and Hopkins and Kozen's definition for commutative semirings, and could have some interest of its own.

Organization and Contributions of This Paper. In Section 2 we define differentials for power series over ω-continuous semirings. Section 3 introduces a generalized Newton's method for approximating the least solution of fixed point equations over arbitrary ω-continuous semirings. In Section 4 (Theorem 4.1) we characterize the iterates of the Newton scheme in terms of the *tree dimension*, a concept generalized from [2]. We apply this result to context-free grammars in Section 5 and prove that the Newton iterates coincide with finite-index approximations. In Section 6 we apply the generalized Newton's method to stochastic context-free grammars.

Missing proofs can be found in a technical report [1].

2   Differentials in ω-Continuous Semirings

The goal of this section is to generalize the notion of differential of a function to ω-continuous semirings. More precisely, we will only define the notion for functions that can be represented as a power series.

Recall the classical notion of differential that can be found in any elementary calculus textbook. Let V be a vector space of finite dimension over the real numbers $\mathbb{R}$. The *dual* of V is the vector space whose elements are the linear functions $V \rightarrow \mathbb{R}$, usually called *linear forms*. Given $f: V \rightarrow \mathbb{R}$, the *differential* Df of f (when it exists) is an application $Df: V \rightarrow \tilde{V}$ that assigns to every vector $\boldsymbol{v} \in V$ a linear form $Df|_{\boldsymbol{v}}$. Loosely speaking, $Df|_{\boldsymbol{v}}$ is the best linear approximation of f at the point $\boldsymbol{v}$.

We wish to generalize the notion of differential to the case in which $\mathbb{R}$ is replaced by an arbitrary ω-continuous semiring, and f is a power series. For this, we first introduce ω-continuous semirings in Section 2.1. In Section 2.2 we generalize the notions of vector and linear form over the reals. In Section 2.3 we introduce power series, and finally in Section 2.4 the notion of differential itself.

2.1 ω-Continuous Semirings

In the following, we work with ω-continuous semirings, as defined in [11].

Definition 2.1. *A semiring S is given by $\langle S, +, \cdot, 0, 1 \rangle$, where S is a set with $0, 1 \in S$, $\langle S, +, 0 \rangle$ is a commutative monoid with neutral element 0, $\langle S, \cdot, 1 \rangle$ is a monoid with neutral element 1, 0 is an annihilator w.r.t. $\cdot$, i.e. $0 \cdot a = a \cdot 0 = 0$ for all $a \in S$, and $\cdot$ distributes over $+$, i.e. $a \cdot (b + c) = a \cdot b + a \cdot c$, and $(a + b) \cdot c = a \cdot c + b \cdot c$. The natural order relation $\sqsubseteq$ on a semiring S is defined by $a \sqsubseteq b \Leftrightarrow \exists d \in S : a + d = b$. The semiring S is naturally ordered if $\sqsubseteq$ is a partial order on S.*

An ω-*continuous semiring* is a naturally ordered semiring extended by an infinite summation-operator $\sum$ that satisfies the following properties[1]:

– For every sequence $a : \mathbb{N} \rightarrow S$ the supremum $\sup\{\sum_{0 \leq i \leq k} a_i \mid k \in \mathbb{N}\}$ exists in S w.r.t. $\sqsubseteq$, and is equal to $\sum_{i \in \mathbb{N}} a_i$. As a consequence, every non-decreasing sequence $a_i \sqsubseteq a_{i+1}$ converges, i.e. $\sup\{a_i\}$ exists.
– It holds

$$\sum_{i \in \mathbb{N}} (c \cdot a_i) = c \cdot \left(\sum_{i \in \mathbb{N}} a_i \right), \quad \sum_{i \in \mathbb{N}} (a_i \cdot c) = \left(\sum_{i \in \mathbb{N}} a_i \right) \cdot c, \quad \sum_{j \in J} \left(\sum_{i \in I_j} a_j \right) = \sum_{i \in \mathbb{N}} a_i$$

for every $a : \mathbb{N} \rightarrow S$, $c \in S$, and every partition $(I_j)_{j \in J}$ of $\mathbb{N}$.

In the following we often omit the dot $\cdot$ in products.

Example 2.1. The *real semiring*, denoted by $\mathcal{S}_{\mathbb{R}}$, has $\mathbb{R}_{\geq 0} \cup \{\infty\}$ as carrier. Sum and multiplication are defined as expected (e.g. $a \cdot \infty = \infty$ for $a \neq 0$). Notice that sum is not idempotent and product is commutative.

The *language semiring* over an alphabet Σ, denoted by $\mathcal{S}_{\Sigma}$, has the set of all languages over Σ as carrier. Sum is union, and product is concatenation of languages. Notice that sum is idempotent and product is not commutative.

---

[1] [11] requires infinite summation for *any* sum, but we need only countable sums here.

2.2   Vectors and Linear Forms

We introduce the notion of vectors and linear forms over an ω-continuous semiring S. Notice that the name vector and linear form have to be taken with a grain of salt, because for instance the set of vectors over S does not build a vector space (since S may not be a field). However, it is useful to keep the names to remember that they generalize the usual notions of vector and linear form.

Definition 2.2. *Let S be an ω-continuous semiring and let $\mathcal{X}$ be a finite set of variables.*

A vector is a mapping $\boldsymbol{v} \colon \mathcal{X} \to S$. The set of all vectors is denoted by V. Given a countable set I and a vector $\boldsymbol{v}_i$ for every $i \in I$, we denote by $\sum_{i \in I} \boldsymbol{v}_i$ the vector given by $\left(\sum_{i \in I} \boldsymbol{v}_i \right)(X) = \sum_{i \in I} \boldsymbol{v}_i(X)$ for every $X \in \mathcal{X}$.

A linear form is a mapping $l \colon V \to S$ satisfying $l(\boldsymbol{v} + \boldsymbol{v}') = l(\boldsymbol{v}) + l(\boldsymbol{v}')$ for every $\boldsymbol{v}, \boldsymbol{v}' \in V$ and $l(\boldsymbol{0}) = 0$, where $\boldsymbol{0}$ denotes the vector given by $\boldsymbol{0}(X) = 0$ for every $X \in \mathcal{X}$. Given a linear form l and $s, s' \in S$, we denote by $s \cdot l \cdot s'$ the linear form given by $(s \cdot l \cdot s')(\boldsymbol{v}) = s \cdot l(\boldsymbol{v}) \cdot s'$ for every $\boldsymbol{v} \in V$. Given a countable set I and a linear form l_i for every $i \in I$, we denote by $\sum_{i \in I} l_i$ the linear form given by $\left(\sum_{i \in I} l_i \right)(\boldsymbol{v}) = \sum_{i \in I} l_i(\boldsymbol{v})$ for every $\boldsymbol{v} \in V$.

2.3   Polynomials and Power Series

Definition 2.3. *Let S be an ω-continuous semiring and $\mathcal{X}$ be a finite set of variables. A monomial is a finite expression*

$$a_1 X_1 a_2 \cdots a_k X_k a_{k+1}$$

where $k \geq 0$, $a_1, \ldots, a_{k+1} \in S$ and $X_1, \ldots, X_k \in \mathcal{X}$. A polynomial is an expression of the form $m_1 + \ldots + m_k$ where $k \geq 0$ and $m_1, \ldots, m_k$ are monomials. We let $S[\mathcal{X}]$ denote the set of polynomials w.r.t. S and $\mathcal{X}$. Similarly, a power series is an expression of the form $\sum_{i \in I} m_i$, where I is a countable set and m_i is a monomial for every $i \in I$. We use $S[\![\mathcal{X}]\!]$ to denote this set.

Definition 2.4. *Let $f = \alpha_1 X_1 \alpha_2 X_2 \alpha_3 \ldots \alpha_k X_k \alpha_{k+1} \in S[\![\mathcal{X}]\!]$ be a monomial and let $\boldsymbol{v}$ be a vector. We define $f(\boldsymbol{v})$, the evaluation of f at $\boldsymbol{v}$, as*

$$f(\boldsymbol{v}) = \alpha_1 \boldsymbol{v}(X_1) \alpha_2 \boldsymbol{v}(X_2) \alpha_3 \cdots \alpha_k \boldsymbol{v}(X_k) \alpha_{k+1}.$$

We extend this to any power series $f = \sum_{i \in I} f_i \in S[\![\mathcal{X}]\!]$ by $f(\boldsymbol{v}) = \sum_{i \in I} f_i(\boldsymbol{v})$.

Finally, we can also define the product of polynomials and linear forms as follows:

Definition 2.5. *Let $I \subseteq \mathbb{N}$, let $f, g \in S[\mathcal{X}]$ be polynomials, and let l be a linear form. The expression flg denotes the mapping $T \colon V \to V \to S$ given by*

$$T(\boldsymbol{u}, \boldsymbol{v}) = f(\boldsymbol{u}) l(\boldsymbol{v}) g(\boldsymbol{u}) .$$

We denote by $T|_{\boldsymbol{u}} \colon V \to S$ the linear form given by $T|_{\boldsymbol{u}}(\boldsymbol{v}) = T(\boldsymbol{u}, \boldsymbol{v})$.

2.4   Differential of a Power Series

Recall that in the real case the differential of a function $f\colon V \to \mathbb{R}$ is a mapping $Df\colon V \to \widetilde{V}$ that assigns to every vector $\boldsymbol{v} \in V$ a linear form $Df|_{\boldsymbol{v}}$, the best linear approximation of f at the point $\boldsymbol{v}$. Given the basis $\{\boldsymbol{e}_1, \dots, \boldsymbol{e}_n\}$ of unit vectors of V, the dual basis $\{\mathrm{dX}_1, \dots, \mathrm{dX}_n\}$ of $\widetilde{V}$ is defined by $\mathrm{dX}_i(a_1\boldsymbol{e}_1 + \dots + a_n\boldsymbol{e}_n) = a_i$ for every $a_1, \dots, a_n \in \mathbb{R}$. Since $\{\mathrm{dX}_1, \dots, \mathrm{dX}_n\}$ is a basis of $\widetilde{V}$ there are functions $\lambda_1, \dots, \lambda_n\colon V \to \mathbb{R}$ such that

$$Df|_{\boldsymbol{v}} = \lambda_1(\boldsymbol{v})\,\mathrm{dX}_1 + \cdots + \lambda_n(\boldsymbol{v})\,\mathrm{dX}_n$$

for every $\boldsymbol{v} \in V$ (here λ_i is the partial derivative of f w.r.t. X_i). If for every variable X_i we define $D_{X_i}f\colon V \to \widetilde{V}$ as the mapping that assigns to every vector $\boldsymbol{v}$ the linear form $D_{X_i}f|_{\boldsymbol{v}} = \lambda_i(\boldsymbol{v})\,\mathrm{dX}_i$, then we have $Df = D_{X_1}f + \dots + D_{X_n}f$.

Definition 2.7 below generalizes the linear forms $D_{X_i}f|_{\boldsymbol{v}}$ to the case in which $\mathbb{R}$ is replaced by an ω-continuous semiring. We start by generalizing the dX_i:

Definition 2.6. *For every $X \in \mathcal{X}$, we denote by dX the linear form defined by $\mathrm{dX}(\boldsymbol{v}) = \boldsymbol{v}(X)$ for every $\boldsymbol{v} \in V$.*

Definition 2.7. *Let f be a power series and let $X \in \mathcal{X}$ be a variable. The differential of f w.r.t. X is the mapping $D_Xf\colon V \to V \to S$ that assigns to every vector $\boldsymbol{v}$ the linear form $D_Xf|_{\boldsymbol{v}}\colon V \to S$ inductively defined as follows:*

$$D_Xf|_{\boldsymbol{v}} = \begin{cases} 0 & \text{if } f \in S \text{ or } f \in \mathcal{X} \setminus \{X\} \\ \mathrm{dX} & \text{if } f = X \\ D_Xg|_{\boldsymbol{v}} \cdot h + g \cdot D_Xh|_{\boldsymbol{v}} & \text{if } f = g \cdot h \\ \sum_{i \in I} D_Xf_i|_{\boldsymbol{v}} & \text{if } f = \sum_{i \in I} f_i. \end{cases}$$

Further, we define the differential *of f as the linear form*

$$Df := \sum_{X \in \mathcal{X}} D_Xf.$$

In the real case the differential is used to approximate the value of a differentiable function $f(\boldsymbol{v} + \boldsymbol{u})$ in terms of $f(\boldsymbol{v})$ and $Df|_{\boldsymbol{v}}(\boldsymbol{u})$. The following lemma goes in the same direction.

Lemma 2.1. *Let f be a power series and let $\boldsymbol{v}, \boldsymbol{u}$ be two vectors. We have*

$$f(\boldsymbol{v}) + Df|_{\boldsymbol{v}}(\boldsymbol{u}) \sqsubseteq f(\boldsymbol{v} + \boldsymbol{u}) \sqsubseteq f(\boldsymbol{v}) + Df|_{\boldsymbol{v}+\boldsymbol{u}}(\boldsymbol{u}).$$

3   Solving Systems of Fixed Point Equations

The partial order $\sqsubseteq$ on the semiring S can be lifted to an order on vectors, also denoted by $\sqsubseteq$, given by $\boldsymbol{v} \sqsubseteq \boldsymbol{v}'$ iff $\boldsymbol{v}(X) \sqsubseteq \boldsymbol{v}'(X)$ for every $X \in \mathcal{X}$.

In the following, let $\boldsymbol{F}$ be a *vector of power series*, i.e., a mapping that assigns to each variable $X \in \mathcal{X}$ a power series $\boldsymbol{F}(X)$. For convenience we denote $\boldsymbol{F}(X)$

by F_X. Given a vector v, we define $F(v)$ as the vector satisfying $(F(v))(X) = F_X(v)$ for every $X \in \mathcal{X}$, i.e., $F(v)$ is the vector that assigns to X the result of evaluating the power series F_X at v. So, F can be seen as a mapping $F: V \to V$.

Given a vector of power series F, we are interested in the least fixed point of F, i.e., the least vector v w.r.t. $\sqsubseteq$ satisfying $v = F(v)$.

3.1   Kleene's Iteration Scheme

Recall that a mapping $f: \mathcal{S} \to \mathcal{S}$ is *monotone* if $a \sqsubseteq b$ implies $f(a) \sqsubseteq f(b)$, and ω-*continuous* if for any infinite chain $a_0 \sqsubseteq a_1 \sqsubseteq a_2 \sqsubseteq \ldots$ we have $\sup\{f(a_i)\} = f(\sup\{a_i\})$. The definition can be extended to mappings $F: V \to V$ from vectors to vectors in the obvious way (componentwise). Then we may formulate the following proposition (cf. [11]).

Proposition 3.1. *Let F be a vector of power series. The mapping induced by F is monotone and continuous. Hence, by Kleene's theorem, F has a unique least fixed point μF. Further, μF is the supremum (w.r.t. $\sqsubseteq$) of the Kleene sequence given by $\kappa^{(0)} = F(0)$, and $\kappa^{(i+1)} = F(\kappa^{(i)})$.*[2]

Kleene's iteration scheme converges very slowly. Consider for instance the equation $X = aXb + 1$ over the semiring of languages over $\{a, b\}$ (where $0 = \emptyset$ and $1 = \{\lambda\}$). The i-th iteration $\kappa^{(i)}$ is the language $\{a^j b^j \mid j \leq i\}$, so the scheme needs an infinite number of iterations to reach μF. Newton's iteration scheme, introduced below, can be seen as an "acceleration" of Kleene's scheme.

3.2   Newton's Iteration Scheme

Let F be a vector of power series, and v any vector. Then $DF|_v$ denotes the mapping $V \to V$ with $(DF|_v(u))_X = DF_X|_v(u)$. So $DF|_v$ can be seen as the evaluation of a mapping $DF: V \to V \to V$ at v (cf. Definition 2.7). Lemma 2.1 then becomes $F(v) + DF|_v(u) \sqsubseteq F(v + u) \sqsubseteq F(v) + DF|_{v+u}(u)$.

Newton's scheme uses DF to obtain a sequence that converges more quickly to the least fixed point than Kleene's sequence. In order to introduce it we first define the Kleene star of an arbitrary mapping $V \to V$:

Definition 3.1. *Let $F: V \to V$ be an arbitrary mapping. The mapping $F^i: V \to V$ is inductively defined by $F^0(v) = v$ and $F^{i+1}(v) = F(F^i(v))$. The Kleene star of F, denoted by F^*, is the mapping $F^*: V \to V$ given by $F^*(v) = \sum_{i \geq 0} F^i(v)$.*

Now we can define Newton's scheme.

Definition 3.2. *Let $F: V \to V$ be a vector of power series. We define the Newton sequence $(\nu^{(i)})_{i \in \mathbb{N}}$ as follows:*
$$\nu^{(0)} = F(0) \quad \text{and} \quad \nu^{(i+1)} = \nu^{(i)} + DF|_{\nu^{(i)}}^*(\delta^{(i)}),$$
where $\delta^{(i)}$ has to satisfy $F(\nu^{(i)}) = \nu^{(i)} + \delta^{(i)}$.

In words, $\nu^{(i+1)}$ is obtained by adding to $\nu^{(i)}$ the result of evaluating the Kleene star of $DF|_{\nu^{(i)}}$ at the point $\delta^{(i)}$.

---

[2] In [2] we define $\kappa^{(0)} = 0$, but $\kappa^{(0)} = F(0)$ is slightly more convenient for this paper.

The name "Newton's method" is justified as follows: Consider a univariate equation $X = F(X)$ over $\mathcal{S}_\mathbb{R}$ with $F'|_x \in (-1,1)$ for $x \in [0, \mu F)$. Applying Newton's method as described above to $X = F(X)$, yields $\nu^{(i+1)} = \nu^{(i)} + F'|^*_{\nu^{(i)}}(F(\nu^{(i)}) - \nu^{(i)}) = \nu^{(i)} + \sum_{k \in \mathbb{N}} F'|^k_{\nu^{(i)}}(F(\nu^{(i)}) - \nu^{(i)}) = \nu^{(i)} + \frac{F(\nu^{(i)}) - \nu^{(i)}}{1 - F'|_{\nu^{(i)}}} = \nu^{(i)} - \frac{G(\nu^{(i)})}{G'|_{\nu^{(i)}}}$. But this is exactly Newton's method for finding a zero of $G(X) = F(X) - X$. This can be generalized to equation systems (see [2,5]).

The following theorem summarizes the properties of the Newton sequence.

Theorem 3.1. *Let $F \in \mathcal{S}[\![\mathcal{X}]\!]^{\mathcal{X}}$ be a vector of power series. For every $i \in \mathbb{N}$:*
$$\kappa^{(i)} \sqsubseteq \nu^{(i)} \sqsubseteq F(\nu^{(i)}) \sqsubseteq \mu F = \sup_j \kappa^{(j)}.$$

In particular, this theorem ensures the existence of a suitable $\delta^{(i)}$ (because $\nu^{(i)} \sqsubseteq F(\nu^{(i)})$), and the convergence of the Newton sequence to the same value as the Kleene sequence. Moreover, since $\kappa^{(i)} \sqsubseteq \nu^{(i)}$, the Newton sequence converges "at least as fast" as the Kleene sequence.

Example 3.1. In the following examples we set $\mathcal{X} = \{X\}$. Since in this case vectors only have one component, given an element s of a semiring we also use s to denote the vector v given by $v(X) = s$.

Consider the language semiring $\mathcal{S}_{\{a,b\}}$ over the alphabet $\{a,b\}$. One can show that by taking $\delta^{(i)} = F(\nu^{(i)})$ Newton's sequence can be simplified to

$$\nu^{(0)} = F(0) \quad \text{and} \quad \nu^{(i+1)} = DF|^*_{\nu^{(i)}}(F(\nu^{(i)})).$$

(1) Consider again the polynomial $f(X) = aXb+1$. As already mentioned above, the Kleene sequence needs ω iterations to reach the fixed point $\{a^n b^n \mid n \geq 0\}$. As a warm-up we show that the Newton sequence converges after one step. We have $Df|_v = a\,\mathrm{dX}\,b$ for every $v \in V$, and so

$$\nu^{(1)} = (a\,\mathrm{dX}\,b)^*(1) = \sum_{j \geq 0} a^j\,\mathrm{dX}(1)b^j = \{a^j b^j \mid j \geq 0\}.$$

The next example shows a more interesting case.

(2) Consider the polynomial $f(X) = aXX + b$. We have:

$$Df|_v = av(X)\,\mathrm{dX} + a\,\mathrm{dX}\,v(X)$$
$$\nu^{(0)} = b$$
$$\nu^{(1)} = Df|^*_b(abb + b) = (ab\,\mathrm{dX} + a\,\mathrm{dX}\,b)^*(abb + b)$$
$$= L(X \to abX \mid aXb \mid abb \mid b)$$
$$\nu^{(i+1)} = Df|^*_{\nu^{(i)}}(f(\nu^{(i)})) = (a\nu^{(i)}\,\mathrm{dX} + a\,\mathrm{dX}\,\nu^{(i)})^*(f(\nu^{(i)}))$$

In this case the Newton sequence also needs ω iterations. We shall see in Section 5 that $\nu^{(i)}$ contains the words generated by the grammar $X \to aXX$, $X \to b$ via derivations of index at most $i+1$, i.e., derivations in which no intermediate word contains more than $i + 1$ occurrences of variables.

(3) Consider the same polynomial as in (2), but over a semiring where product is *commutative* (and + is still idempotent). In this case we have:

$$Df|_v = av(X)\,dX$$

$$\nu^{(0)} = b$$
$$\nu^{(1)} = Df|_b^*(ab^2 + b) = (ab\,dX)^*(ab^2 + b) = (ab)^*(ab^2 + b) = (ab)^*b$$
$$\nu^{(2)} = Df|_{\nu^{(1)}}^*(f(\nu^{(1)})) = (a\nu^{(1)})^*(f(\nu^{(1)})) = a(ab)^*b(a((ab)^*b)^2 + b) = (ab)^*b$$

So the Newton sequence reaches the fixed point at $\nu^{(1)}$. The language $(ab)^*b$ is a regular language having the same Parikh mapping as the context-free language generated by $X \rightarrow aXX$, $X \rightarrow b$.

4 Derivation Trees and the Newton Iterates

In language theory, given a grammar G, one associates derivations and derivation trees with G. The language $L(G)$ can be seen as the set of all words that can be derived by a derivation tree of G. On the other hand, if G is context-free, then $L(G)$ is the least solution of a fixed point equation $x = F(x)$ over a language semiring, where the equation $x = F(x)$ is essentially the production set of G.

In this section we extend the notion of derivation trees to fixed point equations $x = F(x)$ over any ω-continuous semiring. It will be easy to see that the Kleene iterates $\kappa^{(i)}$ correspond to the derivation trees of height at most i. We will show that the Newton iterates $\nu^{(i)}$ correspond to the derivation trees of *dimension* at most i, generalizing the concept of dimension introduced in [2]. This gives valuable insight into the generalized Newton's method from the previous section and establishes a strong link between two apparently disconnected concepts, one from language theory (finite-index languages, see Section 5) and the other from numerical mathematics (Newton's method).

Definition 4.1 (derivation tree). *Let F be a vector of power series. A derivation tree of F is defined inductively as follows. Let $a_1X_1a_2\cdots X_sa_{s+1}$ be a summand of F_X ($X \in \mathcal{X}$) and let v be a node labelled by*

$$\lambda(v) = (\lambda_1(v), \lambda_2(v)) = (X, a_1X_1a_2\cdots X_sa_{s+1}).$$

Let $t_1,\ldots,t_s$ be derivation trees with $\lambda_1(t_r) = X_r$ ($1 \le r \le s$). Then the tree whose root is v and whose (ordered) children are $t_1,\ldots,t_s$ is a derivation tree.

We identify a derivation tree and its root from now on and often simply write *tree* when we mean *derivation tree*.

Remark to multiplicities. Let F be a system of power series. If, for a variable $X \in \mathcal{X}$, the same monomial m occurs more than once as a summand of F_X, and there is a node v in a tree of F s.t. $\lambda_1(v) = X$ and $\lambda_2(v) = m$, then it is not clear "which" summand m of F_X was used at v. But we assume in the following that $\lambda_2(v)$ is a particular *occurrence* of m in F_X. Hence, two trees which are equal up to different occurrences are regarded as *different* in the following. However, we do not make that explicit in our definition to avoid notational clutter.

Definition 4.2 (height, yield). *The height $h(t)$ of a derivation tree t is the length of a longest path from the root to a leaf of t. The yield $Y(t)$ of a derivation tree t with $\lambda_2(t) = a_1 X_1 a_2 \cdots X_s a_{s+1}$ is inductively defined to be*

$$Y(t) = a_1 Y(t_1) a_2 \cdots Y(t_s) a_{s+1}.$$

We can characterize the Kleene sequence $(\kappa^{(i)})_{i \in \mathbb{N}}$ using the height as follows.

Proposition 4.1. *For all $i \in \mathbb{N}$ and $X \in \mathcal{X}$, we have that $(\kappa^{(i)})_X$ is the sum of yields of all derivation trees with $\lambda_1(t) = X$ and $h(t) \leq i$.*

Now we aim at characterizing the Newton sequence $(\nu^{(i)})_{i \in \mathbb{N}}$ in terms of derivation trees. To this end we need use another property of a tree, the tree *dimension*. As does the height, it depends only on the graph structure of a tree.

Definition 4.3 (dimension). *For a tree t, define $dl(t) = (d(t), l(t))$ as follows.*

1. *If $h(t) = 0$, then $dl(t) = (0,0)$.*
2. *If $h(t) > 0$, let $\{t_1, \ldots, t_s\}$ be the children of t where $d(t_1) \geq \ldots \geq d(t_s)$. Let $d_1 = d(t_1)$. If $s > 1$, let $d_2 = d(t_2)$, otherwise let $d_2 = 0$. Then*

$$dl(t) = \begin{cases} (d_1 + 1, 0) & \text{if } d_1 = d_2 \\ (d_1, l(t_1) + 1) & \text{if } d_1 > d_2. \end{cases}$$

We call $d(t)$ the dimension *of the tree t.*

The following Theorem 4.1 defines a concrete Newton sequence $(\nu^{(i)})_{i \in \mathbb{N}}$ which allows for the desired tree characterization of $\nu^{(i)}$ (cf. Prop. 4.1).

Theorem 4.1. *Let F be a vector of power series. Define the sequence $(\nu^{(i)})_{i \in \mathbb{N}}$ as follows:*

$$\nu^{(0)} = F(0) \quad \text{and} \quad \nu^{(i+1)} = \nu^{(i)} + DF|_{\nu^{(i)}}^{*}(\delta^{(i)}),$$

where $\delta_X^{(i)}$ is the sum of yields of all derivation trees t with $\lambda_1(t) = X$ and $dl(t) = (i+1, 0)$. Then for all $i \geq 0$:

(1) $F(\nu^{(i)}) = \nu^{(i)} + \delta^{(i)}$, so $(\nu^{(i)})_{i \in \mathbb{N}}$ is a Newton sequence as defined in Def. 3.2;

(2) $\nu_X^{(i)}$ is the sum of yields of all derivation trees t with $\lambda_1(t) = X$ and $d(t) \leq i$.

5   Languages with Finite Index

In this section we study fixed point equations $x = F(x)$ over language semirings. Let $\mathcal{S}_\Sigma$ be the language semiring over a finite alphabet Σ. Let F be a vector of polynomials over $\mathcal{X}$ whose coefficients are elements of Σ. Then, for each $X_0 \in \mathcal{X}$, there is a naturally associated context-free grammar $G_{F,X_0} = (\mathcal{X}, \Sigma, P, X_0)$, where the set of productions is $P = \{(X_i \to \alpha) \mid \alpha \text{ is a summand of } F_{X_i}\}$. It is well-known that $L(G_{F,X_0}) = (\mu F)_{X_0}$ (see e.g. [11]). Analogously, each grammar is naturally associated with a vector of polynomials. In the following we use grammars and vectors of polynomials interchangeably.

We show in this section that the approximations $\boldsymbol{\nu}^{(i)}$ obtained from our generalized Newton's method are strongly linked with the *finite-index* approximations of $L(G)$. Finite-index languages have been extensively investigated under different names [13,8,14,7,6] (see [6] for historical background).

Definition 5.1. *Let G be a grammar, and let D be a derivation $X_0 = \alpha_0 \Rightarrow \cdots \Rightarrow \alpha_r = w$ of $w \in L(G)$, and for every $i \in \{0, \ldots, r\}$ let β_r be the projection of α_r onto the variables of G. The index of D is the maximum of $\{|\beta_0|, \ldots, |\beta_r|\}$. The index-$i$ approximation of $L(G)$, denoted by $L_i(G)$, contains the words derivable by some derivation of G of index at most i.*

We show that for a context-free grammar G in Chomsky normal form (CNF), the Newton approximations to $L(G)$ coincide with the finite-index approximations.

Theorem 5.1. *Let $G = (\mathcal{X}, \Sigma, P, X_0)$ be a context-free grammar in CNF and let $(\boldsymbol{\nu}^{(i)})_{i \in \mathbb{N}}$ be the Newton sequence associated with G. Then $(\boldsymbol{\nu}^{(i)})_{X_0} = L_{i+1}(G)$ for every $i \geq 0$.*

In particular, it follows from Theorem 5.1 that the (first component of the) Newton sequence for a context-free grammar G converges in finitely many steps if and only if $L(G) = L_i(G)$ for some $i \in \mathbb{N}$.

6   Stochastic Context-Free Grammars

In this section we show how the link between finite-index approximations and (the classical version of) Newton's method can be exploited for the analysis of *stochastic context-free grammars (SCFGs)*, a model that combines the language semiring and the real semiring.

A SCFG is a CFG where every production is assigned a probability. SCFGs are widely used in natural language processing and in bioinformatics (see also the example at the end of the section). We use the grammar G^{ex} with productions $X \xrightarrow{1/6} X^6, X \xrightarrow{1/2} X^5, X \xrightarrow{1/3} a$ as running example.

SCFGs can be seen as systems of polynomials over the *direct product* of the semiring $\mathcal{S}_\Sigma$ of languages over Σ and the semiring of non-negative reals ($\mathbb{R}_{\geq 0} \cup \{\infty\}, +, \cdot, 0, 1$). The system for the grammar G has one polynomial, namely $F_X = (\lambda, \frac{1}{6})X^6 + (\lambda, \frac{1}{2})X^5 + (a, \frac{1}{3})$.

Given an SCFG it is often important to compute the *termination probability* $T(X)$ of a given variable X (see [4,3] for applications to program verification). $T(X)$ is the probability that a derivation starting at X "terminates", i.e., generates some word. For G^{ex} we have $0.3357037075 < T(X) < 0.3357037076$. It is easy to see that the termination probabilities are given by (the real part of) the least fixed point of the corresponding system of polynomials [4,5], and that they may be irrational and not representable by radicals (in fact, G^{ex} is an example). Therefore, they must be numerically approximated. This raises the problem that whenever other parameters are calculated from the termination probabilities it is necessary to conduct an error propagation analysis.

A solution to this problem is to replace the original SCFG G by another one, say G', generating "almost" the same language, and for which all termination probabilities are 1. "Almost" means that (1) the probability of the derivations of G that are not derivations of G' is below a given bound, and (2) that the quotient of the probabilities of any two derivations that appear in both G and G' is the same in G and G'. G' can be obtained as follows. Since $\boldsymbol{\delta}^{(i)}$ is uniquely determined by the equation $\boldsymbol{F}(\boldsymbol{\nu}^{(i)}) = \boldsymbol{\nu}^{(i)} + \boldsymbol{\delta}^{(i)}$ over $S_{\mathbb{R}}$, we know that $\boldsymbol{\nu}^{(K)}$ is the probability of the derivation trees of dimension at most K. Hence, we can approximate the terminating runs by choosing $G' = G_K$, where G_K generates the derivation trees of G of dimension at most K. Assuming that G is in Chomsky normal form, G_K has variables $X_0, X_1, X_{\leq 1}, \ldots, X_K, X_{\leq K}$ for every variable X of G, with $X_{\leq K}$ as axiom. Its productions are constructed so that X_i ($X_{\leq i}$) generates the derivation trees of G of dimension i (at most i) with X as root. For this, G_K has: (i) a production $X_0 \to a$ for every production $X \to a$ of G, (ii) productions[3] $X_i \to Y_{i-1}Z_{i-1}$, $X_i \to Y_i Z_{\leq i-1}$ and $X_i \to Y_{\leq i-1}Z_i$ for every production $X \to YZ$ of G and every $i \in \{1, \ldots, K\}$, and (iii) productions $X_{\leq i} \to X_i$ and $X_{\leq i} \to X_{\leq i-1}$ for every variable X and every $i \in \{1, \ldots, K\}$. It remains to define the probabilities of the productions so that the probability that a tree t is derived from $X_{\leq K}$ in G_K is equal to the conditional probability that t is derived from X in G under the condition that a tree of dimension at most K is derived from X. For this we set $p(X_0 \to a) = \frac{p(X \to a)}{\nu_X^{(0)}}$. For $K > 0$, an induction over the tree dimension shows that we have to choose the remaining probabilities as follows (we omit some symmetric cases):

$$p(X_{\leq K} \to X_K) = \frac{\boldsymbol{\Delta}_X^{(K)}}{\nu_X^{(K)}} \qquad p(X_K \to Y_K Z_{\leq K-1}) = \frac{p(X \to YZ)}{\boldsymbol{\Delta}_X^{(K)}} \boldsymbol{\Delta}_Y^{(K)} \nu_Z^{(K-1)}$$

$$p(X_{\leq K} \to X_{\leq K-1}) = \frac{\nu_X^{(K-1)}}{\nu_X^{(K)}} \qquad p(X_K \to Y_{K-1} Z_{K-1}) = \frac{p(X \to YZ)}{\boldsymbol{\Delta}_X^{(K)}} \boldsymbol{\Delta}_Y^{(K-1)} \boldsymbol{\Delta}_Z^{(K-1)}$$

with $\boldsymbol{\Delta}^{(k)} = \boldsymbol{\nu}^{(k)} - \boldsymbol{\nu}^{(k-1)}$ for $k > 0$, and $\boldsymbol{\Delta}^{(0)} = \boldsymbol{\nu}^{(0)}$.

The first iterations of the Newton sequence for our running example G^{ex} are

$$\boldsymbol{\nu}^{(0)} = 1/3, \quad \boldsymbol{\nu}^{(1)} = 0.3357024402, \quad \boldsymbol{\nu}^{(2)} = 0.3357037075, \quad \boldsymbol{\nu}^{(3)} = 0.3357037075$$

(up to machine accuracy). In this case we could replace G^{ex} by G_2^{ex} or even G_1^{ex}.

We finish the section with another example. The following SCFG, taken from [10], is used to describe the secondary structure in RNA:

$$L \xrightarrow{0.869} CL \quad L \xrightarrow{0.131} C \quad S \xrightarrow{0.788} pSp \quad S \xrightarrow{0.212} CL \quad C \xrightarrow{0.895} s \quad C \xrightarrow{0.105} pSp.$$

The following table shows the first iterates of the Newton and Kleene sequences for the corresponding system of polynomials.

i	($\nu_L^{(i)}$,	$\nu_S^{(i)}$,	$\nu_C^{(i)}$)	($\kappa_L^{(i)}$,	$\kappa_S^{(i)}$,	$\kappa_C^{(i)}$)
1	(0.5585,	0.4998,	0.9475)	(0.1172,	0,	0.895)
3	(0.9250,	0.9150,	0.9911)	(0.2793,	0.0571,	0.8973)
5	(0.9972,	0.9968,	0.9997)	(0.3806,	0.1414,	0.9053)

---

[3] where $X_{\leq 0}$ is identified with X_0.

As we can see, the contribution of trees of dimension larger than 5 is negligible. Here, the Newton sequence converges much faster than the Kleene sequence.

7   Conclusions

We have generalized Newton's method for numerically computing a zero of a differentiable function to a method for approximating the least fixed point of a system of power series over an arbitrary ω-continuous semiring. We have characterized the iterates of the Newton sequence in terms of derivation trees: the i-th iterate corresponds to the trees of dimension at most i. Perhaps surprisingly, in the language semiring the Newton iterates turn out to coincide with the classical notion of finite-index approximations. Finally, we have sketched how our approach can help to analyze stochastic context-free grammars.

References

1. Esparza, J., Kiefer, S., Luttenberger, M.: An extension of Newton's method to ω-continuous semirings. Technical report, Universität Stuttgart (2007)
2. Esparza, J., Kiefer, S., Luttenberger, M.: On fixed point equations over commutative semirings. In: Thomas, W., Weil, P. (eds.) STACS 2007. LNCS, vol. 4393, pp. 296–307. Springer, Heidelberg (2007)
3. Esparza, J., Kučera, A., Mayr, R.: Quantitative analysis of probabilistic pushdown automata: Expectations and variances. In: Proceedings of LICS 2005, pp. 117–126. IEEE Computer Society Press, Los Alamitos (2005)
4. Esparza, J., Kučera, A., Mayr, R.: Model checking probabilistic pushdown automata. In: LICS 2004, IEEE Computer Society Press, Los Alamitos (2004)
5. Etessami, K., Yannakakis, M.: Recursive Markov chains, stochastic grammars, and monotone systems of nonlinear equations. In: STACS, pp. 340–352 (2005)
6. Fernau, H., Holzer, M.: Conditional context-free languages of finite index. In: New Trends in Formal Languages, pp. 10–26 (1997)
7. Ginsburg, S., Spanier, E.: Derivation-bounded languages. Journal of Computer and System Sciences 2, 228–250 (1968)
8. Gruska, J.: A few remarks on the index of context-free grammars and languages. Information and Control 19, 216–223 (1971)
9. Hopkins, M.W., Kozen, D.: Parikh's theorem in commutative Kleene algebra. In: Logic in Computer Science, pp. 394–401 (1999)
10. Knudsen, B., Hein, J.: RNA secondary structure prediction using stochastic context-free grammars and evolutionary history. Oxford Journals - Bioinformatics 15(6), 446–454 (1999)
11. Kuich, W.: Handbook of Formal Languages. Semirings and Formal Power Series: Their Relevance to Formal Languages and Automata, vol. 1, ch. 9, pp. 609–677. Springer, Heidelberg (1997)
12. Ortega, J.M.: Numerical Analysis: A Second Course. Academic Press, New York (1972)
13. Salomaa, A.: On the index of a context-free grammar and language. Information and Control 14, 474–477 (1969)
14. Yntema, M.K.: Inclusion relations among families of context-free languages. Information and Control 10, 572–597 (1967)

Non-constructive Methods for Finite Probabilistic Automata*

Rūsiņš Freivalds

Institute of Mathematics and Computer Science, University of Latvia,
Raiņa bulvāris 29, Rīga, Latvia

Abstract. Size (the number of states) of finite probabilistic automata
with an isolated cut-point can be exponentially smaller than the size of
any equivalent finite deterministic automaton. The result is presented
in two versions. The first version depends on Artin's Conjecture (1927)
in Number Theory. The second version does not depend on conjectures
but the numerical estimates are worse. In both versions the method of
the proof does not allow an explicit description of the languages used.
Since our finite probabilistic automata are reversible, these results imply
a similar result for quantum finite automata.

1 Introduction

M.O.Rabin proved in [14] that if a language is recognized by a finite probabilistic
automaton with n states, r accepting states and isolation radius δ then there
exists a finite deterministic automaton which recognizes the same language and
the deterministic automaton may have no more than $(1 + \frac{r}{\delta})^n$ states. However,
how tight is this bound? Rabin gave an example of languages in [14] where probabilistic automata indeed had size advantages but these advantages were very
far from the exponential gap predicted by the formula $(1 + \frac{r}{\delta})^n$. Unfortunately,
the advantage proved by Rabin's example was only linear, not exponential. Is it
possible to diminish the gap? Is the upper bound $(1 + \frac{r}{\delta})^n$ tight or is Rabin's
example best possible?

R. Freivalds in [5] constructed an infinite sequence of finite probabilistic automata such that every automaton recognizes the corresponding language with
the probability $\frac{3}{4}$, and if the probabilistic automaton has n states then the language cannot be recognized by a finite deterministic automaton with less than
$\Omega(2^{\sqrt{n}})$ states. This did not close the gap between the lower bound $\Omega(2^{\sqrt{n}})$
and the purely exponential upper bound $(1 + \frac{r}{\delta})^n$ but now it was clear that
the size advantage of probabilistic versus deterministic automata may be super-
polynomial.

A.Ambainis [1] constructed a new sequence of languages and corresponding sequence of finite probabilistic automata such that every automaton recognizes the

---

* Research supported by Grant No.05.1528 from the Latvian Council of Science and
European Commission, contract IST-1999-11234.

T. Harju, J. Karhumäki, and A. Lepistö (Eds.): DLT 2007, LNCS 4588, pp. 169–180, 2007.

corresponding language with the probability $\frac{3}{4}$ and if the probabilistic automaton has n states then the language cannot be recognized by a finite deterministic automaton with less than $\Omega(2^{\frac{n \log \log n}{\log n}})$ states. On the other hand, the languages in [5] were in a single-letter alphabet but for the languages in [1] the alphabet grew with n unlimitedly.

This paper gives the first ever purely exponential distiction between the sizes of probabilistic and deterministic finite automata. Existence of an infinite sequence of finite probabilistic automata is proved such that all of them recognize some language with a fixed probability $p > \frac{1}{2}$ and if the probabilistic automaton has n states then the language cannot be recognized by a finite deterministic automaton with less than $\Omega(a^n)$ states for a certain $a > 1$. This does not end the search for the advantages of probabilistic finite automata over deterministic ones. We still do not know the best possible value of a. Moreover, the best estimate proved in this paper is proved under assumption of the well-known Artin's conjecture in Number Theory. Our final Theorem 3 does not depend on any open conjectures but the estimate is worse, and the description of the languages used is even less constructive. These seem to be the first results in Finite Automata depending on open conjectures in Number Theory.

The essential proofs are non-constructive. Such an approach is not new. A good survey of many impressive examples of non-constructive methods is by J. Spencer [15]. Technically, the crucial improvement over existing results and methods comes from our usage of mirage codes to construct finite probabilistic automata. Along this path of proof, it turned out that the best existing result on mirage codes (Theorem A below) is not strong enough for our needs. The improvement of Theorem A is based on the notion of Kolmogorov complexity. It is well known that Kolmogorov complexity is not effectively computable. It turned out that non-computability of Kolmogorov complexity allows to prove the existence of the needed mirage codes and it is enough for us to prove an exponential gap between the size of probabilistic and deterministic finite automata recognizing the same language. On the other hand, some results of abstract algebra (namely, elementary properties of group homomorphisms) are also used in these proofs.

2   Number-Theoretical Conjectures

By p we denote an odd prime number, i.e. a prime greater than 2. To prove the main theorems we consider several lemmas. Most of them are valid for arbitrary p but we are going to use them only for odd primes of a special type.

Consider the sequence

$$2^0, 2^1, 2^2, \ldots, 2^{p-2}, 2^{p-1}, 2^p, \ldots$$

and the corresponding sequence of the remainders of these numbers modulo p

$$r^0, r^1, r^2, \ldots, r^{p-2}, r^{p-1}, r^p, \ldots \tag{1}$$

$(r_k \equiv 2^k \pmod{p})$. For arbitrary p, the sequence (1) is periodic. Since $r_0 = 1$ and, by the Fermat Little Theorem, $r_{p-1} \equiv 2^{p-1} \equiv 1 \pmod{p}$, one may think that $p - 1$ is the least period of the sequence (1).

This is not the case. For instance, $2^{7-1} \equiv 1 \pmod 7$. but also $2^3 \equiv 1$ (mod 7). However, sometimes $p - 1$ can be the least period of the sequence (1). In this case, 2 is called a primitive root modulo p. More generally, a number a is called a primitive root modulo p if and only if a is a relatively prime to p and $p - 1$ is the least period in the sequence of remainders modulo p of the numbers

$$a^0 = 1, a^1, a^2, \ldots, a^{p-2}, a^{p-1}, a^p, \ldots$$

Emil Artin made in 1927 a famous conjecture the validity of which is still an open problem.

Artin's Conjecture. [3] If a is neither -1 nor a square, then a is a primitive root for infinitely many primes.

Moreover, it is conjectured that density of primes for which a is a primitive root equals $A = 0.373956\ldots$. In 1967, C.Hooley [9] proved that Artin's conjecture follows from the Generalized Riemann hypothesis. D.R.Heath-Brown [10] proved that Artin's conjecture can be wrong no more than for 2 distinct primes a.

3   Linear Codes

Linear codes is the simplest class of codes. The alphabet used is a fixed choice of a finite field $GF(q) = F_q$ with q elements. For most of this paper we consider a special case of $GF(2) = F_2$. These codes are binary codes.

A generating matrix G for a linear $[n, k]$ code over F_q is a k-by-n matrix with entries in the finite field F_q, whose rows are linearly independent. The linear code corresponding to the matrix G consists of all the q^k possible linear combinations of rows of G. The requirement of linear independence is equivalent to saying that all the q^k linear combinations are distinct. The linear combinations of the rows in G are called codewords. However we are interested in something more. We need to have the codewords not merely distinct but also as far as possible in terms of Hamming distance. Hamming distance between two vectors $v = (v_1, \ldots, v_n)$ and $w = (w_1, \ldots, w_n)$ in F_{q^k} is the number of indices i such that $v_i \neq w_i$.

The textbook [7] contains

Theorem A. For any integer $n \geq 4$ there is a $[2n, n]$ binary code with a minimum distance between the codewords at least $n/10$.

However the proof of the theorem in [7] has a serious defect. It is non-constructive. It means that we cannot find these codes or describe them in a useful manner. This is why P.Garret calls them *mirage codes*.

If q is a prime number, the set of the codewords with the operation "component-wise addition" is a group. Finite groups have useful properties. We single out Lagrange's Theorem. The order of a finite group is the number of elements in it.

Lagrange's Theorem. (see e.g. [7]) Let GR be a finite group. Let H be a subgroup of GR. Then the order of H divides the order of G.

Definition 1. *A generating matrix G of a linear code is called* **cyclic** *if along with an arbitrary row $(v_1, v_2, v_3, \ldots, v_n)$ the matrix G contains also a row $(v_2, v_3, \ldots, v_n, v_1)$.*

We would have liked to prove a reasonable counterpart of Theorem A for cyclic mirage codes, but this attempt fails. Instead we consider binary generating matrices of a bit different kind. Let p be an odd prime number, and x be a binary word of length p. The generating matrix $G(p, x)$ has p rows and $2p$ columns. Let $x = x_1 x_2 x_3 \ldots x_p$. The first p columns (and all p rows) make a unit matrix with elements 1 on the main diagonal and 0 in all the other positions. The last p columns (and all p rows) make a cyclic matrix with $x = x_1 x_2 x_3 \ldots x_p$ as the first row, $x = x_p x_1 x_2 x_3 \ldots x_{p-1}$ as the second row, and so on.

Lemma 1. *For arbitrary x, if $h_1 h_2 h_3 \ldots h_p h_{p+1} h_{p+2} h_{p+3} \ldots h_{2p}$ is a codeword in the linear code corresponding to $G(p, x)$, then $h_p h_1 h_2 \ldots h_{p-1} h_{2p} h_{p+1} h_{p+2} \ldots h_{2p-1}$ is also a codeword.*

There are 2^p codewords of the length $2p$. If the codeword is obtained as a linear combination with the coefficients $c_1, c_2, \ldots, c_p$ then the first p components of the codeword equal $c_1 c_2 \ldots c_p$. We denote by $R(x, c_1 c_2 \ldots c_p)$ the subword containing the last p components of this codeword.

Lemma 2. *If $c_1 c_2 \ldots c_p = 000 \ldots 0$, then $R(x, c_1 c_2 \ldots c_p) = 000 \ldots 0$, for arbitrary x.*

Definition 2. *We will call a word* **trivial** *if all its symbols are equal. Otherwise we call the word* **nontrivial**.

Lemma 3. *If $c_1 c_2 \ldots c_p$ is trivial, then $R(x, c_1 c_2 \ldots c_p)$ is trivial for arbitrary x.*

Proof. Every symbol of $R(x, c_1 c_2 \ldots c_p)$ equals $x_1 + x_2 + \cdots + x_p \pmod 2$.

Lemma 4. *If x is trivial, then $R(x, c_1 c_2 \ldots c_p)$ is trivial for arbitrary $c_1 c_2 \ldots c_p$.*

Definition 3. *A word $x = x_1 x_2 \ldots x_p$ is called a* **cyclic shift** *of the word $y = y_1 y_2 \ldots y_p$ if there exists i such that $x_1 = y_i, x_2 = y_{i+1}, \ldots, x_p = y_{i+p}$ where the addition is modulo p. If $(i, p) = 1$, then we say that this cyclic shift is* **nontrivial**.

Lemma 5. *If x is a cyclic shift of y, then $R(x, c_1 c_2 \ldots c_p)$ is a cyclic shift of $R(y, c_1 c_2 \ldots c_p)$.*

Lemma 6. *If p is an odd prime, x is a nontrivial word and y is a nontrivial cyclic shift of x, then $x \neq y$.*

Lemma 7. *If p is an odd prime and $c_1 c_2 \ldots c_p$ is nontrivial, then the set $T_{c_1 c_2 \ldots c_p} = \{R(x, c_1 c_2 \ldots c_p) | x \in \{0, 1\}^p$ and $R(x, c_1 c_2 \ldots c_p)$ nontrivial $\}$ has a cardinality which is a multiple of p.*

Proof. Immediately from Lemmas 5 and 6.

For arbitrary fixed $c_1 c_2 \ldots c_p$, the set $\{R(x, c_1 c_2 \ldots c_p) | x \in \{0,1\}^p\}$ with algebraic operation "component-wise addition modulo z" is a group. We denote this group by B. By D we denote the group of all 2^p binary words of the length p with the same operation.

Lemma 8. *For arbitrary $c_1 c_2 \ldots c_p$, x and y, $R(x, c_1 c_2 \ldots c_p) + R(y, c_1 c_2 \ldots c_p)$ $= R(x + y, c_1 c_2 \ldots c_p)$.*

In other words, for arbitrary $c_1 c_2 \ldots c_p$, the map $D \to B$ defined by $x \to R(x, c_1 c_2 \ldots c_p)$ is a group homomorphism. (Definition and properties of group homomorphisms can be found in every textbook on group theory. See e.g. [4].) The **kernel** of the group homomorphism is the set $ker_0 = \{x | R(x, c_1 c_2 \ldots c_p) = 000 \ldots 0\}$.

The image of the group homomorphism is the set B. For arbitrary $z \in B$, by ker_z we denote the set $ker_z = \{x | R(x, c_1 c_2 \ldots c_p) = z\}$.

From Lemma 8 we easily get

Lemma 9. *For arbitrary $z \in B$, $card(ker_z) = card(ker_0)$.*

Lemma 10. *For arbitrary $z \in B$, $card(ker_z) = \frac{card(D)}{card(B)}$.*

Lemma 11. *If x contains $(p-1)$ zeroes and 1 one, and $c_1 c_2 \ldots c_p$ is nontrivial, then $R(x, c_1 c_2 \ldots c_p)$ is nontrivial.*

Proof. For such an x, the number of ones in $R(x, c_1 c_2 \ldots c_p)$ is the same as the number of ones in $c_1 c_2 \ldots c_p$.

Lemma 12. *If p is an odd prime such that 2 is a primitive root modulo p and $c_1 c_2 \ldots c_p$ is nontrivial, then the set $S_{c_1 c_2 \ldots c_p} = \{R(x, c_1 c_2 \ldots c_p) | x \in \{0,1\}^p\}$ is either of cardinality 1 or of cardinality 2.*

Proof. By Lagrange's Theorem the order 2^p of the group B divides the order of the group D. Hence the order of B is 2^b for some integer b. The neutral element of these groups is the word $000 \ldots 0$. It belongs to every subgroup. There are two possible cases:

1. $111 \ldots 1$ is in B,
2. $111 \ldots 1$ is not in B.

In the case 1 $card(T_{c_1 c_2 \ldots c_p}) = card(B) - 2$, and by Lemmas 7 and 10 $card(T_{c_1 c_2 \ldots c_p})$ is a multiple of p. Hence $2^b = card(B) \equiv 2 \pmod{p}$ and $2^{b-1} \equiv 1 \pmod{p}$. Since 2 is a primitive root modulo p, either $2^{b-1} = 2^{p-1}$ or $2^{b-1} = 2^0$. If $2^{b-1} = 2^{p-1}$, then $2^b = 2^p$ and for this fixed $c_1 c_2 \ldots c_p$ the map $x \to R(x, c_1 c_2 \ldots c_p)$ takes distinct x'es into distinct $R(x, c_1 c_2 \ldots c_p)$'s. If $2^{b-1} = 2^0$, then $2^b = 2$ and $B = \{000 \ldots 0, 111 \ldots 1\}$, but this is impossible by Lemma 11.

In the case 2 $card(T_{c_1 c_2 \ldots c_p}) = card(B) - 1$ and by Lemma 7 $card(T_{c_1 c_2 \ldots c_p})$ is a multiple of p. Hence $2^b \equiv 1 \pmod{p}$. Since 2 is a primitive root modulo p, either $2^b = 2^{p-1}$ or $2^b = 2^0$. If $2^b = 2^{p-1}$, then $card(B) = 2^{p-1}$ and, by Lemma 10, for arbitrary $z \in T_{c_1 c_2 \ldots c_p}$, $card(ker_z) = 2$. If $2^b = 2^0$, then $B = \{000 \ldots 0\}$ but this is impossible by Lemma 11.

4   Kolmogorov Complexity

The theorems in this section are well-known results in spite of the fact that it is not easy to find exact references for all of them.

Definition 4. *We say that the numbering* $\Psi = \{\Psi_0(x), \Psi_1(x), \Psi_2(x), \ldots\}$ *of 1-argument partial recursive functions is* **computable** *if the 2-argument function* $U(n, x) = \Psi_n(x)$ *is partial recursive.*

Definition 5. *We say that a numbering* Ψ *is reducible to the numbering* η *if there exists a total recursive function* $f(n)$ *such that, for all* n *and* x, $\Psi_n(x) = \eta_{f(n)}(x)$.

Definition 6. *We say that a computable numbering* φ *of all 1-argument partial recursive functions is a* **Gödel numbering** *if every computable numbering (of any class of 1-argument partial recursive functions) is reducible to* φ.

Theorem. There exists a Gödel numbering.

Definition 7. *We say that a Gödel numbering* ϑ *is a* **Kolmogorov numbering** *if for arbitrary computable numbering* Ψ *(of any class of 1-argument partial recursive functions) there exist constants* $c > 0, d > 0$, *and a total recursive function* $f(n)$ *such that:*

1. *for all* n *and* x, $\Psi_n(x) = \vartheta_{f(n)}(x)$,
2. *for all* n, $f(n) \leq c \cdot n + d$.

Kolmogorov Theorem. [11] There exists a Kolmogorov numbering.

5   New Mirage Codes

In the beginning of Section 3 we introduced a special type generating matrices $G(p, x)$ where p is an odd prime and x is a binary word of length p. Now we introduce two technical auxiliary functions. If z is a binary word of length $2p$, then $d(z)$ is the subword of z containing the first p symbols, and $e(z)$ is subword of z containing the last p symbols. Then $z = d(z)e(z)$.

There exist many distinct Kolmogorov numberings. We now fix one of them and denote it by η. Since Kolmogorov numberings give indices for all partial recursive functions, for arbitrary x and p, there is an i such that $\eta_i(p) = x$. Let $i(x, p)$ be the minimal i such that $\eta_i(p) = x$. It is easy to see that if $x_1 \neq x_2$, then $i(x_1, p) \neq i(x_2, p)$. We consider all binary words x of the length p and denote by

$x(p)$ the word x such $i(x, p)$ exceed $i(y, p)$ for all binary words y of the length p different from x. It is obvious that $i \geq 2^p - 1$.

Until now we considered generating matrices $G(p, x)$ for independently chosen p and x. From now on we consider only odd primes p such that 2 is a primitive root modulo p and the matrices $G(p, x(p))$. We wish to prove that if p is sufficiently large, then Hamming distances between two arbitrary codewords in this linear code is at least $\frac{4p}{19}$.

We introduce a partial recursive function $\mu(z, \epsilon, p)$ defined as follows. Above when defining $G(p, x)$ we considered auxiliary function $R(x, c_1c_2 \ldots c_p)$. To define $\mu(z, \epsilon, p)$ we consider all 2^p binary words x of the length p. If z is not a binary word of length $2p$, then $\mu(z, \epsilon, p)$ is not defined. If ϵ is not in $\{0, 1\}$, then $\mu(z, \epsilon, p)$ is not defined. If z is a binary word of length $2p$ and $\epsilon \in \{0, 1\}$, then we consider all $x \in \{0, 1\}^p$ such that $R(x, d(z)) = e(z)$. If there are no such x, then $\mu(z, \epsilon, p)$ is not defined. If there is only one such x, then $\mu(z, \epsilon, p) = x$. If there are two such x, then

$$\mu(z, \epsilon, p) = \begin{cases} \text{the first such } x \text{ in the lexicographical order, for } \epsilon = 1 \\ \text{the second such } x \text{ in the lexicographical order, for } \epsilon = 0 \end{cases}$$

If there are more than two such x, then $\mu(z, \epsilon, p)$ is not defined.

Now we introduce a computable numbering of some partial recursive functions. This numbering is independent of p.

For each p (independently from other values of p) we order the set of all the 2^{2p} binary words z of the length $2p$: $z_0, z_1, z_2, \ldots, z_{2^{2p}-1}$. We define z_0 as the word $000 \ldots 0$. The words $z_1, z_2, \ldots, z_{2^{2p}-1}$ are words with exactly one symbol 1. We strictly follow a rule "if the word z_i contains less symbols 1 than the word z_j, then $i < j$". Words with equal number of the symbol 1 are ordered lexicographically. Hence $z_{2^{2p}-1} = 111 \ldots 1$.

For each p, we define

$$\Psi_0(p) = \mu(z_0, 0, p)$$
$$\Psi_1(p) = \mu(z_0, 1, p)$$
$$\Psi_2(p) = \mu(z_1, 0, p)$$
$$\Psi_3(p) = \mu(z_1, 1, p)$$
$$\Psi_4(p) = \mu(z_2, 0, p)$$
$$\Psi_5(p) = \mu(z_2, 1, p)$$
$$\cdots$$
$$\Psi_{2^{2p+1}-2}(p) = \mu(z_{2^{2p}-1}, 0, p)$$
$$\Psi_{2^{2p+1}-1}(p) = \mu(z_{2^{2p}-1}, 1, p)$$

For $j \geq 2^{2p+1}$, $\Psi_j(p)$ is undefined.

We have fixed a Kolmogorov numbering η and we have just constructed a computable numbering Ψ of some partial recursive functions.

Lemma 13. *There exist constants $c > 0$ and $d > 0$ (independent of p) such that for arbitrary i there is a j such that*

1. $\Psi_i(t) = \eta_j(t)$ for all t, and
2. $j \leq ci + d$.

Proof. Immediately from Kolmogorov Theorem.

We consider generating matrices $G(p, x(p))$ for linear codes where p is an odd prime such that 2 is a primitive root modulo p, and, as defined above, $x(p)$ is a binary word of length p such that $\eta_i(p) = x(p)$ implies $i \geq 2^p - 1$. We denote the corresponding linear code by $LC_2(p)$.

Now we prove several lemmas showing that, if p is sufficiently large, then Hamming distances between arbitrary two codewords are no less than $\frac{4p}{19}$.

Lemma 14. *For every linear code, there is a codeword* $000\ldots0$.

Proof. The codeword $000\ldots0$ is obtained by using coefficients $c_1 c_2 \ldots c_p = 000\ldots0$.

Lemma 15. *For every linear code, if there exists a pair of distinct codewords with Hamming distance less than d, then there is a codeword with less than d symbols 1 in it.*

Proof. If x_1 and x_2 are codewords, then $x_1 \oplus x_2$ also is a codeword.

Lemma 16. *If p is sufficiently large, and a codeword in $LC_2(p)$ contains less than $\frac{4p}{19}$ symbols 1, then the codeword is* $000\ldots0$.

Proof. Assume from the contrary that there is a codeword $z \neq 000\ldots0$ containing less than $\frac{4p}{19}$ symbols 1. Above we introduced an ordering $z_0, z_1, z_2, \ldots, z_{2^{2p}-1}$ of all binary words of the length $2p$. Then $z = z_i$ where

$$i \leq \binom{2p}{0} + \binom{2p}{1} + \binom{2p}{2} + \cdots + \binom{2p}{\lfloor \frac{4p}{19} \rfloor}.$$

Hence $i = o(2^p)$. On the other hand, the choice of $x(p)$ implies that $i \geq 2^p - 1$. Contradiction.

Lemma 17. *If p is sufficiently large, then the Hamming distance between any two distinct codewords in $LC_2(p)$ is no less than $\frac{4p}{19}$.*

Proof. By Lemmas 16 and 15.

6   Probabilistic Reversible Automata

M.Golovkins and M.Kravtsev [8] introduced probabilistic reversible automata (PRA) to describe the intersection of two classes of automata, namely, the classes of the 1-way probabilistic and quantum automata. The paper [8] describes several versions of these automata. We concentrate here on the simplest and the least powerful class of PRA.

$\Sigma = \{a_1, a_2, \ldots, a_m\}$ is the input alphabet of the automaton. Every input word is enclosed into end-marker symbols # and \$. Therefore the working alphabet is defined as $\Gamma = \Sigma \cup \{\#, \$\}$. $Q = \{q_1, q_2, \ldots, q_n\}$ is a finite set of states. Q is presented as a union of two disjoint sets: Q_A (accepting states) and Q_R (rejecting states). At every step, the PRA is in some probability distribution $(p_1, p_2, \ldots, p_n)$ where $p_1 + p_2 + \cdots + p_n = 1$. As the result of reading the input #, the automaton enters the initial probability distribution $(p_1(0), p_2(0), \ldots, p_n(0))$. $M_1, M_2, \ldots, M_m$ are doubly-stochastic matrices characterising the evolution of probability distributions.

If at some moment t the probability distribution is

$$(p_1(t), p_2(t), \ldots, p_n(t))$$

and the input symbol is a_u, then the probability distribution

$$(p_1(t+1), p_2(t+1), \ldots, p_n(t+1))$$

equals $(p_1(t), p_2(t), \ldots, p_n(t)) \cdot M_u$. If after having read the last symbol of the input word x the automaton has reached a probability distribution $(p_1, p_2, \ldots, p_n)$ then the probability to accept the word equals

$$prob_x = \Sigma_{i \in Q_A} p_i$$

and the probability to reject the word equals

$$1 - prob_x = \Sigma_{i \in Q_R} p_i.$$

We say that a language L is recognized with bounded error with an interval (p_1, p_2) if $p_1 < p_2$ where $p_1 = sup\{prob_x | x \notin L\}$ and $p_2 = inf\{prob_x | x \in L\}$.

We say that a language L is recognized with a probability $p > \frac{1}{2}$ if the language is recognized with interval $(1 - p, p)$.

In the previous section we constructed a binary generating matrix $G(p, p(x))$ for a linear code. Now we use this matrix to construct a probabilistic reversible automaton $R(p)$.

The matrix $G(p, x(p))$ has $2p$ columns and p rows. The automaton $R(p)$ has $4p + 1$ states, $2p$ of them being accepting and $2p + 1$ being rejecting. The input alphabet consists of 2 letters.

The (rejecting) state q_0 is special in the sense that the probability to enter this state and the probability to exit from this state during the work equals 0. This state always has the probability $\frac{17}{36}$. The states $q_1, q_2, \ldots, q_{4p}$ are related to the columns of $G(p, x(p))$ and should be considered as $2p$ pairs $(q_1, q_2), (q_3, q_4), \ldots,$ $\ldots (q_{4p-1}, q_{4p})$ corresponding to the $2p$ columns of $G(p, x(p))$. The states $q_1, q_3, q_5, q_7, \ldots, q_{4p-1}$ are accepting and the states $q_2, q_4, q_6, q_8, \ldots, q_{4p}$ are rejecting. The initial probability distribution is as follows:

$$\begin{cases} \frac{17}{36}, \text{ for } q_0, \\ \frac{19}{72p}, \text{ for each of } q_1, q_3, \ldots, q_{4p-1} \\ 0, \text{ for each of } q_2, q_4, \ldots, q_{4p}. \end{cases}$$

The processing of the input symbols a, b is deterministic. Under the input symbol a the states are permuted as follows:

$$
\begin{array}{llll}
q_1 \rightarrow q_3 & q_2 \rightarrow q_4 & q_{2p+1} \rightarrow q_{2p+3} & q_{2p+2} \rightarrow q_{2p+4} \\
q_3 \rightarrow q_5 & q_4 \rightarrow q_6 & q_{2p+3} \rightarrow q_{2p+5} & q_{2p+4} \rightarrow q_{2p+6} \\
q_5 \rightarrow q_7 & q_6 \rightarrow q_8 & q_{2p+5} \rightarrow q_{2p+7} & q_{2p+6} \rightarrow q_{2p+8} \\
\cdots & \cdots & \cdots & \cdots \\
q_{2p-3} \rightarrow q_{2p-1} & q_{2p-2} \rightarrow q_{2p} & q_{4p-3} \rightarrow q_{4p-1} & q_{4p-2} \rightarrow q_{4p} \\
q_{2p-1} \rightarrow q_1 & q_{2p} \rightarrow q_2 & q_{4p-1} \rightarrow q_{2p+1} & q_{4p} \rightarrow q_{2p+2}
\end{array}
$$

The permutation of the states under the input symbol b depends on $G(p, x(p))$. Let

$$
G(p, x(p)) = \begin{pmatrix}
g_{11} \ g_{12} \cdots g_{1\ 2p} \\
g_{21} \ g_{22} \cdots g_{2\ 2p} \\
\cdots \ \cdots \cdots \ \cdots \\
g_{p1} \ g_{p2} \cdots g_{p\ 2p}
\end{pmatrix}
$$

For arbitrary $i \in \{1, 2, \ldots, p\}$,

$$
\begin{cases}
q_{2i-1} \rightarrow q_{2i-1} , & \text{if } g_{1i} = 0 \\
q_{2i} \rightarrow q_{2i} , & \text{if } g_{1i} = 0 \\
q_{2i-1} \rightarrow q_{2i} , & \text{if } g_{1i} = 1 \\
q_{2i} \rightarrow q_{2i-1} , & \text{if } g_{1i} = 1.
\end{cases}
$$

In order to understand the language recognized by the automaton $R(p)$ we consider the following auxiliary mapping W from the words in $\{a, b\}^*$ into the set of binary $2p$-vectors defined recursively:

1. $CW(\Lambda) = g_{11}g_{12} \cdots g_{1\ 2p}$
2. if $CW(w) = h_1 h_2 h_3 \ldots h_p h_{p+1} h_{p+2} h_{p+3} \ldots h_{2p}$ then

$$
\begin{cases}
CW(wa) = h_p h_1 h_2 \ldots h_{p-1} h_{2p} h_{p+1} h_{p+2} \ldots h_{2p-1} \text{ and} \\
CW(wb) = (h_1 \oplus g_{11})(h_2 \oplus g_{12})(h_3 \oplus g_{13}) \ldots (h_{2p} \oplus g_{1\ 2p}).
\end{cases}
$$

The next two lemmas can be proved by induction over the length of w.

Lemma 18. *For arbitrary word $w \in \{a, b\}^*$, $CW(w)$ is a codeword in the linear code corresponding to the generating matrix $G(p, x(p))$.*

Lemma 19. *Let w be an arbitrary word in $\{a, b\}^*$, and $CW(w) = h_1 h_2 \ldots h_{2p}$. Then the probability distribution of the states in $R(p)$ is*

$$
\begin{cases}
\frac{17}{36} , & \text{for } g_0, \\
\frac{19}{72p} , & \text{for } g_{2i-1} \text{ if } h_i = 0, \\
0 , & \text{for } g_{2i} \text{ if } h_i = 0, \\
0 , & \text{for } g_{2i-1} \text{ if } h_i = 1, \\
\frac{19}{72p} , & \text{for } g_{2i} \text{ if } h_i = 1.
\end{cases}
$$

We introduce a language

$$
L_{G(p,x(p))} = \{w | w \in \{a, b\}^* \& CW(w) = 000 \ldots 0\}.
$$

Lemma 20. *If 2 is a primitive root modulo p and p is sufficiently large, then the automaton $R(p)$ recognizes the language $L_{G(p,x(p))}$ with the probability $\frac{19}{36}$.*

Lemma 21. *For arbitrary p and arbitrary deterministic finite automaton A recognizing $L_{G(p,x(p))}$ the number of states of A is no less than 2^p.*

Lemmas 20 and 21 imply

Theorem 1. *If 2 is a primitive root for infinitely many distinct primes then there exists an infinite sequence of regular languages $L_1, L_2, L_3, \ldots$ in a 2-letter alphabet and a sequence of positive integers $p(1), p(2), p(3), \ldots$ such that for arbitrary j:*

1. *any deterministic finite automaton recognizing L_j has at least $2^{p(j)}$ states,*
2. *there is a probabilistic reversible automaton with $(4p(j)+1)$ states recognizing L_j with the probability $\frac{19}{36}$.*

7 Without Conjectures

In 1989 D. R. Heath-Brown [10] proved Artin's conjecture for "nearly all integers". We use the following corollary from Heath-Brown's Theorem:

Corollary From Heath-Brown's Theorem. [10] At least one integer a in the set $\{3, 5, 7\}$ is a primitive root for infinitely many primes p.

Above we constructed a binary linear code, the binary generating matrix $G(p, x(p))$ of which incorporated a binary word $x(p)$ with maximum complicity in the Kolmogorov numbering η. Now we are going to modify the construction to get generating matrices $G_3(p, x_3(p))$, $G_5(p, x_5(p))$, $G_7(p, x_7(p))$ for ternary, pentary and septary linear codes $LC_3(p)$, $LC_5(p)$ and $LC_7(p)$, respectively. The constructions remain essentially the same only the words x and $c_1 c_2 \ldots c_p$ now are in $\{0, 1, 2\}^p$, $\{0, 1, 2, 3, 4\}^p$ or $\{0, 1, \ldots, 6\}^p$, resp., and the summation is modulo 3, 5, 7, resp. Recall that by Heath-Brown's Theorem [10] there exists $u \in \{3, 5, 7\}$ such that u is a primitive root for infinetely many distinct primes.

Theorem 1 can be re-formulated as follows.

Theorem 2. *Assume Artin's Conjecture. There exists an infinite sequence of regular languages $L_1, L_2, L_3, \ldots$ in a 2-letter alphabet and an infinite sequence of positive integers $z(1), z(2), z(3), \ldots$ such that for arbitrary j:*

1. *there is a probabilistic reversible automaton with $(z(j))$ states recognizing L_j with the probability $\frac{19}{36}$,*
2. *any deterministic finite automaton recognizing L_j has at least $(2^{1/4})^{z(j)} = (1.1892071115\ldots)^{z(j)}$ states,*

Corollary from Heath-Brown's Theorem allows us to prove the following counterpart of Theorem 2.

Theorem 3. *There exists an infinite sequence of regular languages L_1, L_2, $L_3, \ldots$ in a 2-letter alphabet and an infinite sequence of positive integers $z(1), z(2), z(3), \ldots$ such that for arbitrary j:*

1. *there is a probabilistic reversible automaton with $z(j)$ states recognizing L_j with the probability $\frac{68}{135}$,*
2. *any deterministic finite automaton recognizing L_j has at least $(7^{\frac{1}{14}})^{z(j)} = (1.1149116725\ldots)^{z(j)}$ states,*

References

1. Ambainis, A.: The complexity of probabilistic versus deterministic finite automata. In: Nagamochi, H., Suri, S., Igarashi, Y., Miyano, S., Asano, T. (eds.) ISAAC 1996. LNCS, vol. 1178, pp. 233–237. Springer, Heidelberg (1996)
2. Ambainis, A., Freivalds, R.: 1-way quantum finite automata: strengths, weaknesses and generalizations. Proc. IEEE FOCS'98, pp. 332–341 (1998)
3. Artin, E.: Beweis des allgemeinen Reziprozitätsgesetzes. Mat. Sem. Univ. Hamburg B.5, 353–363 (1927)
4. Aschbacher, M.: Finite Group Theory (Cambridge Studies in Advanced Mathematics), 2nd edn. Cambridge University Press, Cambridge (2000)
5. Freivalds, R.: On the growth of the number of states in result of the determinization of probabilistic finite automata. Avtomatika i Vichislitel'naya Tekhnika (Russian) (3), 39–42 (1982)
6. Gabbasov, N.Z., Murtazina, T.A.: Improving the estimate of Rabin's reduction theorem. Algorithms and Automata, Kazan University, pp. 7–10 (Russian) (1979)
7. Garret, P.: The Mathematics of Coding Theory. Pearson Prentice Hall, Upper Saddle River (2004)
8. Golovkins, M., Kravtsev, M.: Probabilistic Reversible Automata and Quantum Automata. In: Ibarra, O.H., Zhang, L. (eds.) COCOON 2002. LNCS, vol. 2387, pp. 574–583. Springer, Heidelberg (2002)
9. Hooley, C.: On Artin's conjecture. J.ReineAngew.Math. 225, 220–229 (1967)
10. Heath-Brown, D.R.: Artin's conjecture for primitive roots. Quart. J. Math. Oxford 37, 27–38 (1986)
11. Kolmogorov, A.N.: Three approaches to the quantitative definition of information. Problems in Information Transmission 1, 1–7 (1965)
12. Kondacs, A., Watrous, J.: On the power of quantum finite state automata. Proc. IEEE FOCS'97, pp. 66–75 (1997)
13. Paz, A.: Some aspects of probabilistic automata. Information and Control 9(1), 26–60 (1966)
14. Rabin, M.O.: Probabilistic Automata. Information and Control 6(3), 230–245 (1963)
15. Spencer, J.: Nonconstructive methods in discrete mathematics. In: Rota, G.-C. (ed.) Studies in Combinatorics (MAA Studies in Mathematics), vol. 17, pp. 142–178. (1978)

The Unambiguity of Segmented Morphisms

Dominik D. Freydenberger[1,*] and Daniel Reidenbach[2]

[1] Research Group on Mathematical Linguistics, URV, Tarragona, Spain
Institut für Informatik, J. W. Goethe-Universität, Postfach 111932,
D-60054 Frankfurt am Main, Germany
freydenberger@em.uni-frankfurt.de
[2] Fachbereich Informatik, Technische Universität Kaiserslautern,
Kaiserslautern, Germany
reidenba@informatik.uni-kl.de

Abstract. A segmented morphism $\sigma_n : \Delta^* \longrightarrow \{\mathtt{a}, \mathtt{b}\}^*$, $n \in \mathbb{N}$, maps each symbol in Δ onto a word which consists of n distinct subwords in $\mathtt{ab^+a}$. In the present paper, we examine the impact of n on the unambiguity of σ_n with respect to any $\alpha \in \Delta^+$, i.e. the question of whether there does not exist a morphism τ satisfying $\tau(\alpha) = \sigma_n(\alpha)$ and, for some symbol x in α, $\tau(x) \neq \sigma_n(x)$. To this end, we consider the set $U(\sigma_n)$ of those $\alpha \in \Delta^+$ with respect to which σ_n is unambiguous, and we comprehensively describe its relation to any $U(\sigma_m)$, $m \neq n$. Our paper thus contributes fundamental (and, in parts, fairly counter-intuitive) results to the recently initiated research on the ambiguity of morphisms.

1 Introduction

This paper deals with morphisms that map a *pattern*, i.e. a finite string over an infinite alphabet Δ of *variables*, onto a finite *word* over $\{\mathtt{a}, \mathtt{b}\}$; for the sake of convenience, we choose $\Delta := \mathbb{N}$. With regard to such a morphism σ, we ask whether it is *unambiguous* with respect to any pattern α, i.e. there is no morphism $\tau : \mathbb{N}^* \longrightarrow \{\mathtt{a}, \mathtt{b}\}^*$ satisfying $\tau(\alpha) = \sigma(\alpha)$ and, for some symbol x in α, $\tau(x) \neq \sigma(x)$. As recently demonstrated in the initial paper on the ambiguity of morphisms by Freydenberger, Reidenbach and Schneider [5], for every pattern α, there is a particular morphism $\sigma_\alpha^{\mathrm{su}}$ such that $\sigma_\alpha^{\mathrm{su}}$ is unambiguous with respect to α if and only if α is *succinct*, i.e. a shortest generator of its E-pattern language, which, in turn, is equivalent to the fact that α is not a fixed point of a nontrivial morphism $\phi : \mathbb{N}^* \longrightarrow \mathbb{N}^*$. Since there is no single morphism which is unambiguous with respect to all succinct patterns, the morphism $\sigma_\alpha^{\mathrm{su}}$ has to be tailor-made for α. More precisely, for various patterns $\alpha \in \mathbb{N}^+$, $\sigma_\alpha^{\mathrm{su}}$ must be *heterogenous* with respect to α, which means that there exist certain variables x, y in α such that the first (or, if appropriate, the last) letter of $\sigma_\alpha^{\mathrm{su}}(x)$ differs from the first (or last, respectively) letter of $\sigma_\alpha^{\mathrm{su}}(y)$. In addition to this, $\sigma_\alpha^{\mathrm{su}}$ has a second important feature: it maps each variable in α onto a word that consists

---

* Corresponding author. A part of this work was done during the author's stay at the Technische Universität Kaiserslautern.

T. Harju, J. Karhumäki, and A. Lepistö (Eds.): DLT 2007, LNCS 4588, pp. 181–192, 2007.
© Springer-Verlag Berlin Heidelberg 2007

of exactly *three* distinct *segments*, i.e. subwords taken from $\mathtt{ab^+a}$ (or, in order to guarantee heterogeneity, $\mathtt{ba^+b}$).

A closer look at the approach by Freydenberger et al. [5] – which is mainly meant to prove the *existence* of an unambiguous morphism with respect to any succinct pattern – reveals that it is not optimal, as there exist numerous patterns with respect to which there is a significantly less complex unambiguous morphism. For instance, as demonstrated by Reidenbach [11], the standard morphism σ_0 given by $\sigma_0(x) := \mathtt{ab}^x$, $x \in \mathbb{N}$, is unambiguous with respect to *every* pattern α satisfying, for some $m \in \mathbb{N}$ and $e_1, e_2, \ldots, e_m \geq 2$, $\alpha = 1^{e_1} \cdot 2^{e_2} \cdot \ldots \cdot m^{e_m}$ (where the superscripts ^{e_j} refer to the concatenation). With regard to this result, it is noteworthy that, first, σ_0 maps each variable onto a much shorter word than $\sigma_\alpha^{\mathrm{su}}$ and, second, σ_0 is *homogeneous*, i.e. for all variables $x, y \in \mathbb{N}$, $\sigma_0(x)$ and $\sigma_0(y)$ have the same first and the same last letter. Consequently, σ_0 is unambiguous with respect to each pattern in a reasonably rich set, although it does not show any of the two decisive properties of $\sigma_\alpha^{\mathrm{su}}$.

In the present paper, we wish to further develop the theory of unambiguous morphisms. In accordance with the structure of $\sigma_\alpha^{\mathrm{su}}$, we focus on *segmented* morphisms σ_n, which map every variable onto n distinct segments. More precisely, for every $x \in \mathbb{N}$, we define the homogeneous morphism σ_n by $\sigma_n(x) := \mathtt{ab}^{nx-(n-1)}\mathtt{a}\,\mathtt{ab}^{nx-(n-2)}\mathtt{a} \ldots \mathtt{ab}^{nx-1}\mathtt{a}\,\mathtt{ab}^{nx}\mathtt{a}$. With regard to such morphisms, we introduce the set $U(\sigma_n) \subseteq \mathbb{N}^+$ of all patterns with respect to which σ_n is unambiguous, and we give a characterisation of $U(\sigma_m)$ for $m \geq 3$. Furthermore, for every $n \in \mathbb{N}$, we compare $U(\sigma_n)$ with every $U(\sigma_m)$, $m \neq n$, and, since every σ_n is a biprefix code, we complement our approach by additionally considering the set $U(\sigma_0)$ of the suffix code σ_0 as introduced above. Our corresponding results yield comprehensive insights into the relation between any two sets $U(\sigma_m), U(\sigma_n)$, $m, n \in \mathbb{N} \cup \{0\}$.

Our studies are largely motivated by the intrinsic interest involved in the examination of the unambiguity of *fixed* instead of tailor-made morphisms. Thereby, we face a task which gives less definitional leeway than the original setting studied by Freydenberger et al. [5], and therefore our paper reveals new elementary phenomena related to the ambiguity of morphisms that have not been discovered by the previous approach. The choice of segmented morphisms as main objects of our considerations, in turn, is primarily derived from the observation that σ_3 is simply the homogeneous version of $\sigma_\alpha^{\mathrm{su}}$. Hence, the insights gained into $U(\sigma_3)$ immediately yield a deeper understanding of the necessity of the heterogeneity of $\sigma_\alpha^{\mathrm{su}}$ and, thus, of a crucial concept introduced in [5]. In addition to this, our partly surprising results on the relation between the number of segments of a morphism σ_n and the set of patterns for which σ_n is unambiguous suggest that – in a similar manner as the work by, e. g., Halava et al. [6] with respect to the Post Correspondence Problem, which is loosely related to our subject – we deal with a vital type of morphisms that addresses some of the very foundations of the problem field of ambiguity of morphisms. Finally, it is surely worth mentioning that the properties of segmented morphisms have also been studied in the context of *pattern languages* (cf., e. g., Jiang et al. [8]); in particular, recent papers prove the substantial impact of the (un-)ambiguity of such morphisms on pattern *inference* (cf. Reidenbach [10,11]). Thus, our results provide a worthwhile starting point for further considerations in a prominent algorithmic research field related to pattern languages. In the present paper, however, we do not explicitly discuss this aspect of our work.

2 Definitions and Basic Notes

We begin the formal part of this paper with a number of basic definitions. A major part of our terminology is adopted from the research on pattern languages (cf. Mateescu and Salomaa [9]). Additionally, for notations not explained explicitly, we refer the reader to Choffrut and Karhumäki [3].

Let $\mathbb{N} := \{1, 2, 3, ...\}$ and $\mathbb{N}_0 := \mathbb{N} \cup \{0\}$. Let Σ be an *alphabet*, i.e. an enumerable set of symbols. We regard two different alphabets: $\mathbb{N}$ and $\{a, b\}$ with $a \neq b$. Henceforth, we call any symbol in $\mathbb{N}$ a *variable* and any symbol in $\{a, b\}$ a *letter*. A *string (over Σ)* is a finite sequence of symbols from Σ. For the *concatenation* of two strings w_1, w_2 we write $w_1 \cdot w_2$ or simply $w_1 w_2$. The notation $|x|$ stands for the size of a set x or the length of a string x, respectively. We denote the *empty string* by λ, i.e. $|\lambda| = 0$. In order to distinguish between a string over $\mathbb{N}$ and a string over $\{a, b\}$, we call the former a *pattern* and the latter a *word*. We name patterns with lower case letters from the beginning of the Greek alphabet such as α, β, γ. With regard to an arbitrary pattern α, $V(\alpha)$ denotes the set of all variables occurring in α. For every alphabet Σ, Σ^* is the set of all (empty and non-empty) strings over Σ, and $\Sigma^+ := \Sigma^* \setminus \{\lambda\}$. Furthermore, we use the regular operations $+$, $*$ and $\cdot$ on sets and letters in the usual way. For any $w \in \Sigma^*$ and any $n \in \mathbb{N}$, w^n describes the n-fold concatenation of w, and $w^0 := \lambda$. We say that a string $v \in \Sigma^*$ is a *substring* of a string $w \in \Sigma^*$ if and only if, for some $u_1, u_2 \in \Sigma^*$, $w = u_1 v u_2$. Subject to the concrete alphabet considered, we call a substring a *subword* or *subpattern*.

Since we deal with word semigroups, a *morphism* σ is a mapping that is compatible with the concatenation, i.e. for patterns $\alpha, \beta \in \mathbb{N}^+$, a morphism $\sigma : \mathbb{N}^* \longrightarrow \{a, b\}^*$ satisfies $\sigma(\alpha \cdot \beta) = \sigma(\alpha) \cdot \sigma(\beta)$. Hence, a morphism is fully explained as soon as it is declared for all variables in $\mathbb{N}$. Note that we restrict ourselves to total morphisms, even though we normally declare a morphism only for those variables explicitly that, in the respective context, are relevant.

For any pattern $\alpha \in \mathbb{N}^+$ with $\sigma(\alpha) \neq \lambda$, we call $\sigma(\alpha)$ *unambiguous (with respect to α or on α)* if there is no morphism $\tau : \mathbb{N}^* \longrightarrow \{a, b\}^*$ such that $\tau(\alpha) = \sigma(\alpha)$ and, for some $x \in V(\alpha)$, $\tau(x) \neq \sigma(x)$; otherwise, we call σ *ambiguous (with respect to α or on α)*. For a given morphism σ, let $U(\sigma)$ denote the set of all $\alpha \in \mathbb{N}^+$ such that σ is unambiguous on α.

We continue the definitions in this section with a partition of the set of all patterns subject to the following criterion that is due to Freydenberger et al. [5]:

Definition 1. *We call any $\alpha \in \mathbb{N}^+$ prolix if and only if there exists a decomposition $\alpha = \beta_0 \gamma_1 \beta_1 \gamma_2 \beta_2 \ldots \beta_{n-1} \gamma_n \beta_n$ with $n \geq 1$, $\beta_k \in \mathbb{N}^*$ and $\gamma_k \in \mathbb{N}^+$, $k \leq n$, such that*

1. *for every* k, $1 \le k \le n$, $|\gamma_k| \ge 2$,
2. *for every* k, $1 \le k \le n$, *and for every* k', $0 \le k' \le n$, $V(\gamma_k) \cap V(\beta_{k'}) = \emptyset$,
3. *for every* k, $1 \le k \le n$, *there exists an* $x_k \in V(\gamma_k)$ *such that* x_k *occurs exactly once in* γ_k *and, for every* k', $1 \le k' \le n$, *if* $x_k \in V(\gamma_{k'})$ *then* $\gamma_k = \gamma_{k'}$.

We call $\alpha \in \mathbb{N}^+$ succinct *if and only if it is not* prolix.

Succinct and prolix patterns possess several interesting characteristic properties. First, Freydenberger et al. [5] demonstrate that a pattern α is succinct if and only if there is an injective morphism σ_α^{su} such that σ_α^{su} is unambiguous on α. Furthermore, there is no injective morphism that is unambiguous on all succinct patterns, and all nonerasing morphism are ambiguous on all prolix patterns. These results serve as the main fundament of the present work. In addition to this aspect, the set of prolix patterns exactly corresponds to the set of finite fixed points of nontrivial morphisms, i.e. for every prolix pattern α there exists a morphism $\phi : \mathbb{N}^* \longrightarrow \mathbb{N}^*$ such that, for an $x \in V(\alpha), \phi(x) \ne x$ and yet $\phi(\alpha) = \alpha$ (cf., e.g., Hamm and Shallit [7]). Finally, according to Reidenbach [10], the succinct patterns are the shortest generators for their respective E-pattern language – this explains the terms "succinct" and "prolix".

Whithin the scope of the present paper, we call a morphism $\sigma : \mathbb{N}^* \longrightarrow \{a, b\}^*$ *homogeneous* if there exist $p, s \in \{a, b\}^+$ such that for all $x \in \mathbb{N}$, p is a prefix of $\sigma(x)$ and s is a suffix of $\sigma(x)$. Otherwise, σ is *heterogeneous*.

For every $n \in \mathbb{N}$, we define σ_n (the *segmented morphism with* n *segments*) by $\sigma_n(x) := ab^{nx-(n-1)}a\, ab^{nx-(n-2)}a \ldots ab^{nx-1}a\, ab^{nx}a$ for every $x \in \mathbb{N}$ and refer to the subwords ab^+a as *segments*. In this work, we mostly concentrate on the morphisms σ_1, σ_2, σ_3 given by $\sigma_1(x) := ab^x a$, $\sigma_2(x) := ab^{2x-1}a\, ab^{2x}a$ and $\sigma_3(x) := ab^{3x-2}a\, ab^{3x-1}a\, ab^{3x}a$. Although it is not a segmented morphism, we also study the morphism σ_0 given by $\sigma_0(x) := ab^x$, as it is quite similar to σ_1 and often used to encode words over infinite alphabets using only two letters.

There is an interesting property of all σ_n with $n \ge 3$ that can be derived from the proof of Lemma 28 by Freydenberger et al. [5]:

Lemma 1. *Let* $\alpha \in \mathbb{N}^+$ *succinct,* $n \ge 3$ *and* $\tau(\alpha) = \sigma_n(\alpha)$ *for some morphism* $\tau \ne \sigma_n$. *Then, for every* $x \in V(\alpha)$, $\tau(x)$ *contains* $a\, ab^{nx-(n-2)}a \ldots ab^{nx-1}a\, a$.

This lemma is of great use in the next section, and the fact that there is no similar property for $n \le 2$ is the very reason for the existence of Section 4.

3 Homogeneous Morphisms with Three or More Segments

Due to Freydenberger et al. [5], we know that the characteristic regularities in prolix patterns render every injective morphism ambiguous on these patterns. Although succinctness prohibits those regularities, some other structures supporting ambiguity of segmented morphisms can occur. For example, it is easy to see that σ_1 is ambiguous on the succinct pattern $\alpha := 1 \cdot 2 \cdot 1 \cdot 3 \cdot 3 \cdot 2$, e.g. by considering morphisms τ_1 or τ_2 which are given by $\tau_1(1) := ab$, $\tau_1(2) := a\, ab^2 a$

and $\tau_1(3) := \mathsf{a}\,\mathsf{a}\mathsf{b}^3$ and $\tau_2(1) := \mathsf{a}\mathsf{b}\mathsf{a}\,\mathsf{a}$, $\tau_2(2) := \mathsf{b}^2\mathsf{a}$ and $\tau_2(3) := \mathsf{b}^3\mathsf{a}\,\mathsf{a}$. In both cases, the arising ambiguity can be understood (albeit rather metaphorically) as some kind of communication where occurrences of 1 decide which modification is applied to their image under σ_1 and communicate this change to occurrences of 2, where applicable using 3 as a carrier. The patterns that show such a structure can be generalised as follows:

Definition 2. Let $\alpha \in \mathbb{N}^+$. An SCRN-partition for α is a partition of $V(\alpha)$ into pairwise disjoint sets S, C, R and N such that $\alpha \in (N^* SC^* R)^+ N^*$. We call α SCRN-partitionable if and only if it has an SCRN-partition.

As demonstrated by the above example, the existence of an SCRN-partition of a pattern α is a sufficient condition for the ambiguity of any segmented morphism (and σ_0 as well). In fact, it holds for every homogeneous morphism:

Proposition 1. Let $\alpha \in \mathbb{N}^+$. If α is SCRN-partitionable, then every homogeneous morphism σ is ambiguous on α.

Proof. As σ is homogeneous, there exist a $p \in \{\mathsf{a}, \mathsf{b}\}^+$ and, for every $x \in \mathbb{N}$, an $s_x \in \{\mathsf{a}, \mathsf{b}\}^*$ such that $\sigma(x) = p\,s_x$. Let S, C, R, N be an SCRN-partition for α. We define τ by, for all $x \in S$, $\tau(x) := \sigma(x)\,p$, for $x \in R$, $\tau(x) := s_x$, for $x \in C$, $\tau(x) := s_x\,p$. For $x \in N$, we simply define $\tau(x) := \sigma(x)$. As we are using an SCRN-partition, $\alpha \notin N^*$; therefore, $\tau \neq \sigma$ holds. It is easy to see that $\tau(\alpha) = \sigma(\alpha)$. Thus, σ is ambiguous on α. $\square$

We now wish to demonstrate that, for σ_n with $n \geq 3$, this condition is even characteristic. If σ_n is ambiguous on some succinct $\alpha \in \mathbb{N}^+$ (i.e., there is some $\tau \neq \sigma_n$ with $\tau(\alpha) = \sigma_n(\alpha)$), every variable possessing different images under τ and σ_n still keeps all its characteristic inner segments under τ (cf. Lemma 1). Any change is therefore limited to some gain or loss of its (or its neighbours') outer segments and has to be communicated along subpatterns resembling the SC^*R-sequences of a SCRN-partition. This allows to construct an SCRN-partition from τ and leads to the following theorem:

Theorem 1. Let $\alpha \in \mathbb{N}^+$. Then, for every $n \geq 3$, σ_n is ambiguous on α if and only if α is prolix or SCRN-partitionable.

Proof. As mentioned above, [5] demonstrates that we can safely restrict ourselves to succint α, since every injective morphism is ambiguous on every prolix $\alpha \in \mathbb{N}^+$. We begin with the only-if-direction. Assume σ_n is ambiguous on some succinct $\alpha \in \mathbb{N}^+$; then there exists some morphism $\tau \neq \sigma_n$ with $\tau(\alpha) = \sigma_n(\alpha)$. Lemma 1 guarantees that every $\tau(x)$ contains the inner segments of $\sigma_n(x)$. This allows us to distinguish the following cases: For every $x \in V(\alpha)$, let $x \in N$ if and only if $\tau(x) = \sigma_n(x)$. If x has neither lost nor gained to its left, but has lost or gained to the right, let $x \in S$, if its the other way around, let $x \in R$. Finally, if $\tau(x)$ is different from $\sigma_n(x)$ on both sides, let $x \in C$. To show that $\alpha \in (N^* SC^* R)^+ N^*$, we read α from the left to the right. As the first variable has no left neighbour, it cannot have gained or lost some word on its left side; thus, it must belong to

N or S. If it belongs to N, the same is true for the next variable, but as $\alpha \in N^+$ would contradict $\tau \neq \sigma_n$, sooner or later some variable from S must occur. As this variable has a changed right segment, its right neighbour experienced the corresponding change on its left segment. Consequently, that variable must belong to C or R. If it is from C instead, again variables from C must follow until a variable from R is encountered; so α has a prefix from N^*SC^*R. But as variables from R do not change their right segments under τ, we now have the same situation as when we started. We conclude $\alpha \in (N^*SC^*R)^+ N^*$; therefore, α is SCRN-partitionable. The if-direction follows from Proposition 1. □

Consequently, ambiguity of morphisms with at least three segments on succinct patterns is always only a transfer of parts of segments in blocks consisting of a sender, a receiver and possibly some carriers between them.[1] As a sidenote, consider *generalised segmented morphisms with n segments* as morphisms $\sigma_G :$ $\mathbb{N}^* \longrightarrow \Sigma^*$ where $\sigma_G(x) \in (\mathsf{ab^+a})^n$ for all $x \in \mathbb{N}$, and for every $w \in \mathsf{ab^+a}$, there is at most one $x \in \mathbb{N}$ such that w is a subword of $\sigma_G(x)$. It can be shown that if $n \geq 3$, Lemma 1 holds for σ_G as well. Thus, for every generalised segmented morphism σ_G with at least three segments, $U(\sigma_G) = U(\sigma_3)$. Furthermore, as σ_3 is the homogeneous version of the heterogeneous unambiguous morphism $\sigma_\alpha^{\mathrm{su}}$ constructed by Freydenberger et al. [5], Theorem 1 precisely distinguishes the patterns for which there is an unambiguous *homogeneous* morphism from those patterns where an unambiguous morphism has to be *heterogeneous*. Thus, our result significantly contributes to a deeper understanding of the impact of the heterogeneity of a segmented morphism on its unambiguity.

Theorem 1 demonstrates, that for σ_n with $n \geq 3$, ambiguity on succinct patterns is inherently related to the occurrence of global regularities that depend on local interactions between *neighbouring variables* only. In fact, these regularities can be described by the equivalence classes $L_i^\sim$ and $R_i^\sim$ on $V(\alpha)$ introduced by Freydenberger et al. [5] as fundamental tools to construct tailor-made unambiguous morphisms $\sigma_\alpha^{\mathrm{su}}$. In the present paper, we describe these equivalence classes using an equivalent but simpler definition that is based on the adjacency graph of a pattern, a construction that has first been employed by Baker et al. [1] to simplify the Bean-Ehrenfeucht-McNulty-Zimin characterisation of avoidable words, cf. Cassaigne [2]. Like Baker et al., we associate a pattern $\alpha \in \mathbb{N}^+$ with a bipartite graph $\mathrm{AG}(\alpha)$, the *adjacency graph of* α: The vertex set consists of two marked copies of $V(\alpha)$, V^L and V^R (for left and right, respectively); for each $x \in V(\alpha)$, there is an element $x^L \in V^L$ and an element $x^R \in V^R$. There is an edge $x^L - y^R$ for $x, y \in V(\alpha)$ if and only if xy is a subpattern of α.

Unlike Baker et al., we consider a partition of $V^L \cup V^R$ into sets $H_1, \ldots, H_n$ such that each H_i is the set of vertices of a maximal and connected subgraph of $\mathrm{AG}(\alpha)$. We call such a set H_i a *neighbourhood in* α and refer to the set of all neighbourhoods as $H(\alpha)$. For every neighbourhood H_i, the *left neighbourhood class* $L_i^\sim$ denotes the set of all x such that x^L is in H_i and likewise the *right neighbourhood class* $R_i^\sim$ the set of all x such that x^R is in H_i.

---

[1] Hence the letters S, C, R, N stand for sender, carrier, receiver and neutral, respectively.

Example 1. Let $\alpha := 1 \cdot 2 \cdot 3 \cdot 1 \cdot 2 \cdot 2 \cdot 3$. We obtain $H_1 = \{1^L, 2^L, 2^R, 3^R\}$ and $H_2 = \{3^L, 1^R\}$ and therefore, $L_1^\sim = \{1, 2\}$, $L_2^\sim = \{3\}$, $R_1^\sim = \{2, 3\}$ and $R_2^\sim = \{1\}$. In the following figure, we display the adjacency graph of α. Boxes mark the elements of H_1:

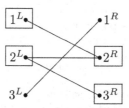

As no injective morphism is unambiguous on a prolix pattern, we mainly deal with succinct patterns. It is useful to note that, apart from patterns of length 1 (like 1), no succinct pattern contains variables that occur only once. Therefore, in succinct patterns every neighbourhood contains elements from V^L and V^R, and every variable belongs to exactly one left and one right neighbourhood class.

Utilising our definition of neighbourhood classes, we now give a second characterisation of $U(\sigma_n)$, $n \geq 3$:

Theorem 2. *For every $\alpha \in \mathbb{N}^+$ with first variable f and last variable l and any $n \geq 3$, σ_n is ambiguous on α if and only if α is prolix or there is a neighbourhood $H_i \in H(\alpha)$ such that $f \notin R_i^\sim$ and $l \notin L_i^\sim$.*

Proof. First, assume that, for some succinct $\alpha := f\alpha'l$ with $\alpha' \in \mathbb{N}^*$, there is some $\tau \neq \sigma_n$ such that $\tau(\alpha) = \sigma_n(\alpha)$. Now we construct an SCRN-partition S, C, R, N of $V(\alpha)$ like in the proof to Theorem 1. Let $x \in S$ and choose i such that $x \in L_i^\sim$. Then $\tau(x)$ can be seen as the result of $\sigma_n(x)$ either loosing a word $\mathtt{b^*a}$ to or gaining some word $\mathtt{ab^*}$ from every right neighbour of an occurrence of x in α. Therefore, all those neighbours must reflect this change on the left side of their image under τ, as anything else would contradict Lemma 1 or $\tau(\alpha) = \sigma_n(\alpha)$. Likewise, all those neighbours' left neighbours must change their right segment in the same way as x. This has to propagate through all of H_i; so all elements of $L_i^\sim$ show the same change to their right segment, and all elements of $R_i^\sim$ show the corresponding change to their left segment. Now assume $f \in R_i^\sim$. As f is the first variable of α and due to Lemma 1, $\tau(f)$ can differ from $\sigma_n(x)$ only to the right of the middle segment, and only by some part of a segment. But this contradicts our previous observation that all elements of $R_i^\sim$ are afflicted by a change to their left segment. This leads to $f \notin R_i^\sim$. Likewise, $l \notin L_i^\sim$, which concludes this direction of the proof. For the other direction, let $\alpha := f\alpha'l$ be succinct with some neighbourhood H_i such that $f \notin R_i^\sim$ and $l \notin L_i^\sim$. Now define S, C, R, N by $S = L_i^\sim \setminus R_i^\sim$, $C = L_i^\sim \cap R_i^\sim$, $R = R_i^\sim \setminus L_i^\sim$ and $N = V(\alpha) \setminus (L_i^\sim \cup R_i^\sim)$. The four sets form a partition of $V(\alpha)$, so it merely remains to be shown that their elements occur in α in the right order. First observe that, by definition, $f \in S \cup N$ and $l \in R \cup N$. Furthermore, for any subpattern xy of α, if $x \in S$ or $x \in C$, then $x \in L_i^\sim$. Therefore, $y \in R_i^\sim$ and thus $y \in C \cup R$. Likewise, $x \in N$ or $x \in R$ implies $x \notin L_i^\sim$ and $y \notin R_i^\sim$, which leads to $y \in N \cup S$ and $\alpha \in (N^* S C^* R)^+ N^*$. By Theorem 1, we conclude that σ_n is ambiguous on α. $\qquad\square$

Consequently, in order to decide ambiguity of σ_n, $n \geq 3$, on a succinct pattern α, it suffices to construct $H(\alpha)$ and check the classes of the first and last variable of α. The construction can be done efficiently, e. g. by using a Union-Find-algorithm.

This theorem provides a useful corollary for a class of patterns first described by Baker et al. [1]. We call a pattern $\alpha \in \mathbb{N}^+$ *locked* if and only if $|H(\alpha)| = 1$ and thus $L_1^{\sim} = R_1^{\sim} = V(\alpha)$. We observe the following consequence:

Corollary 1. *Let $\alpha \in \mathbb{N}^+$. If α is succinct and locked, then $\alpha \in U(\sigma_3)$.*

This corollary is of use in the next section, where we shall see that having less than three segments entails other types of ambiguity than the one described in the previous section.

4   Homogeneous Morphisms with Less Than Three Segments

In this section, we examine the effects caused by reducing the number of segments. One might expect no change in the corresponding sets of unambiguous patterns, or a small hierarchy that reflects the number of segments, but as we shall see, neither is the case. To this end, we construct the following five patterns:

Definition 3. *We define α_0, α_1, α_2 and $\alpha_{0\backslash 2}$ as follows:*

$$\alpha_0 := 1 \cdot 2 \cdot 3 \cdot 1 \cdot 3 \cdot 2,$$
$$\alpha_1 := 1 \cdot 2 \cdot 2 \cdot 3 \cdot 1 \cdot 1 \cdot 3 \cdot 1,$$
$$\alpha_2 := (1 \cdot 2 \cdot 3 \cdot 3 \cdot 4)^2 \cdot 5 \cdot 2 \cdot 6 \cdot 5 \cdot 7 \cdot (8 \cdot 6)^2 \cdot (9 \cdot 7)^2 \cdot 10 \cdot 4 \cdot 11 \cdot 4 \cdot 10 \cdot 12 \cdot$$
$$11 \cdot 12 \cdot (3 \cdot 13)^2 \cdot (14 \cdot 3 \cdot 2 \cdot 15)^2,$$
$$\alpha_{0\backslash 2} := (1 \cdot 2 \cdot 3)^2 \cdot (4 \cdot 5 \cdot 4)^2 \cdot (6 \cdot 7 \cdot 6 \cdot 8)^2 \cdot 1 \cdot 7 \cdot 3 \cdot (9 \cdot 6 \cdot 6 \cdot 10)^2 \cdot (11 \cdot 12)^2 \cdot$$
$$(13 \cdot 7 \cdot 7 \cdot 4 \cdot 14 \cdot 12)^2 \cdot (15 \cdot 14)^2 \cdot 9 \cdot 6.$$

Finally, we define $\alpha_{1\backslash 2}$ by $\alpha_{1\backslash 2} := 1^2 \cdot \delta \cdot 1 \cdot p(\delta) \cdot 1$, where

$$\delta := \beta_1 \cdot 1 \cdot \beta_2 \cdot 1 \cdot \beta_3 \cdot 1 \cdot \beta_4 \cdot 1 \cdot \beta_5 \cdot 1 \cdot \gamma_1 \cdot \beta_6 \cdot 1 \cdot \beta_7 \cdot 1 \cdot \gamma_2 \cdot 1 \cdot \beta_8 \cdot$$
$$1 \cdot \gamma_3 \cdot 1 \cdot \beta_9 \cdot 1 \cdot \beta_{10} \cdot 1 \cdot \beta_{11} \cdot 1 \cdot \beta_{12} \cdot 1 \cdot \beta_{13} \cdot 1 \cdot \beta_{14},$$

and $p(1) := \lambda$, $p(x) := x$ for all $x \in \mathbb{N} \setminus \{1\}$, and furthermore

$\beta_1 := 2 \cdot 3 \cdot 3 \cdot 4,$	$\beta_2 := 3 \cdot 2 \cdot 2 \cdot 5,$
$\beta_3 := 6 \cdot 7,$	$\beta_4 := 8 \cdot 9,$
$\beta_5 := 10 \cdot 11,$	$\gamma_1 := (12 \cdot 1)(13 \cdot 1) \cdot \ldots \cdot (17 \cdot 1),$
$\beta_6 := 18 \cdot 19,$	$\beta_7 := 6 \cdot 20 \cdot 9,$
$\gamma_2 := 21 \cdot 1 \cdot 22,$	$\beta_8 := 6 \cdot 23 \cdot 11,$
$\gamma_3 := 24 \cdot 1 \cdot 25 \cdot 1 \cdot 26,$	$\beta_9 := 27 \cdot 2 \cdot 20 \cdot 2 \cdot 20 \cdot 28 \cdot 2 \cdot 29,$

$\beta_{10} := 30\cdot2\cdot(20\cdot2\cdot23\cdot2)^3\cdot20\cdot(31)^4\cdot32,$ $\beta_{11} := 33\cdot3\cdot34\cdot23\cdot3\cdot23\cdot3\cdot35,$

$\beta_{12} := 36\cdot20\cdot20\cdot28\cdot2\cdot29,$ $\beta_{13} := 33\cdot3\cdot34\cdot23\cdot23\cdot37,$

$\beta_{14} := 18\cdot31\cdot32.$

We begin by establishing the relation between $U(\sigma_3)$ and the other sets:

Theorem 3. *The sets $U(\sigma_0)$, $U(\sigma_1)$ and $U(\sigma_2)$ are strictly included in $U(\sigma_3)$.*

Proof. For all three languages, the inclusion directly follows from Theorem 1: If $\alpha \notin U(\sigma_3)$ then α is prolix or SCRN-partitionable. In the former case, every injective morphism is ambiguous on α, and, due to Proposition 1, the existence of an SCRN-partition is sufficient for ambiguity of segmented morphisms and σ_0. To prove strictness, we show that σ_0, σ_1, σ_2 are ambiguous on the patterns α_0, α_1, α_2, respectively, from Definition 3. All three patterns are succinct and – as demonstrated by their adjacency graphs – have only one neighbourhood class each. Hence, due to Corollary 1, σ_3 is unambiguous on each of the patterns.

We start with α_0 and define τ by $\tau(1) := \sigma_0(1\cdot2)$, $\tau(2) := \sigma_0(2)$ and $\tau(3) := \mathsf{b}$. Then $\tau \neq \sigma_0$, but $\tau(\alpha_0) = \sigma_0(1\cdot2)\cdot\sigma_0(2)\cdot\mathsf{b}\cdot\sigma_0(1\cdot2)\cdot\mathsf{b}\cdot\sigma_0(2) = \sigma_0(\alpha_0)$. Therefore, σ_0 is ambiguous on α_0. For α_1, we set $\tau(1) := \mathsf{a}$, $\tau(2) := \mathsf{baab}$ and $\tau(3) := \mathsf{ba}\,\sigma_1(3)\,\mathsf{ab}$. It is easy to see that $\tau \neq \sigma_1$ and $\tau(\alpha_1) = \sigma_1(\alpha_1)$. With regard to σ_2, we consider the morphism τ given by

$$\tau(1) := \sigma_2(1\cdot2\cdot3)\,\mathsf{ab}^5\mathsf{a}\,\mathsf{ab}^3, \qquad \tau(2) := \mathsf{b}^3\mathsf{a}\,\mathsf{ab}^3,$$

$$\tau(3) := \lambda, \qquad \tau(4) := \mathsf{b}^4\mathsf{a}\,\mathsf{ab}^8\mathsf{a},$$

$$\tau(5) := \sigma_2(5)\,\mathsf{a}, \qquad \tau(6) := \mathsf{ba}\,\sigma_2(6),$$

$$\tau(7) := \mathsf{b}^{13}\mathsf{a}\,\mathsf{ab}^{14}\mathsf{a}, \qquad \tau(8) := \mathsf{ab}^{15}\mathsf{a}\,\mathsf{ab}^{15},$$

$$\tau(9) := \sigma_2(9)\,\mathsf{a}, \qquad \tau(10) := \sigma_2(10)\,\mathsf{ab}^3,$$

$$\tau(11) := \sigma_2(11)\,\mathsf{ab}^3, \qquad \tau(12) := \mathsf{b}^{20}\mathsf{a}\,\mathsf{ab}^{24}\mathsf{a},$$

$$\tau(13) := \sigma_2(3\cdot13), \qquad \tau(14) := \mathsf{ab}^{27}\mathsf{a}\,\mathsf{ab}^{25},$$

$$\tau(15) := \mathsf{b}^2\mathsf{a}\,\mathsf{ab}^6\mathsf{a}\,\sigma_2(2\cdot15).$$

Then $\tau \neq \sigma_2$. Proving $\tau(\alpha_2) = \sigma_2(\alpha_2)$ is less obvious, but straightforward. □

The proof for Theorem 3 is of additional interest as Freydenberger et al. [5] propose to study a morphism $\sigma_\alpha^{2\text{-seg}}$ that maps each variable x in a succinct pattern α onto a word that merely consists of the left and the right segment of $\sigma_\alpha^{\mathrm{su}}(x)$ (recall that $\sigma_\alpha^{\mathrm{su}}$ is a heterogeneous morphism which maps every variable x onto *three* segments). In [5] it is asked whether, for every succinct pattern α, $\sigma_\alpha^{2\text{-seg}}$ is unambiguous on α, thus suggesting the chance for a major improvement of $\sigma_\alpha^{\mathrm{su}}$. With regard to this question, we now consider the pattern α_2. In the above proof, it is stated that α is a locked pattern, which implies that $\sigma_{\alpha_2}^{2\text{-seg}}$ only maps the variable 1 onto a word $\mathsf{b}\ldots\mathsf{b}$ and all other variables in α_2 onto words $\mathsf{a}\ldots\mathsf{a}$. Consequently, for each $x \in V(\alpha_2) \setminus \{1\}$, $\sigma_{\alpha_2}^{2\text{-seg}}(x) = \sigma_2(x)$. Therefore – and since, for the corresponding τ introduced in the proof of Theorem 3, the word $\tau(1)$ completely contains $\sigma_2(1)$ – we can define a morphism τ' by

$\tau'(1) := \sigma_{\alpha_2}^{2\text{-seg}}(1 \cdot 2 \cdot 3)\, \mathsf{ab}^5\mathsf{a}\,\mathsf{ab}^3$ and $\tau'(x) := \tau(x)$, $x \in V(\alpha_2) \backslash \{1\}$, and this definition yields $\tau'(\alpha_2) = \sigma_{\alpha_2}^{2\text{-seg}}(\alpha_2)$. So, there exists a succinct pattern α (namely α_2) such that $\sigma_{\alpha}^{2\text{-seg}}$ is ambiguous on α. Thus, α_2 does not only prove $U(\sigma_2) \subset U(\sigma_3)$, but it also provides a negative answer to an intricate question posed in [5].

Returning to the focus of the present paper, the examples in the proof for Theorem 3 demonstrate ambiguity phenomena that are intrinsic for their respective kind of morphisms and cause ambiguity on patterns that are neither prolix nor SCRN-partitionable: With regard to σ_0, the fact that for each x, y with $x < y$, $\sigma_0(x)$ is a prefix of $\sigma_0(y)$ can be used to achieve ambiguity, as demonstrated by α_0. Concerning σ_1, a variable x can achieve $\tau(x) = \mathsf{a}$ both by giving ab^x to the left or $\mathsf{b}^x\mathsf{a}$ to the right, which can be prefix or suffix of some $\sigma_1(y)$. In α_1, we use this for $\tau(1)$. The situation is less obvious and somewhat more complicated for σ_2, as suggested by the fact that we do not know a shorter pattern serving the same purpose as α_2. Here, a variable x can obtain an image $\tau(x) \in \mathsf{b}^*\mathsf{aab}^*$, which can be used both as a middle part of some $\sigma_2(y)$, and as the borderline between some $\sigma_2(y)$ and some $\sigma_2(z)$. In the proofs for Theorem 5 and Theorem 6 we utilise further examples for complicated cases of σ_2-ambiguity.

It is natural to ask whether these phenomena can be used to find patterns where one of the three morphisms σ_0, σ_1, σ_2 is ambiguous, and another is not. We begin with a comparison of $U(\sigma_0)$ and $U(\sigma_1)$:

Theorem 4. *The sets $U(\sigma_0)$ and $U(\sigma_1)$ are incomparable.*

Proof. We have already established the ambiguity of σ_0 on α_0 and of σ_1 on α_1 in the proof of Theorem 3. The proofs for the unambiguity of σ_1 on α_0 and of σ_0 on α_1 are left out due to space reasons. □

This result is perhaps somewhat counter-intuitive, but the fact that $U(\sigma_0)$ and $U(\sigma_1)$ can be separated by two very short examples might be considered evidence that the two languages are by far not as similar as the two morphisms. We proceed with a comparison of $U(\sigma_0)$ and $U(\sigma_2)$. Surprisingly, the same result holds (although one of the examples is considerably more involved):

Theorem 5. *The sets $U(\sigma_0)$ and $U(\sigma_2)$ are incomparable.*

Proof. Here, we use α_0 and $\alpha_{0\backslash 2}$. In spite of the NP-completeness of the problem (cf. Ehrenfeucht, Rozenberg [4]), $\alpha_{0\backslash 2} \in U(\sigma_0)$ and $\alpha_0 \in U(\sigma_2)$ can be verified by a computer; therefore (and due to space constraints), we omit the corresponding proof. Contrary to this, the length of $\sigma_2(\alpha_{0\backslash 2})$ does not allow for the use of a computer. With regard to the ambiguity of σ_2, we thus refer to the morphism τ given by

$$\tau(1) = \sigma_2(1)\,\mathsf{ab}^2, \qquad\qquad \tau(2) = \mathsf{ba}\,\mathsf{ab}^2,$$
$$\tau(3) = \mathsf{b}^2\mathsf{a}\,\sigma_2(1), \qquad\qquad \tau(4) = \lambda,$$
$$\tau(5) = \sigma_2(4 \cdot 5 \cdot 4), \qquad\qquad \tau(6) = \mathsf{a},$$
$$\tau(7) = \mathsf{b}^{11}\mathsf{a}\,\mathsf{ab}^{12}, \qquad\qquad \tau(8) = \sigma_2(7 \cdot 6 \cdot 8),$$
$$\tau(9) = \sigma_2(9 \cdot 6), \qquad\qquad \tau(10) = \sigma_2(6 \cdot 10),$$

$$\tau(11) = \sigma_2(11)\,\mathsf{ab}^{23}\mathsf{a}\,\mathsf{ab}^{12}, \qquad\qquad \tau(12) = \mathsf{b}^{12}\mathsf{a},$$

$$\tau(13) = \sigma_2(13 \cdot 7 \cdot 7 \cdot 4)\,\mathsf{ab}^{27}\mathsf{a}\,\mathsf{ab}^{17}, \qquad \tau(14) = \lambda,$$

$$\tau(15) = \sigma_2(15 \cdot 14),$$

which yields $\tau(\alpha_{0\backslash2}) = \sigma_2(\alpha_{0\backslash2})$ and, hence, the ambiguity of σ_2 on α. □

We conclude this section by the examination of the last open case, namely the relation between $U(\sigma_1)$ and $U(\sigma_2)$. Again, one might conjecture that the more complex morphism σ_2 is "stronger" than σ_1, but our most sophisticated example pattern $\alpha_{1\backslash2}$ shows that this expectation is not correct:

Theorem 6. *The sets $U(\sigma_1)$ and $U(\sigma_2)$ are incomparable.*

Proof. For this proof we use the patterns α_1 and $\alpha_{1\backslash2}$. Recall that σ_1 is ambiguous on α_1 (cf. proof of Theorem 3). With little effort, it can be seen that σ_2 is unambiguous on α_1. Thus, we know that $\alpha_1 \in U(\sigma_2) \setminus U(\sigma_1)$. The fact that σ_1 is unambiguous on $\alpha_{1\backslash2}$ requires extensive reasoning, which is left out due to space reasons. Showing that σ_2 is ambiguous on $\alpha_{1\backslash2}$ is more straightforward. Let $\tau(x) := \lambda$ for $x \in \{2, 3, 28, 31, 34\}$ and $\tau(x) := \sigma_2(x)$ for $x \in V(\gamma_1) \cup V(\gamma_2) \cup V(\gamma_3)$. For all other $x \in V(\alpha_{1\backslash2})$, define $\tau(x)$ as follows:

$$\tau(4) := \sigma_2(2 \cdot 3 \cdot 3 \cdot 4), \qquad\qquad \tau(5) := \sigma_2(3 \cdot 2 \cdot 2 \cdot 5),$$

$$\tau(6) := \sigma_2(6)\,\mathsf{ab}^{11}, \qquad\qquad \tau(7) := \mathsf{b}^2\mathsf{a}\,\mathsf{ab}^{14}\mathsf{a},$$

$$\tau(8) := \mathsf{ab}^{15}\mathsf{a}\,\mathsf{ab}^3, \qquad\qquad \tau(9) := \mathsf{b}^{13}\mathsf{a}\,\sigma_2(9),$$

$$\tau(10) := \mathsf{ab}^{19}\mathsf{a}\,\mathsf{ab}^8, \qquad\qquad \tau(11) := \mathsf{b}^{12}\mathsf{a}\,\sigma_2(11),$$

$$\tau(18) := \sigma_2(18)\,\mathsf{ab}^{27}, \qquad\qquad \tau(19) := \mathsf{b}^{10}\mathsf{a}\,\mathsf{ab}^{38}\mathsf{a},$$

$$\tau(20) := \mathsf{b}^{28}\mathsf{a}\,\mathsf{ab}^{27}, \qquad\qquad \tau(23) := \mathsf{b}^{34}\mathsf{a}\,\mathsf{ab}^{34},$$

$$\tau(27) := \sigma_2(27 \cdot 2 \cdot 20 \cdot 2)\,\mathsf{ab}^{39}\mathsf{a}\,\mathsf{ab}^{12}, \quad \tau(29) := \mathsf{b}^{29}\mathsf{a}\,\sigma_2(2 \cdot 29),$$

$$\tau(30) := \sigma_2\left(30 \cdot 2 \cdot (20 \cdot 2 \cdot 23 \cdot 2)^3\right)\mathsf{ab}^{39}\mathsf{a}\,\mathsf{ab}^{12}, \quad \tau(32) := \mathsf{b}^{34}\mathsf{a}\,\mathsf{ab}^{62}\mathsf{a}\cdot\sigma_2(32),$$

$$\tau(33) := \sigma_2(33 \cdot 3)\,\mathsf{ab}^{33}, \qquad\qquad \tau(35) := w\,\sigma_2(3 \cdot 23 \cdot 3 \cdot 35),$$

$$\tau(36) := \sigma_2(36 \cdot 20)\,\mathsf{ab}^{39}\mathsf{a}\,\mathsf{ab}^{12}, \qquad \tau(37) := w\,\sigma_2(23 \cdot 37),$$

where $w := \mathsf{b}^{11}\mathsf{a}\,\mathsf{ab}^{46}\mathsf{a}$. Obviously $\tau \neq \sigma_2$. As $\tau(x) = \sigma_2(x)$ for $x \in V(\gamma_1) \cup V(\gamma_2) \cup V(\gamma_3)$, especially for $x = 1$, it suffices to show $\tau(\beta_i) = \sigma_2(\beta_i)$ for all $i \in \{1, 2, \ldots, 14\}$. For β_1 and β_2, the claim holds trivially. For the other β_i, the process is straightforward but somewhat lenghty. Thus, $\alpha_{1\backslash2} \in U(\sigma_1) \setminus U(\sigma_2)$, and therefore $U(\sigma_1)$ and $U(\sigma_2)$ are incomparable. □

Note that we do not know any nontrivial characterisation of $U(\sigma_0)$, $U(\sigma_1)$ and $U(\sigma_2)$. Moreover, we cannot refer to a computationally feasible method to successfully seek for any patterns in $U(\sigma_1) \setminus U(\sigma_2)$, $U(\sigma_0) \setminus U(\sigma_2)$ and $U(\sigma_3) \setminus U(\sigma_2)$. Therefore, we cannot answer the question of whether there exist shorter examples than α_2, $\alpha_{0\backslash2}$ and $\alpha_{1\backslash2}$ suitable for proving Theorems 3, 5 and 6, respectively. The intricacy of the ambiguity phenomena relevant for the construction of such patterns, however, suggests that our examples cannot be shortened significantly.

5   Conclusion and Open Problems

In the present paper, we have studied the unambiguity of an important type of injective morphisms. More precisely, we have examined the impact of the number n of segments of a segmented morphism σ_n on the set $U(\sigma_n)$ of patterns for which σ_n is unambiguous. Our main results show that a change of n, surprisingly, does not give rise to a "real" hierarchy of sets of patterns, as the three pairwise incomparable languages $U(\sigma_0)$, $U(\sigma_1)$ and $U(\sigma_2)$ are all contained in one common superset $U(\sigma_3)$, that is also the maximum any homogeneous morphism can achieve. We have established the result on $U(\sigma_3)$ by two characteristic criteria on $U(\sigma_3)$, which additionally entail a substantial improvement of the main technique introduced in the initial paper [5] on the unambiguity of morphisms.

Contrary to this, a major part of our results on σ_0, σ_1 and σ_2 are not based on criteria, but on example patterns. We regard it as a very interesting problem to find characterisations of $U(\sigma_0)$, $U(\sigma_1)$ and $U(\sigma_2)$. In consideration of the remarkable complexity of the patterns $\alpha_{0\backslash 2}$, $\alpha_{1\backslash 2}$ and α_2, however, we expect this to be an extraordinarily cumbersome task.

References

1. Baker, K.A., McNulty, G.F., Taylor, W.: Growth problems for avoidable words. Theor. Comput. Sci. 69, 319–345 (1989)
2. Cassaigne, J.: Unavoidable patterns. In: Lothaire, M. (ed.) Algebraic Combinatorics on Words, pp. 111–134. Cambridge Mathematical Library (2002)
3. Choffrut, C., Karhumäki, J.: Combinatorics of words. In: Rozenberg, G., Salomaa, A. (eds.) Handbook of Formal Languages, vol. 1, pp. 329–438. Springer, Heidelberg (1997)
4. Ehrenfeucht, A., Rozenberg, G.: Finding a homomorphism between two words is NP-complete. Inform. Process. Lett. 9, 86–88 (1979)
5. Freydenberger, D.D., Reidenbach, D., Schneider, J.C.: Unambiguous morphic images of strings. Int. J. Found. Comput. Sci. 17, 601–628 (2006)
6. Halava, V., Harju, T., Karhumäki, J., Latteux, M.: Post Correspondence Problem for morphisms with unique blocks. In: Proc. Words'05. Publications du LACIM, vol. 36, pp. 265–274 (2005)
7. Hamm, D., Shallit, J.: Characterization of finite and one-sided infinite fixed points of morphisms on free monoids. Technical Report CS-99-17, Dep. of Computer Science, University of Waterloo (1999)
8. Jiang, T., Salomaa, A., Salomaa, K., Yu, S.: Decision problems for patterns. J. Comput. Syst. Sci. 50, 53–63 (1995)
9. Mateescu, A., Salomaa, A.: Patterns. In: Rozenberg, G., Salomaa, A. (eds.) Handbook of Formal Languages, pp. 230–242. Springer, Heidelberg (1997)
10. Reidenbach, D.: A discontinuity in pattern inference. In: Diekert, V., Habib, M. (eds.) STACS 2004. LNCS, vol. 2996, pp. 129–140. Springer, Heidelberg (2004)
11. Reidenbach, D.: A non-learnable class of E-pattern languages. Theor. Comput. Sci. 350, 91–102 (2006)

Commutation of Binary Factorial Languages*

Anna E. Frid

Sobolev Institute of Mathematics SB RAS, Novosibirsk, Russia,
and Novosibirsk State University
frid@math.nsc.ru

Abstract. We solve the commutation equation $AB = BA$ for binary factorial languages A and B. As we show, the situations when such languages commute can be naturally classified. The result is based on the existence and uniqueness of a canonical decomposition of a factorial language, proved by S. V. Avgustinovich and the author in 2005. It continues investigation of the semigroup of factorial languages.

1 Introduction

The commutation equation $XY = YX$ is one of the simplest equations, whatever X and Y are and whatever XY means. Its solution is sometimes also simple: for example, if x and y are finite words (i. e., elements of a free semigroup), and the operation considered is just the catenation, then $xy = yx$ implies $x = z^n$ and $y = z^m$ for some word z and some non-negative integers n and m.

However, if X and Y are languages, and the operation considered is still the catenation (defined naturally by $XY = \{xy | x \in X, y \in Y\}$), the commutation equation becomes very difficult to solve. Much attention has been paid in particular to the *centralizer* of a language X, which is the maximal language commuting with it (this maximal language exists since the set of languages commuting with X is closed under union). Conway [2] conjectured in 1971 that the centralizer of a rational language is rational. Although this conjecture has been recently disproved by Kunc [7] in a very strong sense, positive partial results for prefix codes [9], codes [5] and languages with at most three elements [6] are known. A survey on the history of the problem can also be found in [7].

In this paper, we restrict ourselves to considering *factorial* languages. A language is called factorial if it is closed under taking factors of its elements. It is clear that since each factorial language contains the empty word, its centralizer is always Σ^*, where Σ is the alphabet. Moreover, factorial languages clearly commute in the following two situations:

Word type commutation: $AB = BA$ if $A = C^m$ and $B = C^n$ for some factorial language C and non-negative integers n and m.

Commutation by absorption: Let Σ_A be the subalphabet of all letters occurring in a factorial language A. Then $AB = BA = B$ if $B\Sigma_A \subset B$, $\Sigma_A B \subset B$, and thus $B = B\Sigma_A^* = \Sigma_A^* B$: the language B *absorbs* A.

---

* Supported in part by RFBR grants 05-01-00364 and 06-01-00694.

T. Harju, J. Karhumäki, and A. Lepistö (Eds.): DLT 2007, LNCS 4588, pp. 193–204, 2007.
© Springer-Verlag Berlin Heidelberg 2007

An example of commutation by absorption is given by $A = a^*$ and $B = \{a, b\}^*$. In fact, there exists a continuum of binary factorial languages absorbing A: there is a continuum of binary factorial languages, and for each of them, denoted by C, we can take $B = a^*Ca^*$.

It is also clear that all unary factorial languages on the alphabet $\Sigma = \{a\}$ commute and this commutation is described by one of the two cases above: if the languages are $\{a^k | k \leq n\}$ and $\{a^k | k \leq m\}$ for some $n, m \geq 0$, this is word type commutation, and if at least one of them is equal to a^*, this is absorption.

However, the following binary example shows that commutation of factorial languages in general is not entirely described by these two situations.

Example 1. The languages $A = a^*b^* \cup b^*a^*$ and $B = a^*$ commute since $AB = BA = a^*b^*a^*$.

In this paper, we classify the cases when binary factorial languages commute by defining three types of so-called *unexpected commutation*; an example of one of its types is given above, and the complete description is done in Section 7. The main result of this paper is

Theorem 1. *Two binary factorial languages commute if and only if either they are powers of the same factorial language; or one of them absorbs the other; or their relationship is described below as unexpected commutation I, II, or III.*

Most auxiliary statements of the paper hold or can be generalized to the arbitrary alphabet. The fact that the alphabet is binary is needed only for one of the cases considered in Subsection 7.2. Extending that part of the proof to arbitrary alphabets is a subject of future work.

The proof is based on existence and uniqueness of the *canonical* decomposition of a factorial language proved by Avgustinovich and the author in [1], and on an investigation of canonical decomposition of a catenation of factorial languages taken in [4]. Practically, we examine factorial languages as words on the infinite alphabet of *indecomposable* languages. Although the semigroup of factorial languages is not free, we may consider it almost as a free one and use classical results on word equations.

In Sections 2 and 3, main definitions and techniques are introduced. Section 4 contains technical lemmas conccerning properties of the introduced operators. In all the three sections, factorial languages on arbitrary alphabets are considered.

In Section 5 we list needed classical results on word equations. In Section 6, specific features of the binary alphabet are explained. At last, Section 7 contains the main part of the proof, relying on the results of the previous sections. Cases of unexpected commutation are marked by frames.

2   Factorial Languages and Canonical Decompositions

Let Σ be a finite alphabet; for a while, we not require Σ to be binary. A *language* is an arbitrary subset of the set Σ^* of all finite words on Σ. The empty word is denoted by λ.

A word v is called a *factor* of a word u if $u = svt$ for some words s and t (which can be empty). In particular, λ is a factor of any word. The *factorial closure* $\mathrm{Fac}(L)$ of a language L is the set of all factors of all its elements. Clearly, $\mathrm{Fac}(L) \supseteq L$. If $\mathrm{Fac}(L) = L$, that is, if L is closed under taking factors, we say that L is a *factorial* language.

Typical examples of factorial languages include the set of factors of a finite or infinite word; the set of words avoiding a pattern, etc. Clearly, the factorial closure of an arbitrary language is a factorial language; if the initial language is regular, so is its factorial closure. The family of factorial languages is closed under taking union, intersection and catenation. Factorial languages equipped with catenation constitute a submonoid of the monoid of all languages, and its unit is the language $\{\lambda\}$. We are interested in properties of this submonoid.

A factorial language L is called *indecomposable* if $L = L_1 L_2$ implies $L = L_1$ or $L = L_2$ for any factorial languages L_1 and L_2. In particular, we have the following

Lemma 1. *[1] For each alphabet Σ, the language Σ^* is indecomposable.*

Other examples of indecomposable languages include $a^* \cup b^*$ with $a, b \in \Sigma$, and languages of factors of any recurrent infinite word.

A decomposition $L = L_1 \cdots L_k$ of a factorial language L to catenation of factorial languages is called *minimal* if $L \neq L_1 \cdots L_{i-1} L_i' L_{i+1} \cdots L_k$ for any factorial language $L_i' \subsetneq L_i$. A minimal decomposition to indecomposable languages is called *canonical.*

The following theorem is the starting point of our technique.

Theorem 2. *[1] For each factorial language L, a canonical decmposition exists and is unique.*

Example 2. If L is indecomposable, then its canonical decomposition is just $L = L$. The canonical decomposition of the language $a^* b^* \cup b^* a^*$ is $(a^* \cup b^*)(a^* \cup b^*)$.

3   Important Tools

Let $\mathcal{F}$ be the set of all indecomposable factorial languages. In what follows, the canonical decomposition of a language L considered as a word on the alphabet $\mathcal{F}$ will be denoted by $\overline{L}$. Due to Theorem 2, the equality between two factorial languages L_1 and L_2 always implies the equality of words on $\mathcal{F}$ corresponding to their canonical decompositions. We shall write it as $L_1 \doteq L_2$ (or, more rigorously, $\overline{L_1} \doteq \overline{L_2}$).

In what follows we compare canonical decompositions of factorial languages as words on $\mathcal{F}$: although the alphabet $\mathcal{F}$ is infinite, all considered words on it are finite, and we can use classical results and techniques on word equations.

Another series of results which we shall need concerns canonical decompositions of catenation of factorial languages. Suppose we are given canonical decompositions $\overline{A}$ of a language A and $\overline{B}$ of a language B. What is the canonical decomposition $\overline{AB}$ of their catenation? The answer has been found in [4] and is expressed in terms of subalphabets $\Pi(A)$ and $\Delta(B)$, where

$$\Pi(X) = \{x \in \Sigma \mid Xx \subset X\} \text{ and } \Delta(X) = \{x \in \Sigma \mid xX \subset X\}.$$

So, $\Pi(X)$ is defined as the greatest subalphabet such that $X\Pi^*(X) = X$, and $\Delta(X)$ is the greatest subalphabet such that $\Delta^*(X)X = X$.

Example 3. If $X = \{a,b\}^*$, then $\Delta(X) = \Pi(X) = \{a,b\}$. If $X = a^*b^*$, then $\Delta(X) = \{a\}$ and $\Pi(X) = \{b\}$. If $X = a^* \cup b^*$, then $\Delta(X) = \Pi(X) = \emptyset$. We also have $\Delta(X) = \Pi(X) = \emptyset$ for each finite language X.

Now, given a factorial language X and a subalphabet Δ, let the operators L and R on factorial languages be defined by

$$L_\Delta(X) = \mathrm{Fac}(X \backslash \Delta X) \text{ and } R_\Delta(X) = \mathrm{Fac}(X \backslash X\Delta).$$

The meaning of these sets is described by the following lemma proved in [4].

Lemma 2. *For factorial languages X and Y we have $R_{\Delta(Y)}(X)Y = XY$, and $R_{\Delta(Y)}(X)$ is the minimal factorial set with this property: it is equal to the intersection of all factorial languages Z such that $ZY = XY$. Symmetrically, $YL_{\Pi(Y)}(X) = YX$, and $L_{\Pi(Y)}(X)$ is the minimal factorial language with this property.*

Let us list several staightforward properties of the operators L and R.

Lemma 3. *Suppose that $\Sigma = \{a_1, \ldots, a_q\}$, and X is a factorial language on Σ. Then*

$$\{\lambda\} = L_\Sigma(X) \subseteq L_{\{a_1,\ldots,a_{q-1}\}}(X) \subseteq \cdots \subseteq L_{\{a_1\}}(X) \subseteq L_\emptyset(X) = X, \text{ and}$$
$$\{\lambda\} = R_\Sigma(X) \subseteq R_{\{a_1,\ldots,a_{q-1}\}}(X) \subseteq \cdots \subseteq R_{\{a_1\}}(X) \subseteq R_\emptyset(X) = X. \qquad \square$$

Lemma 4. *For each subalphabet Γ and a factorial language X we have $X\Gamma^* = R_\Gamma(X)\Gamma^*$ and $\Gamma^*X = \Gamma^*L_\Gamma(X)$.*

The two following lemmas show that in many cases we just have $L_\Gamma(X) = X$ and $R_\Gamma(X) = X$.

Lemma 5. *[4] Let X be a factorial language and $\Gamma \subset \Sigma$ be a subalphabet. If $\Pi(X)$ contains a symbol not belonging to Γ, then $R_\Gamma(X) = X$. Symmetrically, if $\Delta(X)$ contains a symbol not belonging to Γ, then $L_\Gamma(X) = X$.*

Lemma 6. *[4] Let X be a factorial language with $\overline{X} \doteq X_1 \cdots X_k$. Then*

$$\overline{L_{\Delta(X)}(X)} \doteq \begin{cases} X_2 \cdots X_k, & \text{if } X_1 = \Delta^*(X), \\ \overline{X}, & \text{otherwise;} \end{cases}$$

and symmetrically,

$$\overline{R_{\Pi(X)}(X)} \doteq \begin{cases} X_1 \cdots X_{k-1}, & \text{if } X_k = \Pi^*(X), \\ \overline{X}, & \text{otherwise.} \end{cases}$$

These two lemmas combined with Lemma 2 lead to the following main result of [4].

Theorem 3. *[4] Let A and B be factorial languages with $\overline{A} \doteq A_1 \cdots A_k$ and $\overline{B} \doteq B_1 \cdots B_m$, where $A_i, B_j \in \mathcal{F}$. Then the canonical decomposition of AB is*

1. $\overline{AB} \doteq \overline{R_{\Delta(B)}(A)} \, \overline{B}$ *if $\Pi(A) \subsetneq \Delta(B)$; symmetrically, $\overline{AB} \doteq \overline{A} \, \overline{L_{\Pi(A)}(B)}$ if $\Delta(B) \subsetneq \Pi(A)$;*
2. $\overline{AB} \doteq A_1 \cdots A_{k-1}\overline{B}$ *if $\Delta(B) = \Pi(A) = \Delta$ and $A_k = \Delta^*$; symmetrically, $\overline{AB} \doteq \overline{A}B_2 \cdots B_m$ if $\Delta(B) = \Pi(A) = \Delta$ and $B_1 = \Delta^*$. Note that the situation of $A_k = B_1 = \Delta^*$ falls into both cases.*
3. $\overline{AB} \doteq \overline{A} \, \overline{B}$ *otherwise.*

Corollary 1. *For all factorial languages A and B, the canonical decomposition of AB is either $\overline{AB} \doteq \overline{R_{\Delta(B)}(A)} \, \overline{B}$ or $\overline{AB} \doteq \overline{A} \, \overline{L_{\Pi(A)}(B)}$.*

PROOF. In fact, in the latter two cases of Theorem 3 the equality also holds as it is listed in Lemmas 5 and 6. For details, see [4].   □

4   Properties of Decompositions and Operators L and R

In this section, the alphabet considered is still arbitrary.

Lemma 7. *[4] If $\overline{X} \doteq X_1 \cdots X_n$, $X_i \in \mathcal{F}$, then $\Pi(X) = \Pi(X_n)$ and $\Delta(X) = \Delta(X_1)$.*

Lemma 8. *Let $\overline{X} \doteq X_1 \cdots X_m$, $X_i \in \mathcal{F}$, be the canonical decomposition of a factorial language X. Then each factor $X_i \cdots X_j$, $i \le j$, of the "word" $X_1 \cdots X_m$ is also a canonical decomposition of the respective language:*

$$\overline{X_i \cdots X_j} \doteq X_i \cdots X_j.$$

PROOF. Suppose that $X_i \cdots X_j$ is not a canonical decomposition. Since all languages X_k are already indecomposable, it is possible only if $X_i \cdots X_j = X_i \cdots X_k' \cdots X_j$ for some factorial language $X_k' \subsetneq X_k$, $i \le k \le j$. But then $X = X_1 \cdots (X_i \cdots X_k' \cdots X_j) \cdots X_m = X_1 \cdots X_k' \cdots X_m$, that is, $\overline{X} \ne X_1 \cdots X_m$, a contradiction.   □

Lemma 9. *Let X be a factorial language. Then for all subalphabets $\Delta, \Pi \subset \Sigma$ the equality $L_\Pi(R_\Delta(X)) = R_\Delta(L_\Pi(X))$ holds.*

PROOF. If a non-empty word $u \in L_\Pi(R_\Delta(X))$, then there exists v (which can be empty) such that vu starts with a symbol from $\Sigma \backslash \Pi$ and belongs to $R_\Delta(X)$. This, in its turn, means that there exists a word w (which can be empty) such that the last symbol of the word vuw belongs to $\Sigma \backslash \Delta$, and $vuw \in X$.

We see that the obtained condition is symmetric with respect to the order of applying the operators L_Π and R_Δ, so, we get it another time if we consider an arbitrary word $u \in R_\Delta(L_\Pi(X))$. Thus, these two sets are equal.   □

Lemma 10. *Let* $\Gamma \subset \Sigma$ *be a subalphabet and* X *be a factorial language with* $R_\Gamma(X) = Y \neq \{\lambda\}$. *Then* $\Delta(Y) \supseteq \Delta(X)$.

PROOF. Let a be a symbol from $\Delta(X)$. Suppose by contrary that $a \notin \Delta(Y)$, that is, that $ay \neq Y$ for some $y \in Y$.

If y is not the empty word, $y \in Y$ means that $yz \in X$ for some z such that the last symbol of yz does not belong to Γ. Since $a \in \Delta(X)$, we have $ayz \in X$ and thus $ay \in Y$, a contradiction.

Now it remains to observe that if $ay \in Y$ for all non-empty words $y \in Y$ (which exist by the assertion), then the same holds for the empty word λ: the symbol $a = a\lambda \in Y$ since Y is factorial. We have shown that $a \in \Delta(X)$ implies $a \in \Delta(Y)$, which was to be proved. □

Lemma 11 (see [4] for another formulation). *Let* X *be a factorial language with* $\overline{X} \doteq X_1 \cdots X_m$, $X_i \in \mathcal{F}$. *Consider a subalphabet* $\Delta \subset \Sigma$ *and the factorial language* $Y = R_\Delta(X)$ *with* $\overline{Y} \doteq Y_1 \cdots Y_n$, $Y_j \in \mathcal{F}$. *Then there exist integers* $0 = i_0 \leq \ldots \leq i_{m-1} \leq i_m = n$ *such that* $Y_{i_{k-1}+1} \cdots Y_{i_k} \subseteq X_k$ *for all* $k = 1, \ldots, m$. *Moreover, if* $Y = R_\Delta(X)$, *then for each* $k < m$ *we have* $Y_1 \cdots Y_{i_k} = R_{\Delta(Y_{i_k+1})}(X_1 \cdots X_k)$ *and* $Y_{i_k+1} \cdots Y_n = R_\Delta(X_{k+1} \cdots X_m)$.

Note that in this lemma, i_k may be equal to i_{k+1}, and thus the sequence of Y_j included into a given X_k can be empty.

Lemmas symmetric to the latter two can be proved as well.

Lemma 12. *Suppose that* $Y = R_\Delta(X)$ *(or* $Y = L_\Delta(X)$*) for some* $\Delta \subset \Sigma$, $\overline{X} \doteq X_1 \cdots X_n$, $X_i \in \mathcal{F}$, *and* $\overline{Y} \doteq X_{\sigma(1)} \cdots X_{\sigma(n)}$ *for some permutation* σ. *Then* $X = Y$.

PROOF. The assertion of the lemma means that each indecomposable factorial language occurs in the canonical decompositions of X and Y an equal number of times.

For the sake of convenience, let us denote $X_{\sigma(i)} = Y_i$. Due to Lemma 11, there exist integers $0 = i_0 \leq \ldots \leq i_{n-1} \leq i_n = n$ such that $Y_{i_{k-1}+1} \cdots Y_{i_k} \subseteq X_k$ for all $k = 1, \ldots, n$. We wish to prove that $i_k = k$ for all k, and all the inclusions are in fact equalities (of the form $Y_i = X_i$).

Suppose the opposite. Then there exists some k_1 such that the corresponding inclusion is of the form $Y_{i_{k_1-1}+1} \cdots Y_{i_{k_1}} \subset X_{k_1}$ (the equality is impossible even if $i_{k_1} - i_{k_1-1} \geq 2$, since all the involved languages are indecomposable, and decompositions are minimal). In particular, neither of the languages $Y_{i_{k_1-1}+1}, \ldots, Y_{i_{k_1}}$ is equal to X_{k_1}. But we know that the language X_{k_1} occurs in $\overline{X}$ and $\overline{Y}$ an equal number of times. So, X_{k_1} is equal to some Y_j, where $i_{k_2-1} + 1 \leq j \leq i_{k_2}$, and $X_{k_1} = Y_j \subsetneq X_{k_2}$. Continuing this argument, we get an infinite sequence $X_{k_1} \subsetneq X_{k_2} \subsetneq \cdots \subsetneq X_{k_m} \subsetneq \cdots$. But there is only a finite number of entries in the canonical decomposition of a factorial language. A contradiction. □

Lemma 13. *Let us fix a subalphabet* $\Delta \subset \Sigma$. *A factorial language* Y *can be equal to* $R_\Delta(X)$ *($L_\Delta(X)$) for some factorial language* X *if and only if* $Y = R_\Delta(Y)$ *(respectively,* $Y = L_\Delta(Y)$*).*

PROOF. Clearly, the "if" part of the proof is just given by $X = Y$. To prove the "only if" part suppose that $Y \neq R_\Delta(Y)$, which means that there exists a word $u \in Y$ such that $u = u'a$, $a \in \Delta$, and u is not a prefix of any word from Y whose last symbol does not belong to Δ. Suppose that $Y = R_\Delta(X)$ for some X. Since $u \in Y$, we must have $uv \in X$ for some word v whose last symbol is from $\Sigma \backslash \Delta$. But then $uv \in Y$, contradicting to our assumption. The proof for the operator L_Δ is symmetric.   □

Lemma 14. *Let Y be a factorial language with $Y = R_\Delta(Y)$ $(Y = L_\Pi(Y))$ for a given $\Delta, \Pi \subset \Sigma$. Then $Y = R_\Delta(X)$ $(Y = L_\Pi(X))$ if and only if for a factorial language X we have $Y \subseteq X \subseteq Y\Delta^*$ (respectively, $Y \subseteq X \subseteq \Pi^* Y$).*

PROOF. Suppose that $Y = R_\Delta(X)$; then $Y \subseteq X$ by the construction (and Lemma 3), and $X \subseteq X\Delta^* = R_\Delta(X)\Delta^* = Y\Delta^*$ due to Lemma 4.

On the other hand, if $Y \subseteq X \subseteq Y\Delta^*$, then $Y \backslash X\Delta \subseteq X \backslash X\Delta \subseteq Y\Delta^* \backslash X\Delta = Y \backslash X\Delta$. So, $Y \backslash X\Delta = X \backslash X\Delta$ and thus $Y = R_\Delta(Y) = R_\Delta(X)$.

The proof for the operator L_Π is symmetric.   □

5   Simple Word Equations

Here we list several classical word equations and their solutions. Words are considered on an alphabet $\mathcal{A}$ which may be infinite since all considered words are finite anyway.

Lemma 15 (Commutation of words, see e.g. [8]). *Let words $x, y \in \mathcal{A}^*$ commute: $xy = yx$. Then $x = z^n$ and $y = z^m$ for some $z \in \mathcal{A}^*$ and $n, m \geq 0$.*   □

Lemma 16 (see, e. g., [3]). *Let $xz = zy$ for some $x, y, z \in \mathcal{A}^*$. Then either $x = y = \lambda$, or $z = \lambda$, or $x = rs$, $y = sr$, and $z = (rs)^k r$ for some $r, s \in \mathcal{A}^*$ with $r \neq \lambda$ and $k \geq 0$.*   □

At last, the following lemma can be easily proved by a standard technique described, e. g., in [3].

Lemma 17. *Let $xay = yax$ for some $x, y \in \mathcal{A}^*$, $a \in \mathcal{A}$. Then $x = (za)^n z$ and $y = (za)^m z$ for some $z \in \mathcal{A}^*$ and $n, m \geq 0$.*   □

6   Binary Alphabet

In this section we pass to considering languages on the binary alphabet $\Sigma_2 = \{a, b\}$ and discuss what changes after this restriction. In what follows we denote an arbitrary letter of the binary alphabet by x and the other letter by y.

Lemma 18. *For each binary factorial language $X \subset \Sigma_2^*$ not equal to Σ_2^*, each of the subalphabets $\Pi(X)$ and $\Delta(X)$ can be equal only to $\{a\}$, $\{b\}$, or $\emptyset$.*

PROOF. The equality $\Pi(X) = \{a, b\}$ or $\Delta(X) = \{a, b\}$ would imply that $X \supseteq \{a, b\}^* = \Sigma_2^*$. So, $X = \Sigma_2^*$, contradicting to the assertion. □

Lemma 19. *Let $\Gamma \subset \Sigma_2$ be a subalphabet and X be a binary factorial language with $Y = R_\Gamma(X) \neq \{\lambda\}$ and $\Delta(X) = \{x\}$. Then $\Delta(Y) = \{x\}$.*

PROOF. We have $\Delta(Y) \supseteq \{x\}$ due to Lemma 10 and $\Delta(Y) \neq \{a, b\}$ due to Lemma 18, since $Y \subseteq X \neq \Sigma_2^*$. □

7   Main Derivations

So, we have the equality $AB = BA$, where A and B are binary factorial languages. Then canonical decompositions of AB and BA are also equal due to Theorem 2. Due to Corollary 1, there are only three possibilities of how the equality for canonical decompositions looks like: either

$$R_{\Delta(B)}(A) \cdot B \doteq B \cdot L_{\Pi(B)}(A), \tag{1}$$

(or $A \cdot L_{\Pi(A)}(B) \doteq R_{\Delta(A)}(B) \cdot A$, which is the same up to renaming A and B); or

$$R_{\Delta(B)}(A) \cdot B \doteq R_{\Delta(A)}(B) \cdot A, \tag{2}$$

or, symmetrically to the previous case,

$$A \cdot L_{\Pi(A)}(B) \doteq B \cdot L_{\Pi(B)}(A). \tag{3}$$

These cases intersect: for example, the situation when $L_{\Pi(B)}(A) = A$ and $R_{\Delta(A)}(B) = B$ falls into both (1) and (2). However, to get a classification of the cases of commutation, we may consider the cases (1) and (2) separately (the case (3) is symmetric to (2)).

7.1   Case of (1)

Suppose that (1) holds. If we denote $R_{\Delta(B)}(A) = A'$ and $L_{\Pi(B)}(A) = A''$, we get a word equation on $\mathcal{F}$:

$$A'B \doteq BA''.$$

Note that the unit element of the semigroup $\mathcal{F}^*$ is the language $\{\lambda\}$. So, according to Lemma 16, the equation has the following solutions:

1. Either $B = \{\lambda\}$; then $A' = A'' = A$ and this is a particular case of absorption.
2. Or $A' = A'' = \{\lambda\}$, and this is again absorption, since $AB = BA = B$.
3. Or $A' \doteq RS$, $A'' \doteq SR$, and $B \doteq (RS)^k R$ for some $R, S \in \mathcal{F}^*$, $R \neq \{\lambda\}$, $k \geq 0$.

Let us consider this third situation in detail.

First, note that due to Lemma 8, the languages R and S are given in canonical decompositions. Due to Lemma 7 (applied several times), we have

$$\Delta(B) = \Delta(R) = \Delta(A') \text{ and } \Pi(B) = \Pi(R) = \Pi(A''); \tag{4}$$

in what follows we denote these subalphabets just by Δ and Π.

Suppose first that one of the subalphabets Δ and Π is empty: say, $\Delta = \emptyset$. Then $A' = R_\emptyset(A) = A = RS$ and $A'' = L_{\Pi(B)}(A) = SR$; due to Lemma 12, $A'' = A$, and the commutation equation (1) is just $AB \doteq BA$. Due to Lemma 15, we have $A \doteq C^n$ and $B = C^m$ for some factorial language $C \in \mathcal{F}^*$, and this is word type commutation.

Note that if $B = \Sigma_2^*$, then $A' = A'' = \{\lambda\}$, and this is absorption. So, due to Lemma 18, we have essentially two situations when anything unexpected may occur: either $\Delta = \{x\}$ and $\Pi = \{y\}$, $y \neq x$, or $\Delta = \Pi = \{x\}$. We shall consider these two situations in succession, but before that, note that in both cases,

$$L_\Pi(A') = R_\Delta(A'') \tag{5}$$

due to Lemma 9 and $A' \cup A'' \subseteq A \subseteq A'\Delta^* \cap \Pi^*A''$ due to Lemma 14, which gives

$$RS \cup SR \subseteq A \subseteq RS\Delta^* \cap \Pi^*SR. \tag{6}$$

<u>Case of $\Delta = \{x\}$ and $\Pi = \{y\}$.</u> Then $L_{\{y\}}(A') = A'$ and $R_{\{x\}}(A'') = A''$ due to Lemma 5. By (5) we see that $A' = A''$, that is, $RS \doteq SR$, and due to Lemma 15, we have $R = Z^n$ and $S = Z^m$ for some $Z \in \mathcal{F}^+$. So, $A' = A'' = Z^{n+m}$ and $B = Z^{k(n+m)+n}$. After renaming variables we can write $A' = A'' = Z^r$ and $B = Z^p$ for some $r, p > 0$ (if p or r is equal to 0, the language A' or B is equal to $\{\lambda\}$, and we have already considered these degenerate situations).

Now (6) can be rewritten as

$$Z^r \subseteq A \subseteq Z^r x^* \cap y^* Z^r. \tag{7}$$

If $Z^r = Z^r x^* \cap y^* Z^r$, then $A = Z^r$ and we have word type commutation. But if $Z^r \subsetneq Z^r x^* \cap y^* Z^r$, consider a word u of minimal length belonging to $(Z^r x^* \cap y^* Z^r) \setminus Z^r$. We see that $u = yvx$, where $yv \in Z^r$, $vx \in Z^r$, but $yvx \notin Z^r$.

If such a word u exists, then we can take *any* factorial set A lying between Z^r and $Z^r x^* \cap y^* Z^r$, and it will commute with any power of Z. This gives us

> **Unexpected commutation I.** Let Z be a binary factorial language with $\Delta(Z) = \{x\}$ and $\Pi(Z) = \{y\}$, $x \neq y$. Then for all $r, p > 0$ the language Z^p commutes with any language A satisfying the inclusion (7). Such a language not equal to Z^r exists if and only if there exists a word v such that $yv \in Z^r$, $vx \in Z^r$, but $yvx \notin Z^r$.

Example 4. Consider $\Sigma = \{a, b\}$ and the languages $F_a = \text{Fac}(\{a, ab\}^*)$ and $F_b = \text{Fac}(\{b, ab\}^*)$: the language F_a contains all words avoiding two successive bs, and the language F_b contains all words avoiding two successive as. Consider $Z = F_b \cdot F_a$; then $\Pi(Z) = \{a\}$ and $\Delta(Z) = \{b\}$. Let us fix $r = 1$. The word $v = ab$ satisfies our conditions since $aab \in Z$, $abb \in Z$, but $aabb \notin Z$. So, any language $A = Z \cup S$, where S is a factorial subset of $a^* b^*$, commutes with any power Z^p of Z.

<u>Case of $\Delta = \Pi = \{x\}$.</u> Suppose first that $\overline{R}$ does not start with x^*. Then we have $L_{\{x\}}(A') = L_{\{x\}}(RS) = RS$ due to Lemma 6, and thus $RS = R_{\{x\}}(SR)$

due to (5). So, due to Lemma 12 we have $RS \doteq SR$, and due to Lemma 15, $RS = SR = A' = A'' = Z^r$ for some factorial language Z with $R \doteq \overline{Z}^n$ and $S \doteq \overline{Z}^m$. Now (6) can be rewritten as

$$Z^r \subseteq A \subseteq Z^r x^* \cap x^* Z^r;$$

but in fact, both inclusions here are equalities: $Z^r x^* = x^* Z^r = Z^r$ since $\Delta(Z) = \Delta(R) = \{x\}$ and $\Pi(Z) = \Delta(R) = \{x\}$ due to Lemma 7. So, $A = Z^r$, $B = (RS)^k R = Z^p$, and this is word type commutation. The same holds if $\overline{R}$ does not end with x^*. So, it remains to check the situation when $R \doteq x^*$ or $R \doteq x^* T x^*$ for some $T \in \mathcal{F}^+$ (note that $T \neq \{\lambda\}$ since $\overline{x^* x^*} \doteq x^*$).

Suppose first that $R \doteq x^*$. Then (6) can be rewritten as

$$x^* S \cup S x^* \subseteq A \subseteq x^* S x^*. \tag{8}$$

Any language A satisfying these inclusions commutes with all languages of the form $(x^* S)^k x^*$. Here S is an arbitrary language which can precede and follow x^* in a canonical decomposition: that is, an arbitrary language such that $L_{\{x\}}(S) = R_{\{x\}}(S) = S$ and $x \notin \Delta(S), \Pi(S)$ (which means that $\Delta(S)$ and $\Pi(S)$ are equal to $\{y\}$ or to $\emptyset$).

Now suppose that $R \doteq x^* T x^*$, $T \in \mathcal{F}^+$. Then $L_{\{x\}}(RS) \doteq T \Delta^* S$ and $R_{\{x\}}(SR) \doteq S \Delta^* T$ due to Lemma 6; due to (5), we have the following word equation on $\mathcal{F}^*$:

$$T x^* S = S x^* T.$$

Due to Lemma 17, the general solution of this equation is $S \doteq (Q x^*)^n Q$ and $T \doteq (Q x^*)^m Q$ for some $Q \in \mathcal{F}^*$ such that $L_{\{x\}}(Q) = R_{\{x\}}(Q) = Q$ and $x \notin \Delta(S), \Pi(S)$, and for $n, m \geq 0$. So, $RS = (x^* Q)^{n+m+2}$, $SR = (Q x^*)^{n+m+2}$, and $B = (x^* Q)^{k(n+m+2)+m+1} x^*$. After renaming variables, we get $RS = (x^* Q)^r$, $SR = (Q x^*)^r$, and $B = (x^* Q)^p x^*$ for some $r \geq 2$ and $p \geq 1$; and (6) takes the form

$$(x^* Q)^r \cup (Q x^*)^r \subseteq A \subseteq (x^* Q)^r x^*. \tag{9}$$

This inclusion together with (8) (which adds the cases of $r = 1$ and $p = 0$) gives

> **Unexpected commutation II.** Let $x \in \Sigma_2$ be a symbol and Q be a binary factorial language with $L_\Delta(Q) = R_\Delta(Q) = Q$ and $\Delta(Q), \Pi(Q)$ equal to $\emptyset$ or $\{y\}$, $y \neq x$. Then for all $p \geq 0$ and $r \geq 1$ the language $(x^* Q)^p x^*$ commutes with any language A satisfying the inclusions (9).

Note that if the second inclusion of (9) turns into equality, that is, if $A = (x^* Q)^r x^*$, then for $p \geq 1$ it is again a word type commutation since $A = (x^* Q x^*)^r$ and $B = (x^* Q x^*)^p$. However, if the equality does not hold, it is a situation of a new type.

An easiest example of commutation of this type has been mentioned above as Example 1: if $x = a$, and $Q = b^*$, we see that $x^* = a^*$ commutes with $a^* b^* \cup b^* a^*$ (and with any factorial language which includes $a^* b^* \cup b^* a^*$ and is included into $a^* b^* a^*$). A more sophisticated "intermediate" example of this family is the following:

Example 5. For each $p \geq 0$, the language $(a^*b^*)^p a^*$ commutes with the language $A = a^*b^*(aa)^*b^*a^* \cup a^*b^*(aa)^*ab^*$, since $a^*b^*a^*b^* \cup b^*a^*b^*a^* \subset A \subset a^*b^*a^*b^*a^*$.

We have considered all situations possible if (1) holds.

7.2   Case of (2)

In this section, we suppose that the canonical decomposition of $AB = BA$ is

$$R_{\Delta(B)}(A) \cdot B \doteq R_{\Delta(A)}(B) \cdot A.$$

In what follows we denote $R_{\Delta(B)}(A) \doteq A'$ and $R_{\Delta(A)}(B) \doteq B'$, so that

$$A'B \doteq B'A. \tag{10}$$

Suppose first that $A' = \{\lambda\}$ or $B' = \{\lambda\}$. Then $AB = B$ or $AB = A$, and this is commutation by absorption. So, in what follows we assume that the canonical decompositions of A' and B' are not empty.

Suppose that $\Delta(A) = \emptyset$. Then $B' \doteq B$ due to Lemma 3, and our case have been considered in the previous subsection (where it has been shown that this is inevitably word type commutation). Due to Lemma 18, we thus have $\Delta(A) = \{x\}$ and $\Delta(B) = \{z\}$ for some $x, z \in \Sigma_2$. Note also that $\Delta(B') \supseteq \{z\}$ and $\Delta(A') \supseteq \{x\}$ due to Lemma 10; but $\Delta(A') = \Delta(B')$ due to Lemma 7 since A' and B' are not equal to $\{\lambda\}$. At the same time, we cannot have $\Delta(A') = \Delta(B') = \Sigma_2$ since it would imply that $A'B \doteq \Sigma_2^*$, but Σ_2^* is indecomposable (see Lemma 1). So, $x = z$, that is,

$$\Delta(A) = \Delta(B) = \Delta(A') = \Delta(B') = \{x\}$$

for some $x \in \Sigma_2$. Note that this is essentially the only point where we require the alphabet to be binary, since all the arguments of the previous subsection could be extended to the general case.

Note that if $A' = B'$, then $A = B$, and this is word type commutation. So, we may assume that one of the "words" A', B' on the alphabet $\mathcal{F}$ is a proper prefix of the other: say, $A' \doteq B'C$ for some $C \in \mathcal{F}^+$. Then $A \doteq CB$ because of (10), and $B'C \doteq A' \doteq R_{\{x\}}(A) \doteq R_{\{x\}}(CB) \doteq R_{\Delta(R_{\{x\}}(B))}(C)R_{\{x\}}(B)$; the latter equality holds due to Lemma 11. But $R_{\{x\}}(B) = B'$, and we already know that $\Delta(B') = \{x\}$; so, $B'C \doteq R_{\{x\}}(C)B'$. Clearly, $\overline{R_{\{x\}}(C)} \doteq C$ since C precedes B in the canonical decomposition of AB, and $\Delta(B) = \{x\}$. Thus, we have $B'C \doteq CB'$, so that $B' = Z^n$, $C = Z^m$ for some $n, m > 0$ due to Lemma 15. Here Z is a factorial language with $\Delta(Z) = \Delta(B') = \{x\}$ and $\overline{ZZ} \doteq \overline{Z}\,\overline{Z}$.

Due to Lemma 14, we have $B' = Z^n \subseteq B \subseteq Z^n x^*$, and B can be equal to any set satisfying these inclusions. Note that B can be not equal to B' only if $\Pi(Z)$ is equal to $\emptyset$ or $\{y\}$, $y \neq x$. After that we just define $A = Z^m B$ and observe that A and B really commute: $AB = BA = Z^{n+m}B$. So, this is the "right-to-left" version of

> **Unexpected commutation III.** Let Z be a binary factorial language such that $\overline{ZZ} \doteq \overline{Z}\,\overline{Z}$ and $\Delta(Z) = \{x\}$. Let B be a factorial language satisfying $Z^n \subseteq B \subseteq Z^n x^*$, $n > 0$. Then B commutes with $Z^m B$ for all $m > 0$.
>
> Symmetrically, if Z is a binary factorial language with $\overline{ZZ} \doteq \overline{Z}\,\overline{Z}$ and $\Pi(Z) = \{x\}$, and if B is a factorial language satisfying $Z^n \subseteq B \subseteq x^* Z^n$, then B commutes with BZ^m for all $n, m > 0$.

Note that the symmetric "left-to-right" version of unexpected commutation III described above can be found and stated symmetrically starting from (3).

Of course unexpected commutation III includes some cases of word type commutation: in particular, if $B = Z^{n-1}D$ for some $Z \subseteq D \subseteq Zx^*$, where $\{x\} = \Delta(Z)$, then $B = D^n$ and $A = D^{m+n}$. But situations when it is not word type commutation also exist.

Example 6. Consider $Z = a^* b^*$ and $B =$Fac$(a^*(bb)^* a^* b^* \cup a^* b(bb)^* a^* b^* a^*)$. Here $\Delta(Z) = \{a\}$ and $Z^2 = a^* b^* a^* b^* \subset B \subset a^* b^* a^* b^* a^* = Z^2 a^*$. We can see that B commutes with all sets A of the form $A = Z^m B$ (even if m is odd): $AB = BA = Z^{m+2} B$.

We have studied all possible cases when binary factorial languages commute. Theorem 1 is proved. □

Acknowledgements. The author is deeply grateful to Julien Cassaigne who found a mistake in the initial text of the paper and made several important suggestions on the presentation of the remaining part of the result.

References

1. Avgustinovich, S.V., Frid, A.E.: A unique decomposition theorem for factorial languages. Internat. J. Algebra Comput. 15, 149–160 (2005)
2. Conway, J.H.: Regular Algebra and Finite Machines. Chapman & Hall, London (1971)
3. Diekert, V.: Makanin's Algorithm. In: Lothaire, M. (ed.) Algebraic combinatorics on words, pp. 387–442. Cambridge Univ. Press, Cambridge (2002)
4. Frid, A.E.: Canonical decomposition of a catenation of factorial languages. Siberian Electronic Mathematical Reports 4, 14–22 (2007),
   http://semr.math.nsc.ru/2007/V4/p12-19.pdf
5. Karhumäki, J., Latteux, M., Petre, I.: Commutation with codes. Theoret. Comput. Sci. 340, 322–333 (2005)
6. Karhumäki, J., Latteux, M., Petre, I.: The commutation with ternary sets of words. Theory Comput. Systems 38, 161–169 (2005)
7. Kunc, M.: The power of commuting with finite sets of words. In: Diekert, V., Durand, B. (eds.) STACS 2005. LNCS, vol. 3404, pp. 569–580. Springer, Heidelberg (2005), Full version currently available at
   http://math.muni.cz/~kunc/math/commutativity.pdf
8. Lothaire, M.: Combinatorics on words. Addison-Wesley, London (1983)
9. Ratoandromanana, B.: Codes et motifs, RAIRO. Inform. Theor. 23, 425–444 (1989)

Inapproximability of Nondeterministic State and Transition Complexity Assuming P ≠ NP

Hermann Gruber[1] and Markus Holzer[2]

[1] Institut für Informatik, Ludwig-Maximilians-Universität München,
Oettingenstraße 67, D-80538 München, Germany
gruberh@tcs.ifi.lmu.de
[2] Institut für Informatik, Technische Universität München,
Boltzmannstraße 3, D-85748 Garching bei München, Germany
holzer@informatik.tu-muenchen.de

Abstract. Inapproximability results concerning minimization of nondeterministic finite automata relative to given deterministic finite automata were obtained only recently, modulo cryptographic assumptions [4]. Here we give upper and lower bounds on the approximability of this problem utilizing only the common assumption $\mathbf{P} \neq \mathbf{NP}$, in the setup where the input is a finite language specified by a truth table. To this end, we derive an improved inapproximability result for the biclique edge cover problem. The obtained lower bounds on approximability can be sharpened in the case where the input is given as a deterministic finite automaton over a binary alphabet. This settles most of the open problems stated in [4]. Note that the biclique edge cover problem was recently studied by the authors as lower bound method for the nondeterministic state complexity of finite automata [5].

1 Introduction

Finite automata are one of the oldest and most intensely investigated computational models. As such, they found widespread use in many other different areas such as circuit design, natural language processing, computational biology, parallel processing, image compression, to mention a few, see [13] and references therein. As some of these applications deal with huge masses of data, the amount of space needed by finite automata is an important research topic.

On the one hand, it is well known that while nondeterministic finite automata and deterministic finite automata are equal in expressive power, nondeterministic automata can be exponentially more succinct than deterministic ones. On the other hand, minimizing deterministic finite automata can be carried out efficiently, whereas the state minimization problem for nondeterministic finite state machines is **PSPACE**-complete, even if the regular language is specified as a deterministic finite automaton [8]. This prompted the authors of the aforementioned paper to ask whether there exist at least polynomial-time approximation algorithms with a reasonable performance guarantee. However, recent work [4] shows that this problem cannot be approximated within $\frac{\sqrt{n}}{\text{polylog}(n)}$ for state minimization and $\frac{n}{\text{polylog}(n)}$ for transition minimization, provided some cryptographic

T. Harju, J. Karhumäki, and A. Lepistö (Eds.): DLT 2007, LNCS 4588, pp. 205–216, 2007.
© Springer-Verlag Berlin Heidelberg 2007

assumption holds. As the result is based on a rather strong assumption, the authors asked for proving approximation hardness results under the weaker (and more familiar) assumption $\mathbf{P} \neq \mathbf{NP}$ as an open problem. Moreover, they asked to determine the approximation complexity when given a regular language specified by a truth table.

In this paper we solve these open problems. To summarize, we have obtained the following results on the minimization problems for nondeterministic finite automata:

- If the input is given as a *nondeterministic* finite automaton with n states (transitions, respectively), the state (transition, respectively) minimization problem is not fixed-parameter tractable (the parameter being the upper bound on the number of states/transitions to be reached) by Theorem 1, unless $\mathbf{P} = \mathbf{PSPACE}$. Earlier work established that this problem is not approximable within $o(n)$, provided $\mathbf{P} = \mathbf{PSPACE}$ [4], and this holds even for unary input alphabets, unless $\mathbf{P} = \mathbf{NP}$ [7].
- If the input is given by a *truth table* specifying a Boolean function of total size N, the state minimization problem is $\mathbf{NP}$-complete (Theorem 4). Moreover we establish a lower bound of $N^{1/6-\varepsilon}$ on the approximability both for state and transition minimization, provided $\mathbf{P} \neq \mathbf{NP}$ (Theorems 7 and 10). These results are nicely contrasted by two simple polynomial-time algorithms achieving ratios of $O(\sqrt{N}/\log N)$ for state minimization, and $O(N/(\log N)^2)$ for the case of transition minimization, respectively (Theorem 5).
- Finally, if the input is given by a *deterministic* finite automaton, Theorem 1 asserts that the corresponding state and transition minimization problems become fixed-parameter tractable. But assuming $\mathbf{P} \neq \mathbf{NP}$, the state minimization problem is not approximable within $n^{1/3-\varepsilon}$ for alphabets of size $O(n)$ (Corollary 14), and not approximable within $n^{1/5-\varepsilon}$ for a binary alphabet, for all $\varepsilon > 0$ (Theorem 15). Under the same assumption, we show that the transition minimization problem for binary input alphabets is not approximable within $n^{1/5-\varepsilon}$, for all $\varepsilon > 0$ (Corollary 16). Before this work, the known inapproximability results for these problems were based on a much stronger assumption [4].

Some of the hardness results are based on a reduction from the biclique edge cover problem, which we prove to be neither approximable within $|V|^{1/3-\varepsilon}$ nor $|E|^{1/5-\varepsilon}$ unless $\mathbf{P} = \mathbf{NP}$ in Theorem 6.

2   Preliminaries

First we recall some definitions from formal language and automata theory. In particular, let Σ be an alphabet and Σ^* the set of all words over the alphabet Σ, containing the empty word λ. The length of a word w is denoted by $|w|$, where $|\lambda| = 0$. The reversal of a word w is denoted by w^R and the reversal of a language $L \subseteq \Sigma^*$ by L^R, which equals the set $\{ w^R \mid w \in L \}$. Furthermore let $\Sigma^n = \{ w \in \Sigma^* \mid |w| = n \}$.

A *nondeterministic finite automaton* (NFA) is a 5-tuple $A = (Q, \Sigma, \delta, q_0, F)$, where Q is a finite set of states, Σ is a finite set of input symbols, $\delta : Q \times \Sigma \to 2^Q$ is the transition function, $q_0 \in Q$ is the initial state, and $F \subseteq Q$ is the set of accepting states. The transition function δ is extended to a function from $\delta : Q \times \Sigma^* \to 2^Q$ in the natural way, i.e., $\delta(q, \lambda) = \{q\}$ and $\delta(q, aw) = \bigcup_{q' \in \delta(q,a)} \delta(q', w)$, for $q \in Q$, $a \in \Sigma$, and $w \in \Sigma^*$. A nondeterministic finite automaton $A = (Q, \Sigma, \delta, q_0, F)$ is a *deterministic* finite automaton (DFA), if $|\delta(q, a)| = 1$ for every $q \in Q$ and $a \in \Sigma$. The *language accepted* by a finite automaton A is $L(A) = \{w \in \Sigma^* \mid \delta(q_0, w) \cap F \neq \emptyset\}$. Two automata are equivalent if they accept the same language.

For a regular language L, the deterministic (nondeterministic, respectively) state complexity of L, denoted by $sc(L)$ ($nsc(L)$, respectively) is the minimal number of states needed by a deterministic (nondeterministic, respectively) finite automaton accepting L. The transition complexity is analogously defined as the state complexity and we abbreviate the deterministic (nondeterministic, respectively) transition complexity of a regular language L by $tc(L)$ ($ntc(L)$, respectively).

Here we are interested in the state (transition, respectively) minimization problem for nondeterministic finite automata. This problem is defined as follows: For a given finite automaton A and an integer k, decide whether there exists a nondeterministic finite automaton B with at most k states (transitions, respectively) such that $L(A) = L(B)$. As already mentioned in the introduction, this problem is **PSPACE**-complete even if the given automaton is guaranteed to be deterministic [8]. However, other computational complexity aspects may vary if the instance to minimize is described as a nondeterministic or deterministic finite automaton. The following theorem describes such a situation—we omit the proof due to lack of space.

Theorem 1. *(i) The problem to determine for a given deterministic finite automaton, whether there exists a nondeterministic finite automaton B with at most k states (transitions, respectively) such that $L(A) = L(B)$ is fixed-parameter tractable w.r.t. parameter k. (ii) Provided* **P** $\neq$ **PSPACE**, *the aforementioned problems are not fixed-parameter tractable, if the input is given as a nondeterministic finite automaton instead.* $\square$

We also need some notions from graph theory. A *bipartite graph* is a 3-tuple $G = (U, V, E)$, where U and V are finite sets of vertices, and $E \subseteq U \times V$ is a set of edges. A bipartite graph $H = (U', V', E')$ is a *subgraph* of G if $U' \subseteq U$, $V' \subseteq V$, and $E' \subseteq E$. The subgraph H is *induced* if $E' = (U' \times V') \cap E$, the subgraph induced by U' and V' is denoted by $G[U', V']$. A set $C = \{H_1, H_2, \dots, H_k\}$ of non-empty bipartite subgraphs of G is an *edge cover* of G if every edge in G is present in at least one subgraph. An edge cover C of the bipartite graph G is a *biclique edge cover* if every subgraph in C is a biclique, where a *biclique* is a bipartite graph $H = (U, V, E)$ satisfying $E = U \times V$. The *bipartite dimension* of G is referred to as $d(G)$ and is defined to be the size of a smallest biclique edge cover of G. The associated decision problem is a classical one [3, GT18], and a reformulation of the biclique edge cover problem in terms of finite sets gives the set basis problem [3, SP7]. The following result was shown in [10]:

Theorem 2. *Deciding whether for a given bipartite graph G and an integer k there exists a biclique edge cover for G consisting of at most k bicliques is* **NP-complete.**

Finally, we assume the reader to be familiar with the basic notations of approximation theory as contained in textbooks such as [12]. In particular, transferring known inapproximability results to new problems is most easily achieved if we use some kind of approximation-preserving reduction. Several such types of reduction have been proposed; our weapon of choice is the S-reduction introduced in [9]: Loosely speaking, for two minimization problems Π and Π' and associated functions $|x|_\Pi$ and $|y|_{\Pi'}$ measuring the size of respective inputs, an S-reduction from Π to Π' with amplification $(a(n), |x|_\Pi, |y|_{\Pi'})$, where $a(n)$ is monotonically increasing, consists of a polynomial-time computable function f which maps each instance x of Π to an instance y of Π' such that $|y|_{\Pi'} \leq a(|x|_\Pi)$, and a polynomial-time computable function g that maps back instance-solution pairs of Π' to instance-solution pairs of Π such that the performance ratios of the solutions are linearly related. This kind of reduction has the following nice property [9, Proposition 1]:

Lemma 3. *Let $b : \mathbb{N} \to \mathbb{R}^+$ be a positive function, and let $\Pi = (I, \mathrm{sol}, m)$, $\Pi' = (I', \mathrm{sol}', m')$ be two minimization problems. Assume Π' is approximable within $b(|y|_{\Pi'})$, for all $y \in I'$, and there is an S-reduction from Π to Π' with amplification $(a(n), |\cdot|_\Pi, |\cdot|_{\Pi'})$. Then Π is approximable within $O(b(a(|x|_\Pi)))$, for all instances x of Π.*

3   Synthesizing a Minimal Nondeterministic Finite Automaton from a Given Truth Table

In this section we investigate the approximation complexity of minimizing nondeterministic finite automata when specifying the input by a truth table, an open question in [4]. First we show that the decision version of the problem of minimizing the number of states is **NP**-complete. Then we present two simple approximation algorithms for minimizing the number of states or transitions. Moreover, we show that the best possible approximation ratio is related to the one of the biclique edge cover problem. In order to formally define the problem we are interested in, we need some more notations.

To each m-bit Boolean function $f : \{0,1\}^m \to \{0,1\}$, where $m \geq 1$ is some natural number, we can naturally associate a finite binary language as follows:

$$L_f = \{\, x_1 x_2 \ldots x_m \in \{0,1\}^m \mid f(x_1, x_2, \ldots, x_m) = 1 \,\}.$$

In [4], the following problem was proposed: Given a truth table specifying a Boolean function $f : \{0,1\}^m \to \{0,1\}$ and an integer k, is there a nondeterministic finite automata with at most k states (transitions, respectively) accepting the language L_f?

For the the above stated problem we are able to show **NP**-completeness in case of state minimization by a reduction from the biclique edge cover problem. But the used reduction does not preserve approximability. The proof of the following Theorem can be found in the full version of the paper.

Theorem 4. *Deciding whether for a given truth table* $f : \{0,1\}^m \to \{0,1\}$ *and a positive integer* k *there is a nondeterministic finite automaton with at most* k *states accepting language* L_f *is* **NP***-complete.* □

Next we consider how well the problem under consideration can be approximated. By very simple algorithms, we obtain the following situation:

Theorem 5. *(i) Given a truth table of size* $N = 2^m$, *specifying an m-bit Boolean function function* f, *then there is a polynomial-time algorithm approximating the number of states of a state minimal nondeterministic (unambiguous, respectively) finite automaton accepting* L_f *within a factor of* $O(\sqrt{N}/\log N)$. *(ii) When considering transition minimization the performance ratio changes to* $O(N/(\log N)^2)$.

Proof (Sketch). First we note that nondeterministic state and transition complexity are both at least $m = \log N$, except when L_f is empty. For state minimization we use a construction given in [6] to obtain a NFA with $O(\sqrt{N})$ states. For transition minimization recall that the minimal deterministic finite automaton accepting L_f can have at most $O(N/\log N)$ transitions [2]. Then the stated bounds easily follow. □

In the remainder of this section we derive an inapproximability result for the problem under consideration. In order to get good inapproximability ratios, the biclique edge cover problem is a natural candidate to reduce from. By combining a recent inapproximability result for the chromatic number problem [14] with the approximation preserving reduction from the minimum clique partition problem given in [11], we see that the problem is not approximable within $|V|^{1/5-\varepsilon}$. But that is not the end of the line:

Theorem 6. *For all* $\varepsilon > 0$, *the biclique edge cover problem cannot be approximated within* $|V|^{1/3-\varepsilon}$ *or* $|E|^{1/5-\varepsilon}$, *unless* **P** = **NP***. This still holds in the restricted case where the input* $G = (U, V, E)$ *is a balanced bipartite graph, that is* $|U| = |V|$, *and has bipartite dimension at least* $\Omega(|V|^{2/3})$.

Proof. Let the clique partition number $\overline{\chi}(I)$ of a graph I be defined as the smallest number k such that the vertex set of I can be covered by at most k cliques. The associated decision problem is **NP**-complete [3, GT15], and as a simple reformulation of the graph coloring problem, not approximable within $|V|^{1-\varepsilon}$, for all $\varepsilon > 0$, unless **P** = **NP** [14]. We briefly recall the construction for reducing the clique partition problem to the biclique edge cover problem given in [11, Theorem 5.1a].

For an undirected graph $I = (V, E)$ with $V = \{v_1, v_2, \ldots, v_n\}$, we construct its bipartite version by setting $I_B = (X_B, Y_B, E_B)$ as set of left vertices

$X_B = \{x_1, x_2, \ldots, x_n\}$, as set of right vertices $Y_B = \{y_1, y_2, \ldots, y_n\}$, and $(x_i, y_j) \in E_B$ if and only if $i = j$ or $\{v_i, v_j\} \in E$. An edge (x_i, y_j) is called ascending if $i < j$, descending if $i > j$, and horizontal if $i = j$.

The biclique edge cover instance $G = (X, Y, \mathcal{E})$ consists of t^2 copies (the number t to be fixed later accordingly) of I_B, which we think of as being arranged in a $t \times t$ grid; and the bipartition of the vertex set is inherited from I_B. The ith left (right, respectively) vertex in copy (p, q) is denoted by (x_i, p, q) $((y_i, p, q)$, respectively). Two vertices (x_i, p, q) and (y_j, r, s) in different copies are connected by an edge if: either they are in the same row, i.e., $p = r$, and (x_i, y_j) is an ascending edge in E_B, or they are in the same column, i.e., $q = s$, and (x_i, y_j) is a descending edge in E_B. Accordingly, we say that an edge in $\mathcal{E}$ connecting vertices (x_i, p, q) and (y_j, r, s) is ascending if $i < j$, descending if $i > j$, and horizontal if $i = j$.

In [11], it is noted that if there is a system of s bicliques covering all horizontal edges in $\mathcal{E}$, then a partition of I into at most s/t^2 cliques can be constructed in polynomial time from this system, and

$$\overline{\chi}(I) \leq d(G)/t^2. \tag{1}$$

Conversely, each partition of I into r cliques corresponds to a system of rt^2 bicliques which cover all the horizontal edges in $\mathcal{E}$, and maybe some non-horizontal edges. However, note that the rt^2 bicliques do not necessarily cover all edges involving only vertices of a single copy of I_B: As an example, consider the partition of the graph I given in Figure 1 into $r = 3$ cliques.

To cover the remaining edges, we can do somewhat better than proposed in the original reduction: For $x_i \in X_B$, define $X_{i,p}$ as the set of ith left vertices in the copies of I_B which are in row p, and define $Y_{i,p}$ as the set of right vertices y in row p such that $((x_i, p, q), y)$ is an ascending edge in $\mathcal{E}$. It is not hard to see that the induced subgraph $G[X_{i,p}, Y_{i,p}]$ is a biclique which covers all ascending edges in row p incident to x_i, see Figure 1 for illustration by example.

By proceeding in this way for each row and each left vertex x_i in X_B, all ascending edges in G can be covered using no more than tn bicliques. The descending edges in G can be covered by tn bicliques in a similar manner. Thus

$$d(G) \leq t^2 \overline{\chi}(I) + 2tn. \tag{2}$$

Suppose now C is a biclique edge cover for G of size s. Then we can construct a clique partition for I of size s/t^2 in polynomial time from C, see [11] for details. Now we fix $t = 4n$, and compare performance ratios using Inequality (2):

$$\frac{s/t^2}{\overline{\chi}(I)} \leq \frac{s}{d(G) - 2tn} \leq 2\frac{s}{d(G)},$$

where the last statement above holds since $2tn = \frac{1}{2}t^2 \leq \frac{1}{2}d(G)$ by Inequality (1). We have established a S-reduction with expansion $(O(n^3), |V|, |X|)$. Then the desired hardness result regarding the measure $|X|$ follows by Lemma 3. Estimating the number of edges in $\mathcal{E}$ gives a total number of at most

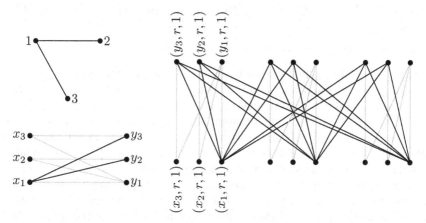

Fig. 1. The original graph I (top left), the bipartite graph I_B (lower left), and the subgraph of G induced by the vertices in row r (right), consisting of $t = 3$ copies of I_B. The induced subgraph $G[X_{1,r}, Y_{1,r}]$ forms a biclique.

$t^2|E_B| + 2t \cdot \binom{t}{2}|E_B| = O(|V|^5)$, so this is equally a S-reduction with expansion $(O(n^5), |V|, |\mathcal{E}|)$. Again by Lemma 3, the claimed inapproximability result follows. Finally, we note that Inequality (1) implies that $d(G) \geq t^2 = \Omega(|X|^{2/3})$, since $|X| = \Theta(n^3)$ and $t = 4n$. □

Theorem 7. *Given a truth table of size N specifying a Boolean function function f, no polynomial-time algorithm can approximate the number of states of a state minimal nondeterministic finite automaton accepting L_f with performance ratio $N^{1/6-\varepsilon}$, for all $\varepsilon > 0$, unless $\mathbf{P} = \mathbf{NP}$.*

Proof. We use the finite language to encode the edges in the graph $G = (X, Y, \mathcal{E})$ from the proof of Theorem 6, and the notations defined therein. Recall X consists of nodes of the form (x_i, p, q) with $x_i \in X_B$, p denotes a row index and q a column index, and similar for y_j, that is (x_i, p, q) and (y_j, p, q) belong to the same copy of I_B. Without loss of generality we assume $V = \{0, 1\}^m$ for some m. The t addresses for rows and columns can be respectively encoded in binary using a fixed length of $\log t = m + 2$. Throughout the rest of the proof, $c_1, c_2, \ldots, c_t$ denote the words encoding the t column addresses, and in a similar manner, $r_1, r_2, \ldots, r_t$ the row addresses. We then encode the edges $((x, p, q), (y, a, b))$ in $\mathcal{E}$ as $xr_pc_q(r_ac_b)^Ry$, and define L_G as the set of all codewords corresponding to an edge in $\mathcal{E}$. In the following, we will use the term "edge" to denote a word encoding an edge in $\mathcal{E}$ if there is no risk of confusion.

Claim 8. The nondeterministic state complexity of L_G is bounded below by the bipartite dimension of G.

Proof. We apply the biclique edge cover technique [5, Theorem 4] to give a lower bound for $\mathrm{nsc}(L_G)$. Let $\Gamma = (A, B, E_{L_G})$ be the bipartite graph given by $A = B = \{0, 1\}^{m+2(m+2)}$, and $E_{L_G} = \{(u, v) \in A \times B \mid uv \in L_G\}$. By an obvious bijection holds $d(G) = d(\Gamma)$, and the latter gives a lower bound for the nondeterministic state complexity of L_G.   □

Claim 9. $\mathrm{nsc}(L_G) = O(d(G)) + O(|X|^{2/3} \log |X|)$.

Proof. We establish the claim by constructing a sufficiently small NFA accepting the language L_G from a minimum biclique edge cover for G. For the horizontal edges in $\mathcal{E}$, we give a construction inspired by the proof of Theorem 6. Let $\{ (X_j, Y_j) \mid 1 \leq j \leq \overline{\chi}(I) \}$ be a minimum set of bicliques covering all horizontal edges in $I_B = (X_B, Y_B, E_B)$. For the ith biclique, we define an auxiliary language H_j as $H_j = X_j \cdot M \cdot Y_j$, where $M = \{ rc(rc)^R \mid r, c \in \{0,1\}^{m+2} \}$ is the language ensuring that the row and column address of $x \in X_j$ is the same as the row and column address of $y \in Y_j$. As there are no horizontal edges between different copies of I_B, language M provides that the union of languages $\bigcup_j H_j$ contains all codewords corresponding to horizontal edges in $\mathcal{E}$, and is a subset of L_G. Each H_j can be described by a nondeterministic finite automaton having $O(t^2)$ states: As all words in the sets X_j and Y_j have length m, each of them can be accepted by a NFA with $O(2^{m/2}) = o(t)$ states. The language M can be accepted by a NFA with $O(2^{2(m+2)}) = O(t^2)$ states. A schematic drawing of such an automaton is given in Figure 2. And a standard construction for nondeterministic finite automata yields an automaton with $O(t^2)$ states for the concatenation of these languages. Finally, the union of these languages can be accepted by a NFA having $O(t^2 \cdot \overline{\chi}(I)) = O(d(G))$ many states.

We use a similar matching language as M to construct a NFA accepting a subset of the codewords of $\mathcal{E}$ which contains all ascending edges. This time, the language has to ensure that the the left and the right vertex share the same row address, that is $M' = \{ rc_1 c_2 r^R \mid r, c_1, c_2 \in \{0,1\}^{m+2} \}$, and this language can be accepted by a NFA with only $O(t \log t)$ states, see Figure 2 for illustration.

Following the idea in the proof of Theorem 6, the graph G has for every row p and every vertex $x_i \in X_B$ a biclique $G[X_{i,p}, Y_{i,p}]$ containing only ascending edges. As we have an ascending biclique for each $x_i \in \{0,1\}^m$, it is more economic to share the states needed for addressing. Thus, a part of the automaton is a binary tree, whose root is the start state and whose leaves address the nodes in X_B. That is, after reading a word x of length m, the automaton is in a unique leaf of the binary tree. In a symmetric manner, we construct an inverted binary tree whose leaves address the nodes in Y_B, and whose transitions are directed towards the root, which is the final state of the automaton. It remains to wire the copies of the automaton accepting M' into these two binary trees appropriately, using no more than $|X_B|$ copies of it: Each leaf x_i of the binary tree, addressing some node in X_B, is identified with the start state of a fresh copy of the NFA. The transitions entering the final state of this copy are replaced with transitions entering the inverted binary tree at the appropriate address. This completes the description of the construction for a NFA having $O(|X_B| + |Y_B| + |X_B| t \log t) = O(|X|^{2/3} \log |X|)$ states accepting a subset of $\mathcal{E}$ including all ascending edges, since $|X| = \Theta(t^3)$ and $|X_B| = |Y_B| = \Theta(t)$.

For the descending edges, we carry out a similar construction, this time using the language $M'' = \{ r_1 cc^R r_2 \mid r_1, c, r_2 \in \{0,1\}^{m+2} \}$ ensuring that the column

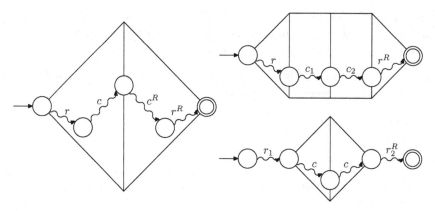

Fig. 2. Schematic drawings of the nondeterministic finite automata accepting M (left), M' (top right), and M'' (bottom right)

addresses match, see Figure 2. Then a similar construction gives a compact NFA describing the codewords of a set of edges including all descending edges in G. Finally, the union of all these languages can be described by a NFA with the desired upper bound on the number of states.   □

Assume now there exists a polynomial-time algorithm approximating $\mathrm{nsc}(L_G)$ within $|X|^{1/3-\varepsilon}$, that is it finds a NFA A of size at most $|X|^{1/3-\varepsilon} \cdot \mathrm{nsc}(L_G)$. By Claim 8, this can be seen equivalently as an algorithm finding a biclique edge cover for G of size at most $|X|^{1/3-\varepsilon} \cdot \mathrm{nsc}(L_G)$. We estimate the performance ratio of the latter algorithm:

$$\frac{|X|^{1/3-\varepsilon} \cdot \mathrm{nsc}(L_G)}{d(G)} = |X|^{1/3-\varepsilon} \cdot \frac{O(d(G)) + O(|X|^{2/3} \log |X|)}{d(G)}$$

by using Claim 9. The latter term is $O(|X|^{1/3-\varepsilon} \log |X|)$ because Theorem 6 asserts that $d(G) \geq |X|^{2/3}$. If we choose a small positive real δ such that $\varepsilon - \delta > 0$, then for $|X|$ large enough, $|X|^{1/3-\varepsilon} \log |X| < |X|^{1/3-(\varepsilon-\delta)}$. Together with the final argument given in Theorem 6, this implies $\mathbf{P} = \mathbf{NP}$.

As the size of the graph and the size of the truth table are related by $N = \Theta(|X|^2)$, the problem is not approximable within $N^{1/6-2\varepsilon}$ for every positive real ε, and the theorem is established.   □

For transition minimization we encounter the following situation.

Theorem 10. *Given a truth table of size N specifying a Boolean function f, no polynomial-time algorithm can approximate the number of transitions of a transition minimal nondeterministic finite automaton accepting L_f with performance ratio $N^{1/6-\varepsilon}$, for all $\varepsilon > 0$, unless $\mathbf{P} = \mathbf{NP}$.*

Proof. The language L_G defined in the proof of Theorem 7 can be accepted by a polynomial-time constructible DFA A having at most $O(m \cdot |L_G|)$ states and transitions. We can mimic the proof of Theorem 7 if we are able to verify inequalities relating nondeterministic transition complexity of L_G to the bipartite dimension of G in a way similar to Claim 8 and Claim 9.

Claim 11. The nondeterministic transition complexity of L_G is bounded below by the bipartite dimension of G minus 1.

For an upper bound on $\mathrm{ntc}(L_G)$, we take a closer look at the NFA constructed in the proof of Claim 9.

Claim 12. $\mathrm{ntc}(L_G) = O(d(G)) + O(|X|^{2/3} \log |X|)$.

The rest of the proof follows along the lines of the proof of Theorem 7.   □

4   Synthesizing a Minimal Nondeterministic Finite Automaton from a Given Deterministic One

This section contains results on the inapproximability of the minimization problem for nondeterministic finite automata, when the input is specified by a (deterministic) finite state automaton. This problem was investigated in [4,8]: Given a finite automaton A and an integer k, is there a nondeterministic finite automaton with at most k states (transitions, respectively) accepting language $L(A)$?

Note that the minimization problems w.r.t. states (transitions, respectively) for nondeterministic finite automata are trivially approximable within $O(n/\log n)$, if the input is given by a deterministic finite automaton. Observe that the minimal deterministic finite automaton equivalent to a given deterministic one is also a feasible solution for the respective problem. The performance ratio of this solution can be bounded using the fact that the blow-up in the number of states or transitions inferred by determinization is at most exponential. While this is only a poor performance guarantee, a strong inapproximability result is obtained in [4], but under a much stronger (cryptographic) assumption than $\mathbf{P} \neq \mathbf{NP}$. We just note their result in passing:

Theorem 13. *(i) Given an n-state deterministic finite automaton A, no polynomial-time algorithm can approximate the number of states of a state minimal nondeterministic finite automaton accepting $L(A)$ with performance ratio better than $\frac{\sqrt{n}}{\mathrm{polylog}(n)}$, if nonuniform logspace contains strong pseudo-random functions. (ii) In case of transition minimization the problem remains inapproximable with the same assumption as above and performance ratio better than $\frac{n}{\mathrm{polylog}(n)}$, where t is the number of transitions of the given deterministic finite state automaton.*

In order to obtain our first inapproximability result on the problem where a DFA is given, we use Theorem 6 and a reduction from the biclique edge cover problem to the nondeterministic finite state automaton minimization problem, where the input is a deterministic finite state automaton, given in [1, Lemma 1]. The noted reduction is an S-reduction with expansion $(O(n), |V|, |Q|)$.

Corollary 14. *Given a n-state deterministic finite automaton A accepting a finite language over an alphabet of size $O(n)$, no efficient algorithm can approximate the number of states of a state minimal nondeterministic finite automaton accepting $L(A)$ with performance ratio $n^{1/3-\varepsilon}$, for all $\varepsilon > 0$, unless $\mathbf{P} = \mathbf{NP}$.* ☐

For fixed alphabet size, we obtain a corresponding result from Theorem 7, as from every truth table, an equivalent DFA of smaller size can be constructed in polynomial time. Exploiting the special structure of the language used in the proof of Theorem 7, we can get an even higher bound.

Theorem 15. *Given a n-state deterministic finite automaton A accepting a finite language over an alphabet of size two, no efficient algorithm can approximate the number of states of a state minimal nondeterministic finite automaton accepting $L(A)$ within a factor of $n^{1/5-\varepsilon}$, for all $\varepsilon > 0$, unless $\mathbf{P} = \mathbf{NP}$.*

Proof (Sketch). To obtain the inapproximability result, we again use the language L_G defined in the proof of Theorem 7. The crucial observation is that this language contains $|\mathcal{E}| = O(2^{5m})$ words of length $O(m)$. Thus, a binary tree-like deterministic finite automaton of size $O(m \cdot |\mathcal{E}|)$ accepting all these words can be constructed in polynomial time—note that this size is much smaller than the truth table specifying L_G. Then one can show, similarly as in the proof of Theorem 7, that the stated inapproximability bound holds. ☐

By combining the observations in Theorems 10 and 15, we obtain for the corresponding problem of minimizing the number of transitions:

Corollary 16. *Given a deterministic finite automaton A with t transitions accepting a finite language over a binary alphabet, no efficient algorithm can approximate the number of transitions of a transition minimal nondeterministic finite automaton accepting $L(A)$ within a factor of $t^{1/5-\varepsilon}$, for all $\varepsilon > 0$, unless $\mathbf{P} = \mathbf{NP}$.* ☐

5   Conclusions

We compared nondeterministic finite automata minimization problems for regular languages, where the language is specified by different means—in decreasing order of succinctness: By a nondeterministic finite automaton, a deterministic automaton, or by a truth table. When given an NFA, the approximability of these problems is already settled [4,7]. When given a DFA as input, approximation hardness was known only modulo cryptographic assumptions [4]. The main contribution of this paper is that we do not need such strong assumptions, that is, the problems are hard to approximate unless $\mathbf{P} = \mathbf{NP}$. This essentially also holds if the input is specified as a truth table, but for the latter case, we were able to provide simple approximation algorithms with nontrivial performance guarantees. This settles most of the research problems suggested in [4].

Acknowledgments. We thank Gregor Gramlich for carefully reading an earlier draft of this work, and to the anonymous referees for many valuable suggestions and corrections.

References

1. Amilhastre, J., Janssen, P., Vilarem, M.-C.: FA minimisation heuristics for a class of finite languages. In: Boldt, O., Jürgensen, H. (eds.) WIA 1999. LNCS, vol. 2214, pp. 1–12. Springer, Heidelberg (2001)
2. Champarnaud, J.-M., Pin, J.-E.: A maxmin problem on finite automata. Discrete Applied Mathematics 23, 91–96 (1989)
3. Garey, M.R., Johnson, D.S.: Computers and Intractability, A Guide to the Theory of NP-Completeness. Freeman, San Francisco (1979)
4. Gramlich, G., Schnitger, G.: Minimizing NFA's and regular expressions. In: Diekert, V., Durand, B. (eds.) STACS 2005. LNCS, vol. 3404, pp. 399–411. Springer, Heidelberg (2005)
5. Gruber, H., Holzer, M.: Finding lower bounds for nondeterministic state complexity is hard (extended abstract). In: Ibarra, O.H., Dang, Z. (eds.) DLT 2006. LNCS, vol. 4036, pp. 363–374. Springer, Heidelberg (2006)
6. Gruber, H., Holzer, M.: On the average state and transition complexity of finite languages. Theoretical Computer Science, Special Issue: Selected papers of Descriptional Complexity of Formal Systems, Accepted for publication (2006)
7. Gruber, H., Holzer, M.: Computational complexity of NFA minimization for finite and unary languages. In: Proceedings of the 1st International Conference on Language and Automata Theory and Applications, Tarragona, Spain, LNCS, Springer. Accepted for publication (March 2007)
8. Jiang, T., Ravikumar, B.: Minimal NFA problems are hard. SIAM Journal on Computing 22(6), 1117–1141 (1993)
9. Kann, V.: Polynomially bounded minimization problems that are hard to approximate. Nordic Journal of Computing 1(3), 317–331 (1994)
10. Orlin, J.: Contentment in graph theory: Covering graphs with cliques. Indagationes Mathematicae 80, 406–424 (1977)
11. Simon, H.U.: On approximate solutions for combinatorial optimization problems. SIAM Journal on Discrete Mathematics 3(2), 294–310 (1990)
12. Vazirani, V.V.: Approximation Algorithms. Springer, Heidelberg (2001)
13. Yu, S.: Regular languages. In: Rozenberg, G., Salomaa, A. (eds.) Handbook of Formal Languages, vol. 1, pp. 41–110. Springer, Heidelberg (1997)
14. Zuckerman, D.: Linear degree extractors and the inapproximability of max clique and chromatic number. Report TR05-100, Electronic Colloquium on Computational Complexity (ECCC) (September 2005)

State Complexity of Union and Intersection of Finite Languages

Yo-Sub Han[1],* and Kai Salomaa[2],**

[1] Intelligence and Interaction Research Center, Korea Institute of Science and Technology, P.O.Box 131, Cheongryang, Seoul, Korea
emmous@kist.re.kr
[2] School of Computing, Queen's University, Kingston, Ontario K7L 3N6, Canada
ksalomaa@cs.queensu.ca

Abstract. We investigate the state complexity of union and intersection for finite languages. Note that the problem of obtaining the tight bounds for both operations was open. We compute the upper bounds based on the structural properties of minimal deterministic finite-state automata (DFAs) for finite languages. Then, we show that the upper bounds are tight if we have a variable sized alphabet that can depend on the size of input DFAs. In addition, we prove that the upper bounds are unreachable for any fixed sized alphabet.

1 Introduction

Regular languages are one of the most important and well-studied topics in computer science. They are often used in various practical applications such as vi, emacs and Perl. Furthermore, researchers developed a number of software libraries for manipulating formal language objects with the emphasis on regular languages; examples are Grail [12] and Vaucanson [2].

The applications and implementations of regular languages motivate the study of the descriptional complexity of regular languages. The descriptional complexity of regular languages can be defined in different ways since regular languages can be defined in different ways. For example, a regular language L is accepted by a deterministic finite-state automaton (DFA) or a nondeterministic finite-state automaton (NFA). L is also described by a regular expression. Yu and his co-authors [1,13,14] regarded the number of states in the minimal DFA for L as the complexity of L and studied the state complexity of basic operations on regular languages and finite languages. Holzer and Kutrib [5,6] investigated the state complexity of NFAs. Recently, Ellul et al. [3] examined the size of the shortest regular expression for a given regular language. There are many other results on state complexity with different viewpoints [4,8,9,10,11]. We focus on the measure of Yu [13]: The *state complexity* of a regular language L is the number of states of the minimal DFA for L. The state complexity of an operation

---

* Han was supported by the KIST Tangible Space Initiative Grant.
** Salomaa was supported by the Natural Sciences and Engineering Research Council of Canada Grant OGP0147224.

T. Harju, J. Karhumäki, and A. Lepistö (Eds.): DLT 2007, LNCS 4588, pp. 217–228, 2007.

on regular languages is a function that associates to the state complexities of the operand languages the worst-case state complexity of the language resulting from the operation. For instance, we say that the state complexity of the intersection of $L(A)$ and $L(B)$ is mn, where A and B are minimal DFAs and the numbers of states in A and B are m and n, respectively. It means that mn is the worst-case number of states of the minimal DFA for $L(A) \cap L(B)$.

Yu et al. [14] gave the first formal study of state complexity of regular language operations. Later, Câmpeanu et al. [1] investigated the state complexity of finite languages. Let A and B be minimal DFAs for two regular languages L_1 and L_2, and m and n be the numbers of states for A and B, respectively.

operation	finite languages	regular languages
$L_1 \cup L_2$	$O(mn)$	mn
$L_1 \cap L_2$	$O(mn)$	mn
$\Sigma^* \setminus L_1$	m	m
$L_1 \cdot L_2$	$(m-n+3)2^{n-2} - 1^{\diamond}$	$(2m-1)2^{n-1}$
L_1^*	$2^{m-3} + 2^{m-4}$, for $m \geq 4^{\diamond}$	$2^{m-1} + 2^{m-2}$
L_1^R	$3 \cdot 2^{p-1} - 1$ if $m = 2p$ $2^p - 1$ if $m = 2p-1$	2^m

Fig. 1. State complexity of basic operations on finite languages and regular languages [1,14]. Note that $\diamond$ refers to results using a two-character alphabet.

Fig. 1 shows the state complexity of basic operations on finite languages and regular languages. All complexity bounds, except for union and intersection of finite languages, in Fig. 1 are tight; namely, there exist worst-case examples that reach the given bounds. Clearly, mn is an upper bound since finite languages are a proper subfamily of regular languages. We also note that Yu [13] briefly mentioned a rough upper bound $mn - (m+n-2)$ for both operations. Therefore, it is natural to investigate the tight bounds for union and intersection of finite languages.

We define some basic notions in Section 2. In Section 3, we obtain an upper bound $mn - (m+n)$ for the union of two finite languages L_1 and L_2 based on the structural properties, where the sizes of L_1 and L_2 are m and n. Then, we prove that the bound is tight if the alphabet size can depend on m and n. We also examine the intersection of L_1 and L_2 in Section 4 and obtain an upper bound $mn - 3(m+n) + 12$. We again demonstrate that the upper bound is reachable using a variable sized alphabet. We conclude the paper in Section 5.

2   Preliminaries

Let Σ denote a finite alphabet of characters and Σ^* denote the set of all strings over Σ. The size $|\Sigma|$ of Σ is the number of characters in Σ. A language over

Σ is any subset of Σ^*. The symbol $\emptyset$ denotes the empty language and the symbol λ denotes the null string. A finite-state automaton (FA) A is specified by a tuple $(Q, \Sigma, \delta, s, F)$, where Q is a finite set of states, Σ is an input alphabet, $\delta : Q \times \Sigma \rightarrow 2^Q$ is a transition function, $s \in Q$ is the start state and $F \subseteq Q$ is a set of final states. Given a DFA A, we assume that A is complete; namely, each state has $|\Sigma|$ out-transitions and, therefore, A may have a sink (or dead) state. Since all sink states are always equivalent, we can assume that A has a unique sink state. Let $|Q|$ be the number of states in Q and $|\delta|$ be the number of transitions in δ. For a transition $\delta(p, a) = q$ in A, we say that p has an *out-transition* and q has an *in-transition*. Furthermore, p is a *source state* of q and q is a *target state* of p. A string x over Σ is accepted by A if there is a labeled path from s to a final state in F such that this path spells out the string x. Thus, the language $L(A)$ of an FA A is the set of all strings that are spelled out by paths from s to a final state in F. We say that A is *non-returning* if the start state of A does not have any in-transitions and A is *non-exiting* if all out-transitions of any final state in A go to the sink state.

Given an FA $A = (Q, \Sigma, \delta, s, F)$, we define the *right language* L_q of a state q to be the set of strings that are spelled out by some path from q to a final state in A; namely, L_q is the language accepted by the FA obtained from A by changing the start state to q. We say that two states p and q are *equivalent* if $L_p = L_q$.

3 Union of Finite Languages

Given two minimal DFAs A and B for non-empty finite languages L_1 and L_2, we can in the standard way construct a DFA for the union of $L(A)$ and $L(B)$ based on the Cartesian product of states.

Proposition 1. *Given two DFAs $A = (Q_1, \Sigma, \delta_1, s_1, F_1)$ and $B = (Q_2, \Sigma, \delta_2, s_2, F_2)$, let $M_\cup = (Q_1 \times Q_2, \Sigma, \delta, (s_1, s_2), F)$, where for all $p \in Q_1$ and $q \in Q_2$ and $a \in \Sigma$, $\delta((p, q), a) = (\delta(p, a), \delta(q, a))$ and $F = \{(p, f_2) \mid p \in Q_1 \text{ and } f_2 \in F_2\} \cup \{(f_1, q) \mid f_1 \in F_1 \text{ and } q \in Q_2\}$. Then, $L(M_\cup) = L(A) \cup L(B)$ and $M_\cup$ is deterministic.*

A crucial observation is that both A and B must be non-returning since L_1 and L_2 are finite. Therefore, as Yu [13] observed, if we apply the Cartesian product for union, all states (s_1, q) for $q \neq s_2$ and all states (p, s_2) for $p \neq s_1$ are not reachable from the start state (s_1, s_2) in $M_\cup$. Thus, we can reduce $(m + n) - 2$ states.

Another observation is that A must have a final state f such that all of f's out-transitions go to the sink state. Consider the right language of a state (i, j) in $M_\cup$.

Proposition 2 (Han et al. [4]). *For a state (i, j) in $M_\cup$, the right language $L_{(i,j)}$ of (i, j) is the union of L_i in A and L_j in B.*

Let d_1 and d_2 be the sink states of A and B and f_1 and f_2 be final states of A and B such that d_1 is the only target state of f_1 in A and d_2 is the only target state of f_2 in B, respectively. Then, by Proposition 2, (f_1, f_2), (d_1, f_2) and (f_1, d_2) are equivalent and, thus, can be merged into a single state. It shows that we can reduce two more states from $M_\cup$. Therefore, we obtain the following result.

Lemma 1. *Given two minimal DFAs A and B for finite languages, $mn-(m+n)$ states are sufficient for the union of $L(A)$ and $L(B)$, where $m = |A|$ and $n = |B|$.*

We next examine whether or not we can reach the upper bound of Lemma 1.

Lemma 2. *The upper bound $mn - (m + n)$ for union cannot be reached with a fixed alphabet when m and n are arbitrarily large.*

Lemma 2 shows that we cannot reach the upper bound in Lemma 1 if $|\Sigma|$ is relatively small compared with the number states of the given DFAs. Then, the next question is what if $|\Sigma|$ is large enough?

Lemma 3. *The upper bound $mn - (m + n)$ for union is reachable if the size of the alphabet can depend on m and n.*

Proof. Let m and n be positive numbers (namely, $m, n \in \mathbb{N}$) and

$$\Sigma = \{b, c\} \cup \{a_{i,j} \mid 1 \le i \le m - 2, 1 \le j \le n - 2 \text{ and } (i,j) \ne (m - 2, n - 2)\}.$$

Let $A = (Q_1, \Sigma, \delta_1, p_0, \{p_{m-2}\})$, where $Q_1 = \{p_0, p_1, \ldots, p_{m-1}\}$ and δ_1 is defined as follows:

- $\delta_1(p_i, b) = p_{i+1}$, for $0 \le i \le m - 2$.
- $\delta_1(p_0, a_{i,j}) = p_i$, for $1 \le i \le m - 2$ and $1 \le j \le n - 2$, $(i,j) \ne (m - 2, n - 2)$.

For all other cases in δ_1 that are not covered above, the target state is the sink state p_{m-1}.

Next, let $B = (Q_2, \Sigma, \delta_2, q_0, \{q_{n-2}\})$, where $Q_2 = \{q_0, q_1, \ldots, q_{n-1}\}$ and δ_2 is defined as follows:

- $\delta_2(q_i, c) = q_{i+1}$, for $0 \le i \le n - 2$.
- $\delta_2(q_0, a_{i,j}) = q_j$, for $1 \le j \le n - 2$ and $1 \le i \le m - 2$, $(i,j) \ne (m - 2, n - 2)$.

Again, for all other cases in δ_2 that are not covered above, the target state is the sink state q_{n-1}. Fig. 2 gives an example of such DFAs A and B.

Let $L = L(A_1) \cup L(A_2)$. We claim that the minimal (complete) DFA for L needs $mn - (m + n)$ states. To prove the claim, it is sufficient to show that there exists a set R consisting of $mn - (m + n)$ strings over Σ that are pairwise inequivalent modulo the right invariant congruence of L, $\equiv_L$.

We show that $R = R_1 \cup R_2 \cup R_3$, where

$R_1 = \{b^i \mid 0 \le i \le m - 1\}$.
$R_2 = \{c^j \mid 1 \le j \le n - 3\}$. (Note that R_2 does not include strings c^0, c^{n-2} and c^{n-1}.)
$R_3 = \{a_{i,j} \mid 1 \le i \le m - 2 \text{ and } 1 \le j \le n - 2 \text{ and } (i,j) \ne (m - 2, n - 2)\}$.

Any string b^i from R_1 cannot be equivalent with a string c^j from R_2 since $c^j \cdot c^{n-2-j} \in L$ but $b^i \cdot c^{n-2-j} \notin L$. Note that $j \ge 1$ and hence also $b^0 \cdot c^{n-2-j} \notin L$.

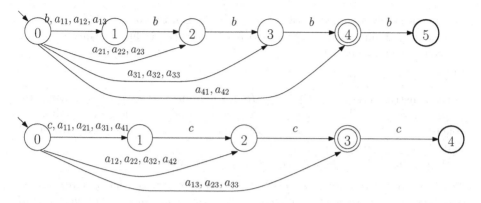

Fig. 2. An example of two minimal DFAs for finite languages whose sizes are 6 and 5, respectively, where state 5 above and state 4 below are sink states. Except for the b-transition to state 5 in A and the c-transition to state 4 in B, we omit all other in-transitions of the sink state.

Next consider a string b^i from R_1 and a string $a_{k,j}$ from R_3. There are four cases to consider.

1. $k \neq i$ and $0 \leq i \leq m - 3$: It means that b^i and $a_{k,j}$ are inequivalent since $b^i \cdot b^{m-2-i} \in L$ but $a_{k,j} \cdot b^{m-2-i} \notin L$.
2. $k \neq i$ and $i = m - 2$: It implies that $k < m - 2$ and, thus, b^i and $a_{k,j}$ are inequivalent since $a_{k,j} \cdot b^{m-2-k} \in L$ but $b^i \cdot b^{m-2-k} \notin L$.
3. $k \neq i$ and $i = m - 1$: The path for $b^i = b^{m-1}$ must end at the sink state for the minimal DFA for L since $b^{m-1} \notin L$. On the other hand, $a_{k,j}$ can be completed to be a string of L by appending zero or more symbols c.
4. $k = i$: Now we have strings b^i and $a_{i,j}$.
   (a) $j < n - 2$: We note that $a_{i,j} \cdot c^{n-2-j} \in L$ but $b^i c^{n-2-j} \notin L$ since no string of L can have both b's and c's. Note that $k = i$ implies that $i \geq 1$.
   (b) $j = n - 2$: Since $j = n - 2$, $i < m - 2$ by the definition of R_3. Now $b^i \cdot \lambda \notin L$ but $a_{i,j} = a_{i,n-2} \cdot \lambda \in L(B) \subseteq L$.
   Therefore, b^i and $a_{i,j}$ are inequivalent.

Symmetrically, we see that any string from R_2 cannot be equivalent with a string from R_3. This case is, in fact, simpler than the previous case since R_2 is more restrictive than R_1.

Finally we need to show that all strings from R_1 (respectively, from R_2 and from R_3) are pairwise inequivalent with each other.

1. R_1: By appending a suitable number of b's, we can always distinguish two distinct strings from R_1.
2. R_2: By appending a suitable number of c's, we can always distinguish two distinct strings from R_2.
3. R_3: Consider two distinct strings $a_{i,j}$ and $a_{x,y}$ from R_3. Without loss of generality, we assume that $i < x$. The other possibility, where j and y differ,

is completely symmetric. Since $a_{i,j} \cdot b^{m-2-i} \in L$ and $a_{x,y} \cdot b^{m-2-i} \notin L$, $a_{i,j}$ and $a_{x,y}$ are inequivalent. Note that $m-2-i > 0$ and, thus, the inequivalence holds even in the case when $y = n - 2$.

This concludes the proof.   □

In the construction of $R = R_1 \cup R_2 \cup R_3$ for Lemma 3, the size of Σ that we use is $mn - 2m - 2n + 5$. By using a more complicated construction, we might be able to reduce the size of Σ. On the other hand, we already know from Lemma 2 that $|\Sigma|$ has to depend on m and n.

We establish the following statement from Lemmas 1 and 3.

Theorem 1. *Given two minimal DFAs A and B for finite languages, $mn - (m + n)$ states are necessary and sufficient in the worst-case for the minimal DFA of $L(A) \cup L(B)$, where $m = |A|$ and $n = |B|$.*

Lemma 2 shows that the upper bound in Lemma 1 is unreachable if $|\Sigma|$ is fixed and m and n are arbitrarily large whereas Lemma 3 shows that the upper bound is reachable if $|\Sigma|$ depends on m and n. These results naturally lead us to examine the state complexity of union with a fixed sized alphabet. For easiness of presentation, we first give the result for a four character alphabet and afterward explain how the construction can be modified for a binary alphabet.

Lemma 4. *Let Σ be an alphabet with four characters. There exists a constant c such that the following holds for infinitely many $m, n \geq 1$, where $\min\{m, n\}$ is unbounded. There exist DFAs A and B, with m and n states respectively, that recognize finite languages over Σ such that the minimal DFA for the union $L(A) \cup L(B)$ requires $c(\min\{m, n\})^2$ states.*

The same result holds for a binary alphabet.

Proof. Let $\Sigma = \{a, b, c, d\}$. We introduce some new notations for the proof. Given an even length string $w \in \Sigma^*$, $\mathsf{odd}(w)$ denotes the subsequence of characters that occur in odd positions in w and, thus, the length of $\mathsf{odd}(w)$ is half the length of w. For example, if $w = adacbcbc$, then $\mathsf{odd}(w) = aabb$. Similarly, $\mathsf{even}(w)$ denotes the subsequence of characters that occur in even positions in w. With the same example above, $\mathsf{even}(w) = dccc$.

Let $s \geq 1$ be arbitrary and $r = \lceil \log s \rceil$. We define the finite language

$$L_1 = \{w_1 w_2 \mid |w_1| = 2r, w_2 = \mathsf{odd}(w_1) \in \{a, b\}^*, \mathsf{even}(w_1) \in \{c, d\}^*\}.$$

The language L_1 can be recognized by a DFA A with at most $10s$ states. For reading a prefix of length $2r$ of an input string, the start state of A has two out-transitions with labels a and b and the two corresponding target states are different. Then, each target state has two out-transitions with labels c and d where the target states are the same. This repeats in A until we finish reading the prefix of length $2r$. All other transitions go to the sink state. Fig. 3 illustrates the construction of A with $r = 3$.

The computations of A, which do not go to the sink state, on inputs of length $2r$ form a tree-like structure that branches into 2^r different states. Each of the 2^r states represents a unique string $\mathsf{odd}(u) \in \{a, b\}^*$, where u is the (prefix

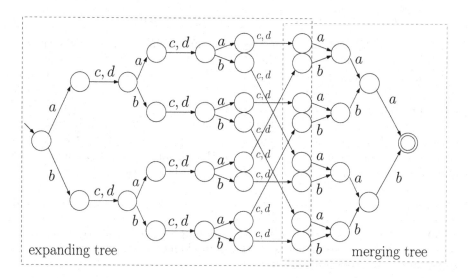

Fig. 3. A DFA A that recognizes L_1 when $r = 3$. We omit the sink state and its in-transitions.

of the) input of length $2r$. Then, the computation from each of these 2^r states verifies whether or not the remaining suffix is identical to the string $\mathsf{odd}(u)$. This can be accomplished using a tree that merges all the computations into a single final state. (See the right part of Fig. 3 for an example.) From each state, there is only one out-transition (either with symbol a or b), if we ignore transitions into the sink state. (The structure looks like a tree when we ignore transitions into the sink state.)

The computations of A on strings of length $2r$ branch into 2^r states. The first "expanding" tree (for instance, the left part of Fig. 3) uses less than $4 \cdot 2^r < 8s$ states[1] since we repeat each level with the c, d transitions in the tree and $s \le 2^r < 2s$.

Finally, consider the number of states in the "merging" tree. (For example, we rotate the right part of Fig. 3.) Similarly, the merging tree has 2^r states and, therefore, the tree needs at most $2 \cdot 2^r < 4s$ states. However, we observe that the the last 2^r states of the expanding tree is the same state to the last 2^r states of the merging tree in A. Therefore, we only need $2s$ states for the merging tree.

$$\text{The total number of states in } A \text{ is less than } 10s. \tag{1}$$

Symmetrically, we define

$$L_2 = \{w_1 w_2 \mid |w_1| = 2r, \mathsf{odd}(w_1) \in \{a, b\}^*, w_2 = \mathsf{even}(w_1) \in \{c, d\}^*\}.$$

The language L_2 consists of strings uv, where $|u| = 2r$, odd characters of u are in $\{a, b\}$, even characters of u are in $\{c, d\}$ and $\mathsf{even}(u)$ coincides with v. Using an argument similar to that for equation (1), we establish that

---

[1] Note that a balanced tree with 2^r leaves has less than $2 \cdot 2^r$ nodes.

$$L_2 \text{ can be recognized by a DFA with less than } 10s \text{ states.} \tag{2}$$

Now let $L = L_1 \cup L_2$. Let u_1 and u_2 be distinct strings of length $2r$ such that $\mathsf{odd}(u_i) \in \{a, b\}^*$ and $\mathsf{even}(u_i) \in \{c, d\}^*$ for $i = 1, 2$.

If $\mathsf{odd}(u_1) \neq \mathsf{odd}(u_2)$, then $u_1 \cdot \mathsf{odd}(u_1) \in L_1 \subseteq L$ but $u_2 \cdot \mathsf{odd}(u_1) \notin L$. Hence, u_1 and u_2 are not equivalent modulo the right invariant congruence of L. Similarly, if $\mathsf{even}(u_1) \neq \mathsf{even}(u_2)$, then, $u_1 \cdot \mathsf{even}(u_1) \in L_2 \subseteq L$ but $u_2 \cdot \mathsf{even}(u_1) \notin L$.

The above implies that the right invariant congruence of L has at least $2^r \cdot 2^r \geq s^2$ different classes. Therefore, if $m = n = 10s$ is the size of the minimal DFAs for the finite languages L_1 and L_2, then from equations (1) and (2) we know that the minimal DFA for $L = L_1 \cup L_2$ needs at least

$$\frac{1}{100} n^2 \text{ states.} \tag{3}$$

Note that $|\Sigma| = 4$. The languages L_1 and L_2 can be straightforwardly encoded over a binary alphabet with the only change that the constant $\frac{1}{100}$ in equation (3) would become smaller. $\qquad \square$

4   Intersection of Finite Languages

We examine the state complexity of intersection of finite languages. Our approach is again based on the structural properties of minimal DFAs of finite languages. We start from the Cartesian product of states for the intersection of two DFAs.

Proposition 3 (Hopcroft and Ullman [7]). *Given two DFAs $A = (Q_1, \Sigma, \delta_1, s_1, F_1)$ and $B = (Q_2, \Sigma, \delta_2, s_2, F_2)$, let $M_\cap = (Q_1 \times Q_2, \Sigma, \delta, (s_1, s_2), F_1 \times F_2)$, where for all $p \in Q_1$ and $q \in Q_2$ and $a \in \Sigma$,*

$$\delta((p, q), a) = (\delta_1(p, a), \delta_2(q, a)).$$

Then, $L(M_\cap) = L(A) \cap L(B)$.

Let $M_\cap$ denote the Cartesian product of states. Let m and n denote the sink states of A and B and $m-1$ and $n-1$ denote the final states whose target states are always the sink states of A and B, respectively. If we regard $M_\cap$ as a $m \times n$ matrix, then all states in the first row and in the first column are unreachable from $(1, 1)$ since A and B are non-returning and, thus, these states are useless in $M_\cap$. Moreover, by the construction, all remaining states in the last row and in the last column are equivalent to the sink state and, therefore, can be merged. Let us examine the remaining states in the second-to-last row and in the second-to-last column except for $(m - 1, n - 1)$.

Lemma 5. *A state $(i, n - 1)$ in the second-to-last column, for $1 \leq i \leq m - 1$, is either*

> *equivalent to $(m - 1, n - 1)$ if state i is a final state in A or*
> *equivalent to (m, n) if state i is not a final state in A.*

We can obtain a similar statement for the states in the second-to-last row in $M_\cap$. Therefore, all the remaining states at the second-to-last row and at the second-to-last column except for $(n-1, m-1)$ can be merged with either $(n-1, m-1)$ or (n, m). Thus, the number of remaining states is

$$mn - \{(m-1) + (n-1)\} - \{(m-2) + (n-2)\} - \{(m-3) + (n-3)\}$$
$$= mn - 3(m+n) + 12,$$

where $\{(m-1) + (n-1)\}$ is from the first row and the first column, $\{(m-2) + (n-2)\}$ is from the last row and the last column and $\{(m-3) + (n-3)\}$ is from the second-to-last row and the second-to-last column. We establish the following lemma from the calculation.

Lemma 6. *Given two minimal DFAs A and B for finite languages, $mn - 3(m+n) + 12$ states are sufficient for the intersection of $L(A)$ and $L(B)$, where $m = |A|$ and $n = |B|$.*

We now show that $mn - 3(m+n) + 12$ states are necessary and, therefore, the bound is tight. Let $m, n \in \mathbb{N}$ and choose $\Sigma = \{a_{i,j} \mid 1 \le i \le m-1 \text{ and } 1 \le j \le n-1\}$.
    Let $A = (Q_1, \Sigma, \delta_1, p_0, \{p_{m-2}\})$, where $Q_1 = \{p_0, p_1, \dots, p_{m-1}\}$ and δ_1 is defined as follows:

- $\delta_1(p_x, a_{i,j}) = p_{x+i}$, for $0 \le x \le m-2$, $1 \le i \le m-1$ and $1 \le j \le n-1$.

If the sum $x + i$ is larger than $m-1$, then p_{x+i} is the sink state $(= p_{m-1})$. For all other cases in δ_1 that are not covered above, the target state is the sink state p_{m-1}.
    Next, let $B = (Q_2, \Sigma, \delta_2, q_0, \{q_{m-2}\})$, where $Q_2 = \{q_0, q_1, \dots, q_{n-1}\}$ and δ_2 is defined as follows:

- $\delta_2(q_x, a_{i,j}) = q_{x+j}$, for $0 \le x \le m-2$, $1 \le j \le n-1$ and $1 \le i \le m-1$.

Similarly, if the sum $x+j$ is larger than $n-1$, then q_{x+j} is the sink state $(= q_{m-1})$. For all other cases in δ_2 that are not covered above, the target state is the sink state q_{m-1}. Fig. 4 shows an example of such DFAs A and B.

Lemma 7. *Let $L = L(A) \cap L(B)$. The minimal (complete) DFA for L needs $mn - 3(m+n) + 12$ states.*

Proof. We prove the statement by showing that there exists a set R of $mn - 3(m+n) + 12$ strings over Σ that are pairwise inequivalent modulo the right invariant congruence of L, $\equiv_L$. We assume that $m \le n$.
    We choose $R = R_1 \cup R_2 \cup R_3 \cup R_4$, where

$R_1 = \{\lambda\}$.
$R_2 = \{a_{m-2,n-2}\}$.
$R_3 = \{a_{m-1,n-1}\}$.
$R_4 = \{a_{i,j} \mid \text{for } 1 \le i \le m-3 \text{ and } 1 \le j \le n-3\}$.

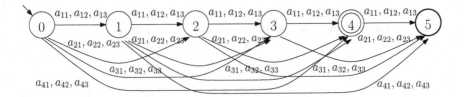

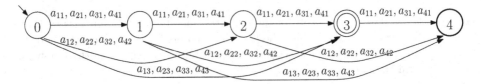

Fig. 4. An example of two minimal DFAs for finite languages whose sizes are 6 and 5, respectively, where state 5 above and state 4 below are sink states. We omit a large number of in-transitions of the sink state.

Any string x from $R_2 \cup R_3 \cup R_4$ cannot be equivalent with λ from R_1 since $\lambda \cdot a_{m-2,n-2} \in L$ but $x \cdot a_{m-2,n-2} \notin L$. Similarly, any string x from $R_1 \cup R_3 \cup R_4$ cannot be equivalent with $a_{m-2,n-2}$ from R_2 since $a_{m-2,n-2} \cdot \lambda \in L$ but $x \cdot \lambda \notin L$. Note that string $a_{m-1,n-1}$ from R_3 is not in L and it can never be in L by appending some string whereas any string x from $R_1 \cup R_2 \cup R_4$ can be in L by appending a suitable string. Therefore, R_1, R_2 and R_3 are inequivalent with each other including R_4.

Finally, we consider two strings $a_{i,j}$ and $a_{x,y}$ in R_4. The two strings are not equivalent since $a_{i,j} \cdot a_{m-2-i,n-2-j} \in L$ but $a_{x,y} \cdot a_{m-2-i,n-2-j} \notin L$ when $(i,j) \neq (x,y)$. Therefore, any two strings from R_4 are not equivalent.

Now we count the number of strings in R. We note that $|R_1| = |R_2| = |R_3| = 1$ and $|R_4| = (m-3)(n-3)$. Therefore, $|R| = mn - 3(m+n) + 12$. It implies that there are at least $mn - 3(m+n) + 12$ states in the minimal DFA for L. $\square$

We obtain the following result from Lemmas 6 and 7.

Theorem 2. *Given two minimal DFAs A and B for finite languages, $mn - 3(m+n) + 12$ states are necessary and sufficient in the worst-case for the intersection of $L(A)$ and $L(B)$, where $m = |A|$ and $n = |B|$.*

Note that the upper bound $mn - 3(m+n) + 12$ is reachable when $|\Sigma|$ depends on m and n as shown in Lemma 7. On the other hand, we can prove that it is impossible to reach the upper bound with a fixed Σ using the same argument as in Lemma 2.

Let us investigate a lower bound for the state complexity of intersection of $L(A)$ and $L(B)$ over a fixed alphabet.

Lemma 8. *Let Σ be an alphabet with four characters. There exists a constant c such that the following holds for infinitely many $m, n \geq 1$, where $\min\{m, n\}$*

unbounded. There exist minimal DFAs A and B that recognize finite languages over Σ such that the minimal DFA for the intersection $L(A) \cap L(B)$ requires $c(\min\{m, n\})^2$ states, where $|A| = m$ and $|B| = n$.
The same result holds for a binary alphabet.

Proof. We omit the proof due to the space limit. The proof is similar to the proof for Lemma 4. $\square$

5   Conclusions

The state complexity of an operation on regular languages is the number of states in the minimal DFA that recognizes the resulting language from the operation. Fig. 1 gives a summary of the results. We have noted that the precise state complexity of union and intersection cases have been open although rough upper bounds were given by Yu [13]. Based on the structural properties of minimal DFAs for finite languages, we have proved that

1. For union, $mn - (m + n)$ states are necessary and sufficient.
2. For intersection, $mn - 3(m + n) + 12$ states are necessary and sufficient.

We have also noted that the bounds are reachable if $|\Sigma|$ depends on m and n, where m and n are the sizes of minimal DFAs for two finite languages. If $|\Sigma|$ is fixed and m and n are arbitrarily large, then we have shown that the upper bounds for both cases are not reachable.

References

1. Câmpeanu, C., Culik II, K., Salomaa, K., Yu, S.: State complexity of basic operations on finite languages. In: Boldt, O., Jürgensen, H. (eds.) WIA 1999. LNCS, vol. 2214, pp. 60–70. Springer, Heidelberg (2001)
2. Claveirole, T., Lombardy, S., O'Connor, S., Pouchet, L.-N., Sakarovitch, J.: Inside Vaucanson. In: Farré, J., Litovsky, I., Schmitz, S. (eds.) CIAA 2005. LNCS, vol. 3845, pp. 116–128. Springer, Heidelberg (2006)
3. Ellul, K., Krawetz, B., Shallit, J., Wang, M.-W.: Regular expressions: New results and open problems. Journal of Automata, Languages and Combinatorics 9, 233–256 (2004)
4. Han, Y.-S., Salomaa, K., Wood, D.: State complexity of prefix-free regular languages. In: Proceedings of DCFS'06, pp. 165–176, Full version is submitted for publication (2006)
5. Holzer, M., Kutrib, M.: Unary language operations and their nondeterministic state complexity. In: Ito, M., Toyama, M. (eds.) DLT 2002. LNCS, vol. 2450, pp. 162–172. Springer, Heidelberg (2002)
6. Holzer, M., Kutrib, M.: Nondeterministic descriptional complexity of regular languages. International Journal of Foundations of Computer Science 14(6), 1087–1102 (2003)
7. Hopcroft, J., Ullman, J.: Introduction to Automata Theory, Languages, and Computation, 2nd edn. Addison-Wesley, Reading, MA (1979)

8. Hricko, M., Jirásková, G., Szabari, A.: Union and intersection of regular languages and descriptional complexity. In: Proceedings of DCFS'05, pp. 170–181 (2005)
9. Jirásek, J., Jiráskov, G., Szabari, A.: State complexity of concatenation and complementation of regular languages. In: Domaratzki, M., Okhotin, A., Salomaa, K., Yu, S. (eds.) CIAA 2004. LNCS, vol. 3317, pp. 178–189. Springer, Heidelberg (2005)
10. Nicaud, C.: Average state complexity of operations on unary automata. In: Kutyłowski, M., Wierzbicki, T., Pacholski, L. (eds.) MFCS 1999. LNCS, vol. 1672, pp. 231–240. Springer, Heidelberg (1999)
11. Pighizzini, G., Shallit, J.: Unary language operations, state complexity and Jacobsthal's function. International Journal of Foundations of Computer Science 13(1), 145–159 (2002)
12. Raymond, D.R., Wood, D.: Grail: A C++ library for automata and expressions. Journal of Symbolic Computation 17, 341–350 (1994)
13. Yu, S.: State complexity of regular languages. Journal of Automata, Languages and Combinatorics 6(2), 221–234 (2001)
14. Yu, S., Zhuang, Q., Salomaa, K.: The state complexities of some basic operations on regular languages. Theoretical Computer Science 125(2), 315–328 (1994)

Bisimulation Minimisation for Weighted Tree Automata*

Johanna Högberg[1], Andreas Maletti[2], and Jonathan May[3]

[1] Department of Computing Science, Umeå University,
S–90187 Umeå, Sweden
johanna@cs.umu.se
[2] Faculty of Computer Science, Technische Universität Dresden,
D–01062 Dresden, Germany
maletti@tcs.inf.tu-dresden.de
[3] Information Sciences Institute, University of Southern California,
Marina Del Rey, CA 90292
jonmay@isi.edu

Abstract. We generalise existing forward and backward bisimulation minimisation algorithms for tree automata to weighted tree automata. The obtained algorithms work for all semirings and retain the time complexity of their unweighted variants for all additively cancellative semirings. On all other semirings the time complexity is slightly higher (linear instead of logarithmic in the number of states). We discuss implementations of these algorithms on a typical task in natural language processing.

1   Introduction

By the Myhill-Nerode theorem there exists, for every regular string language L, a unique (up to isomorphism) minimal deterministic finite automaton (dfa) that recognises L. It was a breakthrough when Hopcroft [1] presented an $O(n \log n)$ minimisation algorithm for dfa where n is the number of states. This still up-to-date bound was obtained by partitioning the state space through a "process the smaller half" strategy. However, in general there exists no unique minimal non-deterministic finite automaton (nfa) recognising a given regular language. Meyer and Stockmeyer [2] proved that minimisation of nfa is PSPACE-complete. The minimisation problem for nfa with n states cannot even be efficiently approximated within the factor $o(n)$, unless P = PSPACE [3]. This meant that the problem had to be simplified; either by restricting the domain to a smaller class of devices, or by surrendering every hope of a non-trivial approximation bound. Algorithms that minimise with respect to a *bisimulation* are examples of the latter approach. The concept of bisimularity was introduced by Milner [4] as a formal tool to investigate transition systems. Simply put, two transition systems are bisimulation equivalent if their behaviour—in response to a sequence of actions—cannot be distinguished by an outside observer. Although bisimulation

---

* This work was partially supported by NSF grant IIS-0428020.

equivalence, as interpreted for various devices, implies language equality, the opposite does not hold in general. We consider weighted tree automata (wta) [5], which are a joint generalisation of tree automata [6,7] and weighted automata [8]. Classical tree automata can then be seen as wta with weights in the Boolean semiring, i.e. a transition has weight *true* if it is present, and *false* otherwise.

One type of bisimulation, called *forward bisimulation* in [9,10], restricts bisimilar states to have identical futures. The future of a state q is the tree series of contexts that is recognised by the wta if the computation starts with the state q and weight 1 at the unique position of the special symbol $\square$ in the context. A similar condition is found in the MYHILL-NERODE congruence for a tree language [11] or even in the MYHILL-NERODE congruence [12] for a tree series. Let us explain it on the latter. Two trees t and u are equal in the MYHILL-NERODE congruence for a given tree series S over the field $(A, +, \cdot, 0, 1)$, if there exist nonzero coefficients $a, b \in A$ such that for all contexts C we observe that $a^{-1} \cdot (S, C[t]) = b^{-1} \cdot (S, C[u])$. The coefficients a and b can be understood as the weights of t and u, respectively. In contrast to the MYHILL-NERODE congruence, a forward bisimulation requires a local condition on the tree representation. The condition is strong enough to enforce equivalent futures, but not too strong which is shown by the fact that, on a deterministic *all-accepting* [13] wta M over a field [14] or a wta M over the Boolean semiring [10], minimisation via forward bisimulation yields the unique (up to isomorphism) minimal deterministic wta that recognises the same tree series as M.

The other type of bisimulation we will consider is called *backward bisimulation* in [9,10]. Backward bisimulation also uses a local condition on the tree representation that enforces that the past of any two bisimilar states is equal. The past of a state is the series that is recognised by the wta if that particular state would be the only final state and its final weight would be 1 (i.e., the past of a state q is the series that maps an input tree t to $h_\mu(t)_q$; see Sect. 2).

The idea behind bisimulation minimisation is to discover and collapse states that in some sense exhibit the same behaviour, thus freeing the input automaton of redundancy. This implies a search for the coarsest relation on the state space that meets the local conditions of the bisimulation relation that we are interested in. The $O(n^2 \log n)$ minimisation algorithm for nfa by Paige & Tarjan [15] could be called a forward bisimulation minimisation. Bisimulation minimisation of tree automata is discussed in [10]. The paper [10] presents two minimisation algorithms that are based on forward and backward bisimulation and run in time $O(rn^{r+1} \log n)$ and $O(r^2 n^{r+1} \log n)$, respectively, where r is the maximal rank of the input symbols and n is the number of states. In this paper, we generalise these results to weighted tree automata and obtain minimisation algorithms that work for arbitrary semirings in $O(rn^{r+2})$ and $O(r^2 n^{r+2})$ for the forward and backward approach, respectively. The counting argument used in [15] and later in [10] is no longer applicable: it was devised for the Boolean semiring and does not generalise. However, when cancellative semirings are considered, we can improve the algorithms to run in $O(rn^{r+1} \log n)$ and $O(r^2 n^{r+1} \log n)$ for the forward and backward approach, respectively, by taking advantage of the "process the smaller half" strategy of Hopcroft. When the forward algorithm is given a deterministic wta, it yields an equivalent deterministic wta in time $O(rn^{r+1})$, which can be optimised to $O(rn^r \log n)$ for additively cancellative semirings.

There are advantages that support having two algorithms. First, forward and backward bisimulation minimisation only yield a minimal wta with respect to the corresponding bisimulation. Thus applying forward and backward bisimulation minimisation in an alternating fashion commonly yields a yet smaller wta. Since both minimisation procedures are very efficient, this approach also works in practice. For the problem of *tree language model* minimisation, discussed in Sect. 5, we minimised our candidate wta in an alternating fashion and found that we were able to get equally small wta after two iterations beginning with backward or three iterations beginning with forward. Our implementation typically ran in $\Theta(rn^{r+1} \log n)^{0.36}$ and $\Theta(r^2 n^{r+1} \log n)^{0.36}$ for forward and backward, respectively; well below the theoretical upper bound.

Second, in certain domains one type of bisimulation minimisation is more effective. For example, backward bisimulation is ineffective on deterministic wta because no two states have the same past[1]. On the other hand, wta recognising languages of trees that vary greatly in the root but little in the leaves (for example, syntax parses of natural language sentences), will benefit more from backward bisimulation minimisation than forward. When presented with an unknown wta, we know no way to say for certain which method of minimisation is superior, so it is beneficial to have both.

The bisimulation introduced in [16] can be seen as a combination of backward and forward bisimulation. Containing the restrictions of both, it is less efficient than backward bisimulation when applied to the minimisation of nondeterministic automata, but just as expensive to calculate, and unlike forward bisimulation it does not yield the standard algorithm when applied to deterministic automata. The pair of algorithms presented in this paper thus supersedes that of [16].

2   Preliminaries

We write $\mathbb{N}$ to denote the set of natural numbers including zero. The subset $\{k, k + 1, \ldots, n\}$ of $\mathbb{N}$ is abbreviated to $[k, n]$, and the cardinality of a set S is denoted by $|S|$. We abbreviate the Cartesian product $S \times \cdots \times S$ with n factors by S^n, and the inclusion $d_i \in D_i$ for all $i \in [1, k]$ as $d_1 \cdots d_k \in D_1 \cdots D_k$.

Let $\mathcal{P}$ and $\mathcal{R}$ be equivalence relations on S. We say that $\mathcal{P}$ is *coarser* than $\mathcal{R}$ (or equivalently: $\mathcal{R}$ is a *refinement* of $\mathcal{P}$), if $\mathcal{R} \subseteq \mathcal{P}$. The *equivalence class* (or *block*) of an element $s \in S$ with respect to $\mathcal{R}$ is the set $[s]_{\mathcal{R}} = \{s' \mid (s, s') \in \mathcal{R}\}$. Whenever $\mathcal{R}$ is obvious from the context, we simply write $[s]$ instead of $[s]_{\mathcal{R}}$. It should be clear that $[s]$ and $[s']$ are equal if s and s' are in relation $\mathcal{R}$, and disjoint otherwise, so $\mathcal{R}$ induces a partition $(S/\mathcal{R}) = \{[s] \mid s \in S\}$ of S.

A *semiring* is a tuple $(A, +, \cdot, 0, 1)$ such that $(A, +, 0)$ is a commutative monoid, $(A, \cdot, 1)$ is a monoid, $\cdot$ distributes (both-sided) over $+$, and 0 is an absorbing element with respect to $\cdot$. We generally assume that $\cdot$ binds stronger

---

[1] The alteration technique is thus useless for deterministic devices.

than $+$, so $a + b \cdot c$ is interpreted as $a + (b \cdot c)$. The semiring $\mathcal{A} = (A, +, \cdot, 0, 1)$ is said to be *cancellative* if $a + b = a + c$ implies that $b = c$ for every $a, b, c \in A$.

A *ranked alphabet* is a finite set of symbols $\Sigma = \bigcup_{k \in \mathbb{N}} \Sigma_{(k)}$ which is partitioned into pairwise disjoint subsets $\Sigma_{(k)}$. The set T_Σ of *trees over* Σ is the smallest set of strings over Σ such that $f\, t_1 \cdots t_k$ in T_Σ for every f in $\Sigma_{(k)}$ and all $t_1, \ldots, t_k$ in T_Σ. We write $f[t_1, \ldots, t_k]$ instead of $f\, t_1 \cdots t_k$ unless k is zero.

A *tree series* over the ranked alphabet Σ and semiring $\mathcal{A} = (A, +, \cdot, 0, 1)$ is a mapping from T_Σ to A. The set of all tree series over Σ and $\mathcal{A}$ is denoted by $\mathcal{A}\langle\langle T_\Sigma \rangle\rangle$. Let $\mathcal{S} \in \mathcal{A}\langle\langle T_\Sigma \rangle\rangle$. We write $(\mathcal{S}, t)$ with $t \in T_\Sigma$ for $\mathcal{S}(t)$. A *weighted tree automaton* M (for short: wta) [17] is a tuple $(Q, \Sigma, \mathcal{A}, F, \mu)$, where Q is a finite nonempty set of *states*; Σ is a ranked alphabet (of *input symbols*); $\mathcal{A} = (A, +, \cdot, 0, 1)$ is a semiring; $F \in A^Q$ is a *final weight distribution*; and $\mu = (\mu_k)_{k \in \mathbb{N}}$ with $\mu_k : \Sigma_{(k)} \to A^{Q^k \times Q}$ is a *tree representation*.

We define $h_\mu : T_\Sigma \to A^Q$ for every $\sigma \in \Sigma_{(k)}$, $q \in Q$, and $t_1, \ldots, t_k \in T_\Sigma$ by

$$h_\mu(\sigma[t_1, \ldots, t_k])_q = \sum_{q_1, \ldots, q_k \in Q} \mu_k(\sigma)_{q_1 \cdots q_k, q} \cdot h_\mu(t_1)_{q_1} \cdot \ldots \cdot h_\mu(t_k)_{q_k} \ .$$

Finally, the tree series *recognised by* M is given by $(\|M\|, t) = \sum_{q \in Q} F_q \cdot h_\mu(t)_q$ for every tree $t \in T_\Sigma$ and denoted by $\|M\|$.

3   Forward Bisimulation

Foundation. Let $M = (Q, \Sigma, \mathcal{A}, F, \mu)$ be a wta. Roughly speaking, a *forward bisimulation on M* is an equivalence relation on Q such that equivalent states react equivalently to future inputs. We enforce this behaviour with only a local condition on μ and F. Let $\square \notin Q$. The set $C_{(k)}^Q$ of *contexts (over Q)* is given by $\{w \in (Q \cup \{\square\})^k \mid w \text{ contains } \square \text{ exactly once}\}$, and for every context c and state q we write $c[\![q]\!]$ to denote the word that is obtained from c by replacing the special symbol $\square$ with q. Henceforth, we assume that the special symbol $\square$ occurs in no set of states of any wta.

Definition 1 (cf. [9, Definition 3.1]). *Let $\mathcal{R} \subseteq Q \times Q$ be an equivalence relation. We say that $\mathcal{R}$ is a* forward bisimulation on M *if for every (p, q) in $\mathcal{R}$ we have (i) $F(p) = F(q)$ and (ii) $\sum_{r \in D} \mu_k(\sigma)_{c[\![p]\!], r} = \sum_{r \in D} \mu_k(\sigma)_{c[\![q]\!], r}$ for every $\sigma \in \Sigma_{(k)}$, block D in $(Q/\mathcal{R})$, and context c of $C_{(k)}^Q$.*

Example 2. Let $\Delta = \Delta_{(0)} \cup \Delta_{(2)}$ be the ranked alphabet where $\Delta_{(0)} = \{\alpha\}$ and $\Delta_{(2)} = \{\sigma\}$. The mapping ZIGZAG from T_Δ to $\mathbb{N}$ is recursively defined for every t_1, t_2, and t_3 in T_Δ by ZIGZAG$(\alpha) = 1$ and ZIGZAG$(\sigma[\alpha, t_2]) = 2$ and ZIGZAG$(\sigma[\sigma[t_1, t_2], t_3]) = 2 + $ ZIGZAG(t_2). Consider the wta $N = (P, \Delta, \mathbb{N}, G, \nu)$ with the semiring $\mathbb{N} = (\mathbb{N}, +, \cdot, 0, 1)$ and $P = \{l, r, L, R, \bot\}$, $G(l) = G(L) = 1$ and $G(p) = 0$ for every $p \in \{r, R, \bot\}$, and

$$1 = \nu_0(\alpha)_{\varepsilon, l} = \nu_0(\alpha)_{\varepsilon, R} = \nu_0(\alpha)_{\varepsilon, \bot} = \nu_2(\sigma)_{r \bot, l} = \nu_2(\sigma)_{\bot l, r} = \nu_2(\sigma)_{\bot \bot, l}$$
$$1 = \nu_2(\sigma)_{R \bot, L} = \nu_2(\sigma)_{\bot L, R} = \nu_2(\sigma)_{\bot \bot, R} = \nu_2(\sigma)_{\bot \bot, \bot} \ .$$

All remaining entries in ν are 0. A straightforward induction shows that N recognises ZIGZAG. Let us consider $\mathcal{P} = \{l, L\}^2 \cup \{r, R\}^2 \cup \{\perp\}^2$. We claim that $\mathcal{P}$ is a forward bisimulation on N. Obviously, $G(l) = G(L)$ and $G(r) = G(R)$. It remains to check Condition (ii) of Definition 1. We only demonstrate the computation on the symbol σ, the context $\perp\square$ and the block $\{r, R\}$.

$$\sum_{p\in\{r,R\}} \nu_2(\sigma)_{\perp l,p} = 1 = \sum_{p\in\{r,R\}} \nu_2(\sigma)_{\perp L,p} \qquad \blacksquare$$

Let $\mathcal{R}$ be a forward bisimulation on M. We identify bisimilar states in order to reduce the size of the wta. Next we present how to achieve this. In essence, we construct a wta $(M/\mathcal{R})$ that uses only one state per equivalence class of $\mathcal{R}$.

Definition 3 (cf. [9, Definition 3.3]). *The* forward aggregated wta $(M/\mathcal{R})$ *is the wta* $((Q/\mathcal{R}), \Sigma, \mathcal{A}, F', \mu')$ *with* $F'([q]) = F(q)$ *for every state q of Q, and* $\mu'_k(\sigma)_{[q_1]\cdots[q_k],D} = \sum_{r\in D} \mu_k(\sigma)_{q_1\cdots q_k,r}$ *for every* $\sigma \in \Sigma_{(k)}$, *word* $q_1 \cdots q_k \in Q^k$, *and block* $D \in (Q/\mathcal{R})$.

Example 4. Recall the wta N and the forward bisimulation $\mathcal{P}$ of Example 2. Let us compute $(N/\mathcal{P}) = (P', \Delta, \mathbb{N}, G', \nu')$. We obtain $P' = \{[l], [r], [\perp]\}$, the final weights $G'([l]) = 1$ and $G'([r]) = G'([\perp]) = 0$ and the nonzero entries

$$1 = \nu'_2(\sigma)_{[r][\perp],[l]} = \nu'_2(\sigma)_{[\perp][l],[r]} = \nu'_2(\sigma)_{[\perp][l],[l]} = \nu'_2(\sigma)_{[\perp][\perp],[r]} = \nu'_2(\sigma)_{[\perp][\perp],[\perp]}$$
$$1 = \nu'_0(\alpha)_{\varepsilon,[l]} = \nu'_0(\alpha)_{\varepsilon,[r]} = \nu'_0(\alpha)_{\varepsilon,[\perp]} \ . \qquad \blacksquare$$

We should verify that the recognised tree series remains the same. The proof of this property is prepared in the next lemma. It essentially states that a collapsed state of $(M/\mathcal{R})$ works like the combination of its constituents in M.

Lemma 5 (cf. [9, Theorem 3.1]). *Let* $(M/\mathcal{R}) = (Q', \Sigma, \mathcal{A}, F', \mu')$. *Then* $h_{\mu'}(t)_D = \sum_{q\in D} h_\mu(t)_q$ *for every tree* $t \in T_\Sigma$ *and block* $D \in (Q/\mathcal{R})$.

The final step establishes that $\|(M/\mathcal{R})\| = \|M\|$. Consequently, collapsing a wta with respect to some forward bisimulation preserves the recognised series.

Theorem 6 (cf. [9, Theorem 3.1]). $\|(M/\mathcal{R})\| = \|M\|$.

The coarser the forward bisimulation $\mathcal{R}$ on M, the smaller $(M/\mathcal{R})$. Our aim is thus to find the coarsest forward bisimulation on M. First we show that a unique coarsest forward bisimulation on M exists.

Theorem 7. *There exists a coarsest forward bisimulation $\mathcal{P}$ on M, and $(M/\mathcal{P})$ admits only the identity as forward bisimulation.*

The previous theorem justifies the name *forward bisimulation minimisation*; given the coarsest forward bisimulation $\mathcal{P}$ on M, the wta $(M/\mathcal{P})$ is minimal with respect to forward bisimulation.

```
┌input:        A wta M = (Q, Σ, A, F, μ);
┌initially:    P₀  := Q × Q;
│              R₀  := ker(F) \ split(Q);
│              i   := 0;
┌while Rᵢ ≠ Pᵢ: choose Sᵢ ∈ (Q/Pᵢ) and Bᵢ ∈ (Q/Rᵢ) such that
│                  Bᵢ ⊂ Sᵢ and |Bᵢ| ≤ |Sᵢ|/2;
│              Pᵢ₊₁ := Pᵢ \ cut(Bᵢ);
│              Rᵢ₊₁ := (Rᵢ \ split(Bᵢ)) \ split(Sᵢ \ Bᵢ);
│              i    := i + 1;
┌return:       (M/Rᵢ);
```

Algorithm 1. A forward bisimulation minimisation algorithm for wta

Algorithm. We now present a minimisation algorithm for wta that draws on the ideas presented in the previous section. Algorithm 1 searches for the coarsest forward bisimulation R on the input wta M by producing increasingly refined equivalence relations $R_0, R_1, R_2, \ldots$. The first of these is the coarsest candidate solution that respects F. The relation R_{i+1} is derived from R_i by removing pairs of states that prevent R_i from being a forward bisimulation. The algorithm also produces an auxiliary sequence of relations $P_0, P_1, P_2, \ldots$ that are used to find these offending pairs. Termination occurs when R_i and P_i coincide. At this point, R_i is the coarsest forward bisimulation on M.

Before we discuss the algorithm, its correctness, and its time complexity, we extend our notation. For the rest of this section, let $M = (Q, \Sigma, A, F, \mu)$ be an arbitrary but fixed wta. We use the following shorthands in Alg. 1.

Definition 8. *Let B be a subset of Q. We write*

- *$cut(B)$ for the subset $(Q^2 \setminus B^2) \setminus (Q \setminus B)^2$ of $Q \times Q$, and*
- *$split(B)$ for the set of all pairs (p, q) in $Q \times Q$ such that $\sum_{r \in B} \mu_k(\sigma)_{c[p], r}$ and $\sum_{r \in B} \mu_k(\sigma)_{c[q], r}$ differ for some $\sigma \in \Sigma_{(k)}$ and $c \in C^Q_{(k)}$.*

Example 9. Let $N = (P, \Delta, \mathbb{N}, G, \nu)$ be the wta of Example 2 that recognises the tree series ZIGZAG. We will show the iterations of the algorithm on this example wta. Let us start with the initialisation: Clearly, P_0 is $P \times P$, and R_0 is the union $\{l, L\}^2 \cup \{r, R\}^2 \cup \{\bot\}^2$. In the first iteration, we select $S_0 = P$ and $B_0 = \{l, L\}$ and thus compute P_1 to be $\{l, L\}^2 \cup \{r, R, \bot\}^2$, and R_1 to be R_0. Obviously, P_1 is still different from R_1, so the algorithm enters a second iteration. We now let $S_1 = \{r, R, \bot\}$ and $B_1 = \{\bot\}$, which yields $R_2 = P_2$, so the algorithm terminates and returns the aggregated wta (N/R_2).   ∎

We henceforth abbreviate $|Q|$ to n, and denote by r the maximum k such that $\Sigma_{(k)}$ is non-empty. As we will later argue, there exists a $t < n$ such that Alg. 1 terminates when $i = t$. We use the notations introduced in the algorithm when we set out to prove correctness and termination.

Lemma 10. *The relation $\mathcal{R}_i$ is a refinement of $\mathcal{P}_i$ for all $i \in [0, t]$.*

Lemma 10 ensures that $\mathcal{R}_i$ is a proper refinement of $\mathcal{P}_i$, for all $i \in [0, t-1]$. Since $\mathcal{P}_{i+1}$ is in turn, by definition, a proper refinement of $\mathcal{P}_i$, termination is guaranteed in less than n iterations. It follows that, up to the termination point t, we can always find blocks $B_i \in (Q/\mathcal{R}_i)$ and $S_i \in (Q/\mathcal{P}_i)$ such that B_i is contained in S_i, and the size of B_i is at most half of that of S_i.

Theorem 11. *Algorithm 1 returns the minimal wta $(M/\mathcal{P})$ with respect to forward bisimulation. Equivalently; $\mathcal{P}$ is the coarsest forward bisimulation on M.*

We now analyse the running time of Alg. 1. We use

$$m = \sum_{k \in [0,r]} |\{(\sigma, q_1 \cdots q_k, q) \in \Sigma_{(k)} \times Q^k \times Q \mid \mu_k(\sigma)_{q_1 \cdots q_k, q} \neq 0\}| \ .$$

to denote the size of μ. In this paper, we assume that the tree representation is not sparse, i.e. that it contains some $\Omega\left(\sum_{k \in [0,r]} n^{k+1}\right)$ entries. For a discussion of how sparse representations affect the performance of the algorithm, see [14]. We also assume that semiring addition can be performed in constant time. We denote by μ_B^{f} the part of μ that contains entries of the form $\mu_k(\sigma)_{q_1 \cdots q_k, q}$, where $q \in B$. The overall time complexity of the algorithm is

$$O\left(\mathrm{INIT}^{\mathrm{f}} + \sum_{i \in [0, t-1]} \left(\mathrm{SELECT}_i + \mathrm{CUT}_i + \mathrm{SPLIT}_i^{\mathrm{f}}\right) + \mathrm{AGGREGATE}^{\mathrm{f}}\right) \ ,$$

where $\mathrm{INIT}^{\mathrm{f}}$, SELECT_i, CUT_i, $\mathrm{SPLIT}_i^{\mathrm{f}}$, and $\mathrm{AGGREGATE}^{\mathrm{f}}$ are the complexity of: (i) the initialisation phase; (ii) the choice of S_i and B_i; (iii) the computation of $\mathcal{P}_{i+1}$; (iv) the computation of $\mathcal{R}_{i+1}$, and (v) the construction of the aggregated automaton $(M/\mathcal{R}_t)$; respectively.

Lemma 12. $\mathrm{INIT}^{\mathrm{f}}$ *and* $\mathrm{AGGREGATE}^{\mathrm{f}}$ *are both in $O(m + n)$, whereas* SELECT_i *is in $O(1)$, CUT_i is in $O(|B_i|)$, and $\mathrm{SPLIT}_i^{\mathrm{f}}$ is in $O(r\,|\mu_{S_i}^{\mathrm{f}}|)$.*

In the worst case, $|S_i|$ equals $n - i$, which means that $|\mu_{S_i}^{\mathrm{f}}|$ is close to m.

Theorem 13. *Algorithm 1 has time complexity $O(rmn)$.*

We now consider a simplification of Alg. 1 for cancellative semirings. In essence, the second split in the computation of $\mathcal{R}_{i+1}$ can be omitted.

Lemma 14. *When the underlying semiring is cancellative, we can replace the computation of $\mathcal{R}_{i+1}$ in Alg. 1 simply by $\mathcal{R}_{i+1} = \mathcal{R}_i \setminus split(B_i)$.*

The optimised algorithm thus only splits against the block B_i, for each $i \in [0, t-1]$. As no state occurs in more than $\log n$ distinct B-blocks, we are able to obtain a lower time complexity:

Theorem 15. *Alg. 1 optimised for cancellative semirings is in $O(rm \log n)$.*

4   Backward Bisimulation

Foundation. Let $M = (Q, \Sigma, \mathcal{A}, F, \mu)$ be a wta. In this section we investigate backward bisimulations [9]. We introduce the following notation. Let Π be a partition of Q. We write $\Pi_{(k)}$ for the set $\{D_1 \times \cdots \times D_k \mid D_1, \ldots, D_k \in \Pi\}$ for every $k \in \mathbb{N}$. Moreover, we write $\Pi_{(\leq k)}$ for the set $\Pi_{(0)} \cup \cdots \cup \Pi_{(k)}$.

Definition 16 (cf. [9, Definition 4.1]). *Let $\mathcal{R}$ be an equivalence relation on Q. If $\sum_{w \in L} \mu_k(\sigma)_{w,p} = \sum_{w \in L} \mu_k(\sigma)_{w,q}$ for every $(p, q) \in \mathcal{R}$, symbol σ in $\Sigma_{(k)}$, and word $L \in (Q/\mathcal{R})_{(k)}$, then we say that $\mathcal{R}$ is a* backward bisimulation *on M.*

Example 17. Let $N = (P, \Delta, \mathbb{N}, G, \nu)$ where $P = \{l, r, L, R, \bot\}$, Δ is as in Example 2, and $G(l) = 1$ and $G(p) = 0$ for every $p \in \{r, L, R, \bot\}$ and

$$1 = \nu_0(\alpha)_{\varepsilon,l} = \nu_0(\alpha)_{\varepsilon,r} = \nu_0(\alpha)_{\varepsilon,L} = \nu_0(\alpha)_{\varepsilon,R} = \nu_0(\alpha)_{\varepsilon,\bot}$$
$$1 = \nu_2(\sigma)_{\bot L,R} = \nu_2(\sigma)_{\bot L,r} = \nu_2(\sigma)_{\bot l,r}$$
$$1 = \nu_2(\sigma)_{R\bot,L} = \nu_2(\sigma)_{R\bot,l} = \nu_2(\sigma)_{r\bot,l} = \nu_2(\sigma)_{\bot\bot,\bot} .$$

All remaining entries in ν are 0. The wta N also recognises ZIGZAG. We propose $\mathcal{P} = \{l\}^2 \cup \{r\}^2 \cup \{L, R, \bot\}^2$ as backward bisimulation. We note that $\nu_0(\alpha)_{\varepsilon,L}$ and $\nu_0(\alpha)_{\varepsilon,R}$ and $\nu_0(\alpha)_{\varepsilon,\bot}$ are all equal and $\sum_{p_1 p_2 \in [\bot][\bot]} \nu_2(\sigma)_{p_1 p_2,p} = 1$ and $\nu_2(\sigma)_{p_1 p_2,p} = 0$ for every $p \in \{L, R, \bot\}$ and $p_1, p_2 \in P$ such that $(p_1, \bot) \notin \mathcal{P}$ and $(p_2, \bot) \notin \mathcal{P}$. ∎

For the rest of this section, let $\mathcal{R}$ be a backward bisimulation on M. Next we define how to collapse M with respect to $\mathcal{R}$.

Definition 18 (cf. [9, Definition 3.3]). *The* backward aggregated *wta $(M/\mathcal{R})$ is the wta $((Q/\mathcal{R}), \Sigma, \mathcal{A}, F', \mu')$ such that (i) $F'(D) = \sum_{q \in D} F(q)$ for every block D of $(Q/\mathcal{R})$ and (ii) $\mu'_k(\sigma)_{D_1 \cdots D_k,[q]} = \sum_{w \in D_1 \cdots D_k} \mu_k(\sigma)_{w,q}$ for every symbol σ in $\Sigma_{(k)}$, word $D_1 \cdots D_k$ of blocks in $(Q/\mathcal{R})$, and state $q \in Q$.*

Example 19. Recall the wta N and the backward bisimulation $\mathcal{P}$ from Example 17. We obtain $(N/\mathcal{P}) = (P', \Delta, \mathbb{N}, G', \nu')$ with $P' = \{[l], [r], [\bot]\}$ and $G'([l]) = 1$ and $G'([r]) = G([\bot]) = 0$ and the nonzero tree representation entries

$$1 = \nu'_2(\sigma)_{[\bot][\bot],[r]} = \nu'_2(\sigma)_{[\bot][l],[r]} = \nu'_2(\sigma)_{[\bot][\bot],[l]} = \nu'_2(\sigma)_{[r][\bot],[l]} = \nu'_2(\sigma)_{[\bot][\bot],[\bot]}$$
$$1 = \nu'_0(\alpha)_{\varepsilon,[l]} = \nu'_0(\alpha)_{\varepsilon,[r]} = \nu'_0(\alpha)_{\varepsilon,[\bot]} .$$
∎

Next we prepare Theorem 21, which will show that M and $(M/\mathcal{R})$ recognise the same series. First we prove that every state q of M recognises the same series as the state $[q]$ of $(M/\mathcal{R})$.

Lemma 20 (cf. [9, Theorem 4.2] and [18, Lemma 5.2]). *Let $(M/\mathcal{R})$ be $(Q', \Sigma, \mathcal{A}, F', \mu')$. Then $h_{\mu'}(t)_{[q]} = h_\mu(t)_q$ for every state $q \in Q$ and tree $t \in T_\Sigma$.*

The previous lemma establishes a nice property of bisimilar states. Namely, $h_\mu(t)_p = h_\mu(t)_q$ for every pair $(p, q) \in \mathcal{R}$ of bisimilar states and every tree $t \in T_\Sigma$.

```
┌input:          A wta M = (Q, Σ, A, F, μ);
┌initially:      P₀  := Q × Q;                 L₀ := (Q/P₀)_(≤r);
└               R₀  := P₀ \ split^b(L₀);   i  := 0;
┌while R_i ≠ P_i: choose S_i ∈ (Q/P_i) and B_i ∈ (Q/R_i) such that
│                   B_i ⊂ S_i and |B_i| ≤ |S_i|/2;
│               P_{i+1}:= P_i \ cut(B_i);
│               L_{i+1}:= (Q/P_{i+1})_(≤r);
│               R_{i+1}:= (R_i \ split^b(L_{i+1}(B_i))) \ split^b(L_{i+1}(S_i \ B_i, ¬B_i));
└               i   := i + 1;
┌return:         (M/R_i);
```

Algorithm 2. A backward bisimulation minimisation algorithm for wta

Theorem 21 (cf. [9, Theorem 4.2] & [18, Lemma 5.3]). $\|(M/R)\| = \|M\|$.

Among all backward bisimulations on M, the coarsest one yields the smallest aggregated wta, and this wta admits only the trivial backward bisimulation.

Theorem 22. *There exists a coarsest backward bisimulation P on M, and the wta (M/P) only admits the identity as backward bisimulation.*

Algorithm. We now show how Alg. 1 can be modified so as to minimise with respect to backward bisimulation. For this we recall the wta $M = (Q, \Sigma, A, F, \mu)$ with $n = |Q|$ states. Intuitively, the sum $\sum_{w \in D_1 \cdots D_k} \mu_k(\sigma)_{w,q}$ captures the extent to which q is reachable from states in $D_1 \cdots D_k$, on input σ, and is thus a local observation of the properties of q (cf. Definition 16). To decide whether states p and q are bisimilar, we compare $\sum_{w \in L} \mu_k(\sigma)_{w,p}$ and $\sum_{w \in L} \mu_k(\sigma)_{w,q}$ on increasing languages L. If we find a pair (σ, L) on which the two sums disagree, then (p, q) can safely be discarded from our maintained set of bisimilar states.

Definition 23. *Let $B, B' \subseteq Q$ and let $\mathcal{L} \subseteq \mathfrak{P}(Q^*)$ be a set of languages.*

- *We write $\mathcal{L}(B)$ to denote $\{L \cap Q^* B Q^* \mid L \in \mathcal{L}\}$.*
- *We write $\mathcal{L}(B, \neg B')$ when we mean $\{L \cap (Q \setminus B')^* \mid L \in \mathcal{L}(B)\}$.*
- *We write $split^b(\mathcal{L})$ for the set of all (p, q) in $Q \times Q$ for which there exist $\sigma \in \Sigma_{(k)}$ and a language $L \in \mathcal{L} \cap \mathfrak{P}(Q^k)$ such that the sums $\sum_{w \in L} \mu_k(\sigma)_{w,p}$ and $\sum_{w \in L} \mu_k(\sigma)_{w,q}$ differ.*

Algorithm 2, as listed above, is obtained from Alg. 1 as follows. The initialisation of R_0 is replaced with the assignment $R_0 = P_0 \setminus split^b((Q/P_0)_{(\leq r)})$, and the computation of R_{i+1} with

$$R_{i+1} = \left(R_i \setminus split^b((Q/P_{i+1})_{(\leq r)}(B_i))\right) \setminus split^b((Q/P_{i+1})_{(\leq r)}(S_i \setminus B_i, \neg B_i)) .$$

Example 24. Consider the execution of the backward bisimulation minimisation algorithm on the wta $N = (P, \Delta, \mathbb{N}, G, \nu)$ of Example 17. Clearly, $\mathcal{P}_0$ is $P \times P$. In the computation of $\mathcal{P}_0 \setminus split^b(\mathcal{L}_0)$, the state space can be divided into $\{L, R, \bot\}$ and $\{l, r\}$, as $\sum_{w \in PP} \nu_k(\sigma)_{w,p}$ is 1 when p is in the former set, but 2, when in the latter. No additional information can be derived by inspecting $\nu_0(\alpha)_{\varepsilon,p}$ because this value equals 1 for every $p \in \{l, r, L, R, \bot\}$, so $\mathcal{R}_0 = \{l, r\}^2 \cup \{L, R, \bot\}^2$.

In Iteration 1, S_0 is by necessity P, and B_0 is $\{l, r\}$, so $\mathcal{P}_1 = \mathcal{R}_0$. The tree representation entries for the nullary symbol α will have no further effect on $\mathcal{R}_0$. On the other hand, we have that $\sum_{w \in [\bot][l]} \nu_2(\sigma)_{w,p}$ is nonzero only when $p = l$, which splits the block $\{l, r\}$. Seeing that ν is such that the block $\{L, R, \bot\}$ is only affected by itself, we know that $\mathcal{R}_1 = \{l\}^2 \cup \{r\}^2 \cup \{L, R, \bot\}^2$, is the sought bisimulation. This means that termination happens in Iteration 3, when $\mathcal{P}_3$ has been refined to the level of $\mathcal{R}_1$.   ∎

Theorem 25. *Algorithm 2 returns the minimal wta $(M/\mathcal{P})$ with respect to backward bisimulation. Equivalently; $\mathcal{P}$ is the coarsest backward bisimulation on M.*

We now compute the time complexity of Alg. 2, using the same assumptions and notations as in Sect. 3. In addition, we denote by μ_L^b, where $L \subseteq \mathfrak{P}(Q^*)$, the part of the tree representation μ that contains entries of the form $\mu_k(\sigma)_{q_1 \cdots q_k, q}$, where $q_1 \cdots q_k$ is in L. The overall complexity of the Alg. 2 can be written as for Alg. 1, with INIT^f, SPLIT_i^f, and AGGREGATE^f, replaced by INIT^b, SPLIT_i^b, and AGGREGATE^b, respectively. By Lemma 26 we thus obtain Theorem 27.

Lemma 26. INIT^b *is in* $O(rm + n)$, *whereas* AGGREGATE^b *is in* $O(m + n)$ *and* SPLIT_i^b *is in* $O\left(r \left|\mu_{\mathcal{L}_i(S_i)}^b\right|\right)$.

Theorem 27. *Algorithm 2 is in* $O(rmn)$.

As in the forward case, we present an optimisation of Alg. 2 for cancellative semirings that reduces the time complexity.

Lemma 28. *When the underlying semiring is cancellative, we can compute the relation $\mathcal{R}_{i+1}$ as $\mathcal{R}_i \setminus split^b(\mathcal{L}_{i+1}(B_i))$ without effect on the overall algorithm.*

Theorem 29. *The optimisation of Alg. 2 is in* $O\left(r^2 m \log n\right)$.

5   Implementation

In this section we present experimental results obtained by applying an implementation (written in Perl) of Alg. 1 and Alg. 2 to the problem of *language modelling* in the natural language processing domain [19]. A language model is a formalism for determining whether a given sentence is in a particular language. Language models are particularly useful in applications of natural language and speech processing such as translation, transliteration, speech recognition, character recognition, etc., where transformation system output must be verified to be an appropriate sentence in the domain language. Typically they are formed

Table 1. Reduction of states and rules by the bisimulation minimisation algorithms

TREES	ORIGINAL		FORWARD		BACKWARD		CONVERGENCE	
	states	rules	states	rules	states	rules	states	rules
25	162	162	141	161	136	136	115	135
45	295	295	248	290	209	209	161	203
85	526	526	436	516	365	365	271	351
165	1087	1087	899	1054	672	672	468	623
305	1996	1996	1630	1924	1143	1143	735	1029

by collecting subsequences of sentences over a large corpus of text and assigning probabilities to the subsequences based on their occurrence counts in the data. To obtain the probability of a sentence one multiplies the probability of subsequences together. It is thus useful to have a data structure for efficiently looking up many subsequences. As effective language models typically have many millions of unique subsequences, but there is considerable similarity between the subsequences, a compressed dictionary of subsequences seems to be a natural choice for such a data structure. A minimisation algorithm is particularly suited for building a compressed dictionary from uncompressed sequence input.

Recent research in natural language processing has focused on using tree-based models to capture syntactic dependencies in applications such as machine translation [20,21]. We thus require a language model of trees, and the subsequences we will represent are subtrees. We prepared a data set by collecting 3-subtrees, i.e. all subtrees of height 3, from sentences taken from the Penn Treebank corpus of syntactically bracketed English news text [22], and collected observation statistics on these subtrees, which we stored as probabilities. In our experiments, we selected at random a subset of these subtrees and constructed an initial wta over the semiring $(\mathbb{R}_+, +, \cdot, 0, 1)$ by representing each 3-subtree in a single path, with an exit weight at the final state equal to the observed probability of the subtree. The sizes of the initial wta are noted in columns 2 and 3 of Table 1. We then performed a single iteration of the forward and backward variants, the results of which are noted in columns 4–7 of Table 1. On average the wta size, taken as $m + n$, is reduced by 10% of original by the forward algorithm and 34% by the backward algorithm. Reduction as a percentage of size by the backward algorithm grew with the size of the wta on this data set, e.g., the largest wta presented in Table 1 was reduced by 42.7%. In contrast, forward minimisation tended to reduce the size of the input by 10% for all wta in our test set. This performance is likely due to the nature of the experimental data used and may differ highly on, e.g., wta with a more densely packed μ, wta representing infinite languages, etc.

As noted in Sect. 1, further minimisation may be obtained by applying the two algorithms in an alternating manner. We found that for the wta in this experiment, two iterations beginning with backward or three iterations beginning with forward resulted in the smallest obtainable wta, the sizes of which are noted in the last two columns of Table 1. On average, the maximal minimisation reduced the size of the input wta by 45% and, as with backward minimisation, the reduction percentage grows with the size of the initial wta, to 55.8% for the largest wta in the sample set.

Acknowledgements. The authors gratefully acknowledge the support of Kevin Knight. We appreciate the sample automata provided by Lisa Kaati and would like to thank Frank Drewes for proof-reading the manuscript. Finally, we would like to extend our thanks to the referees for their insightful comments.

References

1. Hopcroft, J.E.: An *nlogn* algorithm for minimizing states in a finite automaton. In: Kohavi, Z. (ed.) Theory of Machines and Computations, Academic Press, London (1971)
2. Meyer, A.R., Stockmeyer, L.J.: The equivalence problem for regular expressions with squaring requires exponential space. In: Proc. 13th Annual Symp. Foundations of Computer Science, pp. 125–129. IEEE Computer Society, Los Alamitos (1972)
3. Gramlich, G., Schnitger, G.: Minimizing nfas and regular expressions. In: Diekert, V., Durand, B. (eds.) STACS 2005. LNCS, vol. 3404, pp. 399–411. Springer, Heidelberg (2005)
4. Milner, R.: A Calculus of Communicating Systems. Springer Verlag, Heidelberg (1982)
5. Berstel, J., Reutenauer, C.: Recognizable formal power series on trees. Theoretical Computer Science 18, 115–148 (1982)
6. Gécseg, F., Steinby, M.: Tree Automata. Akadémiai Kiadó (1984)
7. Gécseg, F., Steinby, M.: Tree languages. In: Rozenberg, G., Salomaa, A. (eds.) Handbook of Formal Languages, vol. 3, pp. 1–68. Springer Verlag, Heidelberg (1997)
8. Eilenberg, S. (ed.): Automata, Languages, and Machines. Pure and Applied Mathematics, vol. A.59. Academic Press, London (1974)
9. Buchholz, P.: Bisimulation relations for weighted automata. unpublished (2007)
10. Högberg, J., Maletti, A., May, J.: Backward and forward bisimulation minimisation of tree automata. Technical Report ISI-TR-633, U. So. California (2007)
11. Kozen, D.: On the Myhill-Nerode theorem for trees. Bulletin of the EATCS 47, 170–173 (1992)
12. Borchardt, B.: The Myhill-Nerode theorem for recognizable tree series. In: Ésik, Z., Fülöp, Z. (eds.) DLT 2003. LNCS, vol. 2710, pp. 146–158. Springer Verlag, Heidelberg (2003)
13. Drewes, F., Vogler, H.: Learning deterministically recognizable tree series. J. Automata, Languages and Combinatorics (to appear, 2007)
14. Högberg, J., Maletti, A., May, J.: Bisimulation minimisation of weighted tree automata. Technical Report ISI-TR-634, U. So. California (2007)
15. Paige, R., Tarjan, R.: Three partition refinement algorithms. SIAM Journal on Computing 16, 973–989 (1987)
16. Abdulla, P.A., Kaati, L., Högberg, J.: Bisimulation minimization of tree automata. In: Ibarra, O.H., Yen, H.-C. (eds.) CIAA 2006. LNCS, vol. 4094, pp. 173–185. Springer, Heidelberg (2006)
17. Borchardt, B.: The Theory of Recognizable Tree Series. Akademische Abhandlungen zur Informatik. Verlag für Wissenschaft und Forschung (2005)

18. Abdulla, P.A., Kaati, L., Högberg, J.: Bisimulation minimization of tree automata. Technical Report UMINF 06.25, Umeå University (2006)
19. Jelinek, F.: Continuous speech recognition by statistical methods. Proc. IEEE 64, 532–557 (1976)
20. Galley, M., Hopkins, M., Knight, K., Marcu, D.: What's in a translation rule? In: Proc. HLT-NAACL, pp. 273–280 (2004)
21. Yamada, K., Knight, K.: A syntax-based statistical translation model. In: Proc. ACL, pp. 523–530. Morgan Kaufmann, San Francisco (2001)
22. Marcus, M.P., Marcinkiewicz, M.A., Santorini, B.: Building a large annotated corpus of english: The Penn treebank. Comp. Linguistics 19, 313–330 (1993)

Conjunctive Grammars Can Generate Non-regular Unary Languages

Artur Jeż*

Institute of Computer Science,
ul. Joliot-Curie 15, 50-383 Wroclaw, Poland
aje@ii.uni.wroc.pl
http://www.ii.uni.wroc.pl/~aje

Abstract. Conjunctive grammars were introduced by A. Okhotin in [1] as a natural extension of context-free grammars with an additional operation of intersection in the body of any production of the grammar. Several theorems and algorithms for context-free grammars generalize to the conjunctive case. Still some questions remained open. A. Okhotin posed nine problems concerning those grammars. One of them was a question, whether a conjunctive grammar over unary alphabet can generate only regular languages. We give a negative answer, contrary to the conjectured positive one, by constructing a conjunctive grammar for the language $\{a^{4^n} : n \in \mathbb{N}\}$. We then generalise this result—for every set of numbers L such that their representation in some k-ary system is regular set we show that $\{a^{k^n} : n \in L\}$ is generated by some conjunctive grammar over unary alphabet.

Keywords: Conjunctive grammars, regular languages, unary alphabet, non-regular languages.

1 Introduction and Background

Alexander Okhotin introduced conjunctive grammars in [1] as a simple, yet beautiful and powerful extension of context-free grammars. Informally speaking, conjunctive grammars allow additional use of intersection in the body of any rule of the grammar. More formally, conjunctive grammar is defined as a quadruple $\langle \Sigma, N, P, S \rangle$ where Σ is a finite alphabet, N is a set of nonterminal symbols, S is a starting nonterminal symbol and P is a set of productions of the form:

$$A \rightarrow \alpha_1 \& \alpha_2 \& \ldots \& \alpha_k, \quad \text{where } \alpha_i \in (\Sigma \cup N)^* .$$

Word w can be derived by this rule if and only if it can be derived from every string α_i for $i = 1, \ldots, k$.

We can also give semantics of conjunctive grammars with language equations that use sum, intersection and concatenation as allowed operations. Language

---

* Research supported by MNiSW grant number N206 024 31/3826, 2006-2008.

T. Harju, J. Karhumäki, and A. Lepistö (Eds.): DLT 2007, LNCS 4588, pp. 242–253, 2007.
© Springer-Verlag Berlin Heidelberg 2007

generated by conjunctive grammar is a smallest solution of such equations (or rather one coordinate of the solution, since it is a vector of languages).

The usage of intersection allows us to define many natural languages that are not context-free. On the other hand it can be shown [1] that languages generated by the conjunctive grammars are deterministic context-sensitive.

Since in this paper we need only a small piece of theory of conjunctive grammars, we give an example of a grammar, together with language generated by it, instead of formal definitions. The Reader interested in the whole theory of the conjunctive grammars should consult [1] for detailed results or [2] for shorter overview. Also work on the Boolean grammars [3], which extend conjunctive grammars by additional use of negation, may be interesting.

Example 1. Let us consider conjunctive grammar $\langle \Sigma, N, P, S \rangle$ defined by:

$$\Sigma = \{a, b, c\}\,,$$
$$N = \{S, B, C, E, A\}\,,$$
$$P = \{A \rightarrow aA \mid \epsilon, C \rightarrow Cc \mid \epsilon, S \rightarrow (AE)\&(BC),$$
$$B \rightarrow aBb \mid \epsilon, E \rightarrow bEc \mid \epsilon\}\,.$$

The language generated by this grammar is equal to $\{a^n b^n c^n : n \in \mathbb{N}\}$. The associated language equations are:

$$L_A = \{a\}L_A \cup \{\epsilon\}\,,$$
$$L_C = \{c\}L_C \cup \{\epsilon\}\,,$$
$$L_S = (L_A L_E) \cap (L_B L_C)\,,$$
$$L_B = \{a\}L_B\{b\} \cup \{\epsilon\}\,,$$
$$L_E = \{b\}L_E\{c\} \cup \{\epsilon\}\,.$$

Their smallest solution is:

$$L_A = a^*\,,$$
$$L_C = c^*\,,$$
$$L_S = \{a^n b^n c^n : n \in \mathbb{N}\}\,,$$
$$L_B = \{a^n b^n : n \in \mathbb{N}\}\,,$$
$$L_E = \{b^n c^n : n \in \mathbb{N}\}\,.$$

Many natural techniques and properties generalize from context-free grammars to conjunctive grammars. Among them most important are: existence of the Chomsky normal form, parsing using a modification of CYK algorithm *etc.* On the other hand many other techniques do not generalize—there is no Pumping Lemma for conjunctive grammars, they do not have bounded growth property, non-emptiness is undecidable. In particular no technique for showing that a language cannot be generated by conjunctive grammars is known; in fact, as for today, we are only able to show that languages that are not context sensitive lay outside this class of languages.

A. Okhotin in [4] gathered nine open problems for conjunctive and Boolean grammars considered to be the most important in this field. One of those problem was a question, whether conjunctive grammars over unary alphabet generate only regular languages. It is easy to show (using Pumping Lemma), that this is true in case of context-free grammars. The same result was conjectured for conjunctive grammars. We disprove this conjecture by giving conjunctive grammar for a non-regular language $\{a^{4^n} : n \in \mathbb{N}\}$.

The set $\{4^n : n \in \mathbb{N}\}$ written in binary is a regular language. This leads to a natural question, what is the relation between regular (over binary alphabet) languages and unary conjunctive languages. We prove that every regular language (written in some k-ary system) interpreted as a set of numbers can be represented by a conjunctive grammar over an unary alphabet.

2   Main Result—Non-regular Conjunctive Language over Unary Alphabet

Since we deal with an unary alphabet we identify word a^n with number n and work with sets of integers rather than with sets of words.

Let us define the following sets of integers:

$$A_1 = \{1 \cdot 4^n : n \in \mathbb{N}\},$$
$$A_2 = \{2 \cdot 4^n : n \in \mathbb{N}\},$$
$$A_3 = \{3 \cdot 4^n : n \in \mathbb{N}\},$$
$$A_{12} = \{6 \cdot 4^n : n \in \mathbb{N}\}.$$

The indices reflect the fact that these sets written in tetrary system begin with digits $1, 2, 3, 12$, respectively and have only 0'es afterwards. We will show that those sets are the minimal solution of the equations:

$$B_1 = (B_2 B_2 \cap B_1 B_3) \cup \{1\}, \tag{1}$$
$$B_2 = (B_{12} B_2 \cap B_1 B_1) \cup \{2\}, \tag{2}$$
$$B_3 = (B_{12} B_{12} \cap B_1 B_2) \cup \{3\}, \tag{3}$$
$$B_{12} = (B_3 B_3 \cap B_1 B_2). \tag{4}$$

Where in the above equations XY reflects the concatenation of languages:

$$XY := \{x + y : x \in X, y \in Y\}.$$

This set of language equations can be easily transformed to a conjunctive grammar over unary alphabet (we should specify the starting symbol, say B_1). None of the sets A_1, A_3, A_3, A_{12} is a regular language over unary alphabet.

Since solutions are vectors of languages we use notation

$$(A_1, \ldots, A_n) \subset (B_1, \ldots, B_n),$$

meaning, that $A_i \subset B_i$ for $i = 1, \ldots, n$.

Since we often prove theorems be induction on number of digits, it is convenient to use the following notation for language (set) S:

$$S \upharpoonright_n := \{s \in S : s \text{ has } n \text{ digits at the most}\} \, .$$

We shall use it also for the vectors of languages (sets) with an obvious meaning.

Lemma 1. *Every solution* (S_1, S_2, S_3, S_{12}) *of equations* (1)–(4) *satisfies:*

$$(A_1, A_2, A_3, A_{12}) \subset (S_1, S_2, S_3, S_{12}). \tag{5}$$

Proof. We shall prove by induction on n, that

$$(A_1, A_2, A_3, A_{12}) \upharpoonright_n \subset (S_1, S_2, S_3, S_{12}) \, .$$

For $n = 1$ we know that $i \in S_i$ by (1), (2) and (3). This ends induction basis.
For induction step let us assume that

$$(A_1, A_2, A_3, A_{12}) \upharpoonright_{n+1} \subset (S_1, S_2, S_3, S_{12}).$$

We shall prove that this is true also for $(n + 2)$.

Let us start with 4^{n+1} and S_1. By induction assumption $2 \cdot 4^n \in S_2$ and hence $(2 \cdot 4^n) + (2 \cdot 4^n) = 4^{n+1} \in S_2 S_2$. Also by induction assumption $4^n \in S_1$ and $3 \cdot 4^n \in S_3$, hence $(4^n) + (3 \cdot 4^n) = 4^{n+1} \in S_1 S_3$, and so $4^{n+1} \in S_2 S_2 \cap S_1 S_3$ and by (1) we conclude that $4^{n+1} \in S_1$.

Similar calculations can be made for other $(n + 2)$-digit numbers, we present them in simplified way.

For $6 \cdot 4^n$, which is a $(n+2)$-digit number, we can see that $3 \cdot 4^n \in S_3$, $2 \cdot 4^n \in S_2$ by induction hypothesis and $1 \cdot 4^{n+1} = 4 \cdot 4^n \in S_1$, which was proved already in induction step. Hence $6 \cdot 4^n \in S_3 S_3 \cap S_1 S_2$ and by (4) we get $6 \cdot 4^n \in S_{12}$.

For $2 \cdot 4^{n+1}$ notice that $2 \cdot 4^n \in S_2$, $6 \cdot 4^n \in S_{12}$ and $1 \cdot 4^{n+1} \in S_1$ hence $2 \cdot 4^{n+1} \in S_1 S_1 \cap S_{12} S_2$ and by (2) $2 \cdot 4^{n+1} \in S_2$.

For $3 \cdot 4^{n+1}$ notice that $2 \cdot 4^{n+1} \in S_2$, $6 \cdot 4^n \in S_{12}$ and $1 \cdot 4^{n+1} \in S_1$ hence $3 \cdot 4^{n+1} \in S_{12} S_{12} \cap S_1 S_2$ and by (3) $3 \cdot 4^{n+1} \in S_3$.

This ends induction step. $\square$

Lemma 2. *Sets* A_1, A_2, A_3, A_{12} *are a solution of* (1)–(4).

Proof. For every (1)–(4) we have to prove two inclusions: $\subset$ (that is

$$A_1 \subset (A_2 A_2 \cap A_1 A_3) \cup \{1\} \, ,$$
$$A_2 \subset (A_{12} A_2 \cap A_1 A_1) \cup \{2\} \, ,$$
$$A_3 \subset (A_{12} A_{12} \cap A_1 A_2) \cup \{3\} \, ,$$
$$A_{12} \subset (A_3 A_3 \cap A_1 A_2) \, .$$

And $\supset$, that is

$$A_1 \supset (A_2 A_2 \cap A_1 A_3) \cup \{1\} \, ,$$
$$A_2 \supset (A_{12} A_2 \cap A_1 A_1) \cup \{2\} \, ,$$
$$A_3 \supset (A_{12} A_{12} \cap A_1 A_2) \cup \{3\} \, ,$$
$$A_{12} \supset (A_3 A_3 \cap A_1 A_2) \, .$$

For the inclusions $\subset$ the proof is the same as in Lemma 1 and so we skip it.

For the inclusions $\supset$, let us consider m—the smallest number that violates this inclusion, that is m belongs to some right-hand side of one of the (1)–(4), but does not belong to the corresponding left-hand side. First note the easy, but crucial, fact that $m \neq 0$ because all numbers appearing on the right-hand side are strictly greater than 0.

Suppose m violates (1). Then $m \in A_2 A_2$. By definition there are numbers k, l such that $k, l \in A_2$ and $m = k + l$. Since m is the smallest such number then both k, l belong also to the left-hand side, hence $k = 2 \cdot 4^{n_1}$ and $l = 2 \cdot 4^{n_2}$. Then either $k = l$ and $m \in A_1$ and so it does not violate the inclusion or $k \neq l$ and so m have only two non-zero digits, both being 2's. On the other hand $m \in A_1 A_3$, so by definition there are $k' \in A_1$, $l' \in A_3$ such that $m = l' + k'$. Since m is the smallest such number then $l' = 1 \cdot 4^{n_3}$ and $k' = 3 \cdot 4^{n_4}$. And so either $m = 4^{n_3+1}$, if $n_3 = n_4$, or m has only two non-zero digits: 1 and 3. But this is a contradiction with a claim that m has only two non-zero digits, both being $2's$.

We shall deal with other cases in the same manner.

Suppose m violates (2). Then $m \in A_1 A_1$. By definition there are numbers k, l such that $k, l \in A_1$ and $m = k + l$. Since m is the smallest such number then both k, l belong also to the left-hand side, hence $k = 1 \cdot 4^{n_1}$ and $l = 1 \cdot 4^{n_2}$. Then either $k = l$ and $m \in A_2$ and so it does not violate the inclusion or $k \neq l$ and so m have only two non-zero digits, both being 1's. On the other hand $m \in A_{12} A_2$, so by definition there are $k' \in A_{12}$, $l' \in A_2$ such that $m = l' + k'$. Since m is the smallest such number then $l' = 6 \cdot 4^{n_3}$ and $k' = 2 \cdot 4^{n_4}$. And so either $m = 2 \cdot 4^{n_3+1}$, if $n_3 = n_4$, or m has exactly two non-zero digits: 3 and 2 or it has exactly three non-zero digits $1, 2, 2$. But this is a contradiction with a claim that m has only two non-zero digits, both being 1's.

Suppose m violates (3). Then $m \in A_{12} A_{12}$. By definition there are numbers k, l such that $k, l \in A_{12}$ and $m = k + l$. Since m is the smallest such number then both k, l belong also to the left-hand side, hence $k = 6 \cdot 4^{n_1}$ and $l = 6 \cdot 4^{n_2}$. Then either $k = l$ and $m \in A_3$ and so it does not violate the inclusion or $k \neq l$ and so m can have the following multisets of non-zero digits: $\{1, 1, 2, 2\}$ or $\{1, 2, 3\}$. On the other hand $m \in A_1 A_2$, so by definition there are $k' \in A_1$, $l' \in A_2$ such that $m = l' + k'$. Since m is the smallest such number then $l' = 1 \cdot 4^{n_3}$ and $k' = 2 \cdot 4^{n_4}$. And so either $m = 3 \cdot 4^{n_3}$, if $n_3 = n_4$, or m has exactly two non-zero digits: 1 and 2. But this is a contradiction with a previous claim on possible sets of non-zero digits of m.

Suppose it violates (4). Then $m \in A_3 A_3$. By definition there are numbers k, l such that $k, l \in A_3$ and $m = k + l$. Since m is the smallest such number then both k, l belong also to the left-hand side, hence $k = 3 \cdot 4^{n_1}$ and $l = 3 \cdot 4^{n_2}$. Then either $k = l$ and $m \in A_{12}$ and so it does not violate the inclusion or $k \neq l$ and so m have only two non-zero digits, both being 3's. On the other hand $m \in A_1 A_2$, so by definition there are $k' \in A_1$, $l' \in A_2$ such that $m = l' + k'$. Since m is the smallest such number then $k' = 4^{n_3}$ and $l' = 2 \cdot 4^{n_4}$. And so either m has only two non-zero digits: 1 and 2 or it has exactly one non-zero digit—3. But this is a contradiction with a claim that m has only two non-zero digits, both being $3's$. $\square$

Theorem 1. *Sets A_1, A_2, A_3, A_{12} are the smallest solution of* (1)–(4).

Proof. This follows from Lemma 1 and Lemma 2.   □

Corollary 1. *The non-regular language $\{a^{4^n} : n \in \mathbb{N}\}$ can be generated by conjunctive grammar over unary alphabet.*

Corollary 2. *Conjunctive grammars over unary alphabet have more expressive power then context-free grammars.*

3 Additional Results

3.1 Number of Nonterminals Required

The grammar described in the previous section uses four nonterminals. It can be easily converted to Chomsky normal form, we need only to introduce two new nonterminals for languages $\{1\}$ and $\{2\}$ respectively, hence grammar for language $\{4^n : n \in \mathbb{N}\}$ in Chomsky normal form requires only six nonterminals. It is an interesting question, which mechanisms of conjunctive grammars and how many of them are required to generate a non-regular language? How many nonterminals are required? How many of them must generate non-regular languages? How many intersections are needed? Putting this question in the other direction, are there any natural sufficient conditions for a conjunctive grammar to generate regular language?

It should be noted that we are able to reduce the number of nonterminals to three, but we sacrifice Chomsky normal form and introduce also concatenations of three nonterminals in productions. This can be seen as some trade-off between number of nonterminals and length of concatenations. Let us consider a language equation:

$$B_1 = (B_{2,12}B_{2,12} \cap B_1B_3) \cup \{1\}, \tag{6}$$

$$B_{2,12} = ((B_{2,12}B_{2,12} \cap B_1B_1) \cup \{2\}) \cup$$
$$\cup((B_3B_3 \cap B_{2,12}B_{2,12})), \tag{7}$$

$$B_3 = (B_{2,12}B_{2,12} \cap B_1B_1B_1) \cup \{3\}. \tag{8}$$

These are basically the same equations as (1)–(4), except that nonterminals B_2 and B_{12} are identified (or merged) and also conjunct B_2B_1 in (1)–(4) was changed to $B_1B_1B_1$. The Reader can easily check that after only the second change applied to (1)–(4) proofs of Lemma 1 and Lemma 2 can be easily modified to work with new situation.

Theorem 2. *The smallest solution of (6)–(8) is*

$$(A_1, A_2 \cup A_{12}, A_3).$$

Proof. The proof of this theorem is just a slight modification of the proof of Theorem 1, so we shall just sketch it.

The main idea of the proof is to think of nonterminal $B_{2,12}$ that corresponds to the set $A_2 \cup A_{12}$ as two nonterminals: B_2 and B_{12}, corresponding to sets

A_2 and A_{12}, respectively. To implement this approach we should show that every occurrence of $B_{2,12}$ on the right-hand side of any equation can be replaced by exactly one of the nonterminals B_2 or B_{12}, meaning that in each language equation from (6)–(8) if we substitute the variables with intended solution every occurrence of set $A_2 \cup A_{12}$ on the right hand-side can be in fact replaced by exactly one of the sets A_2, A_{12} with keeping the equations true.

Let us consider sets $(A_1,\ A_2 \cup A_{12},\ A_3)$. Using the same arguments as in Lemma 1 we can show that the smallest solutions of (6)–(8) is a pointwise superset of considered sets.

Now we must show that these sets are in fact a solution. Showing $\subset$ is easy, as in Lemma 2. Showing $\supset$ is the same as in Lemma 2 if we show earlier that we can substitute every occurrence of $B_{2,12}$ on the right-hand side with exactly one of B_2 or B_{12}, respectively.

Notice that if we have an intersection of the form $B_i B_j \cap B_k B_l$ then the sum of tetrary digits from B_i and B_j must be equal modulo 3 the sum of tetrary digits from B_k and B_l (this observation is easily generalized to the case with more concatenated nonterminals). We take sums modulo 3 because if we sum two numbers and digits in a column sum up to 4 (or more) then we loose 4 in this column but gain 1 in the next, so the difference is 3. Now if we swap from B_2 to B_{12} then the sum of digits changes by 1. So to get an equation modulo 3 we would have to add and subtract some 1's from both sides. Case inspection shows that this is not possible. And so we can use the same arguments as in Lemma 2. □

3.2   Related Languages

Theorem 3. *For every natural number k, language*

$$\{k^n : n \in \mathbb{N}\}$$

is generated by a conjunctive grammar over unary alphabet.

Proof. For every $k > 4$ we introduce non-terminals $B_{i,j}$, where $i = 1, \ldots k-1$ and $j = 0, \ldots, k-1$, with intention that $B_{i,j}$ defines language of numbers beginning with digits i, j and then only zeroes in k-ary system of numbers. Then we define the productions as:

$B_{1,0} \to B_{2,0}B_{k-2,0}\ \&\ B_{1,0}B_{k-1,0}$,

$B_{1,1} \to B_{1,0}B_{1,0}\ \ \ \&\ B_{k-1,0}B_{2,0}$,

$B_{1,3} \to B_{1,0}B_{3,0}\ \ \ \&\ B_{1,2}B_{1,0}$,

$B_{2,3} \to B_{2,0}B_{3,0}\ \ \ \&\ B_{2,1}B_{2,0}$,

$B_{i,j} \to B_{1,0}B_{i,j-1}\ \&\ B_{2,0}B_{i,j-2}$ for (i,j) not mentioned above ,

$B_{i,0} \to \{i\}$ for $i = 1, \ldots, k-1$.

In the above equations $B_{i,j}$ we use cyclic notation for second lower indices, that is $B_{i,j} = B_{i,j \bmod k}$. It can be shown, using methods as in Lemma 1 and Lemma 2, that the smallest solution of these equations is

$$B_{i,j} = \{(k \cdot i + j) \cdot k^n : n \in \mathbb{N}\}, \text{ for } j \neq 0 \ ,$$
$$B_{i,0} = \{i \cdot k^n : n \in \mathbb{N}\} \ .$$

For $k = 2, 3$ we have to sum up some languages generated in cases of $k = 4, 9$, respectively. The case for $k = 1$ is trivial.

The productions used in this theorem could be simplified for fixed k.   □

4   Regular Languages over k-ary Alphabet

Now we deal with major generalisation of the Theorem 1. We deal with languages $\{a^n : n \in L\}$, where R is some regular language (written in some k-ary system). To simplify the notation, let $\Sigma_k = \{0, \ldots, k - 1\}$. From the following on we consider regular languages over Σ_k for some k that do not have words with leading 0, since this is meaningless in case of numbers.

Definition 1. *Let $w \in \Sigma_k^*$ be a word. We define its unary representation as*

$$f_k(w) = \{a^n : w \text{ read as } k\text{-ary number is } n\} \,.$$

When this does not lead to confusion, we also use f_L applied to languages with an obvious meaning.

The following fact shows that it is enough to consider the k parameter in f_k that are 'large enough'.

Lemma 3. *For every $k = l^n$, $n > 0$ and every unary language L language $f_k^{-1}(L)$ is regular if and only if language $f_l^{-1}(L)$ is regular.*

Proof. An automaton over alphabet Σ_k can clearly simulate reading a word written in l-ary system and *vice-versa*.   □

In the following we shall use 'big enough' k, say $k \geq 100$. We claim, that for regular L language $f_k(L)$ is conjunctive.

We now define the conjunctive grammar for fixed regular language $L \subset \Sigma_k^*$ without leading 0. Let

$$M = \langle \Sigma_k, Q, \delta, Q_{fin}, q_0 \rangle$$

be the (non-deterministic) automaton that recognizes L^r.

We define conjunctive grammar $G = \langle \{a\}, N, P, S \rangle$ over unary alphabet with:

$$N = \{A_{i,j,q}, A_{i,j} : 1 \leq i < k, \ 0 \leq j < k, \ q \in Q\} \cup \{S\} \,.$$

The intended solution is

$$L(A_{i,j}) = \{n \ : \ f_k^{-1}(n) = ij0^k \text{ for some natural } k\} \,, \tag{9}$$

$$L(A_{i,j,q}) = \{n : f_k^{-1}(n) = ijw, \ \delta(q_0, w^r, q)\} \,, \tag{10}$$

$$L(S) = f_k(L) \,. \tag{11}$$

We denote sets defined in (10) by $L_{i,j,q}$.

From Theorem 3 we know, that $A_{i,j}$ can be defined by conjunctive grammars, and so we focus only on productions for $A_{i,j,q}$:

$$A_{i,j,q} \rightarrow A_{i,0}A_{j,x,q'} \ \& \ A_{i,1}A_{j-1,x,q'} \ \& \ A_{i,2}A_{j-2,x,q'} \ \& \ A_{i,3}A_{j-3,x,q'} \,, \tag{12}$$

for every $j > 3$, every i and every x, q' such that $\delta(q', x, q)$.

$$A_{i,j,q} \rightarrow A_{i-1,j+1}A_{k-1,x,q'} \ \& \ A_{i-1,j+2}A_{k-2,x,q'} \ \& $$
$$\& \ A_{i-1,j+3}A_{k-3,x,q'} \ \& \ A_{i-1,j+4}A_{k-4,x,q'} \ , \quad (13)$$

for every $j < 4$ and $i \neq 1$ and every x, q' such that $\delta(q', x, q)$.

$$A_{1,j,q} \rightarrow A_{k-1,0}A_{j+1,x,q'} \ \& \ A_{k-2,0}A_{j+2,x,q'} \ \& $$
$$\& \ A_{k-3,0}A_{j+3,x,q'} \ \& \ A_{k-4,0}A_{j+4,x,q'} \ , \quad (14)$$

for every $j < 4$, every x, q' such that $\delta(q', x, q)$.

$$A_{i,j,q_0} \rightarrow k \cdot i + j \ , \quad (15)$$
$$S \rightarrow A_{i,j,q} \text{ for } q \text{ such that } \exists q' \in Q_{fin}, \ \delta(q, ji, q')$$
$$\text{and } i, j \text{---arbitrary digits} \ , \quad (16)$$
$$S \rightarrow i \ \text{ for } i \in f_k(L \cap \Sigma_k) \ . \quad (17)$$

We shall prove, that $L_{i,j,q}$ are solution of proper language equations and that they are the minimal solution (are included in every other solution). The case of $L(S)$ in (11) will then easily follow. We identify the equations with productions (12)–(14).

Lemma 4. *Every solution $(X_{i,j,q})$ of language equations defining G for non-terminals $(A_{i,j,q})$ is a superset of $(L_{i,j,q})$.*

Proof. We prove by induction that for every $n > 1$

$$L_{i,j,q} \restriction_n \subset X_{i,j,q}.$$

When $|ijw| = 2$ then this is obvious by rule (15).

Suppose we have proven the Lemma for $k < n$, we prove it for n. Choose any $ijw \in L_{i,j,q} \restriction_n$. Let $w = xw'$, suppose $j > 3$. Let p be a state such that $\delta(q_0, w'^r, p)$ and $\delta(p, x, q)$ (choose one if there are many). Consider words $jxw' \in L_{j,x,p} \restriction_{n-1}$, $(j-1)xw' \in L_{j-1,x,p} \restriction_{n-1}$, $(j-2)xw' \in L_{j-2,x,p} \restriction_{n-1}$ and $(j-3)xw' \in L_{j-3,x,p} \restriction_{n-1}$. By induction hypothesis $L_{j,x,p} \restriction_{n-1} \subset X_{j,x,p}$, $L_{j-1,x,p} \restriction_{n-1} \subset X_{j-1,x,p}$, $L_{j-2,x,p} \restriction_{n-1} \subset X_{j-2,x,p}$ and $L_{j-3,x,p} \restriction_{n-1} \subset X_{j-3,x,p}$. And so by (12) $ijw \in X_{i,j,q}$.

Other cases (which use productions (13), (14)) are proved analogously. □

Lemma 5. *languages $L_{i,j,q}$ are a solution of of (12)–(14).*

Proof. For the $\subset$ part the proof is essentially the same as in Lemma 4.

To prove the the $\supset$ part we proceed by induction on number of digits in $w \in L_{i,j,q}$.

We begin with (12). Suppose w belong to the right-hand side. We shall show that it also belongs to the left-hand side. We first proof, that it in fact has the desired two first digits. Then we shall deal with the q index.

Consider the possible positions of the two first digits of each conjunct. Notice, that if j is on the position one to the right of i, then the digits are as desired. And so we may exclude this case from our consideration. The following table summarizes the results:

	i and j are together	j is leading	i is leading
$A_{i,0}A_{j,xq'}$	$(i+j), x$	$j, x\langle +i\rangle$	$i, 0$
$A_{i,1}A_{j-1,xq'}$	$(i+j-1), (x+1)$	$(j-1), x\langle +i\rangle$	$i, 1$
$A_{i,2}A_{j-2,xq'}$	$(i+j-2), (x+2)$	$(j-2), x\langle +i\rangle$	$i, 2$
$A_{i,3}A_{j-3,xq'}$	$(i+j-3), (x+3)$	$(j-3), x\langle +i\rangle$	$i, 3$

The drawback of this table is that it does not include the possibility, that some digits sum up to k (or more) and influence another digit (by carrying 1), we handle with this manually. Also in the second column i may be or may be not on the same position as x, but we deal with those two cases together. This possibility was marked by writing $\langle +i\rangle$. Also there may be an add up to k somewhere to the right, and hence we can add 1 to x.

If we want the intersection to be non-empty we have to choose four items, no two of them in the same row. We show, that this is not possible. We say that some choices fit, if the digits included in the table are the same in those choices.

First of all, no two elements in the third column fit. They have fixed digits and they clearly are different.

Suppose now that we choose two elements from the first column. We show that if in one of them $(i+j-z)$ sums up to k (or more) then the same thing happens in the second choice. If $i+j-z \geq k$ (perhaps by additional 1 carried from the previous position) then the first digit is 1. In the second element the first digit can be 1 (if there is a carrying of 1) or at least $i+j-z'$, but the latter is not possible, since $i+j-z' > 1$. Whatever happens, it is not possible to fit the digits on the column with x.

It is not possible to choose three elements from the second column. Suppose that we have three fitting choices in this column. As before we may argue, that either all of them have $j(-z)$ as the first digit or the digits add up to more than $k-1$ and the first digit is 1. Suppose they add up. Since $j < k$ then this is possible only for the first row. Suppose they do not add up. Then if there are three fitting choices, then one of them must be increased by at least 2. But the maximal value carried from the previous position is 1. Contradiction. Note, that by the same reasoning we may prove, that if there are two fitting choices then the first digit is between $j-3$ and j.

And so we know, that if there are four fitting choices, then exactly one of them is in the first column, one in the third column and two in the middle column. The third column always begins with i. In the first column the leading digit is at least $i+j-3 > i$ or it is 1. Hence $i = 1$. And so the choices in the second column begin with 1 as well. Hence $j < 4$, which is a contradiction.

We now move to (13). The following table summarizes the possible first two digits:

	i and $k-z$ are together	$k-z$ is leading	$i-1$ is leading
$A_{i-1,j+1}A_{k-1,x,q'}$	$(k+i-2),(1+j+x)$	$(k-1),x(+i)$	$i-1,j+1$
$A_{i-1,j+2}A_{k-2,x,q'}$	$(k+i-3),(2+j+x)$	$(k-2),x(+i)$	$i-1,j+2$
$A_{i-1,j+3}A_{k-3,x,q'}$	$(k+i-4),(3+j+x)$	$(k-3),x(+i)$	$i-1,j+3$
$A_{i-1,j+4}A_{k-4,x,q'}$	$(k+i-5),(4+j+x)$	$(k-4),x(+i)$	$i-1,j+4$

As before we may argue, that if there are some fitting entries in some column then on their leading position digits sum up to k in all choices or in all choices do not sum up to k.

We cannot have two choices from the third column (the second digits do not match). We can have at the most two from the second column (to obtain three we would have to ad carry at least 2 to one of them and this is not possible). Fr the same reason there can be at the most two choices from the first column, but in such a case we cannot match the positions with x. Hence there is at the most one choice from the first column. And so we have one choice from the first column, one from the third and two from the second. Since the third and the second column match, then i is a big digit, at least $k-3$. But in such a case in the first column we have at least $k+k-3-5>k$ and so the leading digit is 1. Contradiction.

Consider the last possibility, the (14).The following table summarizes the possible first two digits:

	$k-z$ is leading	$j+z$ is leading
$A_{k-1,0}A_{j+1,x,q'}$	$(k-1),0(+j+1)$	$(j+1),x(+k-1)$
$A_{k-2,0}A_{j+2,x,q'}$	$(k-2),0(+j+2)$	$(j+2),x(+k-2)$
$A_{k-3,0}A_{j+3,x,q'}$	$(k-3),0(+j+3)$	$(j+3),x(+k-3)$
$A_{k-4,0}A_{j+4,x,q'}$	$(k-4),0(+j+4)$	$(j+4),x(+k-4)$

In the first column there are no two fitting choices. So we would have to take at least three from the second one. Since j is small, it cannot add up to more than k. So we cannot choose three different choices there—it would force a carrying of at least 2 from the previous position. Contradiction.

We should also take the indices denoting states of the automaton into our consideration. Consider one production:

$$A_{i,j,q} \to A_{i,0}A_{j,x,q'} \;\&\; A_{i,1}A_{j-1,x,q'} \;\&\; A_{i,2}A_{j-2,x,q'} \&\; A_{i,3}A_{j-3,x,q'}$$

and some w belonging to the right-hand side. We will explain the case of conjunct $A_{i,0}A_{j,x,q'}$. Consider $jxw' \in L_{j,x,q'}$ that was used in derivation of w. By definition of $L_{j,x,q'}$ we obtain $\delta(q_0,w'^r,q')$ and by definition of the Production (12) $\delta(q',x,q)$, and so $\delta(q_0,w'^rx,q)$. By previous discussion w begins with digits i,j, then it inherits its digits from jxw' and hence has digit x and then word w'. And so $w=ijxw'$ and $\delta(q_0,w'^rx,q)$, therefore it belongs to the left-hand side. The case with other productions and conjuncts is proved in the same way.   □

Theorem 4. *For every natural $k > 1$ and every regular $L \subset \Sigma_k^*$ without words with leading 0 language $f_k(L)$ is a conjunctive unary language.*

Proof. By Lemma 3, Lemma 4 and Lemma 5.   □

5   Conclusions and Open Problems

The main result of this paper is an example of a conjunctive grammar over unary alphabet generating non-regular language. This grammar has six nonterminal symbols in Chomsky normal form. Number of nonterminals could be reduced to three if we consider a grammar that is not in a Chomsky normal form. It remains an open question, how many nonterminals, intersection *etc.* are required to generate a non-regular language. In particular, can we give natural sufficient conditions for a conjunctive grammar to generate a regular language? Also, no non-trivial algorithm for recognizing conjunctive languages over unary alphabet is known. An obvious modification of the CYK algorithm requires quadratic time and linear space. Can those bounds be lowered? Closure under complementation of conjunctive languages (both in general and and in case of unary alphabet) remains unknown, with conjectured negative answer. Perhaps dense languages, like

$$\mathbb{N} \setminus \{2^n : n \in \mathbb{N}\}$$

are a good starting point in search for an example.

The second important result is a generalisation of the previous one: for every regular language $R \subset \{0, ..., k-1\}^*$ treated as set of k-ary numbers language $\{a^n : \exists_w \in R \ w$ read as a number is $n\}$ is a conjunctive unary language.

Acknowledgments. The author would like to thank Tomasz Jurdziński and Krzysztof Loryś for introducing to the subject and helpful discussion, Sebastian Bala and Marcin Bieńkowski for helpful discussion. The anonymous referees, who helped in improving the presentation and Alexander Okhotin, who suggested the study of generalisation to unary representations of all regular languages (Theorem 4).

References

[1] Okhotin, A.: Conjunctive grammars. Journal of Automata, Languages and Combinatorics 6(4), 519–535 (2001)

[2] Okhotin, A.: An overview of conjunctive grammars. Formal Language Theory Column. Bulletin of the EATCS 79, 145–163 (2003)

[3] Okhotin, A.: Boolean grammars. Information and Computation 194(1), 19–48 (2004)

[4] Okhotin, A.: Nine open problems on conjunctive and boolean grammars. TUCS Technical Report, p. 794 (2006)

Deterministic Blow-Ups of Minimal Nondeterministic Finite Automata over a Fixed Alphabet*

Jozef Jirásek[1], Galina Jirásková[2], and Alexander Szabari[1]

[1] Institute of Computer Science, P.J. Šafárik University,
Jesenná 5, 041 54 Košice, Slovakia
{jozef.jirasek,alexander.szabari}@upjs.sk
[2] Mathematical Institute, Slovak Academy of Sciences,
Grešákova 6, 040 01 Košice, Slovakia
jiraskov@saske.sk

Abstract. We show that for all integers n and α such that $n \leqslant \alpha \leqslant 2^n$, there exists a minimal nondeterministic finite automaton of n states with a four-letter input alphabet whose equivalent minimal deterministic finite automaton has exactly α states. It follows that in the case of a four-letter alphabet, there are no "magic numbers", i.e., the holes in the hierarchy. This improves a similar result obtained by Geffert for a growing alphabet of size $n + 2$ (*Proc. 7th DCFS*, Como, Italy, 23–37).

1 Introduction

Finite automata and regular languages are among the oldest and the simplest topics in formal language theory. They have been studied for several decades. Despite their simplicity, some important problems are still open. Let us remind the question of how many states are sufficient and necessary for two-way deterministic finite automata to simulate two-way nondeterministic finite automata [3,21].

In recent years, we can observe a renewed interest of researchers in automata theory; see [11,24] for a discussion. Many aspects in this field are now intensively investigated. One such aspect is descriptional complexity which studies the costs of the description of languages by different formal systems. In this paper, we study the relations between the sizes of minimal nondeterministic finite automata and their corresponding minimal deterministic counterparts.

Iwama at al. [13] stated the question whether there always exists a minimal nondeterministic finite automaton (NFA) of n states whose equivalent minimal deterministic finite automaton (DFA) has α states for all integers n and α such that $n \leqslant \alpha \leqslant 2^n$. The question has also been considered in [14]. In these two papers, it is shown that if $\alpha = 2^n - 2^k$ or $\alpha = 2^n - 2^k - 1$, where $0 \leqslant k \leqslant n/2 - 2$, or if $\alpha = 2^n - k$, where $2 \leqslant k \leqslant 2n - 2$ and some coprimality condition holds,

---

* Research supported by the VEGA grants 1/3129/06 and 2/6089/26, and the VEGA grant "Combinatorial Structures and Complexity of Algorithms".

T. Harju, J. Karhumäki, and A. Lepistö (Eds.): DLT 2007, LNCS 4588, pp. 254–265, 2007.
© Springer-Verlag Berlin Heidelberg 2007

then the corresponding binary n-state NFAs requiring α deterministic states do exist. In [15], appropriate NFAs has been described for all values of n and α, however, the size of the input alphabet for these automata grows exponentially with n. Later, in [7], the size of the input alphabet for the witness automata has been reduced to $n + 2$.

In this paper, we continue research on this topic. We reduce the input alphabet to a fixed size. We prove that for all integers n and α such that $n \leqslant \alpha \leqslant 2^n$, there exists a minimal nondeterministic finite automaton of n states with a four-letter input alphabet whose equivalent minimal deterministic finite automaton has exactly α states. Using terminology of [8], this means that in the case of a four-letter alphabet, there are no magic numbers, i.e., the holes in the hierarchy that cannot be reached as the size of the minimal DFA corresponding to a minimal n-state NFA. Let us note that in the case of a unary alphabet, all numbers from $\Theta(F(n))$ to 2^n, where $F(n) \approx e^{\sqrt{n \ln n}}$, are known to be magic since any n-state unary NFA can be simulated by an $O(F(n))$-state DFA [6,18]. Moreover, it has been recently shown in [8] that there are much more magic than non-magic numbers in the range from n to $e^{\sqrt{n \ln n}}$ in the unary case. The question of whether or not there are some magic numbers for binary and ternary alphabets seems to be a challenging open problem.

The paper consists of four sections, including this introduction. The next section contains basic definitions, notations and preliminary results. Section 3 presents the main results of the paper. The last section contains concluding remarks and open problems.

2   Preliminaries

In this section, we give some basic definitions and notations used throughout the paper. For further details, we refer to [22,23].

Let Σ be a finite alphabet and Σ^* the set of all strings over the alphabet Σ including the empty string ε. The length of a string w is denoted by $|w|$. A language is any subset of Σ^*. The cardinality of a finite set A is denoted by $|A|$ and its power-set by 2^A.

A *deterministic finite automaton* (DFA) is a 5-tuple $M = (Q, \Sigma, \delta, q_0, F)$, where Q is a finite set of states, Σ is a finite input alphabet, δ is the transition function that maps $Q \times \Sigma$ to Q, q_0 is the initial state, $q_0 \in Q$, and F is the set of accepting states, $F \subseteq Q$. In this paper, all DFAs are assumed to be complete, i.e., the next state $\delta(q, a)$ is defined for each state q in Q and each symbol a in Σ. The transition function δ is extended to a function from $Q \times \Sigma^*$ to Q in the natural way. A string w in Σ^* is accepted by the DFA M if the state $\delta(q_0, w)$ is an accepting state of the DFA M. The *language accepted by* the DFA M, denoted $L(M)$, is the set of strings $\{w \in \Sigma^* \mid \delta(q_0, w) \in F\}$.

A *nondeterministic finite automaton* (NFA) is a 5-tuple $M = (Q, \Sigma, \delta, q_0, F)$, where Q, Σ, q_0, and F are defined in the same way as for a DFA, and δ is the nondeterministic transition function that maps $Q \times \Sigma$ to 2^Q. The transition function can be naturally extended to the domain $Q \times \Sigma^*$. A string w in Σ^* is accepted by the NFA M if the set $\delta(q_0, w)$ contains an accepting state of the NFA M. The *language accepted by* the NFA M, denoted $L(M)$, is the set of strings $\{w \in \Sigma^* \mid \delta(q_0, w) \cap F \neq \emptyset\}$.

Two automata are said to be *equivalent* if they accept the same language. A DFA (an NFA) M is called *minimal* if all DFAs (all NFAs, respectively) that are equivalent to M have at least as many states as M. It is known that each regular language has a unique minimal DFA, up to isomorphism. However, the same result does not hold for minimal NFAs.

The *(deterministic) state complexity* of a regular language is the number of states in its minimal DFA. The *nondeterministic state complexity* of a regular language is defined as the number of states in a minimal NFA accepting this language. A regular language with nondeterministic state complexity n is called an n-state NFA language.

Every nondeterministic finite automaton $M = (Q, \Sigma, \delta, q_0, F)$ can be converted to an equivalent deterministic finite automaton $M' = (2^Q, \Sigma, \delta', q_0', F')$ using an algorithm known as the "subset construction" [20] in the following way. Every state of the DFA M' is a subset of the state set Q. The initial state of the DFA M' is the set $\{q_0\}$. The transition function δ' is defined by $\delta'(R, a) = \bigcup_{r \in R} \delta(r, a)$ for each state R in 2^Q and each symbol a in Σ. A state R in 2^Q is an accepting state of the DFA M' if it contains at least one accepting state of the NFA M.

To prove that a NFA is minimal we use a fooling-set lower-bound technique known from communication complexity theory [2,10]. Although the lower bounds obtained using fooling sets may sometimes be exponentially smaller than the size of minimal NFAs for the corresponding language [1,12], this technique has been successfully used in the field of regular languages several times [4,5,9,16]. We first define a fooling set. Then we give the lemma from [4] describing a fooling-set lower-bound technique. For the sake of completeness, we recall its proof here.

Definition 1. *A set of pairs of strings* $\{(x_i, y_i) \mid i = 1, 2, \ldots, n\}$ *is said to be a fooling set for a regular language* L *if for every* i *and* j *in* $\{1, 2, \ldots, n\}$,

(1) the string $x_i y_i$ *is in the language* L, *and*
(2) if $i \neq j$, *then at least one of the strings* $x_i y_j$ *and* $x_j y_i$ *is not in* L.

Lemma 1 (Birget [4]). *Let a set of pairs of strings* $\{(x_i, y_i) \mid i = 1, 2, \ldots, n\}$ *be a fooling set for a regular language* L. *Then every NFA for the language* L *needs at least* n *states.*

Proof. Let $M = (Q, \Sigma, \delta, q_0, F)$ be any NFA accepting the language L. Since the string $x_i y_i$ ($1 \leqslant i \leqslant n$) is in L, there is a state p_i in Q such that $p_i \in \delta(q_0, x_i)$ and $\delta(p_i, y_i) \cap F \neq \emptyset$, i.e., p_i is a state on an accepting computation of M on $x_i y_i$ that is reached after reading the string x_i. Assume that a fixed choice of p_i has been made for every i in $\{1, 2, \ldots, n\}$. We prove that $p_i \neq p_j$ if $i \neq j$. Suppose by contradiction that $p_i = p_j$ and $i \neq j$. Then the NFA M accepts both strings $x_i y_j$ and $x_j y_i$ which contradicts the assumption that the set $\{(x_i, y_i) \mid 1 \leq i \leq n\}$ is a fooling set for the language L. Hence the NFA M has at least n states. $\square$

3   Deterministic Blow-Ups of Minimal NFAs over a 4-Letter Alphabet

In this section, we present the main results of our paper. We start by describing two nondeterministic finite automata that we will use later in our constructions. We prove several properties concerning these two automata in the two lemmata below.

First, let us consider a k-state NFA $A_k = (Q_A, \{a, b\}, \delta_A, 1, F_A)$, where $Q_A = \{1, 2, \ldots, k\}$, $F_A = \{k\}$, and for each q in Q_A,

$$\delta_A(q, a) = \begin{cases} \{1, q+1\}, \text{ if } 1 \leqslant q \leqslant k-1, \\ \emptyset, \qquad\quad\text{ if } q = k, \end{cases}$$

$$\delta_A(q, b) = \begin{cases} \{q+1\}, & \text{ if } 1 \leqslant q \leqslant k-1, \\ \emptyset, & \text{ if } q = k. \end{cases}$$

The automaton A_k is depicted in Fig. 1. The next lemma shows that every subset of the state set Q_A is a reachable state in the deterministic finite automaton obtained from the NFA A_k by the subset construction.

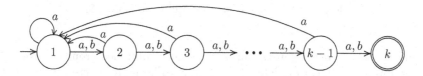

Fig. 1. The nondeterministic finite automaton A_k

Lemma 2. Let $A'_k = (2^{Q_A}, \{a, b\}, \delta'_A, \{1\}, F'_A)$ be the DFA obtained from the NFA A_k by the subset construction. Then every subset of the state set Q_A is reachable in the DFA A'_k.

Proof. The proof is by induction on the cardinality of subsets. The empty set and all the singletons are reachable because

$$\emptyset = \delta'_A(\{1\}, b^k) \text{ and } \{q\} = \delta'_A(\{1\}, b^{q-1}) \text{ for all } q = 1, 2, \ldots, k.$$

Now let $2 \leqslant t \leqslant k$ and assume by induction that every subset of Q_A of size $t-1$ is reachable. Let $\{q_1, q_2, \ldots, q_t\}$ be a subset of size t such that $1 \leqslant q_1 < q_2 < \cdots < q_t \leqslant k$. Then

$$\{q_1, q_2, \ldots, q_t\} = \delta'_A(\{q_2 - q_1, q_3 - q_1, \ldots, q_t - q_1\}, ab^{q_1 - 1}),$$

where the latter subset of size $t-1$ is reachable by induction. Thus the set $\{q_1, q_2, \ldots, q_t\}$ is reachable and our proof is complete.   □

Now, consider the following $(k+1)$-state NFA $B_k = (Q_B, \{a, b\}, \delta_B, 0, \{k\})$, where $Q_B = \{0, 1, 2, \ldots, k\}$, and for each q in Q_B,

$$\delta_B(q, a) = \begin{cases} \{0, 1\}, & \text{if } q = 0, \\ \{q + 1\}, & \text{if } 1 \leqslant q \leqslant k - 1, \\ \{1, 2, \ldots, k\}, & \text{if } q = k, \end{cases}$$

$$\delta_B(q, b) = \begin{cases} \{0\}, & \text{if } q = 0, \\ \{q + 1\}, & \text{if } 1 \leqslant q \leqslant k - 1, \\ \{1, 2, \ldots, k\}, & \text{if } q = k. \end{cases}$$

The automaton B_k is shown in Fig. 2. Note that if we would omit all the transitions defined in the final state k, then the resulting automaton would accept all strings containing a symbol a in the k-th position from the end.

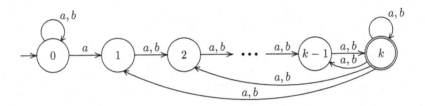

Fig. 2. The nondeterministic finite automaton B_k

Let B'_k be the deterministic finite automaton obtained from the NFA B_k by the subset construction. The DFA B'_4 (or, to be more precise, its reachable states, each of which contains the initial state 0 of the NFA B_4) is shown in Fig. 3. The automaton in the figure looks like a binary tree whose leaves go to state $\{0, 1, 2, 3, 4\}$ on a and b.

In the following, we will consider states $\{0, 2\}, \{0, 2, 3\}, \ldots, \{0, 2, 3, \ldots, r\}, \ldots, \{0, 2, 3, \ldots, k\}$ of the DFA B'_k. Notice that

$$\{0, 2\} \subset \{0, 2, 3\} \subset \ldots \subset \{0, 2, 3, \ldots, r\} \subset \ldots \subset \{0, 2, 3, \ldots, k\}$$

which is a property that will play a crucial role in the proof of our main result. Before stating the next lemma we introduce some notation. Let $2 \leqslant r \leqslant k$. Let

$$\mathcal{R}_{1,1} = \{R \subseteq Q_B \mid R = \delta'_B(\{0, 1\}, w) \text{ for some } w \text{ in } \{a, b\}^*\},$$

$$\mathcal{R}_{1,r} = \{R \subseteq Q_B \mid R = \delta'_B(\{0, 1, 2, 3, \ldots, r\}, w) \text{ for some } w \text{ in } \{a, b\}^*\},$$

$$\mathcal{R}_{2,r} = \{R \subseteq Q_B \mid R = \delta'_B(\{0, 2, 3, 4, \ldots, r\}, w) \text{ for some } w \text{ in } \{a, b\}^*\},$$

i.e., $\mathcal{R}_{1,1}$, $\mathcal{R}_{1,r}$, and $\mathcal{R}_{2,r}$ are the sets of states of the DFA B'_k that are reachable from states $\{0, 1\}, \{0, 1, 2, 3, \ldots, r\}$, and $\{0, 2, 3 \ldots, r\}$, respectively. For example, in our Figure 3, we have $\mathcal{R}_{1,3} = \{\{0, 1, 2, 3\}, \{0, 1, 2, 3, 4\}, \{0, 2, 3, 4\}\}$ and $\mathcal{R}_{2,3} = \{\{0, 2, 3\}, \{0, 1, 3, 4\}, \{0, 3, 4\}, \{0, 1, 2, 3, 4\}\}$.

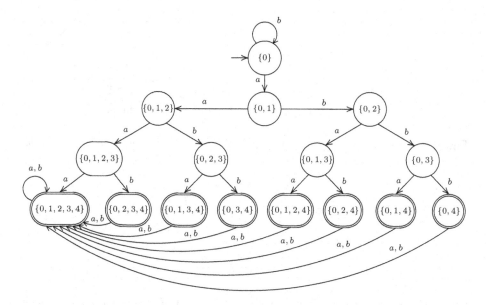

Fig. 3. The deterministic finite automaton B'_4

Lemma 3. *Let* $2 \leqslant r \leqslant s < t \leqslant k$ *and let* $\mathcal{R}_{1,1}$, $\mathcal{R}_{1,r}$, *and* $\mathcal{R}_{2,r}$ *be as above.
Then we have:*

(i) *State* $\{0, 1, 2, 3, \ldots, k\}$ *is a member of the sets* $\mathcal{R}_{1,1}$, $\mathcal{R}_{1,r}$, *and* $\mathcal{R}_{2,r}$.
(ii) *The size of the set* $\mathcal{R}_{1,1}$ *is* $2^k - 1$.
(iii) *The size of the set* $\mathcal{R}_{1,r}$ *and of the set* $\mathcal{R}_{2,r} \setminus \{\{0, 1, 2, 3, \ldots, k\}\}$ *is* $2^{k-r+1} - 1$.
(iv) *The sets* $\mathcal{R}_{1,r}$ *and* $\mathcal{R}_{2,s}$ *have only state* $\{0, 1, 2, 3, \ldots, k\}$ *in common.*
(v) *The sets* $\mathcal{R}_{2,s}$ *and* $\mathcal{R}_{2,t}$ *have only state* $\{0, 1, 2, 3, \ldots, k\}$ *in common.*

Proof. First, notice that every reachable state of the DFA B'_k contains the initial
state 0 of the NFA B_k since state 0 goes to itself on a and b in B_k.

To prove *(i)* note that state k of the NFA B_k is reachable from every state of
this NFA and the transitions on a and b from state k go to $\{1, 2, \ldots, k\}$.

To prove the rest of the lemma let us see how the sets $\mathcal{R}_{1,r}$ and $\mathcal{R}_{2,r}$ look like.
Figures 4 and 5 show what states are reachable from state $\{0, 1, 2, 3, \ldots, r\}$ and
from state $\{0, 2, 3, 4, \ldots, r\}$ after reading a string of length at most two.

It can be shown by induction on the length of strings that after reading a
string w of length i, where $0 \leqslant i \leqslant k - r$ and $1 \leqslant r \leqslant k$, we can reach state
$\{0\} \cup S \cup \{i+1\} \cup \{i+2, i+3, \ldots, i+r\}$ for each subset S of the set $\{1, 2, \ldots, i\}$ from
state $\{0, 1, 2, 3, \ldots, r\}$. Note that $\delta'_B(\{0, 1, 2, 3, \ldots, r\}, a^{k-r}) = \{0, 1, 2, 3, \ldots, k\}$
and that each set reachable from state $\{0, 1, 2, 3, \ldots, r\}$ by a string of length
$k - r$ contains state k which goes to $\{1, 2, 3, \ldots, k\}$ on a and b in the NFA B_k.
It follows that after reading a string of length more then $k - r$ we are always in
state $\{0, 1, 2, 3, \ldots, k\}$. Thus, the size of the set $\mathcal{R}_{1,r}$ is $1 + 2 + 4 + \cdots + 2^{k-r}$.

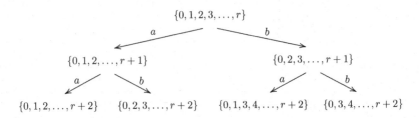

Fig. 4. The states reachable from state $\{0, 1, 2, 3, \ldots, r\}$ by strings $\varepsilon, a, b, aa, ab, ba, bb$

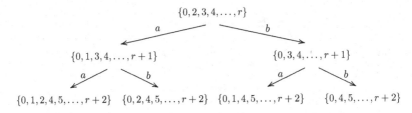

Fig. 5. The states reachable from state $\{0, 2, 3, 4, \ldots, r\}$ by strings $\varepsilon, a, b, aa, ab, ba, bb$

Similarly, we can reach every state $\{0\} \cup S \cup \{i+2, i+3, \ldots, i+r\}$, where S is a subset of $\{1, 2, \ldots, i\}$, from state $\{0, 2, 3, 4, \ldots, r\}$ after reading the strings of length i ($0 \leqslant i \leqslant k - r$ and $2 \leqslant r \leqslant k$). After reading a string of length more then $k - r$ we are again always in state $\{0, 1, 2, 3, \ldots, k\}$. Thus,

$$|\mathcal{R}_{2,r} \setminus \{\{0, 1, 2, 3, \ldots, k\}\}| = 1 + 2 + 4 + \cdots + 2^{k-r} = 2^{k-r+1} - 1,$$

and we have shown *(ii)* and *(iii)*.

Finally, to show *(iv)* and *(v)* let $2 \leqslant r \leqslant s < t \leqslant k$. Then each member of the set $\mathcal{R}_{2,s}$ has a "tail" of size $s - 1$, while each member of the set $\mathcal{R}_{2,t}$ has a "tail" of size $t - 1$, and each member of the set $\mathcal{R}_{1,r}$ has a "tail" of size at least r (here, by a "tail" of size s for a subset S of $\{0, 1, 2, 3, \ldots, k\}$ we mean a sequence $q_1, q_2, \ldots, q_s$ such that $q_i \in S, q_{i+1} = q_i + 1, q_1 - 1 \notin S$, and $S \cap \{q_s + 1, q_s + 2, \ldots, k\} = \emptyset$). This proves the last two items of the lemma and completes our proof. $\quad\square$

We are now ready to prove our main result showing that in the case of a four-letter alphabet, there are no "magic numbers", i.e., each value in the range from n to 2^n can be obtained as the deterministic state complexity of an n-state NFA language over a four-letter alphabet.

Theorem 1. *For all integers n and α such that $n \leqslant \alpha \leqslant 2^n$, there exists a minimal nondeterministic finite automaton of n states with a four-letter input alphabet whose equivalent minimal deterministic finite automaton has α states.*

Proof. Let n and α be arbitrary but fixed integers such that $n \leqslant \alpha \leqslant 2^n$. If $\alpha = n$, then we can consider a unary n-state NFA that counts the numbers of

a's modulo n. On the other hand, the n-state NFAs that need 2^n deterministic states are well-known [17, 19, 15].

If $n < \alpha < 2^n$, then we have $n - k + 2^k \leqslant \alpha < n - (k+1) + 2^{k+1}$ for some integer k such that $1 \leqslant k \leqslant n - 1$. It follows that $\alpha = n - (k+1) + 2^k + m$, where m is an integer such that $1 \leqslant m < 2^k$. For this integer m one of the following cases holds

$$m = (2^{k_1} - 1) + (2^{k_2} - 1) + \cdots + (2^{k_{\ell-1}} - 1) + (2^{k_\ell} - 1) \tag{1}$$
$$m = (2^{k_1} - 1) + (2^{k_2} - 1) + \cdots + (2^{k_{\ell-1}} - 1) + 2 \cdot (2^{k_\ell} - 1) \tag{2}$$

where $1 \leqslant \ell \leqslant k - 1$ and $k \geqslant k_1 > k_2 > \cdots > k_{\ell-1} > k_\ell \geqslant 1$, which can be seen from the following considerations. If $m = 2^k - 1$, then $m = 2^{k_\ell} - 1$, where $\ell = 1$ and $k_\ell = k$. If $m = 2^k - 2$, then $m = 2(2^{k_\ell} - 1)$, where $\ell = 1$ and $k_\ell = k - 1$. Otherwise, $m < 2^k - 2$, and let k_1 be the greatest integer such that $2^{k_1} - 1 \leqslant m$. Then $m = 2^{k_1} - 1 + m_1$, where $m_1 \leqslant 2^{k_1} - 1$. If $m_1 = 2^{k_1} - 1$, then $m = 2 \cdot (2^{k_1} - 1)$, otherwise $m = (2^{k_1} - 1) + (2^{k_2} - 1) + m_2$, where $k_1 > k_2$ and $m_2 \leqslant 2^{k_2} - 1$, and we can continue by induction.

Define an n-state NFA $C = C_{n,k,m} = (Q, \{a, b, c, d\}, \delta, q_0, \{k\})$, where $Q = \{0, 1, 2, \ldots, n - 1\}$, $q_0 = n - 1$ if $k < n - 1$ and $q_0 = 1$ if $k = n - 1$, and for each q in Q,

$$\delta(q, a) = \begin{cases} \{0, 1\}, & \text{if } q = 0, \\ \{1, q + 1\}, & \text{if } 1 \leqslant q \leqslant k - 1, \\ \{1, 2, \ldots, k\}, & \text{if } q = k, \\ \emptyset & \text{if } k + 1 \leqslant q \leqslant n - 1, \end{cases}$$

$$\delta(q, b) = \begin{cases} \{0\}, & \text{if } q = 0, \\ \{q + 1\}, & \text{if } 1 \leqslant q \leqslant k - 1, \\ \{1, 2, \ldots, k\}, & \text{if } q = k, \\ \{1\}, & \text{if } q = k + 1, \\ \{q - 1\}, & \text{if } k + 2 \leqslant q \leqslant n - 1, \end{cases}$$

$$\delta(q, c) = \begin{cases} \{q + 1\}, & \text{if } 0 \leqslant q \leqslant k - 1, \\ \emptyset, & \text{if } k \leqslant q \leqslant n - 1, \end{cases}$$

$$\delta(q, d) = \begin{cases} \{0, 2, 3, 4, \ldots, k - k_q + 1\}, & \text{if } 1 \leqslant q \leqslant \ell - 1, \\ \{0, 1, 2, 3, \ldots, k - k_\ell + 1\}, & \text{if } q = \ell \text{ and (1) holds for } m, \\ \{0, 2, 3, 4, \ldots, k - k_\ell + 1\}, & \text{if } q = \ell \text{ and (2) holds for } m, \\ \{0, 1, 2, 3, \ldots, k - k_\ell + 1\}, & \text{if } q = \ell + 1 \text{ and (2) holds for } m, \\ \emptyset, & \text{otherwise.} \end{cases}$$

The NFA $C_{n,k,m}$ for $m = 2^{k-1} - 1$ is shown in Fig. 6; notice that in this case transitions on d are defined only in state 1 and go to $\{0, 1, 2\}$. Fig. 7 shows transitions on b and d in the NFA $C_{n,k,m}$ for $m = (2^{k-1} - 1) + 2 \cdot (2^{k-2} - 1)$; here, transitions on d are defined in states 1, 2, and 3 and go to $\{0, 2\}$, $\{0, 2, 3\}$, and $\{0, 1, 2, 3\}$, respectively. Next, note that the transitions on a and b in states $1, 2, \ldots, k - 1$ are defined in the same way as for the automaton A_k, while the transitions on these two letters in states $0, 1, 2, \ldots, k$ are defined as for the automaton B_k except for transitions on a going to state 1 from states $1, 2, \ldots, k-1$. Transitions on d are defined in states $1, 2, \ldots, \ell$ in case (1) and in states $1, 2, \ldots, \ell, \ell+1$ in case (2), and go either to a set $\{0, 2, 3, 4, \ldots, r\}$ or to a set $\{0, 1, 2, 3, \ldots, r\}$. As we will see later, this assures that all subsets of the set $\{1, 2, \ldots, k\}$ and m subsets of the set $\{0, 1, 2, \ldots, k\}$ containing state 0 are reachable in the DFA C' obtained from the NFA C by the subset construction. Transitions on symbol c will be used to prove the inequivalence of these reachable subsets.

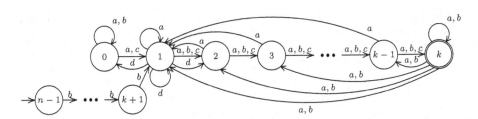

Fig. 6. The nondeterministic finite automaton $C_{n,k,m}$, where $m = 2^{k-1} - 1$

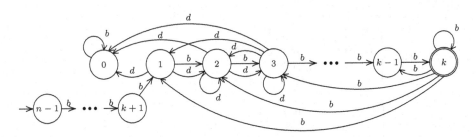

Fig. 7. Transitions on b and d in the NFA $C_{n,k,m}$, where $m = (2^{k-1}-1) + 2 \cdot (2^{k-2}-1)$

Now, we are going to prove that:

(i) The NFA C is minimal.
(ii) The DFA C' obtained from the NFA C by the subset construction has $n - (k+1) + 2^k + m$ reachable states.
(iii) The reachable states of the DFA C' are pairwise inequivalent.

Then, since $\alpha = n - (k+1) + 2^k + m$, the theorem follows immediately.

To prove (i) consider the following sets of pairs of strings

$$\mathcal{A} = \{(b^{i-1}, b^{n-k-i}c^{k-1}) \mid i = 1, 2, \ldots, n-k\},$$
$$\mathcal{B} = \{(b^{n-k-1}c^i, c^{k-1-i}) \mid i = 1, 2, \ldots, k-1\} \cup \{(b^{n-k-1}d, c^k)\}.$$

The set $\mathcal{A} \cup \mathcal{B}$ is a fooling set for the language $L(C)$ because

(1) the strings $b^{n-k-1}c^{k-1}$ and $b^{n-k-1}dc^k$ are in the language $L(C)$ and
(2) if $1 \leqslant i < j \leqslant n-k$, then the string $b^{i-1}b^{n-k-j}c^{k-1}$ is not in $L(C)$,

if $1 \leqslant i < j \leqslant k - 1$, then the string $b^{n-k-1}c^j c^{k-1-i}$ is not in $L(C)$,

if $1 \leqslant i \leqslant n - k$ and $1 \leqslant j \leqslant k - 1$, then the strings $b^{i-1}c^{k-1-j}$, $b^{i-1}c^k$, and $b^{n-k-1}c^j c^k$ are not in the language $L(C)$.

By Lemma 1, every NFA for the language $L(C)$ needs at least n states and so the NFA C is minimal.

To prove *(ii)* let $C' = (2^Q, \{a, b, c, d\}, \delta', q_0', F')$ be the DFA obtained from the NFA C by the subset construction. Consider the following systems $\mathcal{S}_1$ and $\mathcal{S}_2$ of sets of states of the NFA C (remind that the sets $\mathcal{R}_{i,r}$ were defined above Lemma 3):

$$\mathcal{S}_1 = \{\{n - 1\}, \{n - 2\}, \ldots, \{k + 1\}\} \cup 2^{\{1,2,\ldots,k\}} \cup$$
$$\cup \mathcal{R}_{2,k-k_1+1} \cup \mathcal{R}_{2,k-k_2+1} \cup \cdots \cup \mathcal{R}_{2,k-k_\ell-1+1} \cup \mathcal{R}_{1,k-k_\ell+1},$$

$$\mathcal{S}_2 = \{\{n - 1\}, \{n - 2\}, \ldots, \{k + 1\}\} \cup 2^{\{1,2,\ldots,k\}} \cup$$
$$\cup \mathcal{R}_{2,k-k_1+1} \cup \mathcal{R}_{2,k-k_2+1} \cup \cdots \cup \mathcal{R}_{2,k-k_\ell-1+1} \cup \mathcal{R}_{2,k-k_\ell+1} \cup \mathcal{R}_{1,k-k_\ell+1}.$$

We are going to show that $\mathcal{S}_1$ ($\mathcal{S}_2$) is the system of all reachable states of the DFA C' if (1) holds for m (if (2) holds for m, respectively). We prove the case (1); the second case is similar.

Let $m = (2^{k_1} - 1) + (2^{k_2} - 1) + \cdots + (2^{k_\ell-1} - 1) + (2^{k_\ell} - 1)$. We need to show that each set in $\mathcal{S}_1$ is a reachable state of the DFA C' and that no other subset of the state set Q is reachable in C'. The singletons $\{n - 1\}, \{n - 2\}, \ldots, \{k + 1\}$ are reachable since they can be reached from the initial state of C' by reading an appropriate numbers of b's. By Lemma 2, every nonempty subset of the set $\{1, 2, \ldots, k\}$ is reachable because state $\{1\}$ is reachable and the transitions on a and b in states $1, 2, \ldots, k - 1$ (that were used in the proof of Lemma 2) are the same as in the NFA A_k. The empty set is reachable since $\emptyset = \delta'(\{1\}, c^k)$. Next, we have

$$\delta'(q_0', b^{n-k-1+q-1}d) = \delta'(\{q\}, d) = \{0, 2, 3, \ldots, k - k_q + 1\} \text{ for } q = 1, 2, \ldots, \ell - 1$$

and

$$\delta'(q_0', b^{n-k-1+\ell-1}d) = \delta'(\{\ell\}, d) = \{0, 1, 2, 3, \ldots, k - k_\ell + 1\}.$$

The reachability of sets $\mathcal{R}_{2,k-k_1+1}, \mathcal{R}_{2,k-k_2+1}, \ldots, \mathcal{R}_{2,k-k_\ell-1+1}$, and $\mathcal{R}_{1,k-k_\ell+1}$ then follows from their definition and the fact that the transitions on a and b in states $0, 1, 2, \ldots, k$ are almost the same as in the NFA B_k (notice that the transitions on a to state 1 do not mind since the sets in $\mathcal{R}_{i,r}$ always contain state 0 and this state goes to state 1 on a in the NFA B_k). Thus, each set in the system $\mathcal{S}_1$ is a reachable state in the DFA C'. By Lemma 3, the size of the set $\mathcal{R}_{2,k-k_i+1} \setminus \{\{0, 1, 2, 3, \ldots, k\}\}$ is $2^{k_i} - 1$ ($1 \leqslant i \leqslant \ell - 1$), the size of the set $\mathcal{R}_{1,k-k_\ell+1}$ is $2^{k_\ell} - 1$, and these sets are pairwise disjoint accept for they all contain state $\{0, 1, 2, 3, \ldots, k\}$. It follows that

$$|\mathcal{S}_1| = n - k - 1 + 2^k + (2^{k_1} - 1) + (2^{k_2} - 1) + \cdots + (2^{k_\ell-1} - 1) + (2^{k_\ell} - 1)$$
$$= n - (k + 1) + 2^k + m = \alpha.$$

Hence the DFA C' has α reachable states. To see that no other subset of the state set Q is reachable note that for each set S in the system $\mathcal{S}_1$, the sets $\delta'(S, a), \delta'(S, b), \delta'(S, c)$, and $\delta'(S, d)$ are again in the system $\mathcal{S}_1$. This is quite straightforward for symbols a, b, c. In the case of symbol d, it is important to notice that

$$\delta(1, d) \subset \delta(2, d) \subset \ldots \subset \delta(\ell - 1, d) \subset \delta(\ell, d).$$

It follows that $\delta'(S, d)$ is either the empty set or is equal to $\delta(q, d)$ for the greatest integer q in $\{1, 2, \ldots, \ell\}$ which is in S, and so $\delta'(S, d)$ is in the system $\mathcal{S}_1$.

To prove *(iii)* let S and T be two reachable states of the DFA C'. If both S and T are some subsets of $\{0, 1, 2, \ldots, k\}$, then, w.l.o.g., there is a state q in $\{0, 1, 2, \ldots, k\}$ such that $q \in S \setminus T$. Then the string c^{k-q} distinguishes these two states. If S is a subset of $\{0, 1, 2, \ldots, k\}$ and T is one of the singletons $\{k+1\}, \{k+2\}, \ldots, \{n-1\}$, then a string from c^* distinguishes S and T in the case that $S \neq \emptyset$ and a string from $b^+ c^{k-1}$ otherwise. If $S = \{p\}$ and $T = \{q\}$, where $k + 1 \leqslant p < q \leqslant n - 1$, then the string $b^{p-k} c^{k-1}$ is accepted by the DFA C from state S but not from state T. This completes our proof.   □

4   Conclusions

In this paper, we have continued investigation of the relations between the sizes of minimal nondeterministic and deterministic finite automata. We have shown that for all integers n and α such that $1 \leqslant n \leqslant \alpha \leqslant 2^n$, there exists a minimal nondeterministic finite automaton of n states with a four-letter input alphabet whose equivalent minimal deterministic counterpart has α states. This improves the results of [7] that has been obtained using a growing alphabet of size $n + 2$. On the other hand, it has been recently shown [8] that in the case of a unary alphabet there are a lot of values in the range from n to $O(e^{\sqrt{n \ln n}})$ that cannot be reached as the deterministic state complexity of an n-state unary NFA language. It remains open whether the whole range of complexities from n to 2^n can be produced in the case of a binary or ternary alphabet.

References

1. Adorna, H.N.: 3-Party Message Complexity is Better than 2-Party Ones for Proving Lower Bounds on the Size of Minimal Nondeterministic Finite Automata. J. Autom. Lang. Comb. 7, 419–432 (2002)
2. Aho, A.V., Ullman, J.D., Yannakakis, M.: On notions of informations transfer in VLSI circuits. In: Proc.15th ACM STOC, pp. 133–139. ACM, New York (1983)
3. Berman, P., Lingas, A.: On the complexity of regular languages in terms of finite automata. Technical Report 304, Polish Academy of Sciences (1977)
4. Birget, J.C.: Intersection and union of regular languages and state complexity. Inform. Process. Lett. 43, 185–190 (1992)
5. Birget, J.C.: Partial orders on words, minimal elements of regular languages, and state complexity. Theoret. Comput. Sci. 119, 267–291 (1993)

6. Chrobak, M.: Finite automata and unary languages. Theoret. Comput. Sci. 47, 149–158 (1986)
7. Geffert, V. (Non)determinism and the size of one-way finite automata. In: Proc. 7th DCFS, Como, Italy, pp. 23–37 (2006)
8. Geffert, V.: Magic numbers in the state hierarchy of finite automata. In: Královič, R., Urzyczyn, P. (eds.) MFCS 2006. LNCS, vol. 4162, pp. 412–423. Springer, Heidelberg (2006)
9. Glaister, I., Shallit, J.: A lower bound technique for the size of nondeterministic finite automata. Inform. Process. Lett. 59, 75–77 (1996)
10. Hromkovič, J.: Communication Complexity and Parallel Computing. Springer-Verlag, Berlin, Heidelberg (1997)
11. Hromkovič, J.: Descriptional complexity of finite automata: concepts and open problems. J. Autom. Lang. Comb. 7, 519–531 (2002)
12. Hromkovič, J., Seibert, S., Karhumäki, J., Klauck, H., Schnitger, G.: Communication complexity method for measuring nondeterminism in finite automata. Inform. and Comput. 172, 202–217 (2002)
13. Iwama, K., Kambayashi, Y., Takaki, K.: Tight bounds on the number of states of DFAs that are equivalent to n-state NFAs. Theoret. Comput. Sci. 237, 485–494 (2000)
14. Iwama, K., Matsuura, A., Paterson, M.: A family of NFAs which need $2^n - \alpha$ deterministic states. Theoret. Comput. Sci. 301, 451–462 (2003)
15. Jirásková, G.: Note on minimal finite automata. In: Sgall, J., Pultr, A., Kolman, P. (eds.) MFCS 2001. LNCS, vol. 2136, pp. 421–431. Springer, Heidelberg (2001)
16. Jirásková, G.: State complexity of some operations on binary regular languages. Theoret. Comput. Sci. 330, 287–298 (2005)
17. Lupanov, O.B.: A comparison of two types of finite automata. Problemy Kibernetiki 9, 321–326 (in Russian) (1963)
18. Lyubich, Y.I.: Estimates for optimal determinization of nondeterministic autonomous automata. Sibirskii Matematichskii Zhurnal 5, 337–355 (in Russian) (1964)
19. Moore, F.R.: On the bounds for state-set size in the proofs of equivalence between deterministic, nondeterministic, and two-way finite automata. IEEE Trans. Comput. 20, 1211–1214 (1971)
20. Rabin, M., Scott, D.: Finite automata and their decision problems. IBM Res. Develop. 3, 114–129 (1959)
21. Sakoda, W.J., Sipser, M.: Nondeterminism and the size of two-way finite automata. In: Proc. 10th Annual ACM Symp. on Theory of Computing, pp. 275–286 (1978)
22. Sipser, M.: Introduction to the theory of computation. PWS Publishing Company, Boston (1997)
23. Yu, S.: Regular languages. In: Rozenberg, G., Salomaa, A. (eds.) Handbook of Formal Languages, vol. I, ch. 2, pp. 41–110. Springer-Verlag, Berlin Heidelberg (1997)
24. Yu, S.: A renaissance of automata theory? Bull. Eur. Assoc. Theor. Comput. Sci. EATCS 72, 270–272 (2000)

Reduced Languages as ω-Generators

Sandrine Julia[1] and Tran Vinh Duc[2]

[1] Université de Nice - Sophia Antipolis,
Laboratoire I3S - CNRS, B.P. 121,
06903 Sophia Antipolis Cedex, France
Sandrine.Julia@unice.fr
[2] Thang Long School of Technology,
Hanoï, Viet-Nam
tvduc@ifi.edu.vn

Abstract. We consider the following decision problem: "Is a rational ω-language generated by a code ?" Since 1994, the codes admit a characterization in terms of infinite words. We derive from this result the definition of a new class of languages, the reduced languages. A code is a reduced language but the converse does not hold. The idea is to "reduce" easy-to-obtain minimal ω-generators in order to obtain codes as ω-generators.

Introduction

Our research deals with the classical theory of automata and languages. We particularly focus on the rational languages of infinite words (ω-languages) which are recognized by Büchi or Muller automata [16]. A rational ω-language may be ω-generated by a language. The operation $^\omega$ stands for the infinite concatenation and maps a language L into an ω-language L^ω. L^ω is then called an ω-power and L is one of its ω-generators. One can decide if a rational ω-language admits an ω-generator. If so, a rational ω-generator exists [14]. Various decision problems arise from the set of ω-generators of a given rational ω-language.

Here is the open decision problem on which we focus: "Is a rational ω-language generated by a code ?" A language L is a code if and only if every non-empty word in L^* has a unique factorization over L [2]. The similar problem for the Kleene closure * instead of $^\omega$ has a simple solution. The monoid L^* is its own greatest generator and is generated by a code if and only if its root $L^* \backslash (L^* \backslash \{\varepsilon\})^2$ is a code. By analogy, we wonder when a rational ω-language L^ω is the ω-power of a code.

Unfortunately, the set of ω-generators of a rational ω-language does not admit one but a finite number of maximal ω-generators [14]. Even if the greatest ω-generator exists, the ω-power can be generated by a code without the root of its greatest ω-generator being a code. For instance, consider the ω-power L^ω with the root of its greatest ω-generator equals to $L = a + ab + ba$, which is not a code. Surprisingly, L^ω is ω-generated by the infinite code $C = a + (ab)^*ba$.

Our approach of the problem consists of the definition of a new class of ω-languages called the *reduced languages*. It is known that a code is minimal (with

T. Harju, J. Karhumäki, and A. Lepistö (Eds.): DLT 2007, LNCS 4588, pp. 266–277, 2007.
© Springer-Verlag Berlin Heidelberg 2007

respect to inclusion) in the set of ω-generators. Our new class of reduced languages is useful here because it contains the codes and is included in the set of minimal ω-generators. Usual approaches restrict the problem to subclasses of codes: prefix codes [13], ω-codes [9], codes with delay [5]. In addition, the problem is solved for prefix codes in [13]. Here, we decide to widen the problem by considering a notable superclass: the class of *reduced languages*.

The graph of a Büchi automaton reveals that it is possible to sligthly modify an ω-generator without changing the ω-power expressed. But in practice, we had no idea about how to modify a minimal ω-generator towards another which could be a code. Hopefully, to get a reduced language is possible, and may provide codes.

The paper is divided in five main sections. The two first ones set preliminary definitions and useful results, in particular, the characterization of codes by means of infinite words [6]. The third introduces the concept of reduced languages, followed by the study of the whole class and its decidability. Different cases are then detailed in the fourth section to convince that, despite of their great similitary, the reduced languages do not behave exactly as codes when taken as ω-generators. At last, the fifth section explores the ability of the construction of reduced ω-generators to reach codes.

1 Preliminaries

Let Σ be a finite alphabet. A *word* (resp. ω-*word*) is a finite (resp. infinite) concatenation of letters in Σ. We note ε the empty word. Σ^* is the set of words over Σ, $\Sigma^+ = \Sigma^* \setminus \{\varepsilon\}$. Σ^ω is the set of ω-words. Any subset of Σ^* is called a *language* and any subset of Σ^ω is called an ω-*language*.

A word u is a *prefix* of v if $v \in u(\Sigma^* \cup \Sigma^\omega)$ and we write: $u < v$. The induced order is the *prefix order*. For $v \in (\Sigma^* \cup \Sigma^\omega)$, *Pref*$(v)$ stands for the set of all prefixes of v. Hence, for every $L \subseteq (\Sigma^* \cup \Sigma^\omega)$, *Pref*$(L)$ is the set of the prefixes of the words in L.

Let $L \subseteq \Sigma^*$ be a language, the language L^* is the set of words built with words in L: $L^* = \{\varepsilon\} \cup \{a_1 \dots a_n \mid \forall i\ 1 \le i \le n,\ a_i \in L\}$. In the same way, the ω-*power* L^ω is the set of ω-words: $L^\omega = \{a_1 \dots a_n \dots \mid \forall i > 0,\ a_i \in L \setminus \{\varepsilon\}\}$. L^* (resp. L^ω) is generated (resp. ω-generated) by L, and so L is called a *generator* (resp. ω-*generator*). Henceforth, *minimality* or *maximality* are specifically used with respect to inclusion over the set of ω-generators.

Both of following languages are useful: $Prem(L) = (L \setminus \{\varepsilon\}) \setminus (L \setminus \{\varepsilon\})(L \setminus \{\varepsilon\})^+$ and, whenever M is a monoid, $Root(M) = Prem(M) = (M \setminus \{\varepsilon\}) \setminus (M \setminus \{\varepsilon\})^2$.

The *stabilizer* of L^ω is the language: $Stab(L^\omega) = \{u \in \Sigma^+ \mid uL^\omega \subseteq L^\omega\}$. $Stab(L^\omega)$ is a semigroup in which every ω-generator of L^ω is included. So, when $Stab(L^\omega)$ is an ω-generator of L^ω, it is the greatest [14]. The *characteristic language* of L^ω is the language: $\chi(L^\omega) = \{u \in \Sigma^+ \mid uL^\omega \subseteq L^\omega \text{ and } u^\omega \in L^\omega\}$. $\chi(L^\omega)$ is not anymore a semigroup. Every ω-generator of L^ω is still included in $\chi(L^\omega)$ and so, when $\chi(L^\omega)$ is an ω-generator of L^ω, it is also the greatest [14]. L^ω rational implies that $Stab(L^\omega)$ and $\chi(L^\omega)$ are also rational.

Let $L = L \setminus \{\varepsilon\}$ (abusively written $L = L \setminus \varepsilon$). A *L-factorization* of a word u in L^+ is a finite sequence of words in L: $(u_1, u_2, \ldots, u_n)$ such that $u = u_1 u_2 \ldots u_n$. A L-factorization of an ω-word w in L^ω is an infinite sequence: $(w_1, w_2, \ldots, w_n, \ldots)$ such that $w = w_1 w_2 \ldots w_n \ldots$. We will say indifferently L-factorization or factorization over L. A language L is a *code* (resp. an ω-*code*) if every word $u \in \Sigma^*$ (resp. every ω-word $w \in \Sigma^\omega$) has at most one L-factorization [2][17]. Any ω-code is *a fortiori* a code. Later, rational languages and ω-languages can be denoted by their regular (ω)-expressions.

Let L be a language, the *adherence* of L is the ω-language $Adh(L) = \{w \in \Sigma^\omega \mid Pref(w) \subseteq Pref(L)\}$ and an ω-language A is an adherence if $A = Adh(L)$ for some language L.

A language L is said to have a *bounded (deciphering) delay* if: $\exists d \geq 0 \; \forall u \in L$ $(uL^d \Sigma^\omega \cap L^\omega) \subseteq uL^\omega$.

A *finite automaton* $\mathcal{A}$ (resp. *Büchi automaton* $\mathcal{B}$) is specified by $(\Sigma, Q, \delta, I, T)$ where Σ denotes the finite alphabet, Q the finite set of states, $I \subseteq Q$ the set of initial states and $T \subseteq Q$ the set of recognition states. A *run* of a word m in $\mathcal{A}$ (resp. a *run* of an ω-word w in $\mathcal{B}$) is a finite sequence $l = (q_i)_{0 \leq i \leq n}$ (resp. an infinite sequence $l = (q_i)_{i \geq 0}$) of states in Q such that $q_0 \in I$ and $\forall i \; \delta(q_i, m_{i+1}) = q_{i+1}$, with m_i the ith letter of m (resp. w). A word m (resp. an ω-word w) belongs to the language recognized by $\mathcal{A}$ (resp. ω-language recognized by $\mathcal{B}$) if there exists a run $(q_i)_{0 \leq i \leq n}$ (resp. a run $(q_i)_{i \geq 0}$) such that $q_0 \in I$ and $q_n \in T$ (resp. $q_0 \in I$ and $Inf(w) \cap T \neq \varnothing$, where $Inf(w) = \{q \in Q / Card(\{i/q_i = q\})$ is infinite$\})$. We note $L(\mathcal{A})$ (resp. $L(\mathcal{B})$) the language (resp. ω-language) recognized by $\mathcal{A}$ (resp. $\mathcal{B}$). The set of recognized languages coincide with the set of rational languages. A rational ω-language is of the form: $L = \bigcup_{i=1}^{n} A_i B_i^\omega$ with $n \geq 1$ such that for every i, A_i and B_i are rational languages respectively included in Σ^* and Σ^+. Their class coincides with the class of ω-languages recognized by Büchi automata [18].

2   Useful Results

In this section, we present some preliminary results. The first one is very important for our purpose. It gives an elegant characterization of codes based on periodic infinite words.

Proposition 1. *[6] Let $L \subseteq \Sigma^+$. The language L is a code if and only if for every word $u \in L^+$, u^ω has a unique L-factorization.*

Below, we recall two results about adherence needed in the sequel to justify the hypothesis taken.

Proposition 2. *[3][11] Let L be a language. If L^ω is an adherence, then*

$$L^\omega = Adh(L^*) = Adh(Pref(L^*)).$$

Proposition 3. *[14] Let L be a language. If L^ω is an adherence then $\chi(L^\omega)$ and $Stab(L^\omega)$ coincide with the greatest ω-generator of L^ω.*

Example 1. Consider $L = a + ab + b^2$. L^ω is finitely ω-generated so it is an adherence and $\chi(L^\omega) = Stab(L^\omega) = L^+$ is the greatest ω-generator. Let $K = a^*b$. K^ω is not an adherence, $Stab(L^\omega) = \Sigma^+$ is not ω-generator of K^ω but $\chi(L^\omega) = \Sigma^* b \Sigma^*$ is the greatest ω-generator. Finally, let $M = \Sigma^*(aa + bb)$. There are two maximal ω-generators $M_1 = \Sigma^*(aa + bb)\Sigma^* + a(ba)^*$ and $M_2 = \Sigma^*(aa + bb)\Sigma^* + b(ab)^*$. Neither $Stab(L^\omega) = \Sigma^+$ nor $\chi(L^\omega) = M_1 \cup M_2$ are ω-generators of M^ω.

The following result is about languages with a bounded delay. Usually, this deciphering property is linked to codes, not here.

Proposition 4. *[7] Let L be a language with a bounded delay such that L^+ is the greatest ω-generator of L^ω. If L^ω is an adherence, then every ω-generator code of L^ω is necessarily a finite ω-code.*

We point out here the result called *lemma of infinite iteration* frequently used to prove the equality between two ω-powers.

Lemma 1. *[14] Let L and $R \subseteq \Sigma^+$ be two rational languages, $L^\omega \subseteq RL^\omega \Rightarrow L^\omega \subseteq R^\omega$.*

The language $L \setminus LStab(L^\omega)$ is still an ω-generator of L^ω and will be useful to finally simplify ω-generators.

Proposition 5. *[12] Let L be a language, the following properties hold:*

(i) $L \setminus LStab(L^\omega) \subseteq Prem(L)$.
(ii) $L \setminus LStab(L^\omega)$ and $Prem(L)$ are ω-generators of L^ω.

At last, let us recall now a classic result on words.

Lemma 2. *[15] Two words $u, v \in \Sigma^+$ commute, i.e. $uv = vu$, if and only if there exists a word $z \in \Sigma^+$ and two different integers i and $j \geq 1$ such that $u = z^i$ and $v = z^j$.*

3   Reduced Languages

In this section, we present a new class of languages based on a property particularly relevant when refering to ω-generators. This class lies between the class of codes and the class of minimal ω-generators. We call it the class of *reduced languages*.

3.1   Presentation

In the sequel, we present the definition of reduced langages which involves periodic ω-words. Then, we state a characterization of them in order to locate reduced ω-generators among minimal ω-generators.

Definition 1. *A language* $R \subseteq \Sigma^+$ *is called reduced if:*

$$\forall u \in R \quad u^\omega \notin (R \setminus u)^\omega$$

Proposition 6. *Every reduced ω-generator is a minimal ω-generator.*

Proof. Let L be a language. If L is not minimal, then there exists a word $u \in L$ such that $u^\omega \in L^\omega = (L \setminus u)^\omega$. Hence L is not reduced. $\qquad \square$

The converse does not hold. For instance, the language $L = a + ab + ba$ is minimal but is not reduced. Clearly, $(ab)^\omega \in (L \setminus ab)^\omega$.

Proposition 7. *A language L is reduced if and only if for each word $u \in L$, the periodic ω-word u^ω has a unique L-factorization.*

Proof. The second condition clearly implies L reduced. Conversely, assume there exists $u \in L$ such that u^ω has two L-factorizations with different first steps: $(u, u, \ldots)$ and $(v_0, v_1, \ldots)$. Two cases arise:

- either, for each integer $i \geq 0$, $v_i \neq u$, hence $u^\omega \in (L \setminus u)^\omega$.
- either there exists a smallest integer $k > 0$ verifying $v_k = u$.

Hence, there exist two words $\alpha, \beta \in \Sigma^*$ and $n > 0$ such that:

- $v_0 \ldots v_{k-1}\alpha = u^n$
- $v_0 \ldots v_{k-1}u = u^n\beta$
- $v_0 \ldots v_{k-1}u\alpha = u^n\beta\alpha = u^{n+1}$

then, $u = \alpha\beta = \beta\alpha$. According to Lemma 2, there exists z verifying $\alpha = z^i$ and $\beta = z^j$. We obtain $u^\omega = z^\omega = (v_0 \ldots v_{k-1})^\omega$ and so, $u^\omega \in (L \setminus u)^\omega$.

In both cases, a contradiction appears with L reduced. $\qquad \square$

Table 1 gives the maximal number of factorizations of different kinds of ω-words, like in [6]. The asterisk $*$ attests that the column property characterizes the corresponding class of languages. Consequently:

Proposition 8. *A code is a reduced language.*

The converse does not hold. For instance, the language $L = a + ab + bc + c$ is a reduced language but is not a code since the ω-word $(abc)^\omega$ has two distinct L-factorizations. We summarize below the relations between the different classes of ω-generators we consider:

Code ω-generator $\Rightarrow$ reduced ω-generator $\Rightarrow$ minimal ω-generator.

Table 1. Maximal number of factorizations over ω-codes, codes, reduced languages

Language L	u^ω $(u \in L)$	u^ω $(u \in L^+)$	any
ω-code	1	1	1$*$
code	1	1$*$	∞
reduced language	1$*$	∞	∞

3.2 Decidability

The aim of this part is to ensure that the property of reduced language is decidable over the set of rational languages. Four preliminary lemmas are needed before stating the main result.

Let $L \subseteq \Sigma^+$ a language. We use the set $Amb(L)$ introduced in [10]. We restrict $Amb(L)$ to ω-words, so the set $Amb(L)$ contains ω-words in L^ω with several L-factorizations with different first steps.

$$Amb(L) = \{w \in L^\omega \mid \exists (w_i)_{i \in \mathbb{N}}, (w_j')_{j \in \mathbb{N}}$$
$$\text{two } L\text{-factorizations of } w \text{ with } w_0 \neq w_0'\}$$

Lemma 3. *[6][9] If a language $L \subseteq \Sigma^+$ is rational, then the set $Amb(L)$ is rational too.*

Proof. If L is rational, the congruence defined as $u \simeq v \Leftrightarrow u^{-1}L = v^{-1}L$ has a finite index. Let us write $\langle u \rangle$ the equivalence class of the word u. The set $Amb(L)$ is obtained as: $Amb(L) = \bigcup_{\langle u \rangle \subseteq L} \langle u \rangle \, (L^\omega \cap (u^{-1}L \setminus \varepsilon)L^\omega)$. $\qquad\square$

The following lemma is a consequence of Definition 1:

Lemma 4. *Let $L \subseteq \Sigma^+$ be a language, L is reduced if and only if L verifies:*

$$\forall u \in L \qquad u^\omega \notin Amb(L)$$

We present here some notation concerning the Büchi congruence [4] in order to prove our result. Let $\mathcal{A} = (\Sigma, Q, I, \delta, T)$ a complete Büchi automaton. For each state $q \in Q$, and for every word $u \in \Sigma^*$, we write:

$$\delta_T(q, u) = \{q' \in Q \mid \text{ exists } t \in T \text{ and } u_1, u_2 \in \Sigma^*$$
$$\text{with } u = u_1 u_2 \text{ and } t \in \delta(q, u_1) \text{ and } q' \in \delta(t, u_2)\}$$

The Büchi congruence $\approx$ is defined by:

$$u \approx v \quad \Leftrightarrow \quad \forall q \in Q, \begin{cases} \delta(q, u) = \delta(q, v) \\ \delta_T(q, u) = \delta_T(q, v) \end{cases}$$

for every $u, v \in \Sigma^+$. Let $[u] = \{w \in \Sigma^+ \mid w \approx u\}$ be the equivalence class of u. As $\approx$ has a calculable finite index, we obtain:

Lemma 5. *[4] For each $u \in \Sigma^+$, its equivalence class $[u]$ is a constructible rational language.*

Lemma 6. *If $v_1 \approx v_2$, then $v_1{}^\omega \in L_\omega(\mathcal{A}) \Leftrightarrow v_2{}^\omega \in L_\omega(\mathcal{A})$.*

It is time to state the main result.

Theorem 1. *One can decide whether a rational language is a reduced language.*

Proof. Let L be a rational language. It is effective to:

- construct the automaton $\mathcal{A}$ which recognizes the set $Amb(L)$ (according to Lemma 3);
- compute the equivalence classes $[u_1], \ldots, [u_k]$ (according to Lemma 5);
- verify if there exists $[u_i]$ such that $[u_i] \cap L \neq \emptyset$ and $u_i{}^\omega \in Amb(L)$.

If so, L is not reduced, otherwise, L is reduced (according to lemmas 4 and 6).

$\qquad\square$

4   Reduced Languages as ω-Generators

The class of reduced languages comes from considerations on the set of ω-generators. This set contains or not a reduced language. If so, this set contains or not a code. Both subsections illustrate the main two cases, the first revealing incidently that a rational ω-power is not necessarily ω-generated by a code.

4.1   No Reduced ω-Generator

We show that there exists an ω-power that cannot be generated by a reduced language. Consequently, this implies that a rational ω-power is not necessarily generated by a code. Previously, this result has been proved in [19] and clearly reinforces the interest in the decision problem we study.

Proposition 9. *Some rational ω-powers do not admit reduced ω-generators.*

Proof. Consider $L = a^2 + a^3 + ba + b$. Notice that $L^\omega = \Sigma^\omega \setminus ab\Sigma^\omega$ and that $L = \chi(L^\omega)$ is the greatest ω-generator of L^ω. Assume that there exists a reduced ω-generator R of L^ω. Let us prove the following two facts:

Fact 1. *Let $w \in L^\omega$. If $w \in a\Sigma^\omega$ then $aw \in L^\omega$.*

Proof (Fact 1). Clearly, $w \in (a^2L^\omega \cup a^3L^\omega)$. If $w \in a^2L^\omega$ then $aw \in a^3L^\omega \subseteq L^\omega$. If $w \in a^3L^\omega$ then $aw \in (a^2)^2L^\omega \subseteq L^\omega$.   □

Fact 2. *For all $k \geq 1$ and $u \in \Sigma^*$, we obtain $\{a^ku, a^kua\} \not\subseteq R$.*

Proof (Fact 2). If $\{a^ku, a^kua\}$ is included in R, using Fact 1, we get:

$$(a^kua)^\omega = (a^ku)\underbrace{(a^{k+1}u)^\omega}_{\in R^\omega}$$

Consequently, $(a^kua)^\omega$ has two R-factorizations: $(\alpha_i)_{i\geq 0}$ and $(\beta_j)_{j\geq 0}$ with $\alpha_0 = a^kua$ and $\beta_0 = a^ku$. Hence, R is not reduced.   □

- as $a^\omega \in L^\omega$, there exists a unique $i_0 > 1$ such that $a^{i_0} \in R$ (according to Fact 2 with $u = \varepsilon$).
- as $a^{i_0}aba^\omega \in L^\omega$ and $aba^\omega \notin L^\omega$, there exists a unique integer $i_1 \geq 0$ such that $a^{i_0}aba^{i_1} \in R$.
- as $a^{i_0}aba^{i_1}aba^\omega \in L^\omega$ and $a^{i_0}aba^{i_1}a \notin R$ (according to Fact 2), then, there exists a unique integer $i_2 \geq 0$ such that $a^{i_0}aba^{i_1}aba^{i_2} \in R$.
- and so forth, we define a unique infinite sequence $(i_j)_{j\geq 0}$.

Now, let us consider the following ω-word: $w = a^{i_0}aba^{i_1}ab\ldots aba^{i_n}\ldots$. This word w belongs to L^ω but lacks a factorization over R. We deduce that R is not an ω-generator of L^ω. We conclude that L^ω does not have any reduced ω-generator.   □

Corollary 1. *Some rational ω-powers do not admit codes as ω-generators.*

4.2 Reduced vs Code ω-Generator

In this section, we show that an ω-power generated by a reduced language is not necessarily generated by a code.

Proposition 10. *A rational language ω-generated by a reduced language is not necessarily ω-generated by a code.*

Proof. $L = a+ab+bc+c$ is a language with a bounded delay 1, studied in [1]. L^ω is an adherence since L is finite and then $\chi(L^\omega) = L^+$ is its greatest ω-generator. It is clear that L is reduced. Let us show that L^ω cannot be generated by a code. Assume firstly that C is an ω-code ω-generator of L^ω. As $(ua^i)a^\omega = (u)\underbrace{a^\omega}_{\in L^\omega}$ and

$(ub)c^\omega = (u)\underbrace{bc^\omega}_{\in L^\omega}$, we obtain the property P: $\{ua^i, u\} \not\subseteq C$ and $\{ub, u\} \not\subseteq C$, for

every $u \in \Sigma^+$ and $i > 0$. We intend to construct an infinite sequence of elements from C:

- as $a^\omega \in L^\omega$, there exists a unique integer $i_0 > 0$ (according to P) such that $a^{i_0} \in C$;
- $i_0 > 0$, then $a^{i_0}ba^\omega \in L^\omega$, but $a^{i_0}b \notin C$ (according to P), and there exists a unique integer $i_1 > 0$ such that $a^{i_0}ba^{i_1} \in C$;
- and so on; we define a unique infinite sequence $(i_j)_{j\geq0}$.

The cardinality of C is necessarily infinite, so there is no finite ω-code ω-generating L^ω. According to Prop. 4, there is no code C ω-generating L^ω.   □

5 Reducing ω-Generator

For the moment, the interest of the new class of reduced languages is not proven. However, some minimal ω-generators which are not codes are prevented from being codes *essentially* because there are not reduced. So, we present a method in order to make ω-generators reduced without affecting their ω-power.

5.1 Reduction

The reduction mixes two ideas: the first is a transformation required to aim at the uniqueness of factorizations of specific periodic ω-words, according to the characterization of reduced languages (Prop. 7). The second one is a simplification to guarantee the minimality of reduced ω-generators (Prop. 6).

Let us call $A(L)$ the language of words in L which prevents L from being reduced.

$$A(L) = \{u \in L \mid u^\omega \in (L \setminus u)^\omega\}$$

A step of reduction consists in the elimination of an element from $A(L)$, eventually compensated by the apparition of other elements. A first way to do this is described below:

Proposition 11. *Let $L \subseteq \Sigma^+$ be a rational language. For every $u \in A(L)$, both languages $G = u^*(L \setminus u)$ and especially $\Gamma = G \setminus GStab(L^\omega)$ are ω-generators of L^ω.*

Proof. Let $G = u^*(L \setminus u)$. As $u \in L$, we get that $G \subseteq L^+$ and then $G^\omega \subseteq L^\omega$. Conversely, let $w \in L^\omega$. There are two cases:

- either $w = u^\omega$ and then $w \in (L \setminus u)^\omega \subseteq (u^*(L \setminus u))^\omega$.
- either there exists $n \geq 0$ such that $w = xyw'$ where $x = u^n$, $y \in (L \setminus u)$ and $w' \in L^\omega$. Hence, $w \in (u^*(L \setminus u))L^\omega$. From Lemma 1, $w \in (u^*(L \setminus u))^\omega$.

The equality $L^\omega = G^\omega$ is proved. $L^\omega = \Gamma^\omega$ follows from Prop. 5.   □

Example 2. Let $L = a + ab + ba$. As $(a, ba, ba, \ldots)$ is a L-factorization of the word $(ab)^\omega$, we know that $A(L) = ab$. According to Prop. 11, the languages $G = (ab)^*(a + ba)$ and $\Gamma = a + (ab)^*ba$ are ω-generators of L^ω. Here, the latter language is an ω-code. It is necessarily a code and a reduced language too.

To increase the possibility to find a code when reducing a language, we have to treat separately the case where $A(L)$ contains two words sharing the same primitive root. So, here is a second way to remove an element from $A(L)$.

Proposition 12. *Let $L \subseteq \Sigma^+$ be a rational language. If $A(L)$ contains two non-empty words u and v such that u and v commute, then both languages $G = u + v^*(L \setminus \{u, v\})$ and especially $\Gamma = G \setminus GStab(L^\omega)$ are ω-generators of L^ω.*

Proof. From Lemma 2, $uv = vu$ implies that there exist two different integers $i \geq 1$ and $j \geq 1$ and a word $z \in \Sigma^+$ such that $u = z^i$ and $v = z^j$. As $\{z^i, z^j\} \subseteq A(L)$, according to Prop. 11, $G' = (z^j)^*(L \setminus z^j)$ is an ω-generator of L^ω. Moreover, $(z^j)^+z^i = z^i(z^j)^+ \subseteq G'Stab(L^\omega)$. Then, we obtain: $G' \setminus G'Stab(L^\omega) \subseteq G = z^i + (z^j)^*(L \setminus \{z^i, z^j\}) \subseteq G'$ and we deduce from Prop. 5 that G is an ω-generator of L^ω. From Prop. 5 again, $\Gamma = G \setminus GStab(L^\omega)$ is an ω-generator of L^ω.   □

Example 3. Let $L = a^2 + a^3 + b$. $A(L) = a^2 + a^3$ and we choose to remove a^3. We deduce from Prop. 12 that $G = a^2 + (a^3)^*b$ and $\Gamma = a^2 + a^3b + b$ are ω-generators of L^ω. So Γ is an ω-code. The other choice would have lead to the ω-code $\Gamma' = a^3 + a^2b + a^4b + b$.

Obviously, the ω-generators computed by a step of reduction are not necessarily reduced. Perhaps the problem has just been moved. We study in the next section the use of the reduction, its range, and of course, its limit.

5.2   Experimentation

This section explains how to use the reduction in order to find reduced ω-generators, possibly codes. We limit ourselves to rational ω-powers which are adherences. Indeed, from Prop. 2 and Prop. 3, such ω-powers verify:

$$L^\omega = (\chi(L^\omega))^\omega = Adh(\chi(L^\omega))$$

Moreover, every finitely ω-generated language is an adherence [11]. From now on, L will denote the root of $\chi(L^\omega)$ which is ever characteristic of L^ω [8] and is, in addition here, the greatest ω-generator. The reduction principle consists in applying recursively either Proposition 11 or Proposition 12 to L and the languages obtained, while it is possible.

As an illustration, we make here a digression towards automata. Let $\mathcal{A}$ be the minimal (deterministic) automaton which recognizes L^*, we note $L^* = L(\mathcal{A})$. Let $\mathcal{B}$ be the same automaton in its Büchi version, L^ω is recognized by $\mathcal{B}$ and we note $L^\omega = L_\omega(\mathcal{B})$. So, we intend to apply the reduction to L which is already minimal whenever it is not reduced. How does a reduction step operate on a deterministic Büchi automaton ? It suppresses a recognition state from one cycle. To do this, it induces a dilation of others as shown in Figure 1.

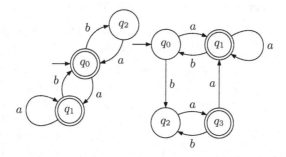

Fig. 1. Two automata for $L^* = (a + ab + ba)^*$ and $C^* = (a + (ab)^*ba)^*$ coupled with two Büchi automata for $L^\omega = C^\omega = (a + ab + ba)^\omega$

Let us come back to the implementation of reduction. Thus, three different cases arise when we apply a step of reduction:

(i) the process halts and gives an ω-generator code;
(ii) the process halts on a reduced ω-generator which is not a code;
(iii) the process does not halt.

In the sequel, we discuss the three cases from examples. The last two cases explore the actual limit of the reduction principle.

Case (i). In this case, the reduction provides a code, as illustrated in the following example. Note that the examples from Section 5.1 would have be convenient here. However, we give another example to show that, sometimes, it is also possible to get a code which is not ω-code.

Example 4. Let $L = ab + aba + baba$. The set $A(L) = ab$ since, in particular, $(aba, baba, baba, \ldots)$ is a $(L \setminus ab)$-factorization of the word $(ab)^\omega$. According to Prop. 11, the language $G = (ab)^*(aba + baba)$ is an ω-generator of L^ω. Hence, $\Gamma = aba + ababa + (ab)^*baba$ is a reduced ω-generator of L^ω. It is a code but not an ω-code. It was already known that L^ω has no ω-code as ω-generator [1].

Case (ii). This time, the process of reduction provides a reduced ω-generator which is not a code.

Example 5. Let $L = a + ab + bab$. We have $A(L) = ab$. According to Prop. 11, $\Gamma = a + bab + aba + ab^2ab$ is an ω-generator of L^ω. It is reduced but it is still not a code. However, there is no way to continue the reduction because Γ is reduced: one can easily verify that $A(\Gamma) = \emptyset$. In addition, L is a language with a bounded delay 1 and it was already known that L^ω cannot be ω-generated by an ω-code [1]. According to Prop. 4, L^ω has no code among its ω-generators.

Every time the process halts on a reduced ω-generator which is not a code, we succeed in finding a proof more or less *ad hoc* that the concerned ω-power is not ω-generated by a code. But examples are quickly difficult to handle and we do not know more. Nevertheless, it is not excluded that a generalization would be possible. We have to investigate for instance deciphering delays.

Case (iii). How do we interpret the third case ? Our process does not halt. It clearly contains the case where L^ω has no reduced ω-generator, nor code.

Example 6. Let $L = a^2 + a^3 + ba + b$. We get $A(L) = a^2 + a^3$. According to Prop. 12, $\Gamma = a^2 + a^3ba + a^3b + ba + b$ and $\Gamma' = a^3 + a^2ba + a^2b + a^4ba + a^4b + ba + b$ are both ω-generators of L^ω. Neither Γ nor Γ' are reduced since $A(\Gamma) = a^3ba + a^3b$ and $A(\Gamma') = a^2ba + a^4ba + a^4b$. So neither Γ nor Γ' are codes. We can continue the process for a while ... But it necessarily continues without halting (nor looping) because we proved there is no reduced ω-generator for such an ω-language (in the proof of Prop. 9).

Is the condition sufficient ? We have no counterexample, nor proof. The only certitude is that the process cannot halt if there is no reduced languages among the ω-generators.

To decide if a rational adherence admits an ω-generator code, it is not sufficient to test whether the root of the greatest ω-generator is a code. The technique of reduction can help but obscure areas remain. Finally, several significative examples are recapitulated in Table 2 like in [19]. It is not worth exhibiting complicated examples to illustrate the complexity of the problem.

Table 2. Examples in brief

$L =$ $Root(\chi(L^\omega))$	L^ω has an ω-generator ...		
	reduced	code	ω-code
$a + ab$	$a + ab$		
$a^2 + a^3 + b$	$a^2 + a^3b + b$		
$a + ab + ba$	$a + (ab)^*ba$		
$a + ab + b^2$	$a + ab + b^2$		no
$ab + aba + baba$	$aba + ababa + (ab)^*baba$		no
$a + ab + bc + c$	$a + ab + bc + c$	no	
$a + ab + bab$	$a + bab + aba + ab^2ab$	no	
$a^2 + a^3 + ba + b$	no		

6 Conclusion

The research of ω-generators codes lead us to define the new class of reduced languages, strongly connected with periodic ω-words. Particularly, we have explained its remarkable position between code and minimal ω-generators though this area is not so large. In the rational case, the definition of reduced languages allows an algorithmic approach to search and sometimes find ω-generators codes. Up to what point does our method produce a code whenever it exists ? An intensive work of experimentation is needed to understand where is exactly the limit of such a method.

References

1. Augros, X.: Etude des générateurs de langages de mots infinis. Master's thesis, Univ. de Nice - Sophia Antipolis (June 1997)
2. Berstel, J., Perrin, D.: Theory of codes. Academic Press, London (1985)
3. Boasson, L., Nivat, M.: Adherences of languages. Journal of Computer and System Sciences 20, 285–309 (1980)
4. Büchi, J.R.: On a decision method in restricted second order arithmetics. In: International Congress on Logic, Methodology and Philosophy of Science, pp. 1–11. Stanford University Press (1960)
5. Devolder, J.: Generators with bounded deciphering delay for rational ω-languages. Journal of Automata, Languages and Combinatorics 4(3), 183–204 (1999)
6. Devolder, J., Latteux, M., Litovsky, I., Staiger, L.: Codes and infinite words. Acta. Cybernetica 11(4), 241–256 (1994)
7. Julia, S.: Sur les codes et les ω-codes générateurs de langages de mots infinis. PhD thesis, Université de Nice - Sophia Antipolis (1996)
8. Julia, S.: A characteristic language for rational ω-powers. In: Proc. 3rd Int. Conf. Developments in Language Theory, pp. 299–308, Thessaloniki (1997)
9. Julia, S., Litovsky, I., Patrou, B.: On codes, ω-codes and ω-generators. Information Processing Letters 60(1), 1–5 (1996)
10. Karhumäki, J.: On three-element codes. Theoret. Comput. Sc. 40, 3–11 (1985)
11. Latteux, M., Timmerman, E.: Finitely generated ω-langages. Information Processing Letters 23, 171–175 (1986)
12. Litovsky, I.: Free submonoids and minimal ω-generators of R^ω. Acta. Cybernetica 10(1-2), 35–43 (1991)
13. Litovsky, I.: Prefix-free languages as ω-generators. Information Processing Letters 37, 61–65 (1991)
14. Litovsky, I., Timmerman, E.: On generators of rational ω-power languages. Theoretical Computer Science 53, 187–200 (1987)
15. Lothaire, M.: Algebraic Combinatorics on Words. Cambridge (2002)
16. Perrin, D., Pin, J.E.: Infinite words. Elsevier Academic Press, Amsterdam (2004)
17. Staiger, L.: On infinitary finite length codes. Theoretical Informatics and Applications 20(4), 483–494 (1986)
18. Thomas, W.: Automata on infinite objects. In: Handbook of Theoretical Computer Science, vol. B, ch. 4, Elsevier Science Publishers, Amsterdam (1990)
19. Tran Vinh Duc. A la recherche des codes générateurs de langages de mots infinis. Master's thesis, IFI Hanoï - Univ. de Nice - Sophia Antipolis (2006)

Avoiding Approximate Squares

Dalia Krieger[2], Pascal Ochem[1], Narad Rampersad[2], and Jeffrey Shallit[2]

[1] LaBRI — Université Bordeaux 1
351, cours de la Libération
33405 Talence Cedex, France
ochem@labri.fr
[2] David R. Cheriton School of Computer Science
University of Waterloo
Waterloo, ON, N2L 3G1 Canada
{d2kriege,nrampersad}@cs.uwaterloo.ca
shallit@graceland.uwaterloo.ca

Abstract. As is well-known, Axel Thue constructed an infinite word over a 3-letter alphabet that contains no squares, that is, no nonempty subwords of the form xx. In this paper we consider a variation on this problem, where we try to avoid *approximate squares*, that is, subwords of the form xx' where $|x| = |x'|$ and x and x' are "nearly" identical.

1 Introduction

A hundred years ago, Norwegian mathematician Axel Thue initiated the study of combinatorics on words [14,15,2]. One of his achievements was the construction of an infinite word over a three-letter alphabet that contains no squares, that is, no nonempty subwords of the form xx.

Many variations on this problem have been considered. For example, Brandenburg [3] and Dejean [6] considered the problem of avoiding *fractional powers*. A word w is an α-*power* if it can be written in the form $w = x^n y$, where y is a prefix of x and $\alpha = |w|/|x|$. A word z *contains an* α-*power* if some subword is a β-power, for $\beta \geq \alpha$; otherwise it *avoids* α-*powers*. Similarly, a word z *contains an* α^+-*power* if some subword is a β-power for $\beta > \alpha$; otherwise it *avoids* α^+-*powers*. We say α-powers (resp. α^+-powers) are *avoidable* over a k-letter alphabet if there exists an infinite word over that alphabet avoiding α-powers (resp., α^+-powers).

Dejean [6] improved Thue's result by showing how to avoid $(7/4)^+$-powers over a 3-letter alphabet; this result is optimal, as every ternary word of length ≥ 39 contains a $7/4$-power. Pansiot [13] showed how to avoid $(7/5)^+$-powers over a 4-letter alphabet. Again, this is optimal, as every quaternary word of length ≥ 122 contains a $7/5$-power.

Dejean also proved that for $k \geq 5$, one cannot avoid $k/(k-1)$-powers over a k-letter alphabet. She conjectured that it was possible to avoid $(k/(k-1))^+$-powers over a k-letter alphabet. This conjecture was proved for $5 \leq k \leq 11$ by Moulin-Ollagnier [11], for $7 \leq k \leq 14$ by Mohammad-Noori and Currie [10], and for $k \geq 38$ by Carpi [4]. The cases $15 \leq k \leq 37$ remain open.

T. Harju, J. Karhumäki, and A. Lepistö (Eds.): DLT 2007, LNCS 4588, pp. 278–289, 2007.
© Springer-Verlag Berlin Heidelberg 2007

Another variation is to avoid not all α-powers, but only sufficiently long ones. Entringer, Jackson, and Schatz [7] showed how to construct a word over a 2-letter alphabet that avoids squares xx with $|x| \geq 3$; here the number 3 is best possible.

In this paper we consider yet another variation, but one that seems natural: we consider avoiding *approximate* squares, that is, subwords of the form xx' where x' is "almost the same" as x. The precise definitions are given below. One of our main results is a further strengthening of Dejean's improvement on Thue for 3 letters.

Approximate squares (also known as approximate *tandem repeats* in the biological literature) have been studied before, but from the algorithmic point of view. Landau and Schmidt [8] and Kolpakov and Kucherov [9] both gave efficient algorithms for finding approximate squares in a string.

Notation: we use Σ_k to denote the alphabet of k letters $\{0, 1, 2, \ldots, k-1\}$.

2   Approximate Squares

There are at least two natural notions of approximate square. We define them below.

For words x, x' of the same length, define the *Hamming distance* $d(x, x')$ as the number of positions in which x and x' differ. For example, $d(01203, 11002) = 3$. We say that a word xx' with $|x| = |x'|$ is a *c-approximate square* if $d(x, x') \leq c$. Using this terminology, for example, a 0-approximate square is a square, and a 1-approximate square is either a square or differs from a square in exactly one position.

To avoid c-approximate squares, we would like to enforce the condition $d(x, x') > c$ for all x, x' of the same length, but clearly this is impossible if $|x| \leq c$. To avoid this technicality, we say a word z *avoids c-approximate squares* if for all its subwords xx' where $|x| = |x'|$ we have $d(x, x') \geq \min(c+1, |x|)$.

This definition is an "additive" version; there is also a "multiplicative" version. Given two words x, x' of the same length, we define their *similarity* $s(x, x')$ as the fraction of the number of positions in which x and x' agree. Formally,

$$s(x, x') := \frac{|x| - d(x, x')}{|x|}.$$

Thus for example, $s(123456, 101406) = 1/2$. The *similarity* of a finite word z is defined to be $\alpha = \max_{\substack{xx' \text{ a subword of } z \\ |x|=|x'|}} s(x, x')$; we say such a word is α-similar. Thus, a 1-similar finite word contains a square.

For infinite words, the situation is slightly more subtle. We say an infinite word $\mathbf{z}$ is α-*similar* if $\alpha = \sup_{\substack{xx' \text{ a subword of } \mathbf{z} \\ |x|=|x'|}} s(x, x')$ and there exists at least one subword xx' with $|x| = |x'|$ and $s(x, x') = \alpha$. Otherwise, if $\alpha = \sup_{\substack{xx' \text{ a subword of } \mathbf{z} \\ |x|=|x'|}} s(x, x')$, but α is not attained by any subword xx' of $\mathbf{z}$, then we say $\mathbf{z}$ is α^--*similar*.

As an example, consider the infinite word over Σ_3, $\mathbf{c} = 21020121012021020120\cdots$, defined to be the length of contiguous blocks of 1's between consecutive 0's in the Thue-Morse sequence $\mathbf{t}$. As is well-known, $\mathbf{c}$ is square-free, so it cannot be 1-similar. However, since $\mathbf{t}$ contains arbitrarily large squares, it follows that $\mathbf{c}$ must contain arbitrarily large 1-approximate squares, and so $\mathbf{c}$ is 1^--similar.

Another definition of approximate square was given in [5], but we do not consider it here.

3   Words of Low Similarity

The main problem of interest is, given an alphabet Σ of size k, what is the smallest similarity possible over all infinite words over Σ? We call this the *similarity coefficient* of k.

Answering this question has two aspects. We can explicitly construct an infinite word that is α-similar (or α^--similar). To show that α is best possible, we can construct a tree of all finite words that are β-similar for $\beta < \alpha$. The root of this tree is labeled 0 (which suffices by symmetry), and if a node is labeled w, its children are labeled wa for all $a \in \Sigma$. If a node is β-similar for some $\beta \geq \alpha$, it becomes a leaf and no children are added. We can then use depth-first or breadth-first search to explore this tree. The number of leaves of this tree equals the (finite) number of words beginning with 0 that are β-similar for $\beta < \alpha$, and the height h of the tree is the length of the longest words with this property. The number of leaves at depth h equals the number of maximal words beginning with 0 that are β similar for some $\beta < \alpha$.

We performed this computation for various alphabet sizes k, and the results are reported below in Table 1. For $k = 8$, our method took advantage of some symmetries to speed up the computation, and as a result, we did not compute the number of leaves or maximal strings. For the reported values, these computations represent a proof that the similarity coefficient is at least as large as the α reported.

For alphabet size $k = 2$, every infinite word is 1-similar. We now report on larger alphabet sizes.

Table 1. Similarity bounds

Alpha bet Size k	Similarity Coefficient α	Height of Tree	Number of Leaves	Number of Maximal Words	Lexicographically First Maximal Words
2	1	3	4	1	010
3	3/4	41	2475	36	010201202101201021201210201202101201020101
4	1/2	9	382	6	012310213
5	2/5	75	3902869	48	01230431034204134012031421041234201240341023042031234032102432041034214024 3
6	1/3	17	342356	480	01234150325143012
7	?	?	?	?	?
8	1/4	71	—	—	01234056731460251647301275634076213574102364075120435674103271564073142

Theorem 1. *There exists an infinite 3/4-similar word* **w** *over* $\{0, 1, 2\}$.

Proof. Let h be the 24-uniform morphism defined by

$$0 \rightarrow 012021201021012102120210$$
$$1 \rightarrow 120102012102120210201021$$
$$2 \rightarrow 201210120210201021012102.$$

The following lemma may be verified computationally.

Lemma 1. *Let* $a, b, c \in \{0, 1, 2\}$, $a \neq b$. *Let* w *be any subword of length 24 of* $h(ab)$. *If* w *is neither a prefix nor a suffix of* $h(ab)$, *then* $h(c)$ *and* w *mismatch in at least 9 positions.*

Let $\mathbf{w} = h^\omega(0)$. We shall show that **w** has the desired property. We argue by contradiction. Suppose that **w** contains a subword yy' with $|y| = |y'|$ such that y and y' match in more than $3/4 \cdot |y|$ positions. Let us suppose further that $|y|$ is minimal.

We may verify computationally that **w** contains no such subword yy' where $|y| \leq 72$. We therefore assume from now on that $|y| > 72$.

Let $w = a_1 a_2 \cdots a_n$ be a word of minimal length such that $h(w) = xyy'z$ for some $x, z \in \{0, 1, 2\}^*$. By the minimality of w, we have $0 \leq |x|, |z| < 24$.

For $i = 1, 2, \ldots, n$, define $A_i = h(a_i)$. Then since $h(w) = xyy'z$, we can write

$$h(w) = A_1 A_2 \cdots A_n = A_1' A_1'' A_2 \cdots A_{j-1} A_j' A_j'' A_{j+1} \cdots A_{n-1} A_n' A_n''$$

where

$$A_1 = A_1' A_1''$$
$$A_j = A_j' A_j''$$
$$A_n = A_n' A_n''$$
$$x = A_1'$$
$$y = A_1'' A_2 \cdots A_{j-1} A_j'$$
$$y' = A_j'' A_{j+1} \cdots A_{n-1} A_n'$$
$$z = A_n'',$$

and $|A_1''|, |A_j''| > 0$. See Figure 1.

If $|A_1''| > |A_j''|$, then, writing y and y' atop one another, as illustrated in Figure 2, one observes that for $t = j + 1, j + 2, \ldots, n - 1$, each A_t "lines up" with a subword, say B_t, of $A_{t-j}A_{t-j+1}$. We now apply Lemma 1 to conclude that each A_t mismatches with B_t in at least 9 of 24 positions. Consequently, y and y' mismatch in at least $9(j-2)$ positions. Since $j \geq |y|/24 + 1$, we have that $9(j-2) \geq 9(|y|/24 - 1)$. However, $9(|y|/24 - 1) > |y|/4$ for $|y| > 72$, so that y and y' mismatch in more than $1/4 \cdot |y|$ positions, contrary to our assumption.

If $|A_1''| < |A_j''|$, as illustrated in Figure 3, then a similar argument shows that y and y' mismatch in more than $1/4 \cdot |y|$ positions, contrary to our assumption.

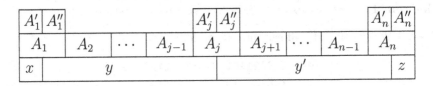

Fig. 1. The string $xyy'z$ within $h(w)$

$$y = \boxed{\begin{array}{c} A_1'' \end{array} \begin{array}{c} A_2 \end{array}} \cdots \boxed{\begin{array}{c} A_{j-1} \end{array} \begin{array}{c} A_j' \end{array}}$$

$$y' = \boxed{A_j'' \;\; A_{j+1} \;\; A_{j+2}'} \cdots \boxed{A_{n-1} \;\; A_n'}$$

Fig. 2. The case $|A_1''| > |A_j''|$

$$y = \boxed{A_1'' \;\; A_2 \;\; A_3'} \cdots \boxed{A_{j-1} \;\; A_j'}$$

$$y' = \boxed{A_j'' \;\; A_{j+1}} \cdots \boxed{A_{n-1} \;\; A_n'}$$

Fig. 3. The case $|A_1''| < |A_j''|$

Therefore $|A_1''| = |A_j''|$. We first observe that any pair of words taken from $\{h(0), h(1), h(2)\}$ mismatch at every position. We now consider several cases.

Case 1: $A_1 = A_j = A_n$. Then letting $u = A_1 A_2 \cdots A_{j-1}$ and $u' = A_j A_{j+1} \cdots A_{n-1}$, we see that u and u' match in exactly the same number of positions as y and y'.

Case 2: $A_1 = A_j \neq A_n$. Then letting $u = A_1 A_2 \cdots A_{j-1}$ and $u' = A_j A_{j+1} \cdots A_{n-1}$, we see that u and u' match in at least as many positions as y and y'.

Case 3: $A_1 \neq A_j = A_n$. Then letting $u = A_2 A_3 \cdots A_j$ and $u' = A_{j+1} A_{j+2} \cdots A_n$, we see that u and u' match in at least as many positions as y and y'.

Case 4: $A_1 = A_n \neq A_j$. Then letting $u = A_1 A_2 \cdots A_{j-1}$ and $u' = A_j A_{j+1} \cdots A_{n-1}$, we see that u and u' match in exactly the same number of positions as y and y'.

Case 5: A_1, A_j, and A_n are all distinct. Then letting $u = A_1 A_2 \cdots A_{j-1}$ and $u' = A_j A_{j+1} \cdots A_{n-1}$, we see that u and u' match in exactly the same number of positions as y and y'.

We finish the argument by considering the word uu'. First observe that either

$$uu' = h(a_1 a_2 \cdots a_{j-1}) h(a_j a_{j+1} \cdots a_{n-1})$$

or

$$uu' = h(a_2 a_3 \cdots a_j) h(a_{j+1} a_{j+2} \cdots a_n).$$

Without loss of generality, let us assume that the first case holds.

Recall our previous observation that the words $h(0)$, $h(1)$, and $h(2)$ have distinct letters at every position. Suppose then that there is a mismatch between u and u' occurring within blocks A_t and A_{t+j} for some t, $1 \leq t \leq j$. Then A_t and A_{t+j} mismatch at every position. Moreover, we have $a_j \neq a_{j+t}$. Conversely, if A_t and A_{t+j} match at any single position, then they match at every position, and we have $a_t = a_{t+j}$.

Let $v = a_1 a_2 \cdots a_{j-1}$ and $v' = a_j a_{j+1} \cdots a_{n-1}$. Let m be the number of matches between u and u'. From our previous observations we deduce that the number of matches m' between v and v' is $m/24$, but since $|v| = |u|/24$, $m'/|v| = m/|u|$. Thus, if $m/|u| > 3/4$, as we have assumed, then $m'/|v| > 3/4$. But the set $\{h(0), h(1), h(2)\}$ is a code, so that vv' is the unique pre-image of uu'. The word vv' is thus a subword of $\mathbf{w}$, contradicting the assumed minimality of yy'. We conclude that no such yy' occurs in $\mathbf{w}$, and this completes the argument that $\mathbf{w}$ is $3/4$-similar. $\qquad\Box$

Next, we consider the case of alphabet size $k = 4$.

Theorem 2. *There exists an infinite $1/2$-similar word $\mathbf{x}$ over $\{0, 1, 2, 3\}$.*

Proof. Let g be the 36-uniform morphism defined by

$$0 \rightarrow 012132303202321020123021203020121310$$
$$1 \rightarrow 123203010313032131230132310131232021$$
$$2 \rightarrow 230310121020103202301203021202303132$$
$$3 \rightarrow 301021232131210313012310132313010203.$$

Then $\mathbf{x} = g^\omega(0)$ has the desired property. The proof is entirely analogous to that of Theorem 1 and is omitted. $\qquad\Box$

In our last result of this section, we show that we can obtain infinite words of arbitrarily low similarity, provided the alphabet size is sufficiently large. The main tool is the following [1, Lemma 5.1.1]:

Lemma 2 (Lovász Local Lemma; asymmetric version). *Let I be a finite set, and let $\{A_i\}_{i \in I}$ be events in a probability space. Let E be a set of pairs $(i, j) \in I \times I$ such that A_i is mutually independent of all the events $\{A_j : (i, j) \notin E\}$. Suppose there exist real numbers $\{x_i\}_{i \in I}$, $0 \leq x_i < 1$, such that for all $i \in I$,*

$$\mathrm{Prob}(A_i) \leq x_i \prod_{(i,j) \in E} (1 - x_j).$$

Then

$$\mathrm{Prob}\left(\bigcap_{i \in I} \overline{A_i}\right) \geq \prod_{i \in I}(1 - x_i) > 0.$$

We now state our result.

Theorem 3. *Let $c > 1$ be an integer. There exists an infinite $1/c$-similar word.*

Proof. Let k and N be positive integers, and let $w = w_1 w_2 \cdots w_N$ be a random word of length N over a k-letter alphabet Σ. Here each letter of w is chosen uniformly and independently at random from Σ.

Let

$$I = \{(t, r) : 0 \le t < N, 1 \le r \le \lfloor (N - t)/2 \rfloor\}.$$

For $i = (t, r) \in I$, write $y = w_t \cdots w_{t+r-1}$ and $y' = w_{t+r} \cdots w_{t+2r-1}$. Let A_i denote the event $s(y, y') > 1/c$. We can derive a crude overestimate of $\mathrm{Prob}(A_i)$ by first choosing a fraction $\ge \frac{1}{c}$ of the positions in the first half that will match the second half, and letting the remaining positions be chosen arbitrarily. We get

$$\mathrm{Prob}(A_i) \le \frac{\binom{r}{\lceil r/c \rceil + 1} k^{\lceil r/c \rceil + 1} k^{2r - 2(\lceil r/c \rceil + 1)}}{k^{2r}}$$

$$\le \binom{r}{\lceil r/2 \rceil} k^{-r/c} \le 2^r k^{-r/c},$$

where the last inequality comes from Stirling's approximation.

For all positive integers r, define $\xi_r = 2^{-2r}$. For any real number $\alpha \le 1/2$, we have $(1 - \alpha) \ge e^{-2\alpha}$. Hence, $(1 - \xi_r) \ge e^{-2\xi_r}$. For $i = (t, r) \in I$, define $x_i = \xi_r$. Let E be as in the local lemma. Note that a subword of length $2r$ of w overlaps with at most $2r + 2s - 1$ subwords of length $2s$. Thus, for all $i = (t, r) \in I$, we have

$$x_i \prod_{(i,j) \in E} (1 - x_j) \ge \xi_r \prod_{s=1}^{\lfloor N/2 \rfloor} (1 - \xi_s)^{2r+2s-1} \ge \xi_r \prod_{s=1}^{\infty} (1 - \xi_s)^{2r+2s-1}$$

$$\ge \xi_r \prod_{s=1}^{\infty} e^{-2\xi_s(2r+2s-1)} \ge 2^{-2r} \prod_{s=1}^{\infty} e^{-2(2^{-2s})(2r+2s-1)}$$

$$\ge 2^{-2r} \exp\left[-2\left(2r \sum_{s=1}^{\infty} \frac{1}{2^{2s}} + \sum_{s=1}^{\infty} \frac{2s-1}{2^{2s}} \right) \right]$$

$$\ge 2^{-2r} \exp\left[-2\left(2r\left(\frac{1}{3}\right) + \frac{5}{9} \right) \right]$$

$$\ge 2^{-2r} \exp\left(-\frac{4}{3}r - \frac{10}{9} \right).$$

The hypotheses of the local lemma are met if $2^r k^{-r/c} \le 2^{-2r} \exp\left(-\frac{4}{3}r - \frac{10}{9}\right)$. Taking logarithms, we require $r \log 2 - \frac{r}{c} \log k \le -2r \log 2 - \frac{4}{3}r - \frac{10}{9}$. Rearranging terms, we require $c\left(3 \log 2 + \frac{4}{3} + \frac{10}{9r}\right) \le \log k$. The left side of this inequality is largest when $r = 1$, so we define $d_1 = 3 \log 2 + \frac{4}{3} + \frac{10}{9}$, and insist that $c \cdot d_1 \le \log k$. Hence, for $k \ge e^{c \cdot d_1}$, we may apply the local lemma to conclude that with positive probability, w is $1/c$-similar. Since $N = |w|$ is arbitrary, we conclude that there are arbitrarily large such w. By König's Infinity Lemma, there exists an infinite $1/c$-similar word, as required. $\qquad\square$

4   Words Avoiding c-Approximate Squares

In this section we consider the "additive" version of the problem. Table 2 reflects our results using a backtracking algorithm: there is no infinite word over a k-letter alphabet that avoids c-approximate squares, for the k and c given below.

Table 2. Lower bounds on avoiding c-approximate squares

Alphabet Size k	c	Height of Tree	Number of Leaves	Number of Maximal Words	Lexicographically First Maximal Words
2	0	4	3	1	010
3	1	5	23	2	01201
4	2	7	184	6	0123012
5	2	11	3253	24	01234102314
6	3	11	35756	960	01234051230
7	4	13	573019	6480	0123450612340
8	5	15	-	-	012345607123450

Theorem 4. *There is an infinite word over a 3-letter alphabet that avoids 0-approximate squares, and the 0 is best possible.*

Proof. Any ternary word avoiding squares, such as the fixed point, starting with 2, of $2 \to 210$, $1 \to 20$, $0 \to 1$, satisfies the conditions of the theorem. The result is best possible, from Table 2.   □

Theorem 5. *There is an infinite word over a 4-letter alphabet that avoids 1-approximate squares, and the 1 is best possible.*

Proof. Let **c** be any squarefree word over $\{0, 1, 2\}$, and consider the image under the 48-uniform morphism γ defined by

$$0 \to 012031023120321031201321032013021320123013203123$$
$$1 \to 012031023120321023103213021032013210312013203123$$
$$2 \to 012031023012310213023103210231203210312013203123$$

The resulting word $\mathbf{d} = \gamma(\mathbf{c})$ avoids 1-approximate squares. The result is best possible, from Table 2.

The proof is similar to that of Theorem 1. Suppose to the contrary that **d** contains a 1-approximate square yy', $|y| = |y'|$. We may verify computationally that **d** contains no such subword yy' where $|y| \leq 96$. We therefore assume from now on that $|y| > 96$.

Let $w = a_1 a_2 \cdots a_n$ be a word of minimal length such that $\gamma(w) = xyy'z$ for some $x, z \in \{0, 1, 2, 3\}^*$. By the minimality of w, we have $0 \leq |x|, |z| < 48$.

For $i = 1, 2, \ldots, n$, define $A_i = \gamma(a_i)$. Just as in the proof of Theorem 1, we write

$$\gamma(w) = A_1 A_2 \cdots A_n = A_1' A_1'' A_2 \cdots A_{j-1} A_j' A_j'' A_{j+1} \cdots A_{n-1} A_n' A_n'',$$

so that the situation illustrated in Figure 1 applies to $xyy'z$ within $\gamma(w)$. We now make the following observations regarding the morphism γ:

1. Let $a, b, c \in \{0, 1, 2\}$, $a \neq b$. Let u be any subword of length 48 of $\gamma(ab)$. If u is neither a prefix nor a suffix of $\gamma(ab)$, then $\gamma(c)$ and u mismatch in at least 18 positions.
2. Let $a, b \in \{0, 1, 2\}$, $a \neq b$. Then $\gamma(a)$ and $\gamma(b)$ mismatch in at least 18 positions.
3. Let u, u', v, v' be words satisfying the following:
   - $|u| = |u'|$, $|v| = |v'|$, and $|uv| = |u'v'| = 48$;
   - each of u and u' is a suffix of a word in $\{\gamma(0), \gamma(1), \gamma(2)\}$; and
   - each of v and v' is a prefix of a word in $\{\gamma(0), \gamma(1), \gamma(2)\}$.
   Then either $uv = u'v'$ or uv and $u'v'$ mismatch in at least 18 positions.
4. Let $a \in \{0, 1, 2\}$. Then $\gamma(a)$ is uniquely determined by either its prefix of length 17 or its suffix of length 17.

From the first observation, we deduce, as in the proof of Theorem 1, that the cases illustrated by Figures 2 and 3 cannot occur. In particular, we have that $|A_1''| = |A_j''|$ and $|A_j'| = |A_n'|$.

From the second observation, we deduce that for $i = 2, 3, \ldots, j - 1$, $A_i = A_{i+j-1}$, and consequently, $a_i = a_{i+j-1}$.

From the third observation, we deduce that $A_1'' = A_j''$ and $A_j' = A_n'$.

From the fourth observation, we deduce that either $A_1 = A_j$ or $A_j = A_n$. If $A_1 = A_j$, then $a_1 = a_j$; if $A_j = A_n$, then $a_j = a_n$. In the first case, $a_1 a_2 \cdots a_{j-1} a_j a_{j+1} \cdots a_{n-1}$ is a square in $\mathbf{c}$, contrary to our assumption. In the second case, $a_2 a_3 \cdots a_j a_{j+1} a_{j+2} \cdots a_n$ is a square in $\mathbf{c}$, contrary to our assumption.

We conclude that $\mathbf{d}$ contains no 1-approximate square yy', as required. $\square$

Theorem 6. *There is an infinite word over a 6-letter alphabet that avoids 2-approximate squares, and the 2 is best possible.*

Proof. Let $\mathbf{c}$ be any squarefree word over $\{0, 1, 2\}$, and consider the image under the 6-uniform morphism β defined by

$$0 \to 012345; \quad 1 \to 012453; \quad 2 \to 012534.$$

The resulting word avoids 2-approximate squares. The result is best possible, from Table 2.

The proof is similar to that of Theorem 5, so we only note the properties of the morphism β needed to derive the result:

1. Let $a, b, c \in \{0, 1, 2\}$, $a \neq b$. Let u be any subword of length 6 of $\beta(ab)$. If u is neither a prefix nor a suffix of $\beta(ab)$, then $\beta(c)$ and u mismatch in at least 3 positions.
2. Let $a, b \in \{0, 1, 2\}$, $a \neq b$. Then $\beta(a)$ and $\beta(b)$ mismatch in at least 3 positions.

3. Let u, u', v, v' be words satisfying the following:
   - $|u| = |u'|$, $|v| = |v'|$, and $|uv| = |u'v'| = 6$;
   - each of u and u' is a suffix of a word in $\{\beta(0), \beta(1), \beta(2)\}$; and
   - each of v and v' is a prefix of a word in $\{\beta(0), \beta(1), \beta(2)\}$.
   Then either $uv = u'v'$ or uv and $u'v'$ mismatch in at least 3 positions.
4. Let $a \in \{0, 1, 2\}$. Then $\beta(a)$ is uniquely determined by either its prefix of
   length 4 or its suffix of length 1.   □

Further results on additive similarity are summarized in the next theorem.

Theorem 7. *For each k, n, d given in Table 3, there is an infinite word over a k-letter alphabet that avoids n-approximate squares, and in each case such an infinite word can be generated by applying the given d-uniform morphism to any infinite squarefree word over $\{0, 1, 2\}$. (Note that we have used the coding $A = 10$, $B = 11$, etc.)*

In each case, the proof is similar to the ones given previously, and is omitted.

Theorem 8. *For all integers $n \geq 3$, there is an infinite word over an alphabet of $2n$ letters that avoids $(n-1)$-approximate squares.*

Table 3. Morphisms for additive similarity

k	n	d	Morphism
7	3	14	0 → 01234056132465
			1 → 01234065214356
			2 → 01234510624356
8	4	16	0 → 0123456071326547
			1 → 0123456072154367
			2 → 0123456710324765
9	5	36	0 → 012345607821345062718345670281346578
			1 → 012345607182346750812347685102346578
			2 → 012345607182346510872345681702346578
11	6	20	0 → 012345670A812954768A
			1 → 0123456709A1843576A9
			2 → 01234567089A24365798
12	7	24	0 → 012345678091AB2354687A9B
			1 → 012345678091A3B4257689AB
			2 → 012345678091A2B3465798AB
13	8	26	0 → 01234567890A1BC24635798BAC
			1 → 01234567890A1B3C4257689ABC
			2 → 01234567890A1B2C354687A9BC
14	9	28	0 → 0123456789A0B1DC32465798BDAC
			1 → 0123456789A0B1DC243576A98DBC
			2 → 0123456789A0B1CD325468A79CBD
15	10	30	0 → 0123456789AB0D1CE3246579B8ACDE
			1 → 0123456789AB0D1CE2435768A9DCBE
			2 → 0123456789AB0CED32154687BA9DEC

Proof. Consider the $2n$-uniform morphism $h : \Sigma_3^* \to \Sigma_{2n}^*$ defined as follows:

$$0 \to 012 \cdots (n-1)n \cdots (2n-1)$$
$$1 \to 012 \cdots (n-1)(n+1)(n+2) \cdots (2n-1)n$$
$$2 \to 012 \cdots (n-1)(n+2)(n+3) \cdots (2n-1)n(n+1)$$

We claim that if $\mathbf{w}$ is any squarefree word over Σ_3, then $h(\mathbf{w})$ has the desired properties. The proof is a simple generalization of Theorem 6.   □

5   Another Variation

Yet another variation we can study is trying to avoid xx' where x is very similar to x', but only for sufficiently large x. Let us say that a finite word is (r, α)-similar if $\alpha = \sup_{\substack{xx' \text{ a subword of } z \\ |x|=|x'| \geq r}} s(x, x')$, and analogous definitions for infinite z.

Exercise 5.8.1 of Alon and Spencer [1] asks the reader to show, using the Lovász local lemma that, (in our language) for every $\epsilon > 0$, there exists an infinite binary word $\mathbf{z}$ and an integer c such that $\mathbf{z}$ is (c, α)-similar for some $\alpha \leq \frac{1}{2} + \epsilon$. The following result shows $\frac{1}{2}$ is best possible.

Theorem 9. *There is no infinite binary word $\mathbf{z}$ and integer c such that every subword xx' with $|x| = |x'| \geq c$ satisfies $s(x, x') < \frac{1}{2}$.*

Proof. Suppose such a $\mathbf{z}$ and c exist. Consider a subword of $\mathbf{z}$ of the form $xx'yy'$, with $|x| = |x'| = |y| = |y'| = c$. By our assumption, $s(x, x') < \frac{1}{2}$ and $s(x', y) < \frac{1}{2}$; hence, since $\mathbf{z}$ is defined over a binary alphabet, necessarily $s(x, y) > \frac{1}{2}$. Similarly, we must have $s(x', y') > \frac{1}{2}$. But then by definition of s, $2c \cdot s(xx', yy') = c \cdot s(x, y) + c \cdot s(x', y') > \frac{c}{2} + \frac{c}{2} = c$, and so $s(xx', yy') > \frac{1}{2}$, a contradiction to our assumption.   □

6   Edit Distance

There are many definitions of *edit distance*, but for our purposes, we say the edit distance $e(x, y) = c$ if x can be transformed into y by a sequence of c insertions, deletions, or replacements, and no sequence of $c - 1$ insertions, deletions, or replacements suffices.

We can expand our notion of approximate square to avoid all words that are within edit distance c of all squares. For example, consider the case $c = 1$. Then every word of length 2, say ab, is within edit distance 1 of a square, as we can simply replace the b by a to get aa. Thus we need to restrict our attention to avoiding words that are within edit distance c of all *sufficiently large* squares.

Theorem 10. *There is an infinite word over 5 letters such that all subwords x with $|x| \geq 3$ are neither squares, nor within edit distance 1 of any square. There is no such word over 4 letters.*

Proof. The usual tree traversal technique shows there is no such word over 4 letters. Over 5 letters we can use the 5-uniform morphism h defined by

$$0 \to 01234; \quad 1 \to 02142; \quad 2 \to 03143.$$

We claim the image of every square-free word under h has the desired property. Details will appear in the final paper. □

References

1. Alon, N., Spencer, J.: The Probabilistic Method, 2nd edn. Wiley, Chichester (2000)
2. Berstel, J.: Axel Thue's Papers on Repetitions in Words: a Translation. Number 20 in Publications du Laboratoire de Combinatoire et d'Informatique Mathématique. Université du Québec à Montréal (February 1995)
3. Brandenburg, F.-J.: Uniformly growing k-th power-free homomorphisms. Theoret. Comput. Sci. 23, 69–82 (1983)
4. Carpi, A.: On the repetition threshold for large alphabets. In: Královič, R., Urzyczyn, P. (eds.) MFCS 2006. LNCS, vol. 4162, pp. 226–237. Springer-Verlag, Heidelberg (2006)
5. Cambouropoulos, E., Crochemore, M., Iliopoulos, C.S., Mouchard, L., Pinzon, Y.J.: Algorithms for computing approximate repetitions in musical sequences. Intern. J. Comput. Math. 79, 1135–1148 (2002)
6. Dejean, F.: Sur un théorème de Thue. J. Comb. Theory. Ser. A 13, 90–99 (1972)
7. Entringer, R.C., Jackson, D.E., Schatz, J.A.: On nonrepetitive sequences. J. Combin. Theory. Ser. A 16, 159–164 (1974)
8. Landau, G.M., Schmidt, J.P.: An algorithm for approximate tandem repeats. In: Apostolico, A., Crochemore, M., Galil, Z., Manber, U. (eds.) CPM 93. LNCS, vol. 684, pp. 120–133. Springer-Verlag, Heidelberg (1993)
9. Kolpakov, R., Kucherov, G.: Finding approximate repetitions under Hamming distance. Theor. Comput. Sci. 303, 135–156 (2003)
10. Mohammad-Noori, M., Currie, J.D.: Dejean's conjecture and Sturmian words. European J. Combin. 28, 876–890 (2007)
11. Moulin-Ollagnier, J.: Proof of Dejean's conjecture for alphabets with 5, 6, 7, 8, 9, 10 and 11 letters. Theoret. Comput. Sci. 95, 187–205 (1992)
12. Nagell, T. (ed.): Selected Mathematical Papers of Axel Thue. Universitetsforlaget, Oslo (1977)
13. Pansiot, J.-J.: A propos d'une conjecture de F. Dejean sur les répétitions dans les mots. Discrete Appl. Math. 7, 297–311 (1984)
14. Thue, A.: Über unendliche Zeichenreihen. Norske vid. Selsk. Skr. Mat. Nat. Kl. 7, 1–22 (1906) Reprinted in [12, pp. 139–158]
15. Thue, A.: Über die gegenseitige Lage gleicher Teile gewisser Zeichenreihen. Norske vid. Selsk. Skr. Mat. Nat. Kl. 1, 1–67 (1912) Reprinted in [pp. 413–478]

Duplication Roots

Peter Leupold

Research Group on Mathematical Linguistics
Rovira i Virgili University
Pça. Imperial Tàrraco 1, 43005 Tarragona, Catalunya, Spain
klauspeter.leupold@urv.cat

Abstract. Recently the duplication closure of words and languages has received much interest. We investigate a reversal of it: the duplication root reduces a word to a square-free one. After stating a few elementary properties of this type of root, we explore the question whether or not a language has finite duplication root. For regular languages and uniformly bounded duplication root this is decidable.

The main result then concerns the closure of regular and context-free languages under duplication. Regular languages are closed under bounded and uniformly bounded duplication root, while neither regular nor context-free language are closed under general duplication root.

1 Duplication

A mutation, which occurs in DNA strands, is the duplication of a factor inside a strand. The interpretation of this as an operation on a string has inspired much recent work in Formal Languages, most prominently the duplication closure.

The duplication closure of a word was introduced by Dassow et al. [6], who showed that the languages generated are always regular over two letters. Wang then proved that this is not the case over three or more letters [14]. These results had actually been discovered before in the context of copy systems [8], [2]. Later on, length bounds for the duplicated factor were introduced [12], [11], and also the closure of language classes under the duplication operations was investigated [5].

In the work presented here, we investigate the effect of applying the inverse of duplications, i.e., the effect of undoing duplications leaving behind only half of a square. In this way words are reduced to square-free words, which are in some sense primitive under this notion; this is why we call the set of all square-free words reachable from a given word w the duplication root of w in analogy to concepts like the primitive root or the periodicity root of words. Duplication roots were already introduced in earlier work [11].

As the duplication root is a type of generating set with respect to duplication closure, an immediate question is whether a given language has a finite root. We show that even very simple languages generated only by catenation and iteration can have infinite duplication root. For uniformly bounded duplication roots the decidability is deduced from the fact that regular languages are closed under this operation. Also their closure under bounded duplication roots is proven.

T. Harju, J. Karhumäki, and A. Lepistö (Eds.): DLT 2007, LNCS 4588, pp. 290–299, 2007.

Part of the results exposed here were presented at the Theorietag Formale Sprachen [10].

2 Duplication Roots

For applying duplications to words we use string-rewriting systems. In our notation we mostly follow Book and Otto [1] and define such a *string-rewriting system R on Σ* to be a subset of $\Sigma^* \times \Sigma^*$. Its single-step reduction relation is defined as $u \to_R v$ iff there exists $(\ell, r) \in R$ such that for some u_1, u_2 we have $u = u_1 \ell u_2$ and $v = u_1 r u_2$. We also write simpler just $\to$, if it is clear which is the underlying rewriting system. By $\overset{*}{\to}$ we denote the relation's reflexive and transitive closure, which is called the *reduction relation* or *rewrite relation*. The inverse of a single-step reduction relation $\to$ is $\to^{-1} := \{(r, \ell) : (\ell, r) \in R\}$.

The string-rewriting system we use here is the *duplication relation* $u \heartsuit v :\Leftrightarrow \exists z[z \in \Sigma^+ \wedge u = u_1 z u_2 \wedge v = u_1 z z u_2]$. If we have length bounds $|z| \le k$ or $|z| = k$ on the factors to be duplicated we write $\heartsuit^{\le k}$ or $\heartsuit k$ respectively; the relations are called *bounded* and *uniformly bounded duplication* respectively. $\heartsuit^*$ is the reflexive and transitive closure of the relation $\heartsuit$ and thus it is our reduction relation. The *duplication closure* of a word w is then $w^{\heartsuit} := \{u : w \heartsuit^* u\}$. The languages $w^{\heartsuit \le k}$ and $w^{\heartsuit k}$ are defined analogously.

Further notation that will be used is $IRR(R)$ for the set of words irreducible of a string-rewriting system R. For a word w with $w[i]$ we denote its i-th letter, with $w[i \ldots j]$ the factor from position i to j. We call a word w *square-free* iff it does not contain any non-empty factor of the form u^2, where exponents of words refer to iterated catenation, and thus u^i is the i-fold catenation of the word u with itself. If also w^2 does not contain such a square shorter than the entire w^2, then w is said to be circular square-free. With this we come to our central definition.

Definition 1. The *duplication root* of a non-empty word w is

$$\sqrt[\heartsuit]{w} := IRR(\heartsuit^{-1}) \cap \{u : w \in u^{\heartsuit}\}.$$

As usual, this notion is extended in the canonical way from words to languages such that

$$\sqrt[\heartsuit]{L} := \bigcup_{w \in L} \sqrt[\heartsuit]{w}.$$

The roots $\sqrt[\heartsuit \le k]{w}$ and $\sqrt[\heartsuit k]{w}$ are defined in completely analogous ways, and also these are extended to entire languages in the canonical way. When we want to contrast the duplication (root) without length bound to the bounded variants we will at times call it *general duplication (root)*. First off, we illustrate this definition with an example that also shows that duplication roots are in general not unique, i.e., the set $\sqrt[\heartsuit]{w}$ can contain more than one element.

Example 2. By undoing duplications, i.e., by applying rules from $\heartsuit^{-1}$, we obtain from the word $w = abcbabcbc$ the words in the set $\{abc, abcbc, abcbabc, abcbabcbc\}$

; in a first step either the prefix $(abcb)^2$ or the suffix $(bc)^2$ can be reduced, only the former case results in a word with another square, which can be reduced to abc.

Thus we have the root $\sqrt[\heartsuit]{abcbabcbc} = \{abc, abcbabc\}$. Exhaustive search of all shorter words shows that this is a shortest possible example of a word with more than one root over three letters.

Other examples with different cardinalities of the root are the words $w_3 = babacabacbcabacb$ where

$$\sqrt[\heartsuit]{w_3} = \{bacabacb, bacbcabacb, bacb\},$$

and $w_5 = ababcbabcacbabcabacbabcab$ where

$$\sqrt[\heartsuit]{w_5} = \{abcbabcabacbabcab, abcbabcab, abcacbabcab, abcabacbabcab, abcab\},$$

As the examples have finite length, the bounded duplication root is in general not unique either. The uniformly bounded duplication root, however, is known to be unique over any alphabet [11].

It is not clear at this point, whether there exist words w_k over three letters for every $k > 0$ such that $\sqrt[\heartsuit]{w_k}$ has cardinality k. Further it would be interesting to find a function relating this number k to the length of the shortest such word w_k.

3   Finiteness of the Duplication Root

In some way, duplication roots can be seen as generating sets –via duplication closure– for the given language, though not in a strict sense, because they usually generate larger sets. That is, we have $L \subseteq (\sqrt[\heartsuit]{L})^\heartsuit$. One of the main questions about generating sets in algebra seems especially interesting also here: does there exist a finite generating set? Or in our context: is the root finite? Trivially, duplication roots are finite over two letters.

Proposition 3. *Over a two-letter alphabet for every language L its duplication root $\sqrt[\heartsuit]{L}$ is finite.*

Proof. It is well-known that over an alphabet of two letters there exist only six non-empty square-free words. Since $\sqrt[\heartsuit]{L}$ contains only square-free words, it must be finite. □

Things become more difficult over three or more letters. Let us first define the *letter sequence* seq(u) of a word u as follows: any word u can be uniquely factorized as $u = x_1^{i_1} x_2^{i_2} \cdots x_\ell^{i_\ell}$ for some integers $\ell \geq 0$ and $i_1, i_2, \ldots, i_\ell \geq 1$ and for letters $x_1, x_2, \ldots, x_\ell$ such that always $x_j \neq x_{j+1}$; then seq$(u) := x_1 x_2 \cdots x_\ell$. Intuitively speaking, every block of several adjacent occurrences of the same letter is reduced to just one occurrence. Notice that seq$(u) = \sqrt[\heartsuit 1]{u}$. As usual, seq$(L) := \bigcup_{u \in L}$ seq(u) for languages L.

We now collect a few elementary properties that connect a word's letter sequence with duplication and duplication roots.

Lemma 4. *If for two words $u, v \in \Sigma^*$ we have $\mathsf{seq}(u) = \mathsf{seq}(v)$, then there exists a word w such that $u(\heartsuit^{-1})^* w \heartsuit^* v$, i.e. both u and v are reducible to w via unduplications.*

Proof. This is immediate, since every word can be reduced to its letter sequence via rules (xx, x) for $x \in \Sigma$. Thus our statement can be satisfied by setting $w = \mathsf{seq}(u)$. □

Now we state a result that links the letter sequence and the duplication root of a word in a fundamental way.

Lemma 5. *If for two words $u, v \in \Sigma^*$ we have $\mathsf{seq}(u) = \mathsf{seq}(v)$, then also $\sqrt[\heartsuit]{u} = \sqrt[\heartsuit]{v} = \sqrt[\heartsuit]{\mathsf{seq}(u)}$.*

Proof. Via rules (xx, x) for all $x \in \Sigma$ we can obviously go from u to $\mathsf{seq}(u)$. Therefore we have $\sqrt[\heartsuit]{\mathsf{seq}(u)} \subseteq \sqrt[\heartsuit]{u}$. So it remains to show the converse inclusion, and $\sqrt[\heartsuit]{\mathsf{seq}(u)} = \sqrt[\heartsuit]{u}$ will then imply our statement.

Let us suppose there exists a word $z \in \sqrt[\heartsuit]{u}$, which is not contained in $\sqrt[\heartsuit]{\mathsf{seq}(u)}$. As already stated there exists a reduction from u to $\mathsf{seq}(u)$ using only rules (xx, x) for $x \in \Sigma$. Application of these rules preserves the letter sequence of a word. There is also a reduction from u to z via rules from $\heartsuit^{-1}$. Let us look at one specific reduction of this type. As all possible reductions from u to $\mathsf{seq}(u)$ via rules (xx, x) this reduction starts in u, too. At some point –possibly already in the first step– it uses for the first time a rule (ww, w) with $|w| \geq 2$ and results in a word z'. Here this reduction becomes different from the ones to $\mathsf{seq}(u)$.

If the first and the last letter of w are different, then $\mathsf{seq}(w)^2$ is a subsequence of the letter sequence of the word, to which this rule is applied. Consequently, $\mathsf{seq}(w)^2$ is a factor of $\mathsf{seq}(u)$. Thus we can apply a rule $(\mathsf{seq}(w)^2, \mathsf{seq}(w))$ there and obtain the word $\mathsf{seq}(z')$. If the first and the last letter of w are the same, then $\mathsf{seq}(w^2) = \mathsf{seq}(w) \cdot \mathsf{seq}(w[2 \ldots |w|])$, and we obtain z' by application of the rule $(w[2 \ldots |w|]^2, w[2 \ldots |w|])$.

By Lemma 4 z' is reducible to $\mathsf{seq}(z')$, and it is still reducible to z. So we can repeat our reasoning. Because the reduction from u to z is finite, this process will terminate and show that there is a word v reachable from both z and $\mathsf{seq}(u)$ via rules from $\heartsuit^{-1}$.

But $z \in \sqrt[\heartsuit]{u}$ is irreducible under this relation, and thus we must have $v = z$. Now $\mathsf{seq}(u)(\heartsuit^{-1})^* z$ shows that $z \in \sqrt[\heartsuit]{\mathsf{seq}(u)}$. Since this contradicts our assumption, there can be no word in $\sqrt[\heartsuit]{u} \setminus \sqrt[\heartsuit]{\mathsf{seq}(u)}$, and this concludes our proof. □

In the proof, the word $\mathsf{seq}(z')$ is obtained by rules, whose left sides are not longer than the one of the simulated rule (ww, w). Therefore the same argumentation works for bounded duplication.

Corollary 6. *If for two words $u, v \in \Sigma^*$ and an integer k we have $\mathsf{seq}(u) = \mathsf{seq}(v)$, then also $\sqrt[\heartsuit \leq k]{u} = \sqrt[\heartsuit \leq k]{v} = \sqrt[\heartsuit \leq k]{\mathsf{seq}(u)}$.*

Without further considerations, we also obtain a statement about the finiteness of the root of a language.

Corollary 7. *A language L has finite duplication root, iff $\sqrt[\heartsuit]{seq(L)}$ is finite.*

If a language does not have a finite duplication root, then this root can not be of any given complexity with respect to the Chomsky Hierarchy. There is a gap between finite and context-free languages, in which no duplication root can be situated.

Proposition 8. *If a language has a context-free duplication root, then its duplication root is finite.*

Proof. For infinite regular and context-free languages the respective, well-known pumping lemmata hold. Since a duplication root consists only of square-free words, no such language can fulfill these lemmata.   □

Already for the bounded case this does not hold any more. For example, for any $k \geq 1$ we can use a circular square-free word w of length greater than k; such words exist for all $k \geq 18$ [4]. Then we have $\sqrt[\heartsuit \leq k]{w^+} = w^+$, and this language is regular.

It is quite clear how the iteration of the union of several singleton sets can generate a regular language with infinite root; for the simplest case of this type consider $\{a, b, c\}^+$. We will now illustrate with an example that there are also regular languages constructed exclusively by concatenation and iteration, which have an infinite duplication root.

Example 9. We have seen in Example 2 that the root of $u = abcbabcbc$ consists of the two words $u_1 = abc$ and $u_2 = abcbabc$. Let ρ be the morphism, which simply renames letters according to the scheme $a \to b \to c \to a$. Then $\rho(u)$ has the two roots $\rho(u_1)$ and $\rho(u_2)$; similarly, $\rho(\rho(u))$ has the two roots $\rho(\rho(u_1))$ and $\rho(\rho(u_2))$.

We will now use this ambiguity to construct a word w such that $\sqrt[\heartsuit]{w^+}$ is infinite. This word over the four-letter alphabet $\{a, b, c, d\}$ is

$$w = ud\rho(u)d\rho(\rho(u))d = abcbabcbc \cdot d \cdot bcacbcaca \cdot d \cdot cabacabab \cdot d.$$

Thus the duplication root of w contains among others the three words

$$w_a = abc \cdot d \cdot bca \cdot d \cdot cabacab \cdot d$$
$$w_b = abc \cdot d \cdot bcacbca \cdot d \cdot cab \cdot d$$
$$w_c = abcbabc \cdot d \cdot bca \cdot d \cdot cab \cdot d,$$

which are square-free. We now need to recall that a morphism h is called square-free, iff $h(w)$ is square-free for all square-free words w. Crochemore has shown that a uniform morphism h is square-free iff it is square-free for all square-free words of length 3, [3]. Here uniform means that all images of single letters have the same length, which is given in our case.

The morphism we define now is $\varphi(x) := w_x$ for all $x \in \{a, b, c\}$. Thus to establish the square-freeness of φ, we need to check this property for the images of all square-free words up to length 3. These are

$$\varphi(aba) = abcdbcadcabacabdabcdbcacbcadcabdabcdbcadcabacabd$$
$$\varphi(abc) = abcdbcadcabacabdabcdbcacbcadcabdabcbabcdbcadcabd$$
$$\varphi(aca) = abcdbcadcabacabdabcbabcdbcadcabdabcdbcadcabacabd$$
$$\varphi(acb) = abcdbcadcabacabdabcbabcdbcadcabdabcdbcacbcadcabd$$
$$\varphi(bab) = abcdbcacbcadcabdabcdbcadcabacabdabcdbcacbcadcabd$$
$$\varphi(bac) = abcdbcacbcadcabdabcdbcadcabacabdabcbabcdbcadcabd$$
$$\varphi(bca) = abcdbcacbcadcabdabcbabcdbcadcabdabcdbcadcabacabd$$
$$\varphi(bcb) = abcdbcacbcadcabdabcbabcdbcadcabdabcdbcacbcadcabd$$
$$\varphi(cac) = abcbabcdbcadcabdabcdbcadcabacabdabcbabcdbcadcabd$$
$$\varphi(cab) = abcbabcdbcadcabdabcdbcadcabacabdabcdbcacbcadcabd$$
$$\varphi(cba) = abcbabcdbcadcabdabcdbcacbcadcabdabcdbcadcabacabd$$
$$\varphi(cbc) = abcbabcdbcadcabdabcdbcacbcadcabdabcbabcdbcadcabd,$$

where, of course, the images of all words shorter than three are contained in them. All the twelve words listed here are indeed square-free as an eager reader can check, and thus φ is square-free.

Now let t be an infinite square-free word over the letters a, b and c. Then $\varphi(\mathsf{pref}(t))$ is an infinite set of square-free words. From the construction of φ we know that for any word z of length i we can reach $\varphi(z)$ from w^i by undoing duplications. Therefore $\varphi(\mathsf{pref}(t)) \subseteq \sqrt[\heartsuit]{w^+}$, whence also the latter set is infinite.

Thus even very simple languages can have rather complicated roots. Therefore it seems to be a challenging problem to decide the finiteness for regular languages in general.

4   Closure Under Duplication Root

Now we investigate the closure of regular languages under the three variants of duplication root. We start with the most restricted variant, uniformly bounded duplication.

Proposition 10. *If $L \in REG$, then also $\sqrt[\heartsuit^k]{L} \in REG$ for all $k \geq 1$.*

Proof. If a language L is regular, then it can be generated by a regular grammar $G = (N, \Sigma, S, P)$, which has only rules of the forms (A, xB) and (A, x) for non-terminals A and B and $x \in \Sigma$; for simplicity we ignore the possible rule (S, λ) to generate the empty word. From this grammar we construct another one that generates $\sqrt[\heartsuit^k]{L}$.

The new grammar's set of non-terminals is $N' = \{A_w : A \in N \wedge w \in \Sigma^{\leq 2k}\}$. The rule set is derived from P in the following way. The rules from $\{(A_w, B_{wx}) : (A, B) \in P \wedge |w| < 2k - 1\}$ go in parallel to those of P, but store the letters generated in the non-terminals' index instead of actually generating them. When the index reaches length $2k$, the first letters stored in the index are finally put out, when new ones come in. This is done by rules from the set

$$\{(A_w, w[1]B_{w[2\ldots 2k]x}) : (A, xB) \in P \wedge |w| = 2k \wedge w[2\ldots 2k]x \in IRR((\heartsuit^k)^{-1})\}.$$

Only if the index would become a square u^2 of length k, then this square is reduced to u, instead of putting anything out. The set

$$\{(A_w, B_{w[1...k+1]}) : (A, xB) \in P \wedge |w| = 2k \wedge w[2...2k]x \notin IRR((\heartsuit^k)^{-1})\}$$

provides the rules implementing this. The rules from

$$\{(A_w, B_{w[1...k]}) : (A, xB) \in P \wedge |w| = 2k - 1 \wedge w[1...k] = w[k+1...2k-1]x\}$$

take care of the case that upon filling the index already a k-square is produced. From the terminating rules of P we derive the sets

$$\{(A_w, wx) : (A, x) \in P \wedge wx \text{ is not a } k - \text{square}\}$$

and

$$\{(A_w, w[1...k]) : (A, x) \in P \wedge wx \text{ is a } k - \text{square}\}.$$

This new grammar obviously generates the words that also G generates, only leaving out all squares of length $2k$ that occur when going from left to right. If there are two squares of length $2k$ overlapping in more than k symbols, then the entire sequence has period k, and thus the reduction of either one results in the same word. Therefore this proceeding from left to right will indeed reduce all the squares, and in this way all the words in $\sqrt[\heartsuit^k]{L}$ are reached. □

The grammar constructed for $\sqrt[\heartsuit^k]{L}$ uses a similar idea as the algorithm for deciding the question "$u \in v^{\heartsuit k}$?," which we gave in an earlier article [11]. The effective closure of regular languages under uniformly bounded duplication can be used to decide the problem of the finiteness of the root for the uniformly bounded case.

Corollary 11. *For regular languages it is decidable, whether their uniformly bounded duplication root is finite.*

Proof. From the proof of Proposition 10 we see that from a regular grammar for a language L a regular grammar for the language $\sqrt[\heartsuit^k]{L}$ can be constructed. This construction method is effective. Since the finiteness problem is decidable for regular languages, it can then also be decided for $\sqrt[\heartsuit^k]{L}$. □

Now we turn to bounded duplication. As the bounded duplication closure of even single words is in general not regular, it is not clear whether its inverse will preserve regularity. We will show that this is the case by essentially reducing bounded duplication root to an inverse monadic string-rewriting system. A string-rewriting system is *monadic*, iff for all of its rules (ℓ, r) we have $|\ell| > |r|$ and $|r| \leq 1$. Such a system R is said to preserve regularity (context-freeness) iff for all regular languages L also the language $\bigcup_{w \in L}\{u : w \xrightarrow{*}_R u\}$ is regular (context-free).

Proposition 12. *The class of regular languages is closed under bounded duplication root.*

Proof. For a given regular language L the defining condition for a word w to be in $\sqrt[\heartsuit \leq k]{L}$ can be formulated in the following way: $w^{\heartsuit \leq k} \cap L \neq \emptyset$. This is the way we will decide the word problem for $\sqrt[\heartsuit \leq k]{L}$.

Thus we start from the input word and have to determine, whether from it one can reach another word in a given regular language L. Let the length bound for duplication be k. We will transform words from Σ^+ into a redundant representation, where every letter contains also the information about the $k-1$ following ones. This way rewrite rules from $\heartsuit^{\leq k}$ can be simulated by ones with a left side of length only one, i.e. by context-free ones.

First off we define the mapping $\phi : \Sigma^+ \mapsto ((\Sigma \cup \{\square\})^k)^+$ as follows. We delimit with $(\ldots)$ letters from $(\Sigma \cup \{\square\})^k$ and with $[\ldots]$ factors of a word as usual. The image of a word u is

$$u \mapsto (u[1 \ldots k]) (u[2 \ldots k+1]) \cdots (u[|u| - k + 1 \ldots |u|]) \cdot$$
$$(u[|u| - k + 2 \ldots |u|]\square) \cdots (u[|u|]\square^{k-1}).$$

Thus every letter contains also the information about the k following original ones from the original word u, at the end of the word letters are filled up with the space symbol $\square$. This encoding can be reversed by a letter-to-letter homomorphism h defined as $h(x) := x[1]$ if $x[1] \in \Sigma$, for the other case we select for the sake of completeness some arbitrary letter a and set $h(x) := a$ if $x[1] = \square$; the latter case will never occur in our context. It is clear that $h(\phi(u)) = u$ for words from Σ^*. Both mappings are extended to languages in the canonical way such that $\phi(L) := \{\phi(u) : u \in L\}$ and $h(L) := \{h(u) : u \in L\}$.

Now we define the string-rewriting system R over the alphabet $(\Sigma \cup \{\square\})^k$ as follows:

$$R := \{((uv), \phi(u^2v')[1 \ldots |\phi(u^2v')| - k + 1]) : uv \in \Sigma^k \wedge v' \in \Sigma^* \wedge$$
$$v \in v' \cdot \{\square^*\}\}.$$

A letter $[uv]$ is replaced by the image of u^2v under ϕ minus the suffix of letters that contain $\square$. This way application of rules from R keeps this space symbol only in the last letters of our words. It should be rather clear that $\phi(w^{\heartsuit \leq k}) = \{u : \phi(w)R^*u\}$ or, in other words $w^{\heartsuit \leq k} = h(\{u : \phi(w)R^*u\})$.

As all the rules of R have left sides of length one, the application of this system amounts to substituting each letter of w with a word from a context-free language. Since context-free languages are closed under this type of substitution, the language of all the resulting words is context-free if the original one is.

As all the rules of R have left sides of length one and right sides of length greater than one, their inverses are all monadic, i.e. the system R^{-1} is monadic. Monadic string-rewriting systems preserve regularity, see for example the work of Hofbauer and Waldmann [9].

The language $\sqrt[\heartsuit \leq k]{\Sigma^*}$ is regular as it is the complement of the regular $\Sigma^*\{uu : |u| \leq k\}\Sigma^*$.

Summarizing, we can obtain $\sqrt[\heartsuit \leq k]{L}$ by a series of regularity-preserving operations in the following way:

$$\sqrt[\heartsuit \leq k]{L} = (R^{-1})^*(\phi(L)) \cap \sqrt[\heartsuit \leq k]{\Sigma^*}.$$

Though ϕ is not a morphism, it is a gsm mapping and as such certainly preserves regularity. Thus bounded duplication root preserves regularity.   □

Reducing the set R does not affect the context-freeness of the language. Thus we can take away all rules effecting duplications shorter than the length bound k and obain another proof for Proposition 10, however not a constructive one that lets us conclude the decidability. Unfortunately, the proof can not be generalized to context-free languages, because inverse monadic rewriting systems do not preserve context-freeness.

Under the unbounded root neither the regular nor the context-free languages are not closed, as for example $\sqrt[\heartsuit]{\{a,b,c\}^*}$, the language of all square-free words, is not context-free, see [7] and [13].

Summarizing we state the closure properties determined in this article in a comprehensive form.

Theorem 13. *The closure properties of the class of regular languages under the three duplication roots are as follows:*

	$\sqrt[\heartsuit k]{L}$	$\sqrt[\heartsuit \leq k]{L}$	$\sqrt[\heartsuit]{L}$
REG	Y	Y	N
CF	?	?	N

Here Y stands for closure, N stands for non-closure, and ? means that the problem is open.

5   Outlook

Theorem 13 leaves open the closure of context-free languages under bounded duplication root. The answer might actually depend on the length bound as is the case for bounded duplication. For $k \in \{1, 2\}$ regular languages are closed under bounded duplication, for $k \geq 4$ they are not, while for $k = 3$ the problem remains open, see [5]. Also for context-sensitive languages the only trivial case seems to be uniformly bounded duplication root. For all others the straight-forward approach of the proof of Theorem 12 cannot be used with a LBA, because it would exceed the linear length bound.

The other complex of interesting questions is the investigation of the boundary between decidability and undecidability for the question of the finiteness of duplication roots. For the simplest case, i.e. regular languages and uniformly bounded duplication, Corollary 11 establishes decidability; on the other hand, for context-free languages or even more complicated classes, one would expect undecidability for all variants of duplication. Thus sómewhere inbetween there must be the crucial point, probably at different complexities for the different variants of duplication root.

Finally, the investigation of the duplication root of single words promises to pose some interesting questions, as hinted at the end of Section 2. Especially the possible degrees of ambiguity might follow interesting rules.

Acknowledgements. The words w_3 and w_5 from Example 2 were found by Szilárd Zsolt Fazekas, the fact that w is the shortest example of a word with ambiguous root was established by Artiom Alhazov with a computer.

References

1. Book, R., Otto, F.: String-Rewriting Systems. Springer, Berlin Heidelberg (1988)
2. Bovet, D.P., Varricchio, S.: On the Regularity of Languages on a Binary Alphabet Generated by Copying Systems. Information Processing Letters 44, 119–123 (1992)
3. Crochemore, M.: Sharp Caracterizations of Squarefree Morphisms. Theoretical Computer Science 18, 221–226 (1982)
4. Currie, J.D.: There are Ternary Circular Square-free Words of Length n for $n \geq 18$. Electric Journal of Combinatorics 9(1), N10 (2002)
5. Ito, M., Leupold, P., Shikishima-Tsuji, K.: Closure of Language Classes under Bounded Duplication. In: Ibarra, O.H., Dang, Z. (eds.) DLT 2006. LNCS, vol. 4036, pp. 238–247. Springer, Heidelberg (2006)
6. Dassow, J., Mitrana, V., Păun, G.: On the Regularity of Duplication Closure. Bull. EATCS 69, 133–136 (1999)
7. Ehrenfeucht, A., Rozenberg, G.: On the Separating Power of EOL Systems. RAIRO Informatique Théorique 17(1), 13–22 (1983)
8. Ehrenfeucht, A., Rozenberg, G.: On Regularity of Languages Generated by Copying Systems. Discrete Applied Mathematics 8, 313–317 (1984)
9. Hofbauer, D., Waldmann, J.: Deleting String-Rewriting Systems preserve regularity. Theoretical Computer Science 327, 301–317 (2004)
10. Leupold, P.: Duplication Roots – Extended Abstract. In: Proceedings Theorietag Automaten und Formale Sprachen, TU Wien, Wien, pp. 91–93 (2006)
11. Leupold, P., Martín, C., Mitrana, V.: Uniformly Bounded Duplication Languages. Discrete Applied Mathematics 146(3), 301–310 (2005)
12. Leupold, P., Mitrana, V., Sempere, J.: Formal Languages Arising from Gene Repeated Duplication. In: Jonoska, N., Păun, G., Rozenberg, G. (eds.) Aspects of Molecular Computing. LNCS, vol. 2950, pp. 297–308. Springer, Heidelberg (2003)
13. Ross, R., Winklmann, K.: Repetitive Strings are not Context-Free. RAIRO Informatique Théorique 16(3), 191–199 (1982)
14. Wang, M.-W.: On the Irregularity of the Duplication Closure. Bull. EATCS 70, 162–163 (2000)

Complexity Theory for Splicing Systems

Remco Loos[1,*] and Mitsunori Ogihara[2]

[1] Research Group on Mathematical Linguistics
Rovira i Virgili University
Pça Imperial Tàrraco 1, 43005 Tarragona, Spain
`remcogerard.loos@urv.cat`
[2] Department of Computer Science
University of Rochester
Box 270226, Rochester, NY 14627, USA
`ogihara@cs.rochester.edu`.

Abstract. This paper proposes a notion of time complexity in splicing systems and presents fundamental properties of SPLTIME, the class of languages with splicing system time complexity $t(n)$. Its relations to classes based on standard computational models are explored. It is shown that for any function $t(n)$, SPLTIME$[t(n)]$ is included in 1-NSPACE$[t(n)]$. Expanding on this result, 1-NSPACE$[t(n)]$ is characterized in terms of splicing systems: it is the class of languages accepted by a $t(n)$-space uniform family of extended splicing systems having production time $O(t(n))$ with regular rules described by finite automata with at most a constant number of states. As to lower bounds, it is shown that for all functions $t(n) \geq \log n$, all languages accepted by a pushdown automaton with maximal stack height $t(|x|)$ for a word x are in SPLTIME$[t(n)]$. From this result, it follows that the regular languages are in SPLTIME$[O(\log(n))]$ and that the context-free languages are in SPLTIME$[O(n)]$.

1 Introduction

The splicing system [4] is a computational model that is inspired by the DNA recombination process that consists of splicing and reassembling. Intuitively speaking, a splicing system produces words by cutting, with respect to a given set of rules, two words into two parts and then swapping their second parts. In a splicing system word production starts with an initial collection of words. Then, for each pair of words in the set, every applicable rule is applied independently and each of the resulting two words is added to the collection, if it is not already in the set. Since words never disappear from the collection we can naturally view that a splicing system produces its language in rounds. Indeed, the standard definition of a splicing system uses this view.

The complexity of a splicing system can be studied in terms of the complexity of the initial set and the complexity of the rules in the Chomsky Hierarchy.

* Work done during a research stay at the University of Rochester, supported by Research Grant BES-2004-6316 of the Spanish Ministry of Education and Science.

T. Harju, J. Karhumäki, and A. Lepistö (Eds.): DLT 2007, LNCS 4588, pp. 300–311, 2007.
© Springer-Verlag Berlin Heidelberg 2007

The complexity of splicing systems has been well studied in this regard (see, e.g. [2,11,12,13,14]). In particular, it is known [11] that the extended splicing system with a finite initial language and with rules whose patterns are specified using regular expressions is universal (i.e., as powerful as the recursively enumerable) and that this system with the set of rules restricted to be finite is regular [2,14].

The universality of extended system with finite set of rules states that the model is equivalent to other standard abstract computation models, such as Turing machines and random access machines. Since these standard models are used to define computational complexity by introducing the concept of resources, one may wonder whether there exists a natural concept of computational resources in the extended splicing system and what complexity classes are defined in terms of the resource concept.

Surprisingly, although these questions sound natural, and indeed for other models of DNA computing similar questions have been addressed before [8,9,15], they have never been asked about splicing systems. We propose a notion of computational complexity in the splicing model. Since the words in a language are thought of as being produced in rounds, the proposal here is to consider the minimum number of rounds that it takes for the system to produce the word as the complexity of the word with respect to the system. The complexity of the language produced by the system at length n is then defined to be the maximum of the time complexity of the word with respect to the system for all members of the language having length n. This time complexity concept is reminiscent of the derivational complexity of grammars [1,3], where the complexity of a word with respect to a grammar is defined to be the smallest number of derivational steps for producing the word with respect to the grammar. Although the derivational complexity uses the number of operational steps as a measure, it is fundamentally different from our notion of time complexity because splicing is applied to two words and the two input words for splicing can be produced asynchronously in preceding steps.

In this paper we explore properties of the proposed notion of time complexity. Based on the aforementioned universality result we will use the model with a finite initial set and with a regular set of rules. It is easy to see that if $L(\Gamma)$ is infinite then the time complexity of L with respect to Γ is not $o(\log n)$. This is because the initial set is finite and the length of the longest word produced by Γ increases by a factor of at most 2 at each round. It is also easy to show that a regular language a^+ has splicing time complexity $\Theta(\log n)$ with respect to an extended splicing system. Indeed, we show that every regular language has splicing time complexity of $O(\log n)$. As an upper bound, we show that the languages produced by extended splicing systems with time complexity $t(n)$ are accepted by $t(n)$ space-bounded nondeterministic Turing machines with one-way input head. From this general result, it follows that every language generated by a splicing system with time complexity of $O(\log n)$ is in 1-NL.

Exploring this result further, we show that the class of languages produced by extended splicing systems with time bound $t(n)$ is captured in terms of extended splicing systems: a language L belongs to this class if and only if there exists a $t(n)$-space uniform family of splicing systems $\{\Gamma_n\}_{n \geq 0}$ such that for each n, the length-n portion of L is produced by Γ_n in $O(t(n))$ rounds and the number of states of each automaton appearing in Γ_n is bounded by a constant not depending on n.

As a general lower bound, we show that for all $t(n) \in \Omega(\log n)$, all languages accepted by a pushdown automaton M with maximal stack height $f(|x|)$ for a word x in $L(M)$ are produced by extended splicing systems with time complexity $O(f(n))$. From this result it follows that all context-free languages have splicing time complexity of at most $O(n)$.

2   Basic Definitions and Notation

We assume the reader's familiarity with the basic concepts in complexity theory and formal language theory. Good references to these are respectively [6] and [10].

A *splicing rule* over an alphabet V is a word of the form $u_1 \# u_2 \$ v_1 \# v_2$ such that u_1, u_2, v_1, and v_2 are in V^* and such that $\$$ and $\#$ are two symbols not in V.

For a splicing rule $r = u_1 \# u_2 \$ v_1 \# v_2$ and for $x, y, w, z \in V^*$, we say that r produces (w, z) from (x, y), denoted by $(x, y) \vdash_r (w, z)$, if there exist some $x_1, x_2, y_1, y_2 \in V^*$ such that $x = x_1 u_1 u_2 x_2$, $y = y_1 v_1 v_2 y_2$, $z = x_1 u_1 v_2 y_2$, and $w = y_1 v_1 u_2 x_2$.

We simplify the notation by viewing (x, y) and (w, z) as unordered pairs and write $(x, y) \vdash_r (w, z)$ and $(x, y) \vdash_r (z, w)$ interchangeably.

A *splicing scheme* is a pair (V, R), where V is a finite alphabet and $R \subseteq V^* \# V^* \$ V^* \# V^*$ is a finite (possibly infinite but finitely represented) set of splicing rules. For a splicing scheme $h = (V, R)$ and for a language $L \subseteq V^*$, define

$$\sigma_R(L) = \{z, w \in V^* \mid (\exists u, v \in L, r \in R)[(u, v) \vdash_r (z, w)]\}. \tag{1}$$

Given a splicing scheme $h = (V, R)$ and an initial language L, the splicing language $\sigma_R^*(L)$ is defined as follows.

$$\sigma_R^0(L) = L, \tag{2}$$

$$\sigma_R^{i+1}(L) = \sigma_R^i(L) \cup \sigma_R(\sigma_R^i(L)), i \geq 0, \tag{3}$$

$$\sigma_R^*(L) = \bigcup_{i \geq 0} \sigma_R^i(L). \tag{4}$$

In the following, we omit the subscript R if the omission will not cause confusion.

We consider two types for the set, R, of splicing rules. One is the finite set and the other is the regular language. The definition when R is regular can be given as follows: A set of rules R is regular if there exist some $m \geq 1$ and m quadruples of regular languages (A_i, B_i, C_i, D_i), $1 \leq i \leq m$, such that $R = \cup_{1 \leq i \leq m} \{a_i \# b_i \$ c_i \# d_i \mid a_i \in A_i, b_i \in B_i, c_i \in C_i, d_i \in D_i\}$.

Now we are ready to define splicing systems. A *splicing system* is a triple $\Gamma = (V, A, R)$ such that (V, R) is a splicing scheme and $A \subseteq V^*$ is the initial language. The language generated by Γ, denoted by $L(\Gamma)$, is $\sigma_R^*(A)$.

We say that a splicing system is *finite* if it consists of a finite set of rules and a finite initial language, i.e., both A and R are finite sets. It is shown [2,14] that finite splicing systems generate only regular languages.

An *extended splicing system* is a quadruple $\Gamma = (V, \Sigma, A, R)$ such that $\Gamma' = (V, A, R)$ is a splicing system. The language produced by Γ, denoted by $L(\Gamma)$, is $\Sigma^* \cap L(\Gamma')$.

It is known that extended systems with a finite initial set and a regular set of rules are "universal" [11]. Because of this first universality result, the extended splicing system with a finite initial language and a regular set of rules can be considered to be the "standard" universal splicing system, and this is the model we will consider in this paper. Indeed, many well-studied computational models having universal power can be straightforwardly simulated with these systems, with at most a constant slowdown (see Chapter 8 of [13] for an overview).

3   Time Complexity for Splicing Systems

Let $\Gamma = (V, \Sigma, A, R)$ be an extended splicing system. For each $w \in V^*$, define

$$\text{SplicingTime}_\Gamma(w) = \begin{cases} \min\{i \mid w \in \sigma_R^i(A)\} & \text{if } w \in \sigma_R^*(A) \\ 0 & \text{otherwise.} \end{cases}$$

Let $\mathbb{N}$ denote the set of all natural numbers.

Definition 1. *Let $T(n)$ be a monotonically nondecreasing function from $\mathbb{N}$ to itself. Then we define* $\text{SPLTIME}[T(n)]$ *to be the set of all languages L for which there exists an extended splicing system with regular rules $\Gamma = (V, \Sigma, A, R)$ such that for all $w \in L$, it holds that* $\text{SplicingTime}_\Gamma(w) \leq T(|w|)$.

Definition 2. *For a class $\mathcal{C}$ of functions from $\mathbb{N}$ to itself, define*

$$\text{SPLTIME}[\mathcal{C}] = \cup_{T(n) \in \mathcal{C}} \text{SPLTIME}[T(n)].$$

As first observations we have that all finite languages have zero complexity and that for any extended splicing system $\Gamma = (V, \Sigma, A, R)$, at any step i, the length of the longest word in $\sigma_\Gamma^i(A)$ is at most twice that in $\sigma_\Gamma^{i-1}(A)$. This implies that the length is at most 2^i times the longest word in A. Thus, we have:

Proposition 1. $\text{SPLTIME}[o(\log n)]$ *contains no infinite languages.*

Due to Proposition 1 a time complexity function $T(n)$ is meaningful for extended splicing systems if $T(n) \in \Omega(\log n)$. Thus, the smallest splicing time complexity class is $\text{SPLTIME}[O(\log n)]$. Here we show some fundamental results about this class.

First, it is not hard to see that the regular language a^+ belongs to $\text{SPLTIME}[O(\log n)]$ via the following unextended system $\Gamma = (V, A, R)$:

$$V = \{a\}, A = \{a\}, \text{ and } R = \{ a\#\lambda\$\lambda\#a \}.$$

This splicing system generates $\{\lambda, a, aa\}$ in the first step, $\{\lambda, a, aa, a^3, a^4\}$ in the second and in general $\sigma_\Gamma^i(A) = \{a^x \mid 0 \le x \le 2^i\}$.

Actually, it is not very difficult to show that every regular language belongs to this class.

Theorem 1. REG $\subseteq$ SPLTIME$[O(\log n)]$.

Proof. Let L be an arbitrary regular language. If L is finite, an unextended system whose initial language is L and whose rule set is empty produces L in no rounds, and so $L \in$ SPLTIME$[O(\log n)]$.

Suppose that L is an infinite regular language. We will construct an extended finite splicing system $\Gamma = (V, \Sigma, A, R)$ witnessing that $L \in$ SPLTIME$[O(\log n)]$. Let $M = (Q, \Sigma, \delta, q_0, F)$ be a non-deterministic finite automaton accepting L, where Q is the set of states, Σ the input alphabet, q_0 the initial state, F the set of final states, and δ the transition function. We assume without loss of generality that M has no λ-transitions. We construct Γ as follows:

- $V = \Sigma \cup Q \cup \{Z\}$, where Z is a new symbol not in $\Sigma \cup Q$.
- $A = \{Z\} \cup \{q_i a q_j \mid q_i, q_j \in Q, a \in \Sigma, \delta(q_i, a) = q_j\}$.
- R consists of the following rules:
  - $a\#q\$q\#b$ for all $q \in Q, a, b \in \Sigma$,
  - $q_0\#\lambda\$\lambda\#Z$,
  - $\lambda\#q_f\$Z\#\lambda$ for all $q_f \in F$.

The initial language A contains all the words of the form $q_i a q_j$ such that M transitions from q_i to q_j on a. Thus, A is the set of all valid paths of length 1. The rules of the form $a\#q\$q\#b$ connect two paths sharing the same state in the middle. The last two rules eliminate the initial state appearing at the beginning and the final state appearing at the end.

Production of a word w in L can be in a divide-and-conquer fashion: split w into halves, separately produce them with the corresponding states at each end, and connect them. Thus, the time that it takes to produce a word having length n is $\lceil \log(n+1) \rceil + 2$ (the additive term of 2 is for eliminating the initial and accept states after producing a word of the form $q_0 w q_f$ such that $q_f \in F$). Thus, $L \in$ SPLTIME$[O(\log n)]$. This proves the theorem. $\quad\square$

Note that the set of rules in the above construction is finite. Thus, from the characterization of the class of languages generated by extended splicing systems with a finite set of rules [12] we have:

Corollary 1. *Let $\mathcal{F}$ be an arbitrary class of monotonically nondecreasing functions from $\mathbb{N}$ to itself such that $\mathcal{F} \supseteq O(\log n)$. If the set of rules is restricted to be finite, then* REG $=$ SPLTIME$[O(\log n)] =$ SPLTIME$[\mathcal{F}]$.

We go on to show some closure properties for the class SPLTIME$[O(T(n))]$.

Theorem 2. *Let $T(n)$ be an arbitrary monotonically nondecreasing function such that $T(n) \in \Omega(\log n)$. Then the class* SPLTIME$[O(T(n))]$ *is closed under concatenation, star-operation, and union.*

Proof (sketch). Let L_1 and L_2 be languages in SPLTIME$[O(T(n))]$. For each $i \in \{1, 2\}$, suppose that $L_i \in$ SPLTIME$[T(n)]$ is witnessed by an extended splicing system $H_i = (V_i, \Sigma_i, A_i, R_i)$. Without loss of generality, assume that there is no common nonterminal between H_1 and H_2, that is, $(V_1 - \Sigma_1) \cap (V_2 - \Sigma_2) = \emptyset$. To prove the theorem it suffices to show that $L_1 \cup L_2$, $L_1 L_2$, and $(L_1)^*$ each belong to SPLTIME$[O(T(n))]$.

For all three cases, the idea is to add delimiters α_i to the left and β_i to the right to all words in A_i. The rules of each R_i are adapted to also test for the presence of α_i or β_i. This ensures that there are no interferences between the derivations in the two systems. In this way, if H_i produces w, we obtain $\alpha_i w \beta_i$. Finally, we add rules to perform the operation (for catenation and star) and to remove the delimiters. It is not hard to see that the resulting languages are in SPLTIME$[O(T(n))]$. □

We note here that it is unknown whether for any time function $T(n) \in \Theta(\log n)$, SPLTIME$[O(T(n))]$ is closed under intersection or under complementation.

4   Splicing Systems Versus One-Way Nondeterministic Space-Bounded Computation

In this section we consider an upper bound of splicing time complexity classes. The difficulty here is that, although the extended splicing system is universal, there doesn't appear to exist any immediate connection between the running time of a Turing machine and the number of production rounds required by the splicing system that produces the language recognized by the Turing machine.

A straightforward method for checking the membership of a word w in a language L in SPLTIME$[T(n)]$ would be to simulate the splicing system for at most $T(|w|)$ rounds while keeping the collection of the words that have been produced and then check whether w appears in the final collection. Though correct, the space needed for this algorithm can increase rapidly. Indeed, following this strategy gives rise to a doubly-exponential (!) space upper bound.

This upper bound is, not surprisingly, very naive. By guessing the "components" of the splicing operations that are conducted produce a word w, we can reduce the upper bound to a nondeterministic exponential time.

Theorem 3. *For all monotonically nondecreasing functions $T(n)$,*

$$\text{SPLTIME}[T(n)] \subseteq \cup_{c>0} \text{NTIME}[c^{T(n)}].$$

Proof. Let $L \in$ SPLTIME$[T(n)]$ be witnessed by an extended splicing system $\Gamma = (V, \Sigma, A, R)$. Let d be the length of the longest word in A. For all natural numbers $i \geq 0$, and for all $w \in V^*$ such that SplicingTime$_\Gamma(w) \leq i$, $|w| \leq d2^i$. Also, for all positive integers i and for all $w \in V^*$, SplicingTime$_\Gamma(w) \leq i$ if and only if

- either $w \in \mathrm{SplicingTime}_\Gamma(w) \leq i - 1$ or
- there exist $x, y, z \in V^*$ and a rule $r \in R$ such that $\mathrm{SplicingTime}_\Gamma(u) \leq i - 1$, $\mathrm{SplicingTime}_\Gamma(v) \leq i - 1$, and $(u, v) \vdash_r (z, w)$ (recall that we use $(u, v) \vdash_r (z, w)$ and $(u, v) \vdash_r (w, z)$ interchangeably).

Consider the following nondeterministic algorithm Q that takes as input an integer $i \geq 0$ and a word $w \in V^*$ and tests whether w is produced by Γ within i rounds.

Step 1. If $i = 0$, return 1 if $w \in A$ and 0 otherwise.

Step 2. Nondeterministically select $u, v \in V^*$ having length at most $d2^{i-1}$, $z \in V^*$ having length at most $d2^i$, and a finite automaton quadruple $r \in R$.

Step 3. Test whether $(u, v) \vdash_r (z, w)$ by exhaustively examining all possible positions for aligning the finite automata on u and v. If $(u, v) \vdash_r (z, w)$ doesn't hold, return 0.

Step 4. Make two recursive calls, $Q(i-1, u)$ and $Q(i-1, v)$. Both return with 1 as the value, return 1; otherwise, return 0.

It is not difficult to see that this nondeterministic algorithm works correctly. The total number of recursive calls to Q on input (i, w) is at most $2 + 2^2 + \cdots + 2^i < 2^{i+1}$; the running time for the algorithm excluding the time spent on recursive calls is bounded by polynomial in $d2^i$ on input (i, w). Thus, the total running time is $O(c^i)$ for some constant $c > 0$. Now, to test whether $w \in L$, we have only to execute $Q(T(|w|), w)$. This implies, that $L \in \mathrm{NTIME}[c^{T(n)}]$. This proves the theorem. $\qquad\square$

From the above theorem, we immediately have the following corollary.

Corollary 2. $\mathrm{SPLTIME}[O(\log n)] \subseteq \mathrm{NP}$.

The idea of nondeterministic verification shown in the above can be further explored to tighten the upper bound. For a function $T(n)$ from $\mathbb{N}$ to $\mathbb{N}$, $1\text{-NSPACE}[T(n)]$ is the set of all languages accepted by a $T(n)$ space-bounded nondeterministic Turing machine with one-way input tape [5]. In the one-way nondeterministic space-bounded Turing machine model, since the input head moves from left to right only, the usable amount of space must be communicated to the machine prior to computation. For a $T(n)$ space-bounded machine in this model, this communication is accomplished by assuming that on input of length n on each work tape a blank word of length $T(n)$ is written flanked by end markers with the initial position of the head being at the symbol immediately to the right of the left end marker and that the head never goes beyond the end markers.

Among the many 1-NSPACE classes of particular interest to us is 1-NL, which is $\cup_{c>0} 1\text{-NSPACE}[c(\log n)]$. Hartmanis and Mahaney [5] show that the reachability problem of topologically sorted directed graph is complete for 1-NL under the logarithmic space-bounded many-one reductions.

We show an improved upper-bound of 1-NL for SplicingTime.

Theorem 4. $\mathrm{SPLTIME}[O(\log n)] \subseteq 1\text{-NL}$.

This theorem is straightforwardly derived from the following more general statement.

Theorem 5. *For all $f(n) \geq \log n$, SPLTIME$[f(n)] \subseteq$ 1-NSPACE$[f(n)]$.*

Proof (sketch). Let $f(n) \geq \log n$. Let L be a language in SPLTIME$[f(n)]$. Let $\Gamma = (V, \Sigma, I, R)$ be a splicing system that witnesses $L \in$ SPLTIME$[f(n)]$. We regard each rule in R as a quadruple of finite automata $r = (A, B, C, D)$.

Let n be an arbitrary natural number. The process in which Γ produces a word in at most $f(n)$ rounds can be described as a node-labelled, full binary tree of height at most $f(n)$ with each leaf labelled with a word in I and each non-leaf labelled with a word $w \in V^*$, a rule r, and two natural numbers i and j.

Each non-leaf represents a splicing operation as follows: Let g be a non-leaf with labels w, r, i, j. Let $r = (A, B, C, D)$, and let u and v be the word label of the left and right child respectively. Then u is of the form $u_1 u_2$ and v is of the form $v_1 v_2$ such that

- $w = u_1 v_2$, $|u_1| = i$, $|v_1| = j$,
- $u_1 \in L(A)$, $u_2 \in L(B)$, $v_1 \in L(C)$, and $v_2 \in L(D)$.

Note that given a valid production tree the word label of each non-leaf can be computed from the labels of its proper descendants. The output of the production tree is the word label of the root.

Using this notion of production trees, the membership test of any word $x \in \Sigma^n$ in L can be done by testing whether there is a production tree of height at most $f(n)$ whose output is x. Our goal is to design an $O(f(n))$ space-bounded one-way nondeterministic algorithm for this task. So our algorithm should

1. Encode, as a word over some finite alphabet, the tree structure, the leaf labels, and the rule and the splicing positions at each non-leaf.
2. Test, given such an encoding, whether it is in a valid format and, if so, in the specified production tree whether the word assigned to each leaf is in the initial language I and whether the splicing rule specified at each non-leaf can be successfully applied to the word labels of the children.
3. Test whether x is the output of the encoded tree.

We will not give details of our encoding, but it should be clear that introducing delimiters we can represent the tree structure. After guessing such a production tree, it is not particularly difficult to see that its syntactical correctness can be tested deterministically in space $O(f(n))$ by scanning w from left to right. For checking the correctness of the splicing specified at each non-leaf, we need to check the parts of the encoding against the rule automata. This means moving down the tree until a word over V is reached. Since membership in a regular language can be tested by simply scanning the input from left to right and only nodes appearing in a downward path are considered simultaneously, this test can be done deterministically in space $O(f(n))$ by scanning w from left to right. Finally, checking whether the production tree produces x can be tested by comparing the letters of W of x letter by letter. Thus, by concurrently running these tests while nondeterministically producing an encoding of a production tree, the membership of x in L can be tested in space $O(f(n))$.   □

Theorem 4 immediately raises the question of whether we might have equality between SPLTIME$[O(\log n)]$ and 1-NL. We show that this is unlikely—even allowing the use of larger splicing systems for longer words does not enable logarithmic time-bounded splicing systems to produce anything beyond 1-NL. A family of boolean circuits $\mathcal{F} = \{F_n\}_{n\geq 0}$ is said to be *logarithmic-space uniform* [16] if the function $1^n \mapsto F_n$ is logarithmic-space computable. By analogy we introduce a concept of uniform families of splicing systems.

Let $\Gamma = (V, \Sigma, A, R)$ be a splicing system. We consider a binary encoding of Γ similar to those given for Turing machines (see, e.g. [6]). Here we leave out the details of the encoding.

Definition 3. *Let $f(n) \geq \log n$ be a function from $\mathbb{N}$ to itself. We say that a family of extended splicing systems, $\mathcal{G} = \{\Gamma_n\}_{n\geq 0}$, is $f(n)$-space uniform if the function that maps for each $n \geq 0$ from 1^n to the encoding of Γ_n is computable in deterministic $f(n)$ space.*

Definition 4. *We say that a family of extended splicing systems, $\mathcal{G} = \{\Gamma_n\}_{n\geq 0}$ accepts a language L if the splicing systems in $\mathcal{G}$ have the same terminal alphabet Σ such that $L \subseteq \Sigma^*$ and for all $n \geq 0$, it holds that $L^{=n}$, the length-n portion of L, is equal to that of $L(\Gamma_n)$.*

Now we characterize 1-NSPACE using uniform families of splicing systems.

Theorem 6. *Let $f(n) \geq \log n$ be a function from $\mathbb{N}$ to itself that is (Turing-machine) space-constructible. A language L is in 1-NSPACE$[f(n)]$ if and only if there is an $f(n)$-space uniform family $\mathcal{G} = \{\Gamma_n\}_{n\geq 0}$ of splicing systems that accepts L with the following properties:*

1. *There exists a constant c such that for all $n \geq 0$, each automaton appearing in the rule set of Γ_n has at most c states.*
2. *There exists a constant d such that for all $n \geq 0$ and for all $w \in L^{=n}$, there is a production tree of Γ_n to produce w of height not more than $df(n)$.*

Proof (sketch). We omit the "if"-part, which is based on the nondeterministic one-way algorithm presented in the proof of Theorem 5. To prove the "only if"-part, let L be accepted by a one-way nondeterministic $f(n)$ space-bounded machine M. Let Σ be the input alphabet of M. We assume that a special symbol $\dashv$ not in Σ is appended at the end of the input of M so that M knows the end of the input. We also assume that M accepts only after seeing a $\dashv$. Let Q be the set of states of M. Let q_A be a unique accept state of M. For each symbol $a \in \Sigma$, introduce a new symbol $\hat{a}$. Let $\hat{\Sigma}$ be the collection of all newly introduced symbols and let $\Delta = \Sigma \cup \{\dashv\} \cup \hat{\Sigma}$. Let n be fixed. We construct Γ_n as follows: Let S_n be the set of all configurations of M on an input of length M without the specification of the input head position and without the symbol scanned by the input head. Since M is $f(n)$ space-bounded, each element in S_n

can be encoded using $O(f(n))$ characters. Let α_0 be the initial configuration in S_n and let Θ be the set of all accepting configurations in S_n. Suppose M may nondeterministically transition from a configuration α to another β upon scanning an input symbol a. If the input head moves to the right when M makes the transition, we describe this as $(\alpha, a) \rightarrow \beta$; if the input head does not move when M makes the transition, we describe this as $(\alpha, \hat{a}) \rightarrow \beta$.

Let @ and % be new symbols. We define

$$V_n = \Delta \cup S_n \cup \{@, \%\},$$
$$I_n = \{@, \%\} \cup \{\alpha a \beta \mid (\alpha, a) \rightarrow \beta \text{ is a transition of } M\},$$
$$R_n = \{S_n \Sigma^+ \# a\$\alpha \#(\Sigma \cup \{\dashv\})^+ S_n \mid \alpha \in S_n\} \cup$$
$$\{S_n h \# a \$ \alpha h \# S_n \mid \alpha \in S_n, h \in \hat{\Sigma}\} \cup$$
$$\{S_n \# \hat{h} a \$ \alpha \# h S_n \mid \alpha \in S_n, h \in \Sigma\} \cup$$
$$\{\alpha_0 \Sigma^* \# \dashv \Theta \$ @ \# \lambda, \alpha_0 \# \Sigma^* \$ \lambda \# \&\}.$$

Note that each finite automaton appearing in these rules has at most four states, so its deterministic version has at most 16 states. The constant c thus can be 16. Γ_n produces valid computations of M and gives a string $w \in \sigma^*$ only if it takes M from an initial to an accepting configuration. The system produces a word $x \in L^{=n}$ in a number of rounds bounded by $\lceil \log(h^{f(n)} + 1) \rceil + \lceil \log(n+1) \rceil + 2$ for some fixed constant h. Since $f(n) \geq \log n$, this bound is $\Theta(f(n))$. $\qquad \square$

The following corollary immediately follows from the above theorem.

Corollary 3. *A language L is in 1-NL if and only if there is a logarithmic-space uniform family $\mathcal{G} = \{\Gamma_n\}_{n \geq 0}$ of splicing systems that accepts L with the following properties:*

1. *There exists a constant c such that for all $n \geq 0$, each automaton appearing in the rule set of Γ_n has at most c states and has at most c transitions appearing in the encoding.*
2. *There exists a constant d such that for all $n \geq 0$ and for each $w \in L^{=n}$, there is a production tree of Γ_n to produce w of height not more than $d \log n$.*

5   Splicing Systems Versus Pushdown Automata

Theorem 4 sheds light on the question of whether CFL $\subseteq$ SPLTIME$[O(\log n)]$. Since the closure of 1-NL under the logarithmic-space Turing reducibility (see [7]) is NL, the closure of SPLTIME$[O(\log n)]$ under that reducibility is included in NL. On the other hand, the closure of CFL under the logarithmic-space many-one reducibility, i.e., LOGCFL, is equal to SAC^1, the languages accepted by a logarithmic space uniform, polynomial-size, logarithmic-depth semi-unbounded-fan-in circuits [17]. The class SAC^1 is known to include NL but it is unknown whether the two classes are equal to each other. If CFL $\subseteq$ SPLTIME$[O(\log n)]$, then we have that SAC^1 = NL. Because of this, it appears difficult to settle the question of whether CFL $\subseteq$ SPLTIME$[O(\log n)]$. We show that CFL $\subseteq$ SPLTIME$[O(n)]$. This inclusion follows from the following general result.

Theorem 7. *Let $f(n) \geq \log n$ be an arbitrary function. Let L be a language accepted by a pushdown automaton M with the property that for each member x of L there exists an accepting computation of M on x such that the height of the stack of M never exceeds $f(|x|)$. Then L belongs to SPLTIME$[f(n)]$.*

For lack of space, we omit the proof of this theorem, in which we construct a splicing system which simulates M in a way reminiscent of the system used in Theorem 6.

The standard PDA algorithm for a context-free language uses the Greibach normal form and the stack height is bounded by a linear function of the input size. More precisely, one push operation is executed when a production rule of the form $A \rightarrow BC$ is performed and when alignment of C with the input is postponed until the alignment of B with the input has been completed (see, for example, [6]). Since each nonterminal produces a nonempty word, this property means that the number of symbols in the stack does not exceed the length of the input. This observation immediately yields the following corollary.

Corollary 4. CFL $\subseteq$ SPLTIME$[O(n)]$,

Can we improve the upper bound of SPLTIME$[O(n)]$ to SPLTIME$[O(\log n)]$? This is a hard question to answer. As mentioned earlier, the logarithmic-space reducibility closure of CFL is equal to SAC^1, the languages accepted by polynomial size-bounded, logarithmic depth-bounded, logarithmic-space uniform families of semi-unbounded-fan-in (OR gates have no limits on the number of input signals feeding into them while the number is two for AND gates) circuits. This class clearly solves the reachability problem, so SAC$^1 \subseteq$ NL, but it is not known whether the converse holds. Theorem 4 shows that SPLTIME$[O(\log n)] \subseteq$ 1-NL. Since NL is closed under logarithmic-space reductions, we have that the closure of SPLTIME$[O(\log n)]$ under logarithmic-space reductions is included in NL. So, the hypothesis CFL $\subseteq$ SPLTIME$[O(\log n)]$ implies SAC$^1 =$ NL.

Proposition 2. CFL $\not\subseteq$ SPLTIME$[O(\log n)]$ *unless* SAC$^1 =$ NL.

6   Conclusions and Further Research

In this paper, we laid the foundations of the study of time complexity in splicing systems. We defined the time complexity function in terms of the number of rounds needed to generate a word of length n. Specifically, for each n we define the time complexity of the system at length n to be the maximum of the smallest number of rounds needed to generate the words having length n in the language produced by the system. We showed that the class SPLTIME$[O(\log n)]$ is included in the class 1-NL, and in general SPLTIME$[f(n)] \subseteq$ 1-NSPACE$[f(n)]$. In addition, we saw that SPLTIME$[O(\log n)]$ includes all regular languages and SPLTIME$[O(n)]$ includes all context-free languages. In fact, SPLTIME$[f(n)]$ contains all languages accepted by a pushdown automaton with maximal stack height $f(|x|)$ for a word x.

Our work gives rise to many interesting research questions. Of course, we would like to find an exact characterization of our splicing classes. By our characterization of 1-NSPACE$[f(n)]$ in terms of splicing system we show that equality with SPLTIME$[f(n)]$ is unlikely: it is the class of languages accepted by a $f(n)$-space uniform family of extended splicing systems whose production time is $O(f(n))$ such that each finite automaton appearing in the splicing systems has at most a constant number of states. We believe that our concept will be a useful tool in understanding the intrinsic computational power of splicing systems.

References

1. Book, R.V.: Time-bounded grammars and their languages. Journal of Computer and System Sciences 5(4), 397–429 (1971)
2. Culik II, K., Harju, T.: Splicing semigroups of dominoes and DNA. Discrete Applied Mathematics 31, 261–277 (1991)
3. Gladkiĭ, A.V.: On the complexity of derivations in phase-structure grammars. Algebra i Logika Seminar 3(5-6), 29–44 (1964) (in Russian)
4. Head, T.: Formal language theory and DNA: an analysis of the generative capacity of specific recombinant behaviors. Bulletin of Mathematical Biology 49, 737–759 (1987)
5. Hartmanis, J., Mahaney, S.R.: Languages simultaneously complete for one-way and two-way log-tape automata. SIAM Journal of Computing 10(2), 383–390 (1981)
6. Hopcroft, J E., Ullman, J.D.: Introduction to Automata Theory, Languages and Computation. Addison-Wesley, Reading, MA (1979)
7. Ladner, R.E., Lynch, N.A.: Relativization of questions about logspace computability. Mathematical Systems Theory 10(1), 19–32 (1976)
8. Ogihara, M.: Relating the minimum model for DNA computation and Boolean circuits. Proceedings of the 1999 Genetic and Evolutionary Computation Conference, pp. 1817–1821. Morgan Kaufmann Publishers, San Francisco, CA (1999)
9. Ogihara, M., Ray, A.: The minimum DNA computation model and its computational power. In: Ogihara, M., Ray, A. (eds.) Unconventional Models of Computation, Singapore, pp. 309–322. Springer, Heidelberg (1998)
10. Papadimitriou, C.H.: Computational Complexity. Addison-Wesley, Reading, MA (1994)
11. Păun, G.: Regular extended H systems are computationally universal. Journal of Automata, Languages, Combinatorics 1(1), 27–36 (1996)
12. Păun, G., Rozenberg, G., Salomaa, A.: Computing by splicing. Theoretical Computer Science 168(2), 32–336 (1996)
13. Păun, G., Rozenberg, G., Salomaa, A.: DNA Computing - New Computing Paradigms. Springer-Verlag, Berlin Heidelberg (1998)
14. Pixton, D.: Regularity of splicing languages. Discrete Applied Mathematics 69, 101–124 (1996)
15. Reif, J.H.: Parallel molecular computation. Proceedings of the 7th ACM Symposium on Parallel Algorithms and Architecture, pp. 213–223. ACM Press, New York (1995)
16. Ruzzo, W.: On uniform circuit complexity. Journal of Computer and System Sciences 22, 365–383 (1981)
17. Venkateswaran, H.: Properties that characterize LOGCFL. Journal of Computer and System Sciences 43, 380–404 (1991)

Descriptional Complexity of Bounded Context-Free Languages*

Andreas Malcher[1] and Giovanni Pighizzini[2]

[1] Institut für Informatik, Johann Wolfgang Goethe-Universität
D-60054 Frankfurt am Main, Germany
a.malcher@em.uni-frankfurt.de
[2] Dipartimento di Informatica e Comunicazione, Università degli Studi di Milano
via Comelico 39, I-20135 Milano, Italy
pighizzini@dico.unimi.it

Abstract. Finite-turn pushdown automata (PDA) are investigated concerning their descriptional complexity. It is known that they accept exactly the class of ultralinear context-free languages. Furthermore, the increase in size when converting arbitrary PDAs accepting ultralinear languages to finite-turn PDAs cannot be bounded by any recursive function. The latter phenomenon is known as non-recursive trade-off. In this paper, finite-turn PDAs accepting letter-bounded languages are considered. It turns out that in this case the non-recursive trade-off is reduced to a recursive trade-off, more precisely, to an exponential trade-off. A conversion algorithm is presented and the optimality of the construction is shown by proving tight lower bounds. Furthermore, the question of reducing the number of turns of a given finite-turn PDA is studied. Again, a conversion algorithm is provided which shows that in this case the trade-off is at most polynomial.

Keywords: automata and formal languages, descriptional complexity, finite-turn pushdown automata, recursive trade-offs, bounded languages.

1 Introduction

Finite-turn pushdown automata (PDAs) were introduced in [2] by Ginsburg and Spanier. They are defined by fixing a constant bound on the number of switches between push and pop operations in accepting computation paths of PDAs. The class of languages defined by these models is called the class of *ultralinear languages* and is a proper subclass of the class of context-free languages. It can be also characterized in terms of *ultralinear* and *non-terminal bounded grammars* [2]. (In the special case of 1-turn PDAs, i.e., devices making at most one switch between push and pop operations, we get the class of linear context-free languages).

---

* This work was partially supported by MIUR under the project PRIN "Automi e Linguaggi Formali: aspetti matematici e applicativi".

T. Harju, J. Karhumäki, and A. Lepistö (Eds.): DLT 2007, LNCS 4588, pp. 312–323, 2007.

In [8], descriptional complexity questions concerning finite-turn PDAs were investigated, by showing, among other results, the existence of non-recursive trade-offs between PDAs and finite-turn PDAs. Roughly speaking, this means that for any recursive function $f(n)$ and for arbitrarily large integers n, there exists a PDA of size n accepting an ultralinear language such that any equivalent finite-turn PDA must have at least $f(n)$ states. Thus, a PDA with arbitrary many turns may represent an ultralinear language more succinctly than any finite-turn PDA and the savings in size cannot be bounded by any recursive function.

This phenomenon of non-recursive trade-offs was first observed between context-free grammars and deterministic finite automata (DFAs) in the fundamental paper by Meyer and Fischer [9]. Nowadays, many non-recursive trade-offs are known which are summarized, e.g., in [1] and [7]. In the context of context-free languages non-recursive trade-offs are known to exist between PDAs and deterministic PDAs (DPDAs), between DPDAs and unambiguous PDAs (UPDAs), and between UPDAs and PDAs.

Interestingly, the witness languages used in [9] were defined over an alphabet of two symbols and leave open the unary case which was recently solved in [10] by proving an exponential trade-off. Thus, the non-recursive trade-off in the binary case turns into a recursive trade-off in the unary case. More generally, a careful investigation of the known cases of non-recursive trade-offs reveals that the used witness languages are not bounded resp. word-bounded, i.e., they are not included in some subset of $w_1^* w_2^* \ldots w_m^*$ for some fixed words $w_1, w_2, \ldots, w_m$. So, the question arises whether the above non-recursive trade-offs can be translated to the bounded case or whether the structural limitation on boundedness is one that will allow only recursive trade-offs.

In this paper we tackle this question and restrict ourselves to the case of letter-bounded languages, namely, subsets of $a_1^* \ldots a_m^*$, where $a_1, \ldots, a_m$ are symbols. Our main result shows that for these languages the trade-off between PDAs (or context-free grammars) and finite-turn PDAs becomes recursive. More precisely, in Section 3 we first show that each context-free grammar in Chomsky normal form with h variables generating a letter-bounded set can be converted to an equivalent finite-turn PDA whose size is $2^{O(h)}$. Furthermore, the resulting PDA makes at most $m - 1$ turns where m is the number of letters in the terminal alphabet. In a second step, an exponential trade-off is also shown for arbitrary context-free grammars. We prove (in Section 5) that this result is tight by showing that the size of the resulting PDA and the number of turns cannot be reduced. Note that this result is a generalization of the above-mentioned transformation of unary context-free grammars into finite automata which is presented in [10]. In Section 4 the investigation is further deepened by studying how to reduce the number of turns in a PDA. In particular, given a k-turn PDA accepting a subset of $a_1^* a_2^* \ldots a_m^*$, where $k > m-1$, we show how to build an equivalent $(m-1)$-turn PDA. It turns out that in this case the trade-off is polynomial. This result is also used to prove the optimality of our simulation of PDAs accepting letter-bounded languages by finite-turn PDAs. The results of this paper on letter-bounded languages can be seen as a first step towards proving similar results for the general situation of word-bounded languages. Note that many proofs are omitted in this version of the paper due to space limits.

2   Preliminaries and Definitions

Let Σ^* denote the set of all words over the finite alphabet Σ, with the empty string denoted by ϵ, and $\Sigma^+ = \Sigma^* \setminus \{\epsilon\}$. Given a string $x \in \Sigma^*$, $|x|$ denotes its length. For the sake of simplicity, we will consider languages without the empty word ϵ. However, our results can be easily extended to languages containing ϵ. We assume that the reader is familiar with the common notions of formal language theory as presented in [5].

A *context-free grammar* (CFG, for short) will be denoted as usual as a 4-tuple $G = (V, \Sigma, P, S)$. For productions and derivations, we will use the symbols $\rightarrow$, $\Rightarrow$, $\overset{*}{\Rightarrow}$, and $\overset{*}{\Rightarrow}$, with the usual meanings. If T is a parse tree whose root is labeled with a variable $A \in V$ and such that the labels of the leaves, from left to right, form a string $\alpha \in (V \cup \Sigma)^*$, then we write $T : A \overset{*}{\Rightarrow} \alpha$. Furthermore, we indicate as $\nu(T)$ the set of variables which appear as labels of some nodes in T.

Let $M = (Q, \Sigma, \Gamma, \delta, q_0, Z_0, F)$ be a pushdown automaton [5]. The language accepted by M will be denoted, as usual, as $T(M)$. A *configuration* of M is a triple (q, w, γ) where q is the current state, w the unread part of the input, and γ the current content of the pushdown store. Given two configurations c', c'', we write $c' \vdash c''$ if c'' is an immediate successor of c'. A sequence of configurations on M $(q_1, w_1, \gamma_1) \vdash \ldots \vdash (q_k, w_k, \gamma_k)$ is called *one-turn* if there exists $i \in \{1, \ldots, k\}$ such that

$$|\gamma_1| \leq \ldots \leq |\gamma_{i-1}| \leq |\gamma_i| > |\gamma_{i+1}| \geq \ldots \geq |\gamma_k|$$

A sequence of configurations $c_0 \vdash \ldots \vdash c_m$ is called k-*turn* if there are integers $0 = i_0, \ldots, i_l = m$ with $l \leq k$ such that for $j = 0, \ldots, l-1$ holds: $c_{i_j} \vdash \ldots \vdash c_{i_{j+1}}$ is one-turn. M is a k-turn pushdown automaton if every word $w \in T(M)$ is accepted by a sequence of configurations which is k-turn. Without loss of generality, we make the following assumptions about PDAs (cf. [10]).

(1) at the start of the computation the pushdown store contains only the start symbol Z_0; this symbol is never pushed or popped on the stack;
(2) the input is accepted if and only if the automaton reaches a final state, the pushdown store only contains Z_0 and all the input has been scanned;
(3) if the automaton moves the input head, then no operations are performed on the stack;
(4) every push adds exactly one symbol on the stack.

According to the discussion in [3] the size of a PDA should be defined depending on the number of states, the number of stack symbols, the number of input symbols, and the maximum number of stack symbols appearing in the right hand side of transition rules. In this paper, we consider PDAs in the above defined normal form over a fixed alphabet Σ. Thus, $size(M)$ of a PDA M in normal form is defined as the product of the number of states and the number of stack symbols. The size of a finite automaton is defined to be the number of states. As measures for the size of a context-free grammar $G = (V, \Sigma, P, S)$ we consider the number of non-terminals of G, defined as $\mathrm{Var}(G) = \#V$, and the number of symbols of G, defined as $\mathrm{Symb}(G) = \sum_{(A \to \alpha) \in P}(2 + |\alpha|)$ (cf. [6]). Some general information on descriptional complexity may be found in [1].

3 From Grammars to Finite-Turn PDAs

In this section, we study the transformation of CFGs into finite-turn PDAs. Our main result shows that given a grammar G of size h, we can build an equivalent finite-turn PDA M of size $2^{O(h)}$. Furthermore, if the terminal alphabet of G contains m letters, then M is an $(m-1)$-turn PDA. The tightness of the bounds will be shown in Section 5. For the sake of simplicity, we start by considering CFGs in Chomsky normal form with the measure Var for the size. At the end of the section, we will discuss the generalization to arbitrary context-free grammars, taking into consideration the more realistic measure Symb.

In the following we consider an alphabet $\Sigma = \{a_1, \ldots, a_m\}$ and a CFG $G = (V, \Sigma, P, S)$ in Chomsky normal form with h variables, generating a subset of $a_1^* \ldots a_m^*$. Without loss of generality, we can suppose that each variable of G is *useful*, i.e., for each $A \in V$, there exist terminal strings u, v, w, such that $S \overset{*}{\Rightarrow} uAw \overset{*}{\Rightarrow} uvw$.

It can be proved that with each variable A we can associate at most one pair of indices from $\{1, \ldots, m\}$ as follows:

$$border(A) = \begin{cases} (l, r) & \text{if } A \overset{*}{\Rightarrow} uAv \text{ for some } u \in a_l^+, v \in a_r^+ \\ (l, l) & \text{if } u \in a_l^+ \text{ and } v = \epsilon \text{ for any } A \overset{*}{\Rightarrow} uAv \\ (r, r) & \text{if } u = \epsilon \text{ and } v \in a_r^+ \text{ for any } A \overset{*}{\Rightarrow} uAv \\ \text{undefined} & \text{otherwise.} \end{cases}$$

For the sake of brevity, $border(A)$ will be denoted also as (l_A, r_A).

We now consider the relation $\leq$ on the set of possible borders defined as $(l, r) \leq (l', r')$ if and only if $l \leq l'$ and if $l = l'$ then $r \geq r'$, for all $(l, r), (l', r') \in \{1, \ldots, m\}^2$, with $l \leq r$ and $l' \leq r'$. It is not difficult to verify that $\leq$ is a total order on the set of pairs of indices l, r from $\{1, \ldots, m\}$, such that $l \leq r$.

Actually, we are interested in computing borders of variables belonging to the same derivation tree. In this case, either a variable is a descendant of the other in the tree, and then the interval defined by its border is inside the interval defined by the border of the other variable, or one variable is to the right of the other one, and then the corresponding interval is to the right of the other one. More formally, we can prove the following:

Lemma 3.1. *Given a derivation tree T and two variables $A, B \in \nu(T)$, if $border(A) \leq border(B)$ then either:*

(a) $l_A \leq l_B \leq r_B \leq r_A$, or
(b) $l_A < l_B$, $r_A < r_B$, and $r_A \leq l_B$.

A *partial derivation tree* (or *partial tree*, for short) $U : A \overset{*}{\Rightarrow} vAx$ is a parse tree whose root is labeled with a variable A and all the leaves, with the exception of one whose label is the same variable A, are labeled with terminal symbols.

Given a partial tree $U : A \overset{*}{\Rightarrow} vAx$, any derivation tree $T : S \overset{*}{\Rightarrow} z$ with $A \in \nu(T)$ can be "pumped" using U, by replacing a node labeled A in T with the subtree U. In this way, a new tree $T' : S \overset{*}{\Rightarrow} z'$ is obtained, where $z' = uvwxy$, such that $z = uwy$, $S \overset{*}{\Rightarrow} uAy$, and $A \overset{*}{\Rightarrow} w$. Moreover, $\nu(T') = \nu(T) \cup \nu(U)$.

On the other hand, any derivation tree producing a sufficiently long terminal string can be obtained by pumping a derivation tree of a shorter string with a partial tree. By applying the pumping lemma several times, we can prove that any derivation tree can be obtained by starting from a derivation tree of a "short" string (namely, a string of length at most 2^{h-1}) and iteratively pumping it with "small" partial trees. Furthermore, a sequence of partial trees can be considered such that the sequence of borders of their roots is not decreasing. More precisely:

Lemma 3.2. *Given a derivation tree $T : S \overset{*}{\Rightarrow} z$ of a string $z \in \Sigma^*$, with $|z| > 2^{h-1}$, for some integer $k > 0$ there are:*

- *$k + 1$ derivation trees $T_0, T_1, \ldots, T_k$, where $T_i : S \overset{*}{\Rightarrow} z_i$, $i = 0, \ldots, k$, $0 < |z_0| \leq 2^{h-1}$, $T_k = T$, $z_k = z$;*
- *k partial trees $U_1, \ldots U_k$, where, for $i = 1, \ldots, k$, $U_i : A_i \overset{*}{\Rightarrow} v_i A_i x_i$, $0 < |v_i x_i| < 2^h$, and T_i is obtained by pumping T_{i-1} with U_i.*

Furthermore, $border(A_1) \leq border(A_2) \leq \ldots \leq border(A_k)$.

Example 3.3. The language $L = \{a_1^{n+k} a_2^{k+p} a_3^{p+n} \mid n, k, p > 0\}$ can be generated by a grammar in Chomsky normal form with the following productions:

$$
\begin{array}{lllll}
S \rightarrow A_1 E & S' \rightarrow AB & A \rightarrow A_1 F & B \rightarrow A_2 G & A_1 \rightarrow a_1 \\
E \rightarrow S A_3 & A \rightarrow A_1 A_2 & F \rightarrow A A_2 & G \rightarrow B A_3 & A_2 \rightarrow a_2 \\
E \rightarrow S' A_3 & B \rightarrow A_2 A_3 & & & A_3 \rightarrow a_3
\end{array}
$$

Note that $S \overset{*}{\Rightarrow} a_1 S a_3$, $A \overset{*}{\Rightarrow} a_1 A a_2$, and $B \overset{*}{\Rightarrow} a_2 B a_3$. It is easy to get a tree $T_0 : S \overset{*}{\Rightarrow} a_1^2 a_2^2 a_3^2$ and three partial trees $U' : S \overset{*}{\Rightarrow} a_1 S a_3$, $U'' : A \overset{*}{\Rightarrow} a_1 A a_2$, and $U''' : B \overset{*}{\Rightarrow} a_2 B a_3$. Given integers $n, k, p > 0$, a derivation tree for the string $a_1^{n+k} a_2^{k+p} a_3^{p+n}$ can be obtained considering T_0, and pumping it $n - 1$ times with the tree U', $k - 1$ times with the tree U'', and $p - 1$ with the tree U'''. Note that $border(S) = (1, 3) \leq border(A) = (1, 2) \leq border(B) = (2, 3)$.

At this point we are able to describe an $(m - 1)$-turn PDA recognizing the language generated by the grammar G. Such a PDA implements the following nondeterministic procedure, which builds strings in $L(G)$ starting from short derivation trees and pumping them, by respecting the order stated in Lemma 3.2. The procedure verifies the matching between the generated and the input strings. Its correctness is proved in Lemma 3.4 and Theorem 3.5.

nondeterministically select a tree $T : S \overset{*}{\Rightarrow} a_1^{n_1} a_2^{n_2} \ldots a_m^{n_m}$,
   with $n_1 + n_2 + \ldots + n_m \leq 2^{h-1}$
read $a_1^{n_1}$ from the input
$enabled \leftarrow \nu(T)$
$(l, r) \leftarrow (1, m)$ // the "work context"
$iterate \leftarrow$ nondeterministically choose $true$ or $false$
while $iterate$ **do**
   nondeterministically select a tree $U : A \overset{+}{\Rightarrow} vAx$, with $0 < |vx| < 2^h$,
      $A \in enabled$, and $(l, r) \leq border(A) = (l_A, r_A)$
   if $r < r_A$ **then** //new context to the right of the previous one
      for $j \leftarrow l + 1$ **to** $r - 1$ **do**
         consumeInputAndCounter(j)
      endfor
      for $j \leftarrow r$ **to** l_A **do**
         consumeInputAndCounter(j)
         consumeInputAndStack(j)
      endfor
   else //$r_A \leq r$: new context inside the previous one
      for $j \leftarrow l + 1$ **to** l_A **do**
         consumeInputAndCounter(j)
      endfor
   endif
   $(l, r) \leftarrow (l_A, r_A)$
   read v from the input
   if $r \neq l$ **then** push x on the stack
               else read x from the input
   endif
   $enabled \leftarrow enabled \cup \nu(U)$
   $iterate \leftarrow$ nondeterministically choose $true$ or $false$
endwhile
for $j \leftarrow l + 1$ **to** $r - 1$ **do**
   consumeInputAndCounter(j)
endfor
for $j \leftarrow r$ **to** m **do**
   consumeInputAndCounter(j)
   consumeInputAndStack(j)
endfor
if the end of the input has been reached **then** accept
                        else reject
endif

In the previous procedure and in the following macros, the instruction "read x from the input tape," for $x \in \Sigma^*$, actually means that the automaton verifies whether or not x is a prefix of the next part of the input. If the outcome of this test is positive, then the input head is moved immediately to the right of x, namely, x is "consumed," otherwise the machine stops and rejects.

The macros are defined as follows:

ConsumeInputAndCounter(j):

while $n_j \geq 0$ **do**

    read a_j from the input tape

    $n_j \leftarrow n_j - 1$

endwhile

ConsumeInputAndStack(j):

while the symbol at the top of the stack is a_j **do**

    read a_j from the input tape

    pop

endwhile

In order to prove that the pushdown automaton described in the previous procedure accepts the language $L(G)$ generated by the given grammar G, it is useful to state the following lemma:

Lemma 3.4. *Consider one execution of the previous procedure. Let $T_0 : S \overset{\star}{\Rightarrow} a_1^{n_1} \ldots a_m^{n_m}$ be the tree selected at the beginning of such an execution. At every evaluation of the condition of the while loop, there exists a tree $T : S \overset{\star}{\Rightarrow} a_1^{k_1} \ldots a_m^{k_m}$, for some $k_1, \ldots, k_m \geq 0$, such that*

- *the scanned input prefix is $z = a_1^{k_1} \ldots a_l^{k_l}$;*
- *the pushdown store contains the string $\gamma = a_r^{p_r} \ldots a_m^{p_m}$, where, for $j = r, \ldots, m$, $p_j \geq 0$ and $k_j = p_j + n_j$;*
- *for $l < j < r$, $k_j = n_j$;*
- *enabled $= \nu(T)$.*

As a consequence:

Theorem 3.5. *The pushdown automaton M described by the previous procedure is an $(m-1)$–turn PDA accepting the language $L(G)$.*

Proof. First, we show that the number of turns of the PDA M defined in the above procedure is at most $m - 1$. To this aim we count how many times the automaton can switch from push operations to pop operations.

At each iteration of the while loop, the automaton can perform push operations. Pop operations are possible only by calling the macro consumeInputAndStack. This happens first in the while loop, when the condition $r < r_A$ holds true, i.e., when the new context (l_A, r_A) is to the right of the previous context (l, r), and secondly after the end of the loop.

Let $(l_1, r_1), (l_2, r_2), \ldots (l_k, r_k)$ be the sequence of the contexts which in the computation make the above-mentioned condition hold true. Hence, $1 < l_1 < \ldots < l_k \leq m$, that implies $k \leq m - 1$. If $k < m - 1$, then the PDA M makes at most $k \leq m - 2$ turns in the simulation of the while loop and one more turn after the loop. So the total number of turns is bounded by $m - 1$.

Now, suppose that $k = m - 1$. This implies that $l_k = m = r_k$. Before reaching the context (l_k, r_k), at most $m - 2$ turns can be performed. When the automaton switches to the new context $(l_k, r_k) = (m, m)$, it can make pop operations, by calling the macro consumeInputAndStack(m). This requires one more turn. After that, the automaton can execute further iterations, using the same context (m, m). By reading the procedure carefully, we can observe that it never executes further push operations. Finally, at the exit of the loop, further pop operations can be executed (consumeInputAndStack). Hence, the total number of turns is bounded by $m - 1$.

To prove that the language $L(G)$ and the language accepted by the automaton defined in the above procedure coincide it is enough to observe that given a string $z \in L(G)$, the procedure is able to guess the tree T_0 and the partial trees $U_1, \ldots, U_k$ of Lemma 3.2, recognizing in this way z. Conversely, using Lemma 3.4, it is easy to show that each string accepted by the procedure should belong to $L(G)$. $\square$

Corollary 3.6. *Let $\Sigma = \{a_1, \ldots, a_m\}$. For any context–free grammar G in Chomsky normal form with h variables generating a letter-bounded language $L \subseteq a_1^* \ldots a_m^*$, there exists an equivalent $(m-1)$-turn PDA M with $2^{O(h)}$ states and $O(1)$ stack symbols.*

Proof. The most expensive information that the automaton defined in the previous procedure has to remember in its state are the $m-1$ counters bounded by 2^{h-1}, and the set *enabled*, which is a subset of V. For the pushdown store an alphabet with $m+1$ symbols can be used. With a small modification, the pushdown store can be implemented using only two symbols (one symbol to keep a counter p_j and another one to separate two consecutive counters), and increasing the number of states by a factor m, to remember what input symbol a_j the stack symbol A is representing. $\square$

Using standard techniques, a PDA of size n can be converted to an equivalent CFG in Chomsky normal form with $O(n^2)$ variables. Hence, we easily get:

Corollary 3.7. *Each PDA of size n accepting a subset of $a_1^* \ldots a_m^*$ can be simulated by an equivalent $(m-1)$-turn PDA of size $2^{O(n^2)}$.*

We now consider the situation when the given CFG is not necessarily in Chomsky normal form.

Corollary 3.8. *Let $\Sigma = \{a_1, \ldots, a_m\}$. For any context–free grammar G with $\mathrm{Symb}(G) = h$ and generating a letter-bounded language $L \subseteq a_1^* \ldots a_m^*$, there exists an equivalent $(m-1)$-turn PDA M with $2^{O(h)}$ states and $O(1)$ stack symbols.*

Proof. At first, it can be observed that Lemma 3.2 is true not only for CFGs in Chomsky normal form but also for CFGs whose productions have right hand sides of length at most 2. Thus, all arguments in Section 3 are also true for such "normalized" CFGs. It can be shown that any CFG G can be converted to an equivalent CFG G' such that the length of the right hand side of any production belonging to G' is at most 2 and $\mathrm{Var}(G') \leq \mathrm{Symb}(G)$. With similar arguments as in Corollary 3.6 we obtain the claim. $\square$

4   Reducing the Number of Turns

By the results presented in Section 3, each context-free subset of $a_1^* \ldots a_m^*$ can be accepted by an $(m-1)$-turn PDA. In particular, Corollary 3.7 shows that the size of an $(m-1)$-turn PDA equivalent to a given PDA of size n accepting a subset of $a_1^* \ldots a_m^*$, is at most exponential in the square of n.

In this section, we further deepen this kind of investigation by studying how to convert an arbitrary k-turn PDA accepting a language $L \subseteq a_1^* a_2^* \ldots a_m^*$ to an $(m-1)$-turn PDA. It turns out that the increase in size is at most polynomial.

Let us start by considering the unary case, i.e., $m = 1$, which turns out to be crucial to get the simulation in the general case. All PDAs we consider are in normal form. Then we know that at most one symbol is pushed on the stack in every transition.

Lemma 4.1. *Let M be a PDA accepting a unary language L. Let $L(q_1, A, q_2)$ be the set of all words which are processed by 1-turn sequences π of configurations starting with some stack height h in a state q_1 and having A as topmost stack symbol and ending with the same stack height h in some state q_2 and having A as topmost stack symbol. Then, $L(q_1, A, q_2)$ can be recognized by an NFA M' such that $size(M') \leq n^2$ and $n = size(M)$.*

Proof. Consider the following CFG G with start symbol $[q_1, A, q_2]$ having the following rules. Let $p, p', q, q' \in Q$, $Z \in \Gamma$, and $\sigma \in \{a, \epsilon\}$.

(1) $[p, Z, q] \to \sigma[p', Z, q]$, if $\delta(p, Z, \sigma) \ni (p', -)$,
(2) $[p, Z, q] \to \sigma[p, Z, q']$, if $\delta(q', Z, \sigma) \ni (q, -)$,
(3) $[p, Z, q] \to [p', Z', q']$, if $\delta(p, Z, \epsilon) \ni (p', push(Z'))$ and $\delta(q', Z', \epsilon) \ni (q, pop)$,
(4) $[p, Z, q] \to \epsilon$, if $p = q$.

We want to describe how M' simulates a 1-turn sequence π. We simulate the parts of π with A as topmost stack symbol and stack height h with rules (1) and (2). The first part from the beginning up to the first push operation is simulated using rule (1). The second part starting at the end of the computation and going backwards up to the last pop operation is simulated with rule (2). We may change nondeterministically between rules (1) and (2). This is possible, since the input is unary. Having simulated the parts of π with stack height h it is decided nondeterministically to proceed with simulating the parts of π with stack height $h + 1$. Rule (3) simulates a push operation and the corresponding pop operation. Then, rules (1) and (2) can be again used to simulate the parts of π with stack height $h + 1$. Now, we iterate this behavior and simulate all computational steps in π while the stack height simulated is growing. Finally, we can terminate the derivation when the stack height has reached its highest level and all computational steps have been simulated.

Construct an NFA $M' = (Q', \Sigma, \delta', (q_1, A, q_2), F')$ as follows: $Q' = Q \times \Gamma \times Q$ and $F' = \{(q, Z, q) \mid q \in Q, Z \in \Gamma\}$. For $\sigma \in \{a, \epsilon\}$ the transition function δ' is defined as follows:

(1) $\delta'((p, Z, q), \sigma) \ni (p', Z, q)$, if $\delta(p, Z, \sigma) \ni (p', -)$,
(2) $\delta'((p, Z, q), \sigma) \ni (p, Z, q')$, if $\delta(q', Z, \sigma) \ni (q, -)$,
(3) $\delta'((p, Z, q), \sigma) \ni (p', Z', q')$, if $\delta(p, Z, \epsilon) \ni (p', push(Z'))$ and $\delta(q', Z', \epsilon) \ni (q, pop)$. $\qquad\square$

Corollary 4.2. *Let M be some 1-turn PDA accepting a unary language L. Then, an equivalent NFA M' can be constructed such that $size(M') \leq n^2 + 1$ and $n = size(M)$.*

Proof. We can use the above construction, but additionally have to guess in a first step in which state a computation ends. Therefore, we add a new start symbol S and add rules $S \rightarrow [q_0, Z_0, q_f]$ for all $q_f \in F$. For the NFA construction we add a new initial state q_0' and the following rules $\delta'(q_0', \epsilon) \ni (q_0, Z_0, q_f)$, for all $q_f \in F$ to M'.

It is easy to observe that the parts of π with stack height one can be again simulated with rules (1) and (2). The remaining part of the simulation is identical to the above described construction. □

A subcomputation π' is called *strong* of level A if it starts with some stack height h and topmost stack symbol A, ends with the same stack height h, and in all other configurations of π' the stack height is greater than h.

Lemma 4.3. *Let M be some k-turn PDA accepting a unary language L. Let $L(q_1, A, q_2)$ be the set of all words which are processed by sequences π of strong computations of level A which, additionally, start in some state q_1 and end in some state q_2. It can be observed that all words in $L(q_1, A, q_2)$ are accepted with $j \leq k$ turns. Then, $L(q_1, A, q_2)$ can be accepted by an NFA M' such that $size(M') \in O(n^{2\lfloor \log_2 j \rfloor + 2})$ and $n = size(M)$.*

Proof (Sketch). The construction is very similar to the above described construction. Additionally, we store the number of turns, which have to be simulated, in the fourth component of the non-terminals. There are two cases how π may look like. In the first case (type I, cf. Fig. 1, left) π consists of more than two strong computations of level A. We introduce a new rule (5) which is used to decompose a sequence of strong computations with i turns into two subsequences with i_1 and i_2 turns, respectively. A resulting subsequence is then either again of type I and can be again decomposed with the new rule (5), or it is of type II, i.e., it consists of one strong computation of level A (cf. Fig. 1, right). If this computation is 1-turn, it can be simulated with the rules (1) to (3) and finished with rule (4). If it is not 1-turn, we can reduce it to a sequence of strong computations of level B by using the rules (1) to (3). Then, the same analysis can be made for strong computations of level B. An induction on the number of turns shows that G generates $L(q_1, A, q_2)$. □

Fig. 1. The two cases arising in the construction in Lemma 4.3

Corollary 4.4. *Let M be some k-turn PDA accepting a unary language L. Then, an equivalent NFA M' can be constructed with $size(M') \in O(n^{2\lfloor \log_2 k \rfloor + 2})$ and $n = size(M)$.*

Proof. Observe that an accepting computation in M is a sequence of strong computations of level Z_0 starting in q_0 and ending in some accepting state.   □

Now, we are able to consider the general case, i.e., $m \geq 1$ and claim the following.

Theorem 4.5. *Let M be some k-turn PDA accepting a letter-bounded language $L \subseteq a_1^* a_2^* \ldots a_m^*$. Then, an equivalent $(m-1)$-turn PDA M' can be constructed such that $\text{size}(M') \in O(m^6 n^{4\lfloor \log_2 k \rfloor + 8})$ and $n = \text{size}(M)$.*

It has been shown in the previous section that any $L \subseteq a_1^* a_2^* \ldots a_m^*$ can be accepted by an $(m-1)$-turn PDA. If L is accepted by a k-turn PDA such that $k > m - 1$, then some turns are in a way "not necessary." It is a result of the proof of this theorem that this finite number of additional turns takes place within unary parts of the input, i.e., while reading some input a_i^* with $1 \leq i \leq m$. With the help of the construction of Lemma 4.3 these parts can be accepted by NFAs and hence do not affect the stack height in the construction of an $(m-1)$-turn PDA accepting L.

Corollary 4.6. *The trade-offs between finite-turn pushdown automata that accept letter-bounded languages are at most polynomial.*

5   Lower Bounds

In this section we show the optimality of the simulation of grammars generating letter-bounded languages by finite-turn PDAs (Corollary 3.6), and of some other simulation results presented in the paper. Even in this case, the preliminary investigation of the unary case will be useful to afford the general case.

Theorem 5.1. *For any integer $n \geq 1$, consider the language $L_n = \{a^{2^n}\}$.*

(1) *L_n can be generated by some CFG in Chomsky normal form with $n + 1$ non-terminals.*
(2) *Every NFA accepting L_n needs at least 2^n states.*
(3) *For each $k > 0$, every k-turn PDA accepting L_n is at least of size 2^{cn} for some constant $c > 0$ and any sufficiently large n.*

From Theorem 5.1(3), it turns out that for each integer m the simulation result stated in Corollary 3.6 is optimal. The witness languages are unary. Hence, they can be also accepted by "simpler" devices, i.e., finite automata or PDAs with less than $m-1$ turns. We now show the optimality in a stronger form, by exhibiting, for each integer m, a family of witness languages that cannot be accepted with less than $m-1$ turns.

Theorem 5.2. *Given the alphabet $\Sigma = \{a_1, \ldots, a_m\}$, for any integer $n \geq 1$ consider the language*

$$\tilde{L}_n = \{a_1^{n_0+n_1} a_2^{n_1+n_2} \ldots a_{m-1}^{n_{m-2}+n_{m-1}} a_m^{n_{m-1}} \mid n_0 = 2^n, n_1 \geq 1, \ldots, n_{m-1} \geq 1\}.$$

(1) $\tilde{L}_n$ is generated by some CFG in Chomsky normal form with $n + 4m - 3$ non-terminals.

(2) $\tilde{L}_n$ is accepted by an $(m - 1)$-turn PDA of size $2^{O(n)}$.

(3) For each integer $k \geq m - 1$, every k-turn PDA accepting $\tilde{L}_n$ is at least of size 2^{cn} for some constant $c > 0$ and any sufficiently large n.

(4) $\tilde{L}_n$ cannot be accepted by any PDA which makes less than $m - 1$ turns.

Remark that we have considered so far only CFGs in Chomsky normal form and the measure Var. It is easy to observe that we also obtain exponential trade-offs when considering the measure Symb. This shows that the result of Corollary 3.8 is also optimal. Since $\tilde{L}_n$ can be accepted by a PDA of size $O(n)$, we obtain that the result of Corollary 3.7 is nearly optimal.

We complete this section by considering again the unary case. In particular, we prove that the upper bound stated in Corollary 4.2 is tight.

Theorem 5.3. *Consider the language family*

$$L'_n = \{a^t \mid t \geq 0 \wedge t \equiv 0 \bmod n \wedge t \equiv 0 \bmod n + 1\}$$

for natural numbers $n \geq 2$. Then each L'_n can be accepted by some 1-turn PDA of size $2n + 1$, but every NFA accepting L'_n needs at least $n^2 + n$ states.

References

1. Goldstine, J., Kappes, M., Kintala, C.M.R., Leung, H., Malcher, A., Wotschke, D.: Descriptional complexity of machines with limited resources. Journal of Universal Computer Science 8(2), 193–234 (2002)
2. Ginsburg, S., Spanier, E.H.: Finite-turn pushdown automata. SIAM Journal on Control 4(3), 429–453 (1966)
3. Harrison, M.A.: Introduction to Formal Language Theory. Addison-Wesley, Reading MA (1978)
4. Holzer, M., Kutrib, M.: Nondeterministic descriptional complexity of regular languages. International Journal of Foundations of Computer Science 14(6), 1087–1102 (2003)
5. Hopcroft, J.E., Ullman, J.D.: Introduction to Automata Theory, Languages and Computation. Addison-Wesley, Reading MA (1979)
6. Kelemenova, A.: Complexity of Normal Form Grammars. Theoretical Computer Science 28, 299–314 (1984)
7. Kutrib, M.: The phenomenon of non-recursive trade-offs. International Journal of Foundations of Computer Science 16(5), 957–973 (2005)
8. Malcher, A.: On recursive and non-recursive trade-offs between finite-turn pushdown automata. In: Descriptional Complexity of Formal Systems (DCFS 2005), Università degli Studi di Milano, Rapporto Tecnico 06-05, pp. 215–226 (2005)
9. Meyer, A.R., Fischer, M.J.: Economy of descriptions by automata, grammars, and formal systems. IEEE Symp. on Foundations of Computer Science, pp. 188–191 (1971)
10. Pighizzini, G., Shallit, J., Wang, M.-W.: Unary context-free grammars and pushdown automata, descriptional complexity and auxiliary space lower bounds. Journal of Computer and System Sciences 65, 393–414 (2002)

Definable Transductions and Weighted Logics for Texts

Christian Mathissen[*]

Institut für Informatik, Universität Leipzig
D-04109 Leipzig, Germany
mathissen@informatik.uni-leipzig.de

Abstract. A text is a word together with an additional linear order on it. We study quantitative models for texts, i.e. text series which assign to texts elements of a semiring. We consider an algebraic notion of recognizability following Reutenauer and Bozapalidis and show that recognizable text series coincide with text series definable in weighted logics as introduced by Droste and Gastin. In order to do so, we study certain definable transductions and show that they are compatible with weighted logics. Moreover, we show that the behavior of weighted parenthesizing automata coincides with certain definable series.

1 Introduction

Texts as introduced by Rozenberg and Ehrenfeucht [9] extend the model of words by an additional linear order. The theory of texts originates in the theory of 2-structures (cf. [8]) and it turns out that texts represent an important subclass of 2-structures, namely T-structures [10]. Moreover, Ehrenfeucht and Rozenberg proposed texts as a well-suited model for natural texts that may carry in its tree-like structure grammatical information [10, p.264].

A number of authors [11,14,15] have investigated classes of text languages such as the families of context-free, equational or recognizable text languages and developed a language theory. In particular, the fundamental result of Büchi on the coincidence of recognizable and definable languages has been extended to texts [15]. Recently, Droste and Gastin [5] introduced weighted logics over words and showed a Büchi-type characterization for weighted automata over words. They enrich the language of monadic second order logic with values from a semiring in order to add quantitative expressiveness. Since they define their logic for arbitrary commutative semirings, the framework is very flexible, e.g. one may now express how often a certain property holds, how much execution time a process needs or how reliable it is. The result of Droste and Gastin has been extended to trees, traces, pictures and infinite words [6,7,16,17].

In this paper we consider quantitative aspects of texts and study weighted logics for them. We extend both results, that of Hoogeboom and ten Pas to a weighted setting and that of Droste and Gastin to texts. However, rather than using a combinatorial automaton model we follow Hoogeboom and ten Pas who considered recognizability in the algebraic sense. We regard a weighted algebraic recognizability concept for general algebras following a line of research initiated by Reutenauer [19] and continued

---

* Supported by the GK 446 of the German Research Foundation.

by Bozapalidis [2]. It generalizes weighted automata on words and trees as well as the notion of recognizable languages as defined by Mezei and Wright in the 1960s [18].

In order to show the coincidence of recognizable series with the ones definable by certain sentences in weighted logics, we refine the transductions from texts to terms and vice versa given by Hoogeboom and ten Pas such that they are compatible with weighted logics. Therefore, we study a certain subclass of Courcelle's definable transductions [3] and show that it preserves definability with respect to weighted logics. This tool enables us to easily transfer results on weighted logics to different structures.

An important subclass of texts, the class of alternating texts, forms the free bisemigroup and is isomorphic to the class of the so-called sp-biposet introduced by Ésik and Németh in [12]. In the last section we will generalize the parenthesizing automata of Ésik and Németh to a weighted setting and show that their behaviors are exactly the series definable by certain sentences in weighted logics.

We point out that our method extends to classes of graphs where there are similar pairs of transductions as for texts. This applies e.g. to classes of graphs where the modular decompositions can be defined by certain restricted formulae in the graph itself, i. e. to classes of graphs that are, in terminology of Courcelle, "RMSO-parsable". This will be subject of further research.

2   Recognizable Series over General Algebras

Let Σ be a finite ranked alphabet interpreted as a functional signature and let $\mathrm{rk}(f) \in \mathbb{N}$ denote the rank of f for all $f \in \Sigma$. Let $\mathcal{C}$ be a finitely generated Σ-algebra. We fix a finite generating set $\Delta \subseteq \mathcal{C}$. We recall the following definition:

Definition 2.1 (Mezei & Wright [18]). *A $\mathcal{C}$-language $L \subseteq \mathcal{C}$ is* recognizable *if there is a finite Σ-algebra $\mathscr{A}$ and a homomorphism $\varphi : \mathcal{C} \to \mathscr{A}$ such that $\varphi^{-1}(\varphi(L)) = L$.*

The free Σ-algebra over Δ is denoted $T_\Sigma(\Delta)$ and comprises all Σ-terms or equivalently all Σ-trees over Δ. Let $\eta_\mathcal{C} : T_\Sigma(\Delta) \to \mathcal{C}$ denote the unique epimorphism extending $\mathrm{id}(\Delta)$. Let x be a fresh symbol. The set of contexts $\mathrm{CTX}(\Sigma, \Delta) \subseteq T_\Sigma(\Delta \cup \{x\})$ is the set of trees where x appears at exactly one leaf. For $s \in \mathcal{C}$ and $\tau \in \mathrm{CTX}(\Sigma, \Delta)$, $\tau[s]$ denotes the value of the term function of τ on $\mathcal{C}$ at s.

Similar to Definition 2.1, we introduce a concept of recognizability for (formal) $\mathcal{C}$-series, i. e. for functions from $\mathcal{C}$ to a semiring $\mathbb{K}$. A *semiring* $\mathbb{K}$ is an algebraic structure $(\mathbb{K}, +, \cdot, 0, 1)$ such that $(\mathbb{K}, +, 0)$ is a commutative monoid, $(\mathbb{K}, \cdot, 1)$ is a monoid, multiplication distributes over addition and 0 acts absorbing. If multiplication is commutative, then $\mathbb{K}$ is a *commutative semiring*. If addition is idempotent, then $\mathbb{K}$ is an *idempotent semiring*. We call a semiring *locally finite* if any finitely generated subsemiring is finite. Examples for semirings comprise the trivial Boolean algebra $\mathbb{B} = (\{0, 1\}, \vee, \wedge, 0, 1)$ and the natural numbers $(\mathbb{N}, +, \cdot, 0, 1)$ as well as the tropical semiring $(\mathbb{N} \cup \{\infty\}, \min, +, \infty, 0)$ and the arctical semiring $(\mathbb{N} \cup \{-\infty\}, \max, +, -\infty, 0)$ which are used to model problems in operations research. Important examples are also the probabilistic semiring $([0, 1], \max, \cdot, 0, 1)$ and the semiring of formal languages $(\mathscr{P}(\Delta^*), \cup, \cap, \emptyset, \Delta^*)$. *Let in the sequel $\mathbb{K}$ be a commutative semiring such that $0 \neq 1$.*

A $\mathbb{K}$-*semimodule* M is an Abelian monoid $(M, +)$ together with a scalar multiplication $\cdot : \mathbb{K} \times M \to M$ such that for all $k, l \in \mathbb{K}$ and $m, n \in M$ we have

$$k \cdot (m + n) = k \cdot m + k \cdot n, \quad (k + l) \cdot m = k \cdot m + l \cdot m, \quad (kl) \cdot m = k \cdot (l \cdot m),$$
$$1 \cdot m = m, \qquad\qquad\qquad 0 \cdot m = 0.$$

A submonoid N of M is a *subsemimodule* if $\mathbb{K}N \subseteq N$. It is *finitely generated* if $N = \mathbb{K} \cdot m_1 + \ldots + \mathbb{K} \cdot m_n$ for some $m_1, \ldots, m_n \in M$.

A $\mathbb{K}$-Σ-*algebra* $\mathscr{A} = (\mathscr{A}, (\mu_f)_{f \in \Sigma})$ consists of a $\mathbb{K}$-semimodule $\mathscr{A}$ together with multilinear operations μ_f of rank $\mathrm{rk}(f)$ (cf. [1]). Letting μ_f interpret the function symbol $f \in \Sigma$, $\mathscr{A}$ becomes a Σ-algebra. A $\mathbb{K}$-Σ-algebra is said to have *finite rank* if it is a finitely generated $\mathbb{K}$-semimodule.

Example 2.2. In the following, $\mathbb{K}\langle\langle \mathcal{C} \rangle\rangle$ denotes the set of (formal) $\mathcal{C}$-series. Together with pointwise addition and the scalar multiplication $(k \cdot S, s) = k \cdot (S, s)$ for all $k \in \mathbb{K}$, $S \in \mathbb{K}\langle\langle \mathcal{C} \rangle\rangle$ and $s \in \mathcal{C}$ it is a $\mathbb{K}$-semimodule. The set $\mathbb{K}\langle \mathcal{C} \rangle$ of series P having finite support, i. e. where $\{s \in \mathcal{C} \mid (P, s) \neq 0\}$ is finite, is a subsemimodule of $\mathbb{K}\langle\langle \mathcal{C} \rangle\rangle$. It is the free $\mathbb{K}$-semimodule over $\mathcal{C}$. Hence, any $S : \mathcal{C} \to \mathbb{K}$ extends linearly to $\mathbb{K}\langle \mathcal{C} \rangle$. We will not distinguish between S and its linear extension. $P \in \mathbb{K}\langle \mathcal{C} \rangle$ is called *polynomial*.

We equip the $\mathbb{K}$-semimodule $\mathbb{K}\langle \mathcal{C} \rangle$ with multilinear operations in order to make it a $\mathbb{K}$-Σ-algebra. We define

$$(\mu_f(P_1, \ldots, P_n), s) = \sum_{\substack{s_1, \ldots, s_n \in \mathcal{C} \\ f(s_1, \ldots, s_n) = s}} (P_1, s_1) \cdot \ldots \cdot (P_n, s_n).$$

Note, as the P_i are polynomials, the sum is in fact finite. It is not hard to see that this definition indeed gives multilinear operations μ_f. Hence, $\mathbb{K}\langle \mathcal{C} \rangle$ is a $\mathbb{K}$-Σ-algebra and thus a Σ-algebra. Identifying $s \in \mathcal{C}$ with the polynomial that maps s to 1 and any other element of $\mathcal{C}$ to 0, $\mathcal{C}$ becomes a subalgebra of $\mathbb{K}\langle \mathcal{C} \rangle$.

We interpret $\mathbb{K}$-Σ-algebras as algebras in the sense of universal algebra over the signature $(+, (k \cdot)_{k \in \mathbb{K}}, (\mu_f)_{f \in \Sigma})$. Semimodules are algebras over the signature $(+, (k \cdot)_{k \in \mathbb{K}})$. The notion of a $\mathbb{K}$-Σ-homomorphism and a $\mathbb{K}$-Σ-epimorphism as well as the notion of a congruence are defined as usual in universal algebra.

Remark 2.3. It is not hard to see that $\mathbb{K}\langle T_\Sigma(\Delta) \rangle$ is the free $\mathbb{K}$-Σ-algebra over Δ. Hence, for any $\mathbb{K}$-Σ-algebra $\mathscr{A}$, any mapping $\mu_\mathscr{A} : \Delta \to \mathscr{A}$ extends uniquely to a $\mathbb{K}$-Σ-homomorphism $\mu_\mathscr{A} : \mathbb{K}\langle T_\Sigma(\Delta) \rangle \to \mathscr{A}$.

For any function $h : A \to B$ the *kernel of h* denoted $\ker(h)$ is the set $\{(x, y) \in A^2 \mid h(x) = h(y)\}$. If h is a homomorphism, then $\ker(h)$ is a congruence. Now, we are ready to define a general notion of weighted recognizability.

Definition 2.4. *A $\mathcal{C}$-series $S : \mathcal{C} \to \mathbb{K}$ is recognizable if there is a $\mathbb{K}$-Σ-algebra of finite rank $\mathscr{A}$ and a $\mathbb{K}$-Σ-epimorphism $\varphi : \mathbb{K}\langle \mathcal{C} \rangle \to \mathscr{A}$ such that $\ker(\varphi) \subseteq \ker(S)$.*

Note that the definition is independent of the set of constants, i. e. independent of the symbols of Σ of rank 0. Hence, we may e.g. add constants from Δ to Σ without altering the class of recognizable series.

First, we show that Definition 2.4 generalizes Definition 2.1. For a language $L \subseteq \mathcal{C}$ let $\mathbb{1}_L$ denote the characteristic series of L.

Proposition 2.5. *A language $L \subseteq C$ is recognizable iff $\mathbb{1}_L : C \to \mathbb{B}$ is recognizable.*

Proof. (*If*). Let $L \subseteq C$ and let $\mathbb{1}_L : C \to \mathbb{B}$ be recognized by $\varphi : \mathbb{B}\langle C \rangle \to \mathscr{A}$. Since $\ker(\varphi) \subseteq \ker(S)$, it is easy to see that $\varphi_{|C}$ saturates L.

(*Only if*). Let $L \subseteq C$ be recognized by $\varphi : C \to \mathscr{A}$. We extend φ to a $\mathbb{B}$-Σ-epimorphism $\varphi : \mathbb{B}\langle C \rangle \to \mathbb{B}\langle \varphi(C) \rangle$. We show that $\ker(\varphi) \subseteq \ker(S)$ which concludes the proof. Let $P_1, P_2 \in C$ with $\varphi(P_1) = \varphi(P_2)$. We may interpret P_1, P_2 as finite subsets of C. We have $(S, P_1) = 1$ iff there is $c_1 \in P_1$ with $(S, c_1) = 1$ iff there is $c_1 \in P_1$ with $\varphi(c_1) \in \varphi(L)$ iff there is $c_2 \in P_2$ with $\varphi(c_2) \in \varphi(L)$ iff $(S, P_2) = 1$. Hence, $(S, P_1) = (S, P_2)$. $\qquad\square$

We say a formal power series $S : \Delta^* \to \mathbb{K}$ is regular if it is the behavior of some weighted finite automaton. Reutenauer proved the following for commutative rings. His proof also works for locally finite commutative semirings.

Proposition 2.6 (Reutenauer [19]). *Let $\mathbb{K}$ be a commutative ring or let $\mathbb{K}$ be a locally finite commutative semiring. A formal power series is recognizable iff it is regular.*

Let $S : C \to \mathbb{K}$ and let $\sim_S = \{(P_1, P_2) \in \mathbb{K}\langle C \rangle \times \mathbb{K}\langle C \rangle \mid (S, \tau[P_1]) = (S, \tau[P_2])$ for all $\tau \in \mathrm{CTX}(\Sigma, \Delta)\}$. It is not hard to see that this is a $\mathbb{K}$-Σ-congruence. Let $\sim$ be any congruence contained in $\ker(S)$ and let $(P_1, P_2) \in \sim$. Then $(\tau[P_1], \tau[P_2]) \in \sim$ for any $\tau \in \mathrm{CTX}(\Sigma, \Delta)$ as $\sim$ is a congruence. Therefore, we have $(S, \tau[P_1]) = (S, \tau[P_2])$ for all $\tau \in \mathrm{CTX}(\Sigma, \Delta)$. This shows that $\sim \subseteq \sim_S$ and, hence, that $\sim_S$ is the greatest congruence fully contained in $\ker(S)$. We define $\mathscr{A}_S = \mathbb{K}\langle C \rangle / \sim_S$, the *syntactic $\mathbb{K}$-Σ-algebra* of S. Note this definition is independent of the choice of Δ. We conclude:

Proposition 2.7. *A series $S : C \to \mathbb{K}$ is recognizable iff $\mathscr{A}_S$ is of finite rank.*

Lemma 2.8. *Let C_1, C_2 be finitely generated Σ-algebras, let $\psi : C_1 \to C_2$ be an epimorphism and let $S : C_2 \to \mathbb{K}$. Then $\psi^{-1}(S)$ is recognizable iff S is recognizable.*

Proof. (*If*). Extend ψ linearly to a $\mathbb{K}$-Σ-epimorphism $\psi : \mathbb{K}\langle C_1 \rangle \to \mathbb{K}\langle C_2 \rangle$. As S is recognizable, there is a $\mathbb{K}$-Σ-algebra $\mathscr{A}$ of finite rank and a $\mathbb{K}$-Σ-epimorphism $\varphi : \mathbb{K}\langle C_2 \rangle \to \mathscr{A}$ such that $\ker(\varphi) \subseteq \ker(S)$. Hence, $\ker(\psi \circ \varphi) \subseteq \ker(\psi^{-1}(S))$.

(*Only if*). Let $\Delta_1 \subseteq C_1$ be a finite generating set. Let $\psi^{-1}(S) : C_1 \to \mathbb{K}$ be recognizable. Hence, $\mathscr{A}_{\psi^{-1}(S)}$ is of finite rank. We have

$$
\begin{aligned}
P_1 \sim_{\psi^{-1}(S)} P_2 &\Longleftrightarrow (\psi^{-1}(S), \tau[P_1]) = (\psi^{-1}(S), \tau[P_2]) \text{ for all } \tau \in \mathrm{CTX}(\Sigma, \Delta_1) \\
&\Longleftrightarrow (S, \psi(\tau[P_1])) = (S, \psi(\tau[P_2])) \text{ for all } \tau \in \mathrm{CTX}(\Sigma, \Delta_1) \\
&\Longleftrightarrow (S, \tau[\psi(P_1)]) = (S, \tau[\psi(P_2)]) \text{ for all } \tau \in \mathrm{CTX}(\Sigma, \psi(\Delta_1)) \\
&\Longleftrightarrow \psi(P_1) \sim_S \psi(P_2).
\end{aligned}
$$

There is, hence, an epimorphism from $\mathscr{A}_{\psi^{-1}(S)}$ to $\mathscr{A}_S$. Thus, we conclude that $\mathscr{A}_S$ is of finite rank, too. $\qquad\square$

Corollary 2.9. *A series $S : C \to \mathbb{K}$ is recognizable iff $\eta_C^{-1}(S)$ is recognizable.*

We now show that the proposed notion of recognizability coincides with the well-known notion of the behavior of weighted tree automata (over trees in $T_\Sigma(\Delta)$) (see e.g. [1]). A weighted tree automaton $\mathcal{A}$ is a tuple (Q, δ, κ) where Q is a finite set of states, $\kappa : Q \to \mathbb{K}$ and $\delta = (\delta_f)_{f \in \Sigma \cup \Delta}$ is a family of mappings $\delta_f : Q^{\mathrm{rk}(f)} \to \mathbb{K}^Q$. We extend δ_f to $\delta_f : \underbrace{\mathbb{K}^Q \times \ldots \times \mathbb{K}^Q}_{\mathrm{rk}(f)} \to \mathbb{K}^Q$ by letting

$$\delta_f(v_1, \ldots, v_k)_q = \sum_{q_1, \ldots, q_k \in Q} \delta_f(q_1, \ldots, q_k)_q \cdot (v_1)_{q_1} \cdot \ldots \cdot (v_n)_{q_n}.$$

Note that the δ_f are multilinear. Hence, they turn $\mathbb{K}^Q$ into a $\mathbb{K}\text{-}\Sigma$-algebra. Let $\delta : \mathbb{K}\langle T_\Sigma(\Delta) \rangle \to \mathbb{K}^Q$ be the $\mathbb{K}\text{-}\Sigma$-homomorphism mentioned in Remark 2.3 extending $\delta : \Delta \to \mathbb{K}^Q : a \mapsto \delta_a$. Now, the behavior $\|\mathcal{A}\| : T_\Sigma(\Delta) \to \mathbb{K}$ of $\mathcal{A}$ is defined by $(\|\mathcal{A}\|, t) = \sum_{q \in Q} \delta(t)_q \cdot \kappa_q$. We say a formal tree series is *regular* if it is the behavior of a weighted tree automaton.

Proposition 2.10. *Let* $S : T_\Sigma(\Delta) \to \mathbb{K}$. *Then S is regular if it is recognizable.*

Proof. Let $\mathscr{A}$ be of finite rank generated by $m_1, \ldots, m_n$ and $\varphi : \mathbb{K}\langle T_\Sigma(\Delta) \rangle \to \mathscr{A}$ a $\mathbb{K}\text{-}\Sigma$-epimorphism such that $\ker(\varphi) \subseteq \ker(S)$. We set $Q = [n] := \{1, \ldots n\}$. Let $f \in \Sigma \cup \Delta$ with $\mathrm{rk}(f) = k$ and let $i_1, \ldots i_k \in Q$. Then $\mu_f(m_{i_1}, \ldots, m_{i_k}) = \sum_{1 \le j \le n} \delta_f(i_1, \ldots, i_k)_j m_j$ for some $\delta_f(i_1, \ldots, i_k)_j \in \mathbb{K}$. This defines $\delta_f : Q^k \to \mathbb{K}^Q$. Since $\ker(\varphi) \subseteq \ker(S)$, there is a linear form $\gamma : \mathscr{A} \to \mathbb{K}$ such that $\gamma \circ \varphi = S$. We define $\kappa : Q \to \mathbb{K}$ by setting $\kappa(i) = \gamma(m_i)$ for all $1 \le i \le n$. Let $\mathcal{A} = (Q, \delta, \kappa)$. It is easy to see by induction that $\varphi(t) = \sum_{1 \le j \le n} \delta(t)_j m_j$. Hence, $\|\mathcal{A}\| = S$. □

Similar to the proof of Reutenauer for Proposition 2.6 one shows for trees:

Proposition 2.11. *Let $\mathbb{K}$ be a commutative ring or let $\mathbb{K}$ be a commutative and locally finite semiring. A tree series $S : T_\Sigma(\Delta) \to \mathbb{K}$ is recognizable iff it is regular.*

Remark 2.12. For the proofs of Propositions 2.6 and 2.11 one needs that finitely generated modules over finitely generated rings are Noetherian, i.e. any submodule is finitely generated. It is open whether the propositions hold for arbitrary commutative semirings.

3 Relational Structures and Weighted Logics

Let $\sigma = ((R_i)_{i \in I}, \rho)$ be a relational signature consisting of a family of relation symbols R_i each of which is equipped with an arity through $\rho : I \to \mathbb{N}_+$. Let $s = (V(s), (R_i^s)_{i \in I})$ be a σ-structure consisting of a domain $V(s)$ together with a relation R_i^s of arity $\rho(i)$ for every relation symbol R_i. Subsequently, we assume that the domain is finite. Moreover, we will distinguish relational structures only up to isomorphisms. In the following, let $\mathcal{C}$ be a class of σ-structures.

We review classical MSO logic for relational structures over signature $\sigma = ((R_i)_{i \in I}, \rho)$. Formulae of $\mathrm{MSO}(\sigma)$ are inductively built from the atomic formulae $x = y$, $R_i(x_1 \ldots x_{\rho(i)})$, $x \in X$ using negation $\neg$, the connective $\vee$ and the quantifications $\exists x$. and $\exists X$. where x, y, x_j are first-order variables and X is a second-order variable.

Let $\varphi \in \mathrm{MSO}(\sigma)$ and let $\mathrm{Free}(\varphi)$ denote the set of variables that occur free in φ. Let $\mathcal{V}$ be a finite set of first-order and second-order variables. A $(\mathcal{V}, s)$-*assignment* γ is a mapping from $\mathcal{V}$ to the power set $\mathscr{P}(V(s))$ such that first-order variables are mapped to singletons. For $v \in V(s)$ and $T \subseteq V(s)$ we denote by $\gamma[x \to v]$ and $\gamma[X \to T]$ the $(\mathcal{V} \cup \{x\}, s)$-assignment which equals γ on $\mathcal{V} \setminus \{x\}$ (resp. $\mathcal{V} \setminus \{X\}$) and assumes $\{v\}$ for x (resp. T for X). Now, let $\mathrm{Free}(\varphi) \subseteq \mathcal{V}$ and γ be a $(\mathcal{V}, s)$-assignment. We write $(s, \gamma) \models \varphi$ if φ holds in s under the assignment γ.

We write $\varphi(x_1, \ldots, x_n, X_1, \ldots, X_m)$ if $\mathrm{Free}(\varphi) = \{x_1, \ldots, x_n, X_1, \ldots, X_m\}$. In this case write $s \models \varphi[v_1, \ldots, v_n, T_1, \ldots, T_m]$ when we have $(s, \gamma) \models \varphi$ if $\gamma(x_i) = \{v_i\}$ and $\gamma(X_i) = T_i$. For $\varphi(x_1, \ldots, x_k) \in \mathrm{MSO}(\sigma)$ we define $\varphi^s = \{(v_1, \ldots, v_k) \in V(s)^k \mid s \models \varphi[v_1, \ldots, v_k]\}$. In the sequel, we identify the pair (s, γ) with the relational structure which expands s with additional unary relations $x^s = \gamma(x)$ and $X^s = \gamma(X)$ for each first-order variable $x \in \mathcal{V}$ and each second-order variable $X \in \mathcal{V}$. By $\sigma_{\mathcal{V}}$ we denote the corresponding signature and by $N_{\mathcal{V}}$ the class of all $\sigma_{\mathcal{V}}$-structures (s, γ) for $s \in \mathcal{C}$ and γ a $(\mathcal{V}, s)$-assignment. Let $\varphi \in \mathrm{MSO}(\sigma)$ and $\mathcal{V} \supseteq \mathrm{Free}(\varphi)$ be a finite set of variables, then $\mathscr{L}_{\mathcal{V}}(\varphi) = \{(s, \gamma) \in N_{\mathcal{V}} \mid (s, \gamma) \models \varphi\}$ and $\mathscr{L}(\varphi) = \mathscr{L}_{\mathrm{Free}(\varphi)}(\varphi)$.

Let $Z \subseteq \mathrm{MSO}(\sigma)$. A language $L \subseteq \mathcal{C}$ is Z-*definable* if $L = \mathscr{L}(\varphi)$ for a sentence $\varphi \in Z$. $\mathrm{MSO}(\sigma)$-definable languages are simply called *definable*. Formulae containing no quantification at all are called *propositional*. First-order formulae, i. e. formulae containing only quantification over first-order variables are collected in $\mathrm{FO}(\sigma)$. The class $\mathrm{EMSO}(\sigma)$ consists of all formulae φ of the form $\exists X_1. \ldots. \exists X_m. \psi$ where $\psi \in \mathrm{FO}(\sigma)$.

We now define weighted MSO logic as introduced in [5]. Formulae of $\mathrm{MSO}(\mathbb{K}, \sigma)$ are built from the atomic formulae k (for $k \in \mathbb{K}$), $x = y$, $R_i(x_1 \ldots x_{\rho(i)})$, $x \in X$, $\neg(x = y)$, $\neg R_i(x_1 \ldots x_{\rho(i)})$, $\neg(x \in X)$ using the connectives $\vee$, $\wedge$ and the quantifications $\exists x.$, $\exists X.$, $\forall x.$, $\forall X.$. Let $\varphi \in \mathrm{MSO}(\mathbb{K}, \sigma)$ and $\mathrm{Free}(\varphi) \subseteq \mathcal{V}$. The weighted semantics $[\![\varphi]\!]_{\mathcal{V}}$ of φ is a function which assigns to each pair $(s, \gamma) \in N_{\mathcal{V}}$ an element of $\mathbb{K}$. For $k \in \mathbb{K}$ we put $[\![k]\!]_{\mathcal{V}}(s, \gamma) = k$. For all other atomic formulae φ semantics $[\![\varphi]\!]_{\mathcal{V}}$ is given by the characteristic function $\mathbb{1}_{\mathscr{L}_{\mathcal{V}}(\varphi)}$. Moreover, we define

$$
\begin{aligned}
[\![\varphi \vee \psi]\!]_{\mathcal{V}}(s, \gamma) &= [\![\varphi]\!]_{\mathcal{V}}(s, \gamma) + [\![\psi]\!]_{\mathcal{V}}(s, \gamma), \\
[\![\varphi \wedge \psi]\!]_{\mathcal{V}}(s, \gamma) &= [\![\varphi]\!]_{\mathcal{V}}(s, \gamma) \cdot [\![\psi]\!]_{\mathcal{V}}(s, \gamma), \\
[\![\exists x. \varphi]\!]_{\mathcal{V}}(s, \gamma) &= \sum_{v \in V(s)} [\![\varphi]\!]_{\mathcal{V} \cup \{x\}}(s, \gamma[x \to v]), \\
[\![\exists X. \varphi]\!]_{\mathcal{V}}(s, \gamma) &= \sum_{T \subseteq V(s)} [\![\varphi]\!]_{\mathcal{V} \cup \{X\}}(s, \gamma[X \to T]), \\
[\![\forall x. \varphi]\!]_{\mathcal{V}}(s, \gamma) &= \prod_{v \in V(s)} [\![\varphi]\!]_{\mathcal{V} \cup \{x\}}(s, \gamma[x \to v]), \\
[\![\forall X. \varphi]\!]_{\mathcal{V}}(s, \gamma) &= \prod_{T \subseteq V(s)} [\![\varphi]\!]_{\mathcal{V} \cup \{X\}}(s, \gamma[X \to T]).
\end{aligned}
$$

We put $[\![\varphi]\!] = [\![\varphi]\!]_{\mathrm{Free}(\varphi)}$. We give an example at the end of Section 5.

Remark 3.1

1. A formula $\varphi \in \mathrm{MSO}(\mathbb{K}, \sigma)$ which does not contain a subformula $k \in \mathbb{K}$ can be interpreted as an unweighted formula.
2. Let $\mathbb{K}$ be the boolean semiring. Then it is easy to see that weighted logics and classical MSO logic coincide. In this case k is either 0 (false) or 1 (true).

Lemma 3.2. *Let s be a σ-structure, $\varphi \in \mathrm{MSO}(\mathbb{K}, \sigma)$ and $\mathcal{V} \supseteq \mathrm{Free}(\varphi)$. Moreover, let γ be a $(s, \mathcal{V})$-assignment. Then $[\![\varphi]\!]_{\mathcal{V}}(s, \gamma) = [\![\varphi]\!](s, \gamma_{|\,\mathrm{Free}(\varphi)})$.*

For words examples show that unrestricted application of universal quantification does not preserve recognizability. We follow Droste and Gastin [5] to resolve this.

Definition 3.3. *A function* $S : C \to \mathbb{K}$ *is a* definable step function *if* $S = \sum_{1 \le j \le m} k_j \mathbb{1}_{L_j}$ *for* $k_j \in \mathbb{K}$ *and definable languages* $L_j \subseteq C$.

Lemma 3.4. *Let* $\varphi \in \mathrm{MSO}(\mathbb{K}, \sigma)$ *and* $V \supseteq \mathrm{Free}(\varphi)$. *Then* $\llbracket \varphi \rrbracket$ *is a definable step function iff* $\llbracket \varphi \rrbracket_V$ *is a definable step function.*

Definition 3.5. *A formula* $\varphi \in \mathrm{MSO}(\mathbb{K}, \sigma)$ *is* restricted *if it does not contain universal set quantification and whenever* φ *has subformula* $\forall x.\psi$, *then* $\llbracket \psi \rrbracket$ *is a definable step function.*

Let $Z \subseteq \mathrm{MSO}(\mathbb{K}, \sigma)$. A series $S : C \to \mathbb{K}$ is Z-*definable* if $S = \llbracket \varphi \rrbracket$ for a sentence $\varphi \in Z$. $\mathrm{MSO}(\mathbb{K}, \sigma)$-definable series are simply called *definable*. Let $\mathrm{RMSO}(\mathbb{K}, \sigma)$ comprise all restricted formulae of $\mathrm{MSO}(\mathbb{K}, \sigma)$. Furthermore, let $\mathrm{REMSO}(\mathbb{K}, \Sigma)$ consist of all $\varphi \in \mathrm{RMSO}(\mathbb{K}, \sigma)$ having the form $\exists X_1.\dots.\exists X_m.\psi$ with ψ not containing any set quantification.

The following theorem extends the result of Droste and Gastin [5] to trees in $T_\Sigma(\Delta)$. The domain of a tree is a finite, nonempty, prefix-closed subset of $\mathbb{N}^*$ and it has relations for the node labeling and relations $\mathrm{E}_i(x, y)$ saying that y is the i-th child of x.

Theorem 3.6 (Droste & Vogler [7]). *Let* $\mathbb{K}$ *be a commutative semiring. A tree series* $S : T_\Sigma(\Delta) \to \mathbb{K}$ *is regular iff it is* RMSO-*definable iff it is* REMSO-*definable.*

We will show how to transfer this result to other relational structures using definable transductions. First, we need some preparing definitions.

Definition 3.7. *Let* $\varphi \in \mathrm{MSO}(\sigma)$.

1. *We call* φ +-*disambiguatable (resp.* +-RMSO-*disambiguatable) if there is a formula (resp. restricted formula)* φ^+ *such that* $\llbracket \varphi^+ \rrbracket = \mathbb{1}_{\mathscr{L}(\varphi)}$.
2. *We call* φ −-*disambiguatable (resp.* −-RMSO-*disambiguatable) if there is a formula (resp. restricted formula)* φ^- *such that* $\llbracket \varphi^- \rrbracket = \mathbb{1}_{\mathscr{L}(\neg\varphi)}$.
3. *We call* φ disambiguatable *(resp.* RMSO-*disambiguatable) if it is both* +-*disambiguatable and* −-*disambiguatable (resp.* +-RMSO-*disambiguatable and* −-RMSO-*disambiguatable).*

For any +-*disambiguatable (resp.* +-RMSO-*disambiguatable) formula* φ *we choose an arbitrary but fixed formula (resp. restricted formula)* φ^+ *such that* $\llbracket \varphi^+ \rrbracket = \mathbb{1}_{\mathscr{L}(\varphi)}$. *We define* φ^- *analogously.*

Remark 3.8

1. Every propositional formula is RMSO-disambiguatable. Moreover, if $\mathbb{K}$ is idempotent, then any φ is disambiguatable. If additionally φ does not contain universal set quantification, then φ is RMSO-disambiguatable.
2. Let $\mathbb{K} = \mathbb{N}$. Consider the class of graphs without edges where the vertices are labeled with a or b. Then $\forall x. \mathrm{Lab}_a(x)$ is +-RMSO-disambiguatable but it is not −-disambiguatable.

Using Theorem 3.6 and Doner's famous Büchi-type theorem for trees [4], we obtain

Lemma 3.9. *Let C be the class of trees. Then every formula $\varphi \in \mathrm{MSO}(\sigma)$ is RMSO-disambiguatable.*

The following lemma is a slight modification of Lemma 5.1 in [17].

Lemma 3.10 (Meinecke [17]). *If there is a $+$-RMSO-disambiguatable formula $\varphi(x, y)$ such that $(\varphi^+)^s$ is a linear order for every $s \in C$, then every first-order formula is RMSO-disambiguatable.*

4   Definable Transductions

In model theory it is common to interpret one relational structure in another. Courcelle [3] takes quite a constructive point of view by introducing the notion of definable transductions between classes of relational structures. There one derives a new structure by interpreting it in m copies of a given structure. Here we only regard deterministic definable transductions which, therefore, we call definable functions. Let σ_1 and $\sigma_2 = ((R_i)_{i \in I}, \rho)$ be two relational signatures and let C_1 and C_2 be classes of finite σ_1- and σ_2-structures, respectively.

Definition 4.1. *A (σ_1, σ_2)-m-copying definition scheme (without parameter) is a tuple*

$$\mathcal{D} = (\vartheta, (\delta_j)_{1 \leq j \leq m}, (\varphi_l)_{l \in I \star m}) \quad \text{where } I \star m = \{(i, \bar{\jmath}) \mid i \in I, \bar{\jmath} \in [m]^{\rho(i)}\}$$

of formulae in $\mathrm{MSO}(\sigma_1)$ such that $\mathrm{Free}(\vartheta) = \emptyset$, $\mathrm{Free}(\delta_j) = \{x_1\}$ and $\mathrm{Free}(\varphi_l) = \{x_1, \ldots, x_{\rho(i)}\}$ for $l = (i, \bar{\jmath}) \in I \star m$.

Let $\mathcal{D}$ be a (σ_1, σ_2)-m-copying definition scheme and let $s_1 \in C_1$ such that $s_1 \models \vartheta$. Then define the σ_2-structure $\mathbf{def}_{\mathcal{D}}(s_1) = s_2 = (V(s_2), (R_i^{s_2})_{i \in I})$ where $V(s_2) = \bigcup_{1 \leq j \leq m} \delta_j^{s_1} \times \{j\}$ and $R_i^{s_2} = \{(v_1, j_1), \ldots, (v_r, j_r) \in V(s_2)^r \mid (v_1, \ldots, v_r) \in \varphi_{i, (j_1, \ldots, j_r)}^{s_1}\}$ with $r = \rho(i)$. The function defined by $\mathcal{D}$ is given by $s_1 \mapsto \mathbf{def}_{\mathcal{D}}(s_1)$.

Definition 4.2. *A partial function $\Phi : C_1 \to C_2$ is a definable function if there is a definition scheme $\mathcal{D}$ such that $\Phi = \mathbf{def}_{\mathcal{D}}$. If there is a $\mathcal{D}$ such that ϑ, δ_j and φ_l are disambiguatable, then Φ is an unambiguously definable function. If ϑ, δ_j and φ_l are RMSO-disambiguatable, then Φ is a RMSO-definable function.*

Courcelle [3] showed that the preimage of a definable set under a definable function is again definable. We will show a similar result for series. Let $\Phi : C_1 \to C_2$ be a partial function with domain $\mathrm{dom}(\Phi)$ and let $S : C_2 \to \mathbb{K}$. Define $\Phi^{-1}(S)$ by letting $(\Phi^{-1}(S), s_1) = (S, \Phi(s_1))$ for all $s_1 \in \mathrm{dom}(\Phi)$ and $(\Phi^{-1}(S), s_1) = 0$ otherwise.

Proposition 4.3. *Let $\Phi : C_1 \to C_2$ be a partial function.*

1. *Let Φ be unambiguously definable. If there is a $+$-disambiguatable formula $\varphi(x, y)$ such that $(\varphi^+)^{s_1}$ is a linear order for every $s_1 \in C_1$ and if $S : C_2 \to \mathbb{K}$ is definable, then so is $\Phi^{-1}(S)$.*

2. *Let Φ be RMSO-definable. If $S : C_2 \to \mathbb{K}$ is RMSO-definable, then so is $\Phi^{-1}(S)$.*

Remark 4.4. To show Proposition 4.3 one translates formulae in $\mathrm{MSO}(\sigma_2)$ to formulae in $\mathrm{MSO}(\sigma_1)$ using an appropriate definition scheme $\mathcal{D} = (\vartheta, (\delta_j), (\varphi_l))$. If $\vartheta^+, \delta_j^+, \delta_j^-$, φ_l^+ and φ_l^- can be chosen in FO, then a translation can be given such that REMSO-definability is preserved.

5   Definable and Recognizable Text Series

A text is, roughly speaking, a word with an additional linear order. More precisely:

Definition 5.1. *Let Δ be a finite alphabet. A text* over *Δ is a tuple $(V, \lambda, \leq_1, \leq_2)$ where $\leq_1$ and $\leq_2$ are linear orders over the domain V and $\lambda : V \to \Delta$ is a labeling function.*

We consider texts as relational structures where the relations are given by the labeling and by $\leq_1$ and $\leq_2$. As usual, we identify isomorphic texts.

We now define an algebraic structure on texts following Hoogeboom and ten Pas [15]. A biorder is a pair of two linear orders, i.e. a text without labeling. Each biorder defines an operation – we obtain a new text by substituting given texts into the nodes of the biorder. These texts then become intervals of the new text in both the first and the second order. Subsets being intervals of both orders are called *clans*. A biorder is *primitive* if it has only trivial clans, i.e. the singletons and the domain itself.

Let Σ be a finite set of primitive biorders of cardinality at least two and let $\mathrm{TXT}_\Sigma(\Delta)$ be the set of texts generated from Δ using Σ. Let txt $= \eta_{\mathrm{TXT}_\Sigma(\Delta)}$. Applying the theory of 2-structures developed by Ehrenfeucht and Rozenberg [9] one obtains that $\mathrm{TXT}_\Sigma(\Delta)$ is almost freely generated in the variety of all Σ-algebras from the singleton texts, i.e. from Δ. Only the two biorders of cardinality two satisfy an associative law [15]. Thus, different preimages of a text $\tau \in \mathrm{TXT}_\Sigma(\Delta)$ under txt only differ with respect to these two associativity laws. Let $\mathrm{sh}(\tau)$ be the preimage where the brackets are in the right most form. Clearly, $\mathrm{sh}^{-1}(\mathrm{txt}^{-1}(L)) = L$ for any $L \subseteq \mathrm{TXT}_\Sigma(\Delta)$. Hoogeboom and ten Pas call $\mathrm{sh}(\tau)$ the *r-shape* of τ. They show

Theorem 5.2 (Hoogeboom & ten Pas [15]). *A language $L \subseteq \mathrm{TXT}_\Sigma(\Delta)$ is recognizable iff it is definable.*

To prove it, they show that sh and txt are definable functions. Now, Lemma 3.9 implies:

Proposition 5.3. *The natural epimorphism* txt $: T_\Sigma(\Delta) \to \mathrm{TXT}_\Sigma(\Delta)$ *is an RMSO-definable function.*

Proposition 5.4. *The function* sh $: \mathrm{TXT}_\Sigma(\Delta) \to T_\Sigma(\Delta)$ *is RMSO-definable.*

Proof (Sketch). Again we follow the idea in [15]. There a 2-copying scheme for sh is given. The formulae involved contain nested universal quantification over sets. The formula interpreting the label of an inner node of an r-shape in its text is e.g. in Σ_4. However, analyzing the formulae it turns out that any quantification only concerns clans. Hence, we can transform them into equivalent first-order formulae by identifying a clan with its first and its last element with respect to the first order, say. Now, any formula involved becomes a first-order formula. The result follows then from Lemma 3.10.   □

Theorem 5.5. *Let $\mathbb{K}$ be a commutative ring or let $\mathbb{K}$ be a commutative and locally finite semiring. Let $S : \mathrm{TXT}_\Sigma(\Delta) \to \mathbb{K}$ be a text series. Then the following are equivalent.*

1. *S is recognizable.*
2. *S is RMSO-definable.*
3. *S is REMSO-definable.*

Proof (Sketch). Let S be recognizable. By Proposition 2.10 and Corollary 2.9 $\text{txt}^{-1}(S)$ is regular. By Theorem 3.6 $\text{txt}^{-1}(S)$ is REMSO-definable. From the proofs of Proposition 5.4 and Lemma 3.10 in [17] we obtain a 2-copying definition scheme for sh consisting of restricted first-order formulae only. By Remark 4.4 $\text{sh}^{-1}(\text{txt}^{-1}(S)) = S$ is definable in REMSO and, hence, in RMSO.

It remains to show that S is recognizable if it is RMSO-definable. Let S be RMSO-definable. From Proposition 5.3 using Proposition 4.3(2), $\text{txt}^{-1}(S)$ is RMSO-definable and, hence, regular by Theorem 3.6. By Prop. 2.11 and Cor. 2.9 S is recognizable. $\square$

Note that there is a $+$-disambiguatable formula $\varphi(x, y)$ such that φ^t is the lexicographic order of positions for any $t \in T_\Sigma(\Delta)$. Using the result of Droste and Vogler on the coincidence of regular and definable tree series over commutative and locally finite semirings [7] and Proposition 4.3(1) we obtain the following theorem.

Theorem 5.6. *Let $\mathbb{K}$ be a commutative and locally finite semiring. A text series $S: \text{TXT}_\Sigma(\Delta) \to \mathbb{K}$ is definable iff it is recognizable.*

A *computable field* is a field with computable operations $(+, -, \cdot, ^{-1})$; e.g. the rationals.

Corollary 5.7. *Let $\mathbb{K}$ be a computable field. It is decidable whether two given restricted sentences over texts φ and ψ satisfy $[\![\varphi]\!] = [\![\psi]\!]$.*

Proof. The proof of Proposition 4.3 is effective and gives restricted tree formulae φ' and ψ' such that $[\![\varphi']\!] = \text{txt}^{-1}([\![\varphi]\!])$ and $[\![\psi']\!] = \text{txt}^{-1}([\![\psi]\!])$. Clearly, $[\![\varphi]\!] = [\![\psi]\!]$ iff $[\![\varphi']\!] = [\![\psi']\!]$. The latter can be decided by Corollary 5.9 of [7]. $\square$

Similarly, using Corollary 6.7 of [7] we obtain

Corollary 5.8. *Let $\mathbb{K}$ be a computable locally finite commutative semiring. It is decidable whether two given sentences over texts φ and ψ satisfy $[\![\varphi]\!] = [\![\psi]\!]$.*

The following corollary sharpens one implication of Theorem 5.2.

Corollary 5.9. *A language $L \subseteq \text{TXT}_\Sigma(\Delta)$ is definable iff it is definable in EMSO.*

Example 5.10. Let $\mathbb{K} = \mathbb{Z}$ be the ring of integers. Let $\text{Clan}(x_1, x_2)$ be a first-order formula saying that for a text τ, $\{x \in \tau \mid x_1 \leq_1 x \leq_1 x_2\}$ is a proper clan. Consider

$$\varphi = \exists x_1, x_2. \text{Clan}(x_1, x_2)^+ \wedge \forall x, y. x_1 \leq_1 x, y \leq_1 x_2 \to (x \leq_1 y \leftrightarrow y \leq_2 x).$$

For a text τ, $([\![\varphi]\!], \tau)$ gives the number of proper clans generated only from the biorder of cardinality two having two reversed orders. By Theorem 5.5 $[\![\varphi]\!]$ is recognizable.

6   Alternating Texts and Weighted Parenthesizing Automata

In this section let $\Sigma = \{\circ_h, \circ_v\}$ be the set of the two biorders of cardinality two, where for $\circ_h$ both orders coincide. Then $\text{TXT}_\Sigma(\Delta)$, the set of the so-called alternating texts ([10, p. 261]), is the free bisemigroup generated by Δ; where a bisemigroup is a set together with two associative operations. Several authors have investigated the free bisemigroup as a fundamental, two-dimensional extension of classical automaton theory, see e.g. Ésik and Németh [12] and Hashiguchi et. al. (e.g. [13]). Ésik and Németh consider as a representation for the free bisemigroup the so-called *sp-biposets*. They define parenthesizing automata. Here we define weighted parenthesizing automata.

Definition 6.1. *A* weighted parenthesizing automaton *(wpa for short) over* Δ *is a tuple* $\mathcal{P} = (\mathcal{H}, \mathcal{V}, \Omega, \mu, \mu_{op}, \mu_{cl}, \lambda, \gamma)$ *where*

1. $\mathcal{H}, \mathcal{V}$ *are finite disjoint sets of* horizontal *and* vertical states, *respectively.*
2. Ω *is a finite set of* parentheses, [1]
3. $\mu : (\mathcal{H} \times \Delta \times \mathcal{H}) \cup (\mathcal{V} \times \Delta \times \mathcal{V}) \to \mathbb{K}$ *is the* transition function,
4. $\mu_{op}, \mu_{cl} : (\mathcal{H} \times \Omega \times \mathcal{V}) \cup (\mathcal{V} \times \Omega \times \mathcal{H}) \to \mathbb{K}$ *are the* opening *and* closing parenthesizing functions *and*
5. $\lambda, \gamma : \mathcal{H} \cup \mathcal{V} \to \mathbb{K}$ *are the* initial *and* final weight functions.

A *run* r of $\mathcal{P}$ is a certain word over the alphabet $(\mathcal{H} \cup \mathcal{V}) \times (\Delta \cup \Omega) \times (\mathcal{H} \cup \mathcal{V})$ defined inductively as follows. We also define its *label* $\mathrm{lab}(r)$, its *weight* $\mathrm{wgt}(r)$, its *initial state* $\mathrm{init}(r)$ and its *final state* $\mathrm{fin}(r)$.

1. (q_1, a, q_2) is a run for all $(q_1, q_2) \in (\mathcal{H} \times \mathcal{H}) \cup (\mathcal{V} \times \mathcal{V})$ and $a \in \Delta$. We set

$$\mathrm{lab}((q_1, a, q_2)) = a \in \mathrm{TXT}_\Sigma(\Delta), \qquad \mathrm{wgt}((q_1, a, q_2)) = \mu(q_1, a, q_2),$$
$$\mathrm{init}((q_1, a, q_2)) = q_1 \text{ and } \mathrm{fin}((q_1, a, q_2)) = q_2.$$

2. Let r_1 and r_2 be runs with $\mathrm{fin}(r_1) = \mathrm{init}(r_2) \in \mathcal{H}$ (resp. $\mathcal{V}$). Then $r = r_1 r_2$ is a run having

$$\mathrm{lab}(r) = \mathrm{lab}(r_1) \circ_h \mathrm{lab}(r_2) \quad (\text{resp. } \mathrm{lab}(r) = \mathrm{lab}(r_1) \circ_v \mathrm{lab}(r_2)),$$
$$\mathrm{wgt}(r) = \mathrm{wgt}(r_1) \cdot \mathrm{wgt}(r_2), \mathrm{init}(r) = \mathrm{init}(r_1) \text{ and } \mathrm{fin}(r) = \mathrm{fin}(r_2).$$

3. Let r be a run resulting from 2 such that $\mathrm{fin}(r), \mathrm{init}(r) \in \mathcal{H}$ (resp. $\mathcal{V}$). Let $q_1, q_2 \in \mathcal{V}$ (resp. $\mathcal{H}$) and $s \in \Omega$. Then $r' = (q_1, (_s, \mathrm{init}(r)) \, r \, (\mathrm{fin}(r),)_s, q_2)$ is a run having

$$\mathrm{lab}(r') = \mathrm{lab}(r), \mathrm{wgt}(r') = \mu_{op}((q_1, (_s, \mathrm{init}(r))) \cdot \mathrm{wgt}(r) \cdot \mu_{cl}((\mathrm{fin}(r),)_s, q_2)),$$
$$\mathrm{init}(r') = q_1 \text{ and } \mathrm{fin}(r') = q_2.$$

Let $\tau \in \mathrm{TXT}_\Sigma(\Delta)$. Since we do not allow repeated application of rule 3, there are only finitely many runs with label τ. If r is a run with $\mathrm{lab}(r) = \tau$, $\mathrm{init}(r) = q_1$, $\mathrm{fin}(r) = q_2$, we write $r : q_1 \xrightarrow{\tau} q_2$. The behavior of $\mathcal{P}$ is a series $\|\mathcal{P}\| : \mathrm{TXT}_\Sigma(\Delta) \to \mathbb{K}$ with

$$(\|\mathcal{P}\|, \tau) = \sum_{q_1, q_2 \in \mathcal{H} \cup \mathcal{V}} \lambda(q_1) \cdot \sum_{r : q_1 \xrightarrow{\tau} q_2} \mathrm{wgt}(r) \cdot \gamma(q_2).$$

An alternating text series S is *regular* if there is a wpa $\mathcal{P}$ such that $\|\mathcal{P}\| = S$.

---

[1] Contrary to the definition of Ésik and Németh we let $s \in \Omega$ represent both the opening and the closing parentheses. To help the intuition we write $(_s$ or $)_s$.

Proposition 6.2. *Let* $S : \mathrm{TXT}_\Sigma(\Delta) \to \mathbb{K}$. *Then* S *is regular iff* $\mathrm{txt}^{-1}(S)$ *is regular.*

From Theorem 3.6 and Propositions 5.3 and 5.4 we now conclude the following connection between weighted logics and weighted parenthesizing automata.

Theorem 6.3. *Let* $\mathbb{K}$ *be any commutative semiring. An alternating text series is regular iff it is* RMSO-*definable iff it is* REMSO-*definable.*

Corollary 6.4. *Let* $\mathbb{K}$ *be a commutative ring or let* $\mathbb{K}$ *be a commutative and locally finite semiring. An alternating text series is regular iff it is recognizable.*

Remark 6.5. The class of alternating texts is isomorphic to the class of sp-biposets. There is an isomorphism that can be defined by propositional formulae (see e.g. [12]). Thus, the results of the last two sections hold as well for sp-biposets.

Acknowledgments. The author thanks Pascal Weil, Manfred Droste and Dietrich Kuske for their helpful comments as well as an anonymous referee whose remarks resulted in improvements of the paper.

References

1. Bozapalidis, S.: Equational elements in additive algebras. Theory of Computing Systems 32(1), 1–33 (1999)
2. Bozapalidis, S., Alexandrakis, A.: Représentations matricielles des séries d'arbre reconnaissables. Theoretical Informatics and Applications 23(4), 449–459 (1989)
3. Courcelle, B.: Monadic second-order definable graph transductions: a survey. Theoretical Computer Science 126, 53–75 (1994)
4. Doner, J.: Tree acceptors and some of their applications. Journal of Computer and System Sciences 4, 406–451 (1970)
5. Droste, M., Gastin, P.: Weighted automata and weighted logics. In: Caires, L., Italiano, G.F., Monteiro, L., Palamidessi, C., Yung, M. (eds.) ICALP 2005. LNCS, vol. 3580, pp. 513–525. Springer, Heidelberg (2005)
6. Droste, M., Rahonis, G.: Weighted automata and weighted logics on infinite words. In: Ibarra, O.H., Dang, Z. (eds.) DLT 2006. LNCS, vol. 4036, pp. 49–58. Springer, Heidelberg (2006)
7. Droste, M., Vogler, H.: Weighted tree automata and weighted logics. Theoretical Computer Science 366, 228–247 (2006)
8. Ehrenfeucht, A., Harju, T., Rozenberg, G.: The Theory of 2-structures: A Framework for Decomposition and Transformation of Graphs. World Scientific, Singapore (1999)
9. Ehrenfeucht, A., Rozenberg, G.: Theory of 2-structures. I and II. Theoretical Computer Science 70, 277–342 (1990)
10. Ehrenfeucht, A., Rozenberg, G.: T-structures, T-functions, and texts. Theoretical Computer Science 116, 227–290 (1993)
11. Ehrenfeucht, A., ten Pas, P., Rozenberg, G.: Context-free text grammars. Acta Informatica 31(2), 161–206 (1994)
12. Ésik, Z., Németh, Z.L.: Higher dimensional automata. Journal of Automata, Languages and Combinatorics 9(1), 3–29 (2004)
13. Hashiguchi, K., Ichihara, S., Jimbo, S.: Formal languages over free bionoids. Journal of Automata, Languages and Combinatorics 5(3), 219–234 (2000)

14. Hoogeboom, H.J., ten Pas, P.: Text languages in an algebraic framework. Fundamenta Informaticae 25(3), 353–380 (1996)
15. Hoogeboom, H.J., ten Pas, P.: Monadic second-order definable text languages. Theory of Computing Systems 30, 335–354 (1997)
16. Mäurer, I.: Weighted picture automata and weighted logics. In: Durand, B., Thomas, W. (eds.) STACS 2006. LNCS, vol. 3884, pp. 313–324. Springer, Heidelberg (2006)
17. Meinecke, I.: Weighted logics for traces. In: Grigoriev, D., Harrison, J., Hirsch, E.A. (eds.) CSR 2006. LNCS, vol. 3967, Springer, Heidelberg (2006)
18. Mezei, J., Wright, J.B.: Algebraic automata and context-free sets. Information and Control 11(1/2), 3–29 (1967)
19. Reutenauer, Ch.: Séries formelles et algèbres syntactiques. J. Algebra 66, 448–483 (1980)

A Star Operation for Star-Free Trace Languages[*]

Edward Ochmański[1,2] and Krystyna Stawikowska[1]

[1] Faculty of Mathematics and Computer Science, Nicolaus Copernicus University, Torun
[2] Institute of Computer Science, Polish Academy of Sciences, Warszawa, Poland
{edoch,entropia}@mat.uni.torun.pl

Abstract. The paper deals with star-free languages in trace monoids. We define a constrained star operation, named star-free star, and show a new characterisation of star-free trace languages, using this operation instead of complement. We obtain this characterisation combining a star-free star characterisation for word languages and logical characterisation of trace languages (Ebinger/Muscholl). Moreover, some new, simple proofs of known results are presented in the paper.

Keywords: traces, star operation, star-free languages.

1 Introduction

Traditional rational expressions (with union, concatenation and star) play an important role in the theory of formal languages, since its very beginning (Kleene Theorem), as well as in the general theory of monoids. Also a big role play extended rational expressions (with intersection and complement, beside the traditional ones) and star-free expressions (with union, concatenation and complement). Subsets defined by star-free expressions are called star-free subsets. They are highly valued, as they have nice algebraic (aperiodicity, Schützenberger [13]) and logical (first-order definability, Mc-Naughton/Papert [7]) properties in free monoids.

Trace theory was initiated by Mazurkiewicz [6] (cf. [1,2] for surveys). It provides a mathematical model for behaviour of concurrent systems. The most important results about star-free trace languages are those of Guaiana/Restivo/Salemi [5] (aperiodicity) and Ebinger/Muscholl [4] (first-order definability).

As it was mentioned above, star-free expressions differ from traditional rational expressions by complement replacing star. It is interesting and useful to have a method of describing star-free subsets with (a subclass of) traditional rational expressions. We introduced, in [10], a constrained star operation, named star-free star, and we showed that this operation, together with union and product, is able to build the whole class of star-free word languages (Theorem 5). It does not hold in all monoids (Example 4). A question, if the same holds for trace languages, remained unsolved in [10].

In the present paper we put into work the logical characterisation (Theorem 9) of Ebinger/Muscholl [4]. Combining the star-free-star characterisation of word languages

---

[*] The research supported by Ministry of Science and Higher Education of Poland, grant N206 023 31/3744.

T. Harju, J. Karhumäki, and A. Lepistö (Eds.): DLT 2007, LNCS 4588, pp. 337–345, 2007.

with the logical characterisation of [4], we generalise the star-free-star characterisation to traces. Namely, we have shown that the classes of star-free and star-free star languages coincide in any trace monoid. This is the main result of the paper.

The notion of star-free star seems to be quite useful. Using it, we produce new proofs of results of Guaiana/Restivo/Salemi [5] (Theorem 7, (1)≡(4)) and Ebinger/Muscholl [4] (Theorem 11), both much shorter than originals.

2 Preliminaries

In this section, we recall some basic notions and results, used in the paper.

2.1 Word and Trace Languages

An *alphabet* A is assumed to be finite. The set A^* with concatenation as the product operation form the *free monoid*; subsets of A^* are called (*word*) *languages*. Let $I \subseteq A \times A$ be a symmetric and irreflexive relation on A, called *independency*, its complement $D = A \times A - I$ is named *dependency*.

The couple (A, I) or (A, D) is said to be a *concurrent alphabet*. Given a concurrent alphabet (A, I), the *trace monoid* A^*/I is the quotient of the free monoid A^* by the least congruence on A^* containing the relation $\{ab = ba \mid aIb\}$. Members of A^*/I are called *traces*, and sets of traces (i.e. subsets of A^*/I) are called *trace languages*. Clearly, a trace monoid A^*/I is free iff $I = \emptyset$. Given a monoid M, the complement of a subset $X \subseteq M$ will be denoted by $'$, i.e. $X' = M - X$.

Let (A, I) be a concurrent alphabet. Any word $w \in A^*$ induces a trace $[w] \in A^*/I$ – the congruency class of w; any word language $L \subseteq A^*$ induces a trace language $[L] = \{[w] \mid w \in L\}$ – the set of all traces induced by members of L. Given a trace language $T \subseteq A^*/I$, the *flattening* of T is the word language $\bigcup T = \{w \in A^* \mid [w] \in T\}$ – the union of traces in T, viewed as subsets of A^*. Given a word language $L \subseteq A^*$, the *closure* of L is the word language $\bar{L} = \bigcup[L]$. A word language L is said to be *closed* (w.r.t. I) iff $L = \bar{L}$. The closure operation allows to study trace languages on the level of free monoids. Namely, trace languages can be identified with closed word languages.

2.2 Star-Free and Aperiodic Subsets

Given a monoid M and a subset $L \subseteq M$, the following notions are commonly known:

- *atomic set* (*atoms*, for short): empty set $\emptyset$ and singletons $\{m\}$ for all $m \in M$;
- *syntactic congruence* $\approx_L \subseteq M \times M$ of L : $x \approx_L y$ iff
  $(\forall r, s \in M) \, rxs \in L \Leftrightarrow rys \in L$;
- *syntactic monoid* of L: the quotient monoid $M_L = M/\approx_L$.

A subset $L \subseteq M$ is said to be:

- *rational*   iff   it is built from atoms with union, product and star;
- *recognizable* iff   its syntactic monoid is finite;
- *star-free*   iff   it is built from atoms with union, product and complement;
- *aperiodic*   iff   its syntactic monoid is finite and aperiodic:
  $(\exists n)(\forall x \in M_L) x^n = x^{n+1}$, or equivalently $(\exists n)(\forall w \in M) w^n w^{n+1}$.

Classes of subsets, defined above, will be denoted by $RAT(M)$, $REC(M)$, $SF(M)$ and $AP(M)$, respectively. The argument will be possibly omitted, if it will not lead to a confusion. As $RAT(A^*) = REC(A^*)$ in finitely generated free monoids (Kleene Theorem), the class $RAT(A^*) = REC(A^*)$ will be uniformly denoted by $REG(A^*)$, and its members will be called *regular languages*. By definitions, $AP(M) \subseteq REC(M)$ for any monoid M. Moreover, if M is a trace monoid, there hold the inclusions $SF(M) \subseteq REC(M) \subseteq RAT(M)$.

Theorem 1 (Schützenberger [13]). In any finitely generated free monoid A^*, the classes $SF(A^*)$ of star-free languages and $AP(A^*)$ of aperiodic languages coincide.

2.3   Automata and Their Languages

A *(deterministic) automaton* is a quadruple $\mathbf{A} = < A, Q, q_0, F >$, where A is a finite *alphabet*, Q is a finite set of *states*, $q_0 \in Q$ is an *initial state* and $F \subseteq Q$ is a set of *final states*; states are (partial) functions $q : A \to Q$. A triple qaq' is said to be an *arc* in $\mathbf{A}$ if $qa = q'$; then a is called the *label* of the arc qaq'. A *path* in $\mathbf{A}$, from q_1 to q_n, is any sequence $q_1 a_1 q_2 a_2 \ldots a_{n-1} q_n$, where all $q_i \in Q$ and all $a_i \in A$, such that $(\forall i) q_i a_i q_{i+1}$ is an arc in $\mathbf{A}$; then $a_1 a_2 \ldots a_{n-1}$ is called the *label* of the path. A *computation* in $\mathbf{A}$ is any path $q_1 a_1 q_2 a_2 \ldots a_{n-1} q_n$, from the initial state $q_1 = q_0$ to a final state $q_n \in F$; then its label $a_1 a_2 \ldots a_n$ is said to be a word *accepted* by $\mathbf{A}$, and the set of all words accepted by $\mathbf{A}$ is called the *language accepted* by $\mathbf{A}$ or simply the *language of* $\mathbf{A}$, and denoted by $L(\mathbf{A})$. Any automaton accepting a language $L \subseteq A^*$ is said to be an *automaton for* L. An automaton is a *minimal deterministic automaton* (shortly, *m.d.a.*) iff its number of states is minimal from among all deterministic automata accepting the language $L(\mathbf{A})$. Languages accepted by finite automata are called *regular languages*. An automaton is said to be *aperiodic* (or *counter-free*) iff $(qwp$ and $pw^n q) \Rightarrow p = q$, for any state q and any word w.

The following theorem follows from Schützenberger Theorem 1.

Theorem 2. Given a regular language $L \subseteq A^*$, the following statements are equivalent:

      (1) The language L is star-free;
      (2) The minimal deterministic automaton for L is aperiodic;
      (3) There exists an aperiodic automaton for L.

3   Star-Free Star and Word Languages

In this section we deal with free monoids. We introduce the notion of star-free star, basic for this paper, and characterise the class of star-free languages using this notion. This characterisation will play crucial role in this paper.

Definition 3. *Star-free star operation, expressions and languages*

- The *star-free star* operation $^\times$ is defined as follows:

$$L^\times = \begin{cases} L^* \text{ if } L^* \text{ is star-free} \\ \text{undefined otherwise} \end{cases}$$

- A rational expression is called an *SFS-expression* if and only if it is built from atomic expressions with symbols of union, concatenation and star-free star (if defined);
- A language is said to be an *SFS-language* if and only if it is built from atoms using union, concatenation and star-free star; the class of all SFS-languages will be denoted by SFS.

It follows directly from Definition 3 that, in any monoid, SFS is a subclass of SF. The following example shows that the inverse inclusion does not hold in general.

Example 4. Let us take the monoid $(\mathbb{Z}, +)$ of integers with addition. In this monoid X is star-free if and only if X or X' is finite. The class of SFS-languages in $(\mathbb{Z}, +)$ is the smallest class of languages, built from $\emptyset$, 1, -1 and $\mathbb{Z}$, using only $\cup$ and $+$. The set $\mathbb{Z} - \{0\}$ is SF, but not SFS. See [10] for details.

The following theorem says that in free monoids classes SF and SFS coincide.

Theorem 5. A regular word language $L \subseteq A^*$ is SF if and only if it is SFS. In other words, the class of star free word languages is the smallest class of languages, built from atoms using union, product and star-free star.

Proof. The proof follows the McNaughton/Yamada [8] construction of rational expressions for automata, taking into account Theorem 2. The detailed proof has been presented in [10]. □

4   Star-Free Trace Languages

Now, we start to enter into the world of traces. We consider the operation of partially commutative closure. The following lemma will appear quite effective in studies of star-free trace languages.

Lemma 6. *Closing product of closed star-free languages*
If languages $K, L \subseteq A^*$ are closed and star-free, then the closure $\overline{KL}$ of their product is star-free.

Proof. Let $A' = \{a' \mid a \in A\}$ and $A'' = \{a'' \mid a \in A\}$ be differently coloured, so disjunctive, copies of the alphabet A. Denote $C = A' \cup A''$. Let the independency relation $I_C \subseteq C \times C$ be the smallest relation, such that if aIb then $a'I_Cb$. Since K and L are closed, we have

$$\overline{KL} = h(\overline{K'L''}),$$

where K' and L'' are the coloured copies of the languages K and L, respectively; the morphism $h : C \to A$ is the washing morphism, such that $h(a') = h(a'') = a$, for all $a \in A$; the left closure is with respect to I, and the right one with respect to I_C.

Let us define two basic substitutions $f', f'' : A \to 2^{C^*}$, where $f'(a) = A''^* a' A''^*$ and $f''(a) = A'^* a'' A'^*$. These substitutions colour letters of alphabet A and surround them by words of the other colour.

To complete the proof we need two claims.

Claim 1. If $X \subseteq A^*$ is star-free, then $f'(X)$ and $f''(X)$ are star-free.

Proof of Claim 1. We construct the automaton for $f'(X)$ by rebuilding m.d.a. for X, which is aperiodic, by Theorem 2. First, we colour the labels of the automaton with the colour of the copy A', i.e. if an arc is labelled with $a \in A$, then the new label is $a' \in A'$. This automaton accepts the language X'. Now we join, with each state q, the bunch of self-loops $qa''q$, for all $a \in A$. This way we have built the m.d.a. accepting $f'(X)$. It is aperiodic, because X' was such one. Hence, $f'(X)$ is star-free. And the dual construction for $f''(X)$ shows that $f''(X)$ is star-free. □

Observe that, if aDb, then always $a'D_C b''$. Hence, we have the equality

$$\overline{K'L''} = f'(K) \cap f''(L) - \bigcup \{C^* b'' C^* a' C^* \mid aDb\}.$$

It follows from Claim 1 that the language $\overline{K'L''}$ is star-free, because the class SF is closed under boolean operations.

We have shown that $\overline{K'L''}$ is star-free. Now we will show that, if we wash up the colours, the language remains star-free.

Claim 2. If a language $X \subseteq C^* - \bigcup \{C^* b'' C^* a' C^* \mid aDb\}$ is star-free, then the language $h(X)$ is star-free.

Proof of Claim 2. We show the claim by structural induction on SFS-expressions. For atomic languages – obvious. Let us assume that if $Y, Z \subseteq C^* - \bigcup \{C^* b'' C^* a' C^* \mid aDb\}$ are star-free, then $h(Y)$ and $h(Z)$ are star-free.

- If $X = Y \cup Z$, then $h(X) = h(Y) \cup h(Z)$ is star-free;
- If $X = YZ$, then $h(X) = h(Y)h(Z)$ is star-free;
- To show that $h(Y^*)$ is star-free we need to notice that at most one of letters a' and a'', for any $a \in A$, can occur in $Alph(Y)$, because a'' cannot occur before a' in Y^*. Hence, $h(Y^*)$ is an isomorphic copy of Y^*, thus it is star-free. □

It is easy to see that the language $\overline{K'L''}$ satisfies the assumption of Claim 2. Hence, the language $h(\overline{K'L''})$ is star-free. The proof is completed, because $KL = h(\overline{K'L''})$. □

We show that star-freeness of trace languages is equivalent to star-freeness of their flattenings. This fact will be important for our further considerations.

Theorem 7. Let T be a trace language. The following statements are equivalent:

$\qquad$ **(1)** T is star-free
$\qquad$ **(2)** $\bigcup T$ is star-free
$\qquad$ **(3)** $\bigcup T$ is aperiodic
$\qquad$ **(4)** T is aperiodic

Proof. (1)⇒(2): Structural induction on SF-expressions. For atomic languages – obvious. Assume that, for star-free trace languages X and Y, both $\bigcup X$ and $\bigcup Y$ are star-free. Let us check the three cases:

$\quad$ **(a)** If $Z = X \cup Y$, then $\bigcup Z = \bigcup (X \cup Y) = \bigcup X \cup \bigcup Y$ is star-free;
$\quad$ **(b)** If $Z = XY$, then $\bigcup Z = \bigcup (XY) = \overline{\bigcup X \cdot \bigcup Y}$ is star-free, by Lemma 6;;
$\quad$ **(c)** If $Z = X'$, then $\bigcup Z = \bigcup X' = (\bigcup X)'$ is star-free.

(2)⇒(3): Directly from Schützenberger Theorem 1.

(3)⇒(4): From definitions (subsection 2.2). For any $u, v \in A^*$ we have $u \approx_{\bigcup T} v$ iff $[u] \approx_T [v]$, thus syntactic monoids of T and $\bigcup T$ are equal.

(4)⇒(1): This part of the proof, as it was remarked in [5], is exactly the same as the proof of the analogous part of Schützenberger Theorem in [11].   □

The equivalence (1)≡(4) of Theorem 7 was proved by Guaiana/Restivo/Salemi [5] with quite involved, combinatorial techniques on traces. The present proof, using Lemma 6, is much simpler, and more general, than that of [5].

5   First-Order Logic for Traces

Logical definability of trace languages was described by Thomas [15]. The following notation is based on that of [4] and [1].

Let (A, D) be a concurrent alphabet and $w = a_1 \ldots a_n \in A^*$ be a word. A *trace-graph* of the word w is an acyclic graph $[V, E, \lambda]$, where $V = \{x_1, \ldots, x_n\}$ is a finite set of vertices, $E \subseteq V \times V$ is a set of edges and $\lambda : V \to A$, with $\lambda(x_i) = a_i$ for all i, is a node labelling such that $(x_i, x_j) \in E$ if and only if $i < j$ and $(\lambda(x_i), \lambda(x_j)) \in D$. Two words $u, v \in A^*$ are equivalent (i.e. $[u] = [v]$) iff their trace-graphs are isomorphic. If $I = \emptyset$ (free case), then E is a total order on V.

A model for a trace is a trace-graph. First-order formulas have variables $x, y, \ldots$ ranging over elements of V, and are built up from atomic formulas $x = y$, $(x, y) \in E$ and $\lambda(x) = a$ for $a \in A$, logical connectives $\vee, \wedge, \to, \leftrightarrow, \neg$, and the quantifiers $\forall$ and $\exists$.

The trace language defined by a sentence Ψ is $L(\Psi) = \{\alpha \in A^*/I \mid \Psi$ is satisfied in $\alpha\}$. A trace language $T \subseteq A^*/I$ is first-order definable if and only if a first-order sentence Ψ exists such that $T = L(\Psi)$. A class of first-order definable trace languages in A^*/I will be denoted by $FO(A^*/I)$.

Theorem 8 (McNaughton/Papert [7]). A word language is star-free if and only if it is first-order definable.

A concurrent alphabet, equipped with a strict order on the alphabet, will be called an *oriented concurrent alphabet* and denoted by $(A, <, I)$. Any strict order on A induces the well-known lexicographic order on A^*. A word $w \in A^*$ is said to be *lexicographic* (w.r.t. $<$ and I) iff it is lexicographically first in its closure $\overline{w} \subseteq A^*$. The set of all lexicographic words is denoted by LEX (assuming $<$ and I are unambiguously fixed). Subsets of LEX are called *lexicographic languages*. Given a trace language $T \subseteq A^*/I$, the word language $\text{Lex}(T) = \bigcup T \cap LEX$ is called the lexicographic representation of T.

Observe that, for any oriented concurrent alphabet $(A, <, I)$, the set LEX is star-free, as

$$LEX = A^* - \bigcup \{A^* b I_a^* a A^* \mid a I b \wedge a < b\}, \text{ where } I_a = \{c \in A \mid a I c\},$$

hence, by Theorem 8, LEX is first-order definable.

We assume, from now on, that the alphabets under consideration are equipped with a strict order.

Theorem 9 (Ebinger/Muscholl [4]). Let T be a trace language. The following statements are equivalent:

      (1) T is first-order definable,

      (2) $\bigcup T$ is first-order definable,

      (3) $\text{Lex}(T)$ is first-order definable.

Directly from (2)≡(3) and Theorem 8 we get

Corollary 10. Let T be a trace language. Then $\bigcup T$ is star-free if and only if $\text{Lex}(T)$ is star-free.

Theorem 11. A trace language T is star-free if and only if it is first-order definable.

Proof. From Theorem 7, T is star-free iff $\bigcup T$ is star-free. By Theorem 8, $\bigcup T$ is star-free iff $\bigcup T$ is first-order definable. By Theorem 9, $\bigcup T$ is first-order definable iff T is first-order definable. □

6 Star-Free Star and Trace Languages

We show that Theorem 5 (SF = SFS in free monoids) holds in arbitrary trace monoid. From Theorem 7 and Corollary 10, we directly obtain

Lemma 12. Let L be a word language included in LEX. Then

$$L \text{ is star-free if and only if } [L] \text{ is star-free.}$$

The following lemmas make use of Theorem 5.

Lemma 13. Let L be a word language.

$$\text{If } L \in \text{SFS}(A^*) \text{ and } L \subseteq LEX, \text{ then } [L] \in \text{SFS}(A^*/I).$$

Proof. Structural induction on SFS-expressions. For atomic languages – obvious. Assume for word languages $X, Y \in \text{SFS}(A^*)$ and $X, Y \subseteq LEX$, both $[X], [Y] \in \text{SFS}(A^*/I)$.

    (a) If $L = X \cup Y$, then $[L] = [X] \cup [Y] \in \text{SFS}(A^*/I)$.

    (b) If $L = XY$, then $[L] = [X][Y] \in \text{SFS}(A^*/I)$.

    (c) Let $L = X^*$, where $X^* \subseteq LEX$ and $X^* \in \text{SFS}(A^*)$, so $X^* \in \text{SF}(A^*)$. By Lemma 12, $[X]^* = [X^*] \in \text{SF}(A^*/I)$. Hence $[L] = [X]^* \in \text{SFS}(A^*/I)$. □

Lemma 14. Let T be a trace language.

$$\text{If } \bigcup T \in \text{SFS}(A^*), \text{ then } T \in \text{SFS}(A^*/I).$$

Proof. Let $T \subseteq A^*/I$ and $\bigcup T \in \text{SFS}(A^*)$, so $\bigcup T \in \text{SF}(A^*)$. Then $\text{Lex}(T) = \bigcup T \cap LEX \in \text{SF}(A^*)$. By Theorem 5, $\text{Lex}(T) \in \text{SFS}(A^*)$. From Lemma 13, $T = [\text{Lex}(T)] \in \text{SFS}(A^*/I)$. □

This way, we have proved

Theorem 15. Let T be a trace language.

$$T \in \mathrm{SF}(A^*/I) \text{ if and only if } T \in \mathrm{SFS}(A^*/I)$$

Proof. By Theorem 7, if $T \in \mathrm{SF}(A^*/I)$ then $\bigcup T \in \mathrm{SF}(A^*)$. By Theorem 5, $\bigcup T \in$ $\mathrm{SFS}(A^*)$. By Lemma 14, $T \in \mathrm{SFS}(A^*/I)$. By Definition 3 (of SFS-language), $T \in$ $\mathrm{SF}(A^*/I)$.   □

7 Conclusions and Problems

Let us collect the main characterisations of star-free trace languages.

Theorem 16. Given a trace language T, the following statements are equivalent

    (1) T is star-free

    (2) T is aperiodic

    (3) T is first-order definable

    (4) T is an SFS-language

Main and original result of the paper is (1)≡(4). Its proof uses Theorem 9 of Ebinger/Muscholl [4] and our result SF = SFS in free monoids (Theorem 5).

The equivalence (1)≡(3) was formulated by Ebinger/Muscholl [4] with a short sketch of the proof; more detailed sketch was presented in [1]. Another proof, for a more general class of monoids, was given by Droste/Kuske [3]. All the proofs are quite long and involved; they utilise some advanced techniques in logic. Our proof (Theorem 11), using implicitly star-free star, is much simpler. Let us recall that also the proof of (1)≡(2) (Theorem 7), using star-free star, is shorter than the original combinatorial proof of Guaiana/Restivo/Salemi [5].

It seems to the authors that the new result (1)≡(4) and the new proofs of (1)≡(2) and (1)≡(3) confirm efficiency of the star-free star notion.

Remark 1. We have shown (1)≡(4) with Theorem 9 on the first-order definability. We were able to prove Lemma 12 (in [14]), hence the equivalence (1)≡(4), without logic, but only for trace monoids possessing a transitively oriented alphabet ($I \cap \, <$ transitive). It would be interesting to find a general proof without logic.

Problem 1. We have defined star-free star in a semantic way – star in T^* is star-free star if T^* is star-free. Find a syntactically formulated condition for star-free star in trace monoid, like that for "recognizable star" – star in T^* is recognizable, if T is connected.

Problem 2. Is the question "Is T^* star-free?" decidable for star-free trace languages T? The similar decision problem for recognizability of T^* for recognizable T has been open for over twenty years. The general problem "Is T star-free?" for arbitrary rational T is undecidable (Muscholl/Petersen [9]).

References

1. Diekert, V., Metivier, Y.: Partial Commutation and Traces, in [12], pp. 457–533 (1997)
2. Diekert, V., Rozenberg, G. (eds.): The Book of Traces. World Scientific, Singapore (1995)
3. Droste, M., Kuske, D.: Languages and Logical Definability in Concurrency Monoids. In: Kleine Büning, H. (ed.) CSL 1995. LNCS, vol. 1092, pp. 233–251. Springer, Heidelberg (1996)
4. Ebinger, W., Muscholl, A.: Logical Definability of Infinite Traces. Theoretical Computer Science 154, 67–84 (1996)
5. Guaiana, G., Restivo, A., Salemi, S.: Star-free trace languages. Theoretical Computer Science 97, 301–311 (1992)
6. Mazurkiewicz, A.: Concurrent program schemes and their interpretations. Report DAIMI-PB-78, Aarhus University (1977)
7. McNaughton, R., Papert, S.: Counter-free Automata. MIT Press, Cambridge, MA (1971)
8. McNaughton, R., Yamada, R.: Regular expressions and state graphs for automata. Trans. of IRE EC 9(1), 11–18 (1960)
9. Muscholl, A., Petersen, H.: A note on the commutative closure of star-free languages. Information Processing Letters 57, 71–74 (1996)
10. Ochmański, E., Stawikowska, K.: Star Free-Star and Trace Languages. Fundamenta Informaticae 72(1-3), 323–331 (2006)
11. Perrin, D.: Finite Automata. Handbook of Theoretical Computer Science, vol. B, pp. 1–57. Elsevier, Amsterdam (1990)
12. Rozenberg, G., Salomaa, A. (eds.): Handbook of Formal Languages, vol. 3. Springer, Heidelberg (1997)
13. Schützenberger, M.P.: On finite monoids having only trivial subgroups. Information and Control 8, 190–194 (1965)
14. Stawikowska, K., Ochmański, E.: On Star-Free Trace Languages and their Lexicographic Representations, LATA 2007 (to appear, 2007)
15. Thomas, W.: On Logical Definability of Trace Languages. In: Proc. of ASMICS workshop, Report TUM-I9002, pp. 172–182 (1990)

Finite Automata on Unranked and Unordered DAGs

Lutz Priese

Fachbereich Informatik
Universität Koblenz-Landau, Germany
priese@uni-koblenz.de

Abstract. We introduce linear expressions for unrestricted dags (directed acyclic graphs) and finite deterministic and nondeterministic automata operating on them. Those dag automata are a conservative extension of the $T_{u,u}$-automata of Courcelle on unranked, unordered trees and forests. Several examples of dag languages acceptable and not acceptable by dag automata and some closure properties are given.

Keywords: finite state automata, directed acyclic graphs, regular dag languages.

1 Introduction

Finite automata that operate on finite or infinite words and (ordered and ranked) trees and various equivalent concepts to regularity are known. Those are recognizability with congruences, rationality with magmas, expressibility with regular expressions, definability in certain classes of monadic second order logic, generation by certain right-linear grammars. A good overview on the tree results is given in the TATA book [6]. However, only ranked and ordered trees are considered there . Note, a graph is ranked if the degree of any node is determined by its label, leading to ranked alphabets. This is not the case in unranked graphs and the degree in an infinite set of such unranked graphs may be unbounded. Thus, a finite automaton operating on unranked graphs has to operate on nodes with an unknown number of incoming and/or outgoing arcs (let us call this the "problem of unbounded degree"). An unpublished but well-known report [4] of Brüggemann-Klein, Murata and Wood researches ordered, unranked trees in some detail. Unranked, ordered and unordered trees have been investigated as an algebra by Courcelle in [8]. He presents a very elegant characterization of acceptability by $T_{u,u}$-magmas and frontier-to-root $T_{u,u}$-automata. Unranked and unordered trees with arc labels instead of node labels allow for a simpler algebraic approach and are found in Boneva and Talbot [2]. Brüggemann-Klein, Murata, Wood, and also Boneva, Talbot, solve the problem of unbounded degree for their unranked trees by allowing infinite but regular sets of transitions for their automata while Courcelle uses an associative and commutative transition function on pairs of state that easily extends to unbounded multisets of states.

T. Harju, J. Karhumäki, and A. Lepistö (Eds.): DLT 2007, LNCS 4588, pp. 346–360, 2007.
© Springer-Verlag Berlin Heidelberg 2007

Some generalizations of automata to ranked graphs or to their sub-class of ranked directed acyclic graphs are known: Finite graph automata have been introduced by Thomas [14], automata over planar dags by Kamimura and Slutzki [10]. A Kleene theorem for planar dags has been presented by Bossut, Dauchet and Warin [3]. They describe planar dags by linear expression that follow a graphic lay-out and use seriell and parallel composition. Graph expressions have also been introduced by Courcelle [7] for hyper-graphs to define context-free graph grammars. Charatonik [5] has researched automata on t-dags where no isomorphic sub-trees are allowed. Anantharaman, Narendran and Rusinowitch [1] continue this work, where the dag automata are mainly tree automata that run on dags, and present several interesting properties of such recognizable dag languages.

However, there exists no satisfying concept of automata on unrestricted unranked, unordered graphs. Kaminski and Pinter [11] avoid the problem of unbounded degree as their automata define a bound on the degree of acceptable graphs. However, the language of all graphs over a fixed alphabet now is not accepted any more. Fanchon and Morin [9] define regular pomsets languages over unranked alphabets with auto-concurrency via congruences of finite index. Those congruences mirror a serial-parallel composition of pomsets. They receive a concept of regularity that is closed under union but not under intersection or complement.

We will follow Courcelle's approach towards automata - but without using algebras as a semantics. We introduce linear dag expressions as a syntax and give a set-theoretical semantics as graphs. Finite automata operating as well on (congruence classes of) dag expressions as on abstract dags are introduced. In contrast to trees, dags possess incoming and outgoing arcs and all problems of root-to-frontiers and of frontiers-to-root automata must appear in dags. It is known that on trees deterministic root-to-frontier automata are a proper subclass of nondeterministic ones which are equivalent to deterministic or nondeterministic frontier-to-root automata. As a consequence, the 'root-to-frontier' part of dag automata should be nondeterministic. Our dag automata will therefore contain aspects of deterministic frontier-to-root and nondeterministic root-to-frontier automata.

2   Graphs and DAGs

Set-Theoretic Approach. A set-theoretical approach to graphs is simple: A *graph* γ over an alphabet Σ is a triple $\gamma = (N, E, \lambda)$ of two finite sets N of *nodes* and $E \subseteq N \times N$ of *edges* and a labelling mapping $\lambda : N \to \Sigma$. Two graphs $\gamma_i = (N_i, E_i.\lambda_i)$ are *isomorphic*, $\gamma_1 \sim_{iso} \gamma_2$, if there exists a bijective function $h : N_1 \to N_2$ with $(v, v') \in E_1 \Leftrightarrow (h(v), h(v')) \in E_2$ and $\lambda_2(h(v)) = \lambda_1(v)$ holds for all v, v' in N_1.

Thus, graphs in this paper are directed, unranked, unordered, finite and node labelled. We use the following rather standard notations.

$^\bullet v := \{v' \in V | (v', v) \in E\}$, $v^\bullet := \{v' \in V | (v, v') \in E\}$. For $V' \subseteq V$:
$^\bullet V' := \bigcup_{v \in V'} {}^\bullet v$, $V'^\bullet := \bigcup_{v \in V'} v^\bullet$. Any node in $^\bullet v$ ($v^\bullet$) is a *father (son)* of v.
$|^\bullet v|$ ($|v^\bullet|$) is the *in- (out-)degree* of v. A *root (leaf)* of a graph is a node with
in-degree (out-degree) 0. A *connection* of length n between two nodes v, v' is a
word $w = v_1...v_{n+1}$ s.t. $v = v_1, v' = v_{n+1}$ and $(v_i, v_{i+1}) \in E \cup E^{-1}$ holds for
$1 \leq i \leq n$. If $(v_i, v_{i+1}) \in E$ holds for all i w is called a *directed path* from v to
v'. A *cycle* is a directed path of some length > 0 from one node to itself.

A *dag* (directed acyclic graph) is a graph without cycles. A *forest* is a dag
where any node possesses at most one connection to at most one root. A *tree* is
a forest with exactly one root. Thus, the empty graph $\varepsilon := (\emptyset, \emptyset, \emptyset)$ is a forest but
not a tree. We usually identify isomorphic graphs and thus deal with *abstract
graphs*.

By Σ^t, Σ^f, $\Sigma^\dagger$, and Σ^g we denote the sets of all abstract trees, forests, dags,
and graphs, respectively, over Σ.

A graph is *ranked* if all nodes with the same label must also possess the same
out-degree, and *double ranked* if the label defines both the in- and out-degree. It
is *ordered* if a specific order between all sons of any node is given.

Algebraic Specification for Unranked Trees. Courcelle [8] defines a theory
$\mathcal{T}_{uu}$ for unranked, unordered trees that consists of a syntax of sorts $S_{uu} = \{l, t, f\}$
(for *letter, tree, forest*), operator symbols $Op_{uu} = \{p_{l \times f \to t}, r_{t \to f}, +_{f \times f \to f}, \theta_{\to f}\}$,
and equations E_{uu} :

$$u + v = v + u$$
$$(u + v) + w = u + (v + w)$$
$$u + \theta = u,$$

for variables u, v, w of sort f.

Let Σ denote a set of 0-ary generators of sort l. Any unranked, unordered tree
over Σ now simply becomes an element of sort t of $\mathcal{F}(\mathcal{T}_{uu}, \Sigma) = Term(\mathcal{S}_{uu}, \Sigma)/\equiv_{uu}$, where $Term(\mathcal{S}_{uu}, \Sigma)$ are all terms generated by $Op_{uu} \cup \Sigma$
and $\equiv_{uu}$ is the $\mathcal{S}_{uu}$-algebra congruence induced by the equations E_{uu}. To get a
theory $\mathcal{T}_{uo}$ for unranked, ordered trees just drop the equation for commutativity.

We might try to follow this approach and define a theory $\mathcal{T}_d$ for dags by
adding to $\mathcal{T}_{uu}$ a new sort s for *synchronization point* and a new operator symbol
$q_{s \times t \to t}$ and study $Term(\mathcal{T}_d, \Sigma \cup \mathbb{N})$, where any integer $i \in \mathbb{N}$ is regarded as a
0-ary symbol of sort s. However, as we will not be able to use $\mathcal{T}_d$-algebras as a
semantics for dags such an approach seems to be overloaded. In the following
syntax of linear dag expression we mainly abbreviate $p(a, f)$ by af, $p(a, \theta)$ by a,
$q(i, t)$ by it, and $r(t)$ by t and give a set-theoretic semantics.

Syntax of Graph Expressions. Let Σ denote a finite alphabet with $\Sigma \cap \mathbb{N} = \emptyset$.
We define the sets $E^t_{\Sigma\mathbb{N}}$ and $E^f_{\Sigma\mathbb{N}}$ of *tree* and *forest* expressions over $\Sigma \cup \mathbb{N}$ as
the smallest sets fulfilling the following requirements $\forall x \in \Sigma \cup \mathbb{N}$:

$$E^t_{\Sigma\mathbb{N}} \subseteq E^f_{\Sigma\mathbb{N}}, \ \theta \in E^f_{\Sigma\mathbb{N}}, \ x \in E^t_{\Sigma\mathbb{N}},$$
$$f, g \in E^f_{\Sigma\mathbb{N}} \implies xf \in E^t_{\Sigma\mathbb{N}}, \ (f + g) \in E^f_{\Sigma\mathbb{N}}.$$

Define for $a \in \Sigma, i \in \mathbb{N}, f, g \in E^f_{\Sigma \cup \mathbb{N}}$:
$int(\theta) := last(\theta) := first(\theta) := nol(\theta) := \emptyset,$
$int(a) := last(a) := nol(a) := first(a) := \emptyset,$
$nol(i) := \emptyset, \ int(i) := last(i) = first(i) := \{i\},$
$first(af) := \emptyset, \ X(af) := X(f), \text{ for } X \in \{int, last, nol\},$
$first(if) := \{i\}, \ last(if) := last(f),$
$X(if) := X(f) \cup \{i\}, \text{ for } X \in \{int, nol\},$
$X(f + g) := X(f) \cup X(g), \text{ for } X \in \{int, last, first, nol\}.$

$int(f)$ tells which integers appear in f, $last(f)$ which integers appear as last elements, $nol(f)$ which integers as "no last" elements, and $first(f)$ which integer as first elements of f. The binary relation $\equiv$ on forest expressions is defined $\forall f, f', g, g', h, \in E^f_{\Sigma \cup \mathbb{N}}, \ x \in \Sigma \cup \mathbb{N}$ by

1) $f \equiv f$, $f \equiv g \implies g \equiv f$, $(f \equiv g \wedge g \equiv h) \implies f \equiv h$,
2) $f \equiv f' \implies xf \equiv xf'$, $(f \equiv f' \wedge g \equiv g') \implies (f + g) \equiv (f' + g')$,
3) $(f + g) \equiv (g + f)$, $(f + (g + h)) \equiv ((f + g) + h)$, $(f + \theta) \equiv f$.

By 1) $\equiv$ becomes an equivalence relation, by 2) a congruence on our expressions, and fulfills by 3) the equations E_{uu}. The congruence $\equiv_0$ is defined as above but without the requirement $f + \theta \equiv_0 f$ in 3). $(f_1 + ... + f_n)$ or $\sum_{1 \leq \nu \leq n} f_\nu$ abbreviates $(...(f_1 + f_2) + ...) + f_n$.

Semantics of Graph Expressions. In a first, intermediate step we interpret a graph expression as a forest over $\Sigma \cup \mathbb{N}$.

$\forall x \in \Sigma \cup \mathbb{N}, f, g \in E^g_\Sigma:$
$\Im^o(\theta) := (\emptyset, \emptyset, \emptyset), \ \Im^o(x) := (\{1\}, \emptyset, \lambda(1) := x),$
$\Im^o(f) = (V, E, \lambda) \implies \Im^o(xf) := (V \cup \{v_{new}\}, E \cup \{(v_{new}, v) | v \in V \wedge^\bullet v = \emptyset\}, \lambda \cup \lambda(v_{new}) := x\},$
$\Im^o(f + g) := \Im^o(f) + \Im^o(g)$, where $\alpha + \beta$ is the disjoint union of the two graphs α, β.

To get our intended interpretation as abstract graphs over Σ we regard all integers as synchronization points that must be synchronized (i.e., all occurrences of the same integer are identified) and deleted. Therefor we introduce the operation Syd (for "Synchronize and delete"):

$$Syd_{i_1, ..., i_k}(\alpha) := Syd_{(i_1}(...(Syd_{i_k}(\alpha)...), \text{ with}$$

$$Syd_i(\alpha) := (V', E \cap (V' \times V') \cup E', \lambda_{|V'}), \text{ for}$$

$V' = \{v \in V | \lambda(v) \neq i\}$, $E' = \{(v, v') | \exists v_1, v_2 \in V : \lambda(v_1) = \lambda(v_2) = i \wedge (v, v_1), (v_2, v') \in E\}$, setting all nodes v as a father of all nodes v' if v possesses some son with label i and v' possesses some father with label i, deleting such all melted synchronization points.

The interpretation of an expression as an abstract graph is given as

$$\Im(f) := [Syd_{int(f)}(\Im^0(f))]_{\sim iso}.$$

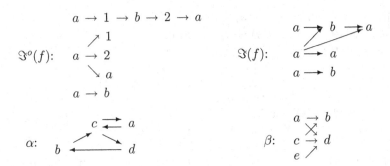

Fig. 1. Some graphs of example 2.1

Example 2.1. $\Im^o(f)$ and $\Im(f)$ for $f = a1b2a + a(1 + 2 + a) + ab$ are shown in figure 1. The abstract graph α is the interpretation of the expressions $1a2c(1 + db2)$, $1c(a1 + db1)$, and $1db2c(a2 + 1)$. $\beta = \Im(d_i)$ for the expressions d_i for $1 \leq i \leq 5$ with

$$d_1 = a(1b + 2d) + c(1 + 2) + e2, \quad d_2 = a(1b + 2) + c(2d + 1) + e2$$
$$d_3 = a1b + c1\,2d + e2, \quad\quad\quad d_4 = a1(b + 2d) + c1 + e2$$
$$d_5 = a1(b + 2d3) + 4c1 + 4e2.$$

Of course, $\equiv$- or $\equiv_0$-congruent expressions describe the same abstract graph.

DAG Expressions. Regard the relation $\trianglelefteq(f) \subseteq int(f) \times int(f)$ defined inductively for $i \in \mathbb{N}$, $a \in \Sigma$, $f, g \in E_{\Sigma \mathrm{UN}}^f$:

$\trianglelefteq(\theta) := \trianglelefteq(a) := \trianglelefteq(i) := \emptyset$,
$\trianglelefteq(af) := \trianglelefteq(f)$, $\trianglelefteq(if) := \trianglelefteq(f) \cup \{(i,j) | j \in int(f)\}$, $\trianglelefteq(f + g) := \trianglelefteq(f) \cup \trianglelefteq(g)$.

Obviously, the interpretation $\Im(f)$ of an expression f is an abstract dag if the transitive closure $\trianglelefteq(f)^+$ of $\trianglelefteq(f)$ is a partial order. Example 2.1 presents 5 rather different expression d_i for the same abstract dag β of figure 1. However, only d_1 and d_2 are "smooth" expressions whilst d_3 to d_5 are against intuition. We need a formal definition for smoothness:

An expression $f \in E_{\Sigma \mathrm{UN}}^f$ is called *smooth* if f contains no sub-expressions

- θ, ij, ig,
- $(g + h)$ with $nol(g) \cap nol(h) \neq \emptyset$,
- it with $i \in nol(t)$,
- i that occurs exactly once in f or with $i \in int(f) - nol(f)$,
- $\sum_{1 \leq \nu \leq n} t_\nu$ with two trees t_{ν_0}, t_{ν_1} for $1 \leq \nu_0 < \nu_1 \leq n$ with $first(t_{\nu_0}) = first(t_{\nu_1}) \neq \emptyset$,

for $a \in \Sigma$, $i, j \in \mathbb{N}$, $t, t_\nu \in E_{\Sigma \mathrm{UN}}^t$, $g, h \in E_{\Sigma \mathrm{UN}}^f$.

In a smooth expression each synchronization point i must occur several times but only once not as a last element. In this case i must precede some tree expression. No two synchronization points must follow each other. No brother trees must be synchronized.

Tree and forest expressions over Σ are expressions without integers, a *graph expression* over Σ is a forest expression over $\Sigma \cup \mathbb{N}$, and a *dag expression* over Σ is a smooth graph expression where $\unlhd^+$ is a partial order:

$$E_\Sigma^t := \{t \in E_{\Sigma \cup \mathbb{N}}^t \mid int(t) = \emptyset\},$$
$$E_\Sigma^f := \{f \in E_{\Sigma \cup \mathbb{N}}^f \mid int(f) = \emptyset\}, \quad E_\Sigma^g := E_{\Sigma \cup \mathbb{N}}^f,$$
$$E_\Sigma^\dagger := \{f \in E_\Sigma^g \mid f \text{ is smooth and } \nexists i, j \in int(f) : (i \unlhd (f)^+ j \wedge j \unlhd (f)^+ i)\}.$$

All abstract trees, forest, dags, and graphs can be expressed:

Lemma 2.1. $\Im(E_\Sigma^t) = \Sigma^t$, $\Im(E_\Sigma^f) = \Sigma^f$, $\Im(E_\Sigma^\dagger) = \Sigma^\dagger$, $\Im(E_\Sigma^g) = \Sigma^g$.

Example 2.2. *Let* $f_1 = a1b2a7 + 6a(1 + 2 + \theta + 2 + 3a4) + 6 + ab(\theta + 7)$, $f_2 = 3a1 + 3 + a(1b2 + 2a4 + a5) + 6ab5$, *and* $d_6 = a1 + a(1b2 + 2a + a) + ab$. $\Im(f_1) = \Im(f_2) = \Im(d_6) = \Im(f)$ *of figure 1.* d_6 *is a dag expression but* f_1, f_2 *are not as they violate smoothness.*

3 DAG-Automata

Let M be a set. A function $f : M \times M \to M$ is *associative* and *commutative* if $f(x, y) = f(y, x)$ and $f(x, f(y, z)) = f(f(x, y), z)$ holds for all $x, y, z \in M$.

Such an associative and commutative function f is easily extended to

$$f^* : (\mathbb{N}^M - \{0\}) \to M$$

operating on nonempty multisets over M: $\forall a \in M : \forall m \in \mathbb{N}^M - \{0\}$:

$$f^*(1 \cdot a) := a, f^*(1 \cdot a + m) := f(a, f^*(m)).$$

A (root-to-frontier) *dag automaton*

$$A = (Q, \Sigma, \delta, \delta_i, \delta_o, \delta_I, \delta_F, I, F)$$

consists of

- a finite set Q of states with $Q \cap \mathbb{N} = \emptyset$,
- a finite alphabet Σ with $Q \cap \Sigma = \emptyset = \Sigma \cap \mathbb{N}$,
- a function $\delta : Q \times \Sigma \to 2^Q$,
- four associative and commutative functions

$$\delta_i, \delta_o, \delta_I, \delta_F : Q \times Q \to Q,$$

- a set $I \subseteq Q$ of initial states, and
- a set $F \subseteq Q$ of final states.

We now introduce the concept of configurations and computations of dag automata. By a simple scan through a dag expression d one can identify as d-*roots* those occurrences of letters in Σ that become roots in $\Im(d)$. A *configuration* C for A is a dag expression over $\Sigma \cup Q$ where exactly the states form the C-roots.

Now, for a given dag d, put in front of each d-root in d some state of Q to get a configuration C_d. C_d is admissible if for the multiset m of added state $\delta_I^*(m) \in I$ holds. C_d denotes the set of all those admissible configurations for d. Each $C \in C_d$ is called a *start configuration* for d. As a configuration is a dag expression itself the congruences $\equiv$ and $\equiv_0$ hold also for configurations.

A configuration C' is a *direct successor* of a configuration C, $C \vdash_A C'$, if there exists a configuration $\hat{C}$ with $C \equiv_0 \hat{C}$ and C' is the result of replacing in $\hat{C}$ a sub-expression

- sa by s', with $s' \in \delta(s, a)$, or
- $s(f_1 + f_2)$ by $(s_1 f_1 + s_2 f_2)$, with $\delta_o(s_1, s_2) = s$, or
- $(s_1 + s_2)$ by s', with $\delta_F(s_1, s_2) = s'$, or
- $(s_1 it + s_2 i)$ by $s' it$, with $\delta_i(s_1, s_2) = s'$, or
- $(s_1 i + s_2 i)$ by $s' i$, with $\delta_i(s_1, s_2) = s'$, or
- si by s, if i occurs exactly once in $\hat{C}$,

for $a \in \Sigma, i \in \mathbb{N}, t \in \Sigma_{\Sigma \cup \mathbb{N}}^t, f_1, f_2 \in \Sigma_\Sigma^g, s, s_1, s_2 \in Q$. In addition, we write $d \vdash C$ for a dag expression d and a start configuration $C \in C_d$. Only here δ_I plays a rôle. δ_o handles the transport of a state to outgoing arcs and δ_i of states from incoming arcs.

$eval^A(d) := \{s \in Q \mid d \vdash_A^* s\}$ is the set of all states into which a dag expression d may evaluate under A.

$D(A) := \{d \in E_\Sigma^\dagger \mid eval^A(d) \cap F \neq \emptyset\}$ is the dag language *accepted* by A. A dag language is *regular* if it is accepted by some dag automaton.

We may regard a dag automaton as operating on dag expressions, $\equiv_0$-congruence classes of them, or on abstract dags, as any dag automaton operates identically on different dag expressions of the same abstract dag:

Theorem 1. *For all dag expressions d_1, d_2 and dag automata A over the same alphabet:*

$$\Im(d_1) = \Im(d_2) \implies eval^A(d_1) = eval^A(d_2).$$

Example 3.1. *Let $\Sigma^\dagger{}_{even}$ denote the language of all dags over Σ with an even number of nodes. Σ_{even} is accepted by A_{even} with $Q := \{0, 1\}$, where we use boldface integers as states, $I := F := \{\mathbf{0}\}$, $\delta(s, a) := s + 1 \bmod 2$ for all $a \in \Sigma$, and $\delta_F := \delta_I := \delta_i := \delta_o := + \bmod 2$.*

A nondeterministic computation with some dag expression d for the dag $\Im(f)$ of figure 1 is, e.g.

$d = a1b2a(3 + \theta) + \left(a\big((1 + 2) + a\big) + ab\right)$

$\vdash 1a1b2a + \left(\mathbf{0}a\big((1 + 2) + a\big) + 1ab\right) (\in C_d)$

$\vdash^*_{(\delta)} 01b2a + 1\big((1 + 2) + a\big) + 0b \vdash^*_{(\delta_o, \delta)} 01b2a + \left(1(1 + 2) + 0a\right) + 1$

$\vdash_{(\delta_o)} 01b2a + \left((11 + 02) + 0a\right) + 1 \equiv_0 \left(\big((01b2a + 11) + 02\big) + 0a\right) + 1$

$\vdash_{(\delta_i)} \left((11b2a + 02) + 0a\right) + 1 \vdash^* (1b2a + 02) + (1 + 1) \vdash^* (02a + 02) + 0$

$\vdash 02a + 0 \vdash^* 1.$

Any computation for d leads to $\mathbf{1}$, thus $d \notin D(A_{even})$.

Example 3.2. *Let $\Sigma^\dagger{}_{dis}$ denote the language of all disconnected dags over Σ and $\Sigma^\dagger{}_{con}$ those of all connected dags. We present an automaton A_d accepting $\Sigma^\dagger{}_{dis}$. When an abstract dag α consists of two disjunct dags α_1, α_2 an accepting computation of A_d guesses a state s_1 to be attached to all roots of α_1 and a different state s_2 to all roots of α_2. A_d passes s_i through α_i. If α_1 and α_2 should have a common node the states s_1 and s_2 will meet and pass an error message to some leaf thats forbids acceptance. Thus simply choose*

$Q := \{s_0, s_1, s_2, \checkmark, \bot\}$ *with a sink state* $\bot$ *(s.t.* $\delta_\bullet(x,y) = \bot$ *if* $x = \bot$ *or* $y = \bot$ *for all transition functions),* $I := \{s_0\}, F := \{\checkmark\}$,

$\delta_I(s_1, s_2) := s_0, \delta_I(s_1, s_1) := s_1, \delta_I(s_2, s_2) := s_2$,

$\delta(s_1, x) := s_1, \delta(s_2, x) := s_2$, *for* $x \in \Sigma$,

$\delta_o(s_1, s_1) := s_1, \delta_o(s_2, s_2) := s_2$,

$\delta_i(s_1, s_1) := s_1, \delta_i(s_2, s_2) := s_2, \delta_i(s_1, s_2) := \bot$,

$\delta_F(s_1, s_1) := s_1, \delta_F(s_2, s_2) := s_2, \delta_F(s_1, s_2) := \checkmark, \delta_F(\checkmark, s_i) := \checkmark$ *for* $i = 1, 2$,

plus all required transitions to get commutative and associative mappings and make $\bot$ *a sink state. A "false" not accepting computation for the above dag expression d is shown in Figure 2.*

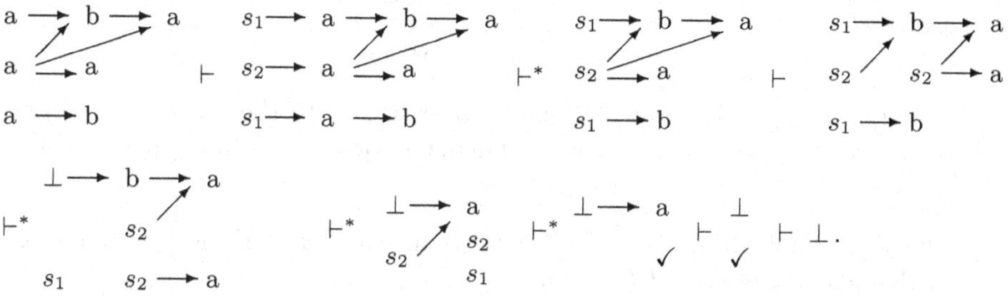

Fig. 2. A not accepting computation of A_{dis}

Although the treatment of incoming and outgoing arcs in dag expressions is completely different (using a simple $+$ for outgoing arcs but an alphabet of infinitely many synchronization points for incoming arcs) they are treated symmetrically in dag automata. This is easily seen with reverse dags.

The *reverse* A^{rev} of an automaton $A = (Q, \Sigma, \delta, \delta_i, \delta_o, \delta_I, \delta_F, I, F)$ is

$$A^{rev} = (Q, \Sigma, \delta^{rev}, \delta_i^{rev}, \delta_o^{rev}, \delta_I^{rev}, \delta_F^{rev}, I^{rev}, F^{rev}),$$

with

$$I^{rev} := F, F^{rev} := I, \delta^{rev}(s, a) := \{s' | s \in \delta(s', a)\},$$

$$\delta_i^{rev} := \delta_o, \delta_o^{rev} := \delta_i, \delta_I^{rev} := \delta_F, \delta_F^{rev} := \delta_I.$$

Any successful computation in A for some α defines in reverse order immediately a successful computation for α^{rev} in A^{rev}. This implies $D(A)^{rev} \subseteq D(A^{rev})$. Further

$$D(A) = (D(A)^{rev})^{rev} \subseteq D(A^{rev})^{rev} \subseteq D((A^{rev})^{rev}) = D(A), \text{ thus}$$

Lemma 3.1. $D(A^{rev}) = D(A)^{rev}$

4 Regular and Nonregular DAG Languages

Simple Regular DAG Languages and Closure Properties. Some typical examples of regular dag languages are - as expected - the language of all dags

- where no node possesses several sons with the same labels,
- where no node possesses several fathers with the same labels,
- where no node possesses several sons (fathers, respectively) with different labels,
- with exactly j roots (j leaves or j nodes, respectively),
- with 0 roots (0 leaves or 0 nodes, respectively) modulo some constant,
- where all nodes have the in-degree i or 0 (out-degree i or 0, in-degree mod i is 0, out-degree mod i is 0, respectively).

  Constructing dag automata that accept those languages is just a simple exercise.

A *maximal path* in a dag is a directed path from some root to some leaf. A path is identified with the word of labels of its nodes. $path(\alpha)$ is the set of all maximal paths in a dag α and $path(D) = \bigcup_{\alpha \in D} path(\alpha)$ for $D \subseteq \Sigma^\dagger$ defines a projection

$$path : 2^{\Sigma^\dagger} \to 2^{\Sigma^*}$$

from languages over dags into languages over words. In the opposite direction there are two canonical ways to embed languages over words into languages over dags:

- the *skinny* embedding of $L \subseteq \Sigma^*$ regards any word $w \in \Sigma^*$ as a path $w \in \Sigma^\dagger$ and is also denoted as $L\,(\subseteq \Sigma^\dagger)$,
- the *fat* embedding D_L of $L \subseteq \Sigma^*$ is the dag language

$$D_L := \{\alpha \in \Sigma^\dagger | path(\alpha) \subseteq L\}.$$

Lemma 4.1. *The skinny embedding L and the fat embedding D_L of a regular word language L are regular dag languages.*

The opposite statement holds for projections of a fat embedding but not for projections of general dag languages.

Lemma 4.2. *If D_L is a regular dag language then L is a regular word language. But there exist regular dag languages whose path projection is not even a context-free word language.*

Lemma 4.3. *The class of regular dag languages is closed under union and intersection.*

The proofs for all mentioned three lemmata are rather straight-forward. Interesting is a counter example for lemma 4.2 of a regular dag language with a noncontext-free path language: It is possible to construct a dag automaton A_D that accepts only dags of the form as shown in Figure 3 with an equal number of

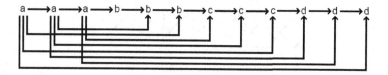

Fig. 3. A dag accepted by A_D

labels a, b, c and d where the order a before b before c before d must be respected. This can be achieved by forcing all nodes with a label a to have four sons, labelled with a, b, c, d, but one who has three sons labelled with b, c and d. All nodes labelled with $x \in \{b, c, d\}$ are forced to have two fathers, one labelled with a and a second labelled with x with an exception for the first label x as shown. Thus, $path(D(A_D)) = \{a^i d^j | 1 \leq i, j\} \cup \{a^i c^j d^n | 1 \leq i, j \leq n\} \cup \{a^i b^j c^n d^n | 1 \leq i, j \leq n\}$, a noncontext-free language. ∎

Some Gaps Between Regular and Nonregular DAG Languages. There are simple example of nonregular dag languages as finite dag automata cannot count above some boundary: the language $D_{r=l}$ of all dags with the same number of roots and leaves, or $D_{n=}$ ($D_{r=}$, $D_{l=}$) over $\{a, b\}$ where equally many nodes (roots, leaves, respectively) are labelled with a and b. However, if one would change the concept of nondeterministic dag automata in such a way that also partial, not associative functions δ_I and δ_F are allowed then $D_{r=}$ and $D_{l=}$ (in contrast to $D_{n=}$) become regular. To accept $D_{l=}$ choose $Q = \{s_0, s_a, s_b, \checkmark\}$, $I = \{s_0\}$, $F = \{\checkmark\}$, $\delta_I(s_0, s_0) = s_0$, $\delta(s_0, x) = s_x$ and $\delta_i(s_x, s_y) = s_0$ and $\delta_o(s_x, s_x) = s_x$ for $x, y \in \{a, b\}$ such that s_x tells that the last node visited has been labelled with x. Now, simply set $\delta_F(s_a, s_b) = \checkmark$, $\delta_F(\checkmark, \checkmark) = \checkmark$ and $\delta_F(., .)$ undefined elsewhere to accept $D_{l=}$. For $D_{r=}$ use the reverse automaton. However, such a trick is impossible with total associative and commutative functions δ_I, δ_F.

Theorem 2. $\Sigma^\dagger_{dis}$ *is regular but* $\Sigma^\dagger_{con}$ *is not. Thus, regular dag languages are not closed under complement.*

Proof. Regularity of $\Sigma^\dagger_{dis}$ was shown in example 3.2. Non-regularity of $\Sigma^\dagger_{con}$ is seen as follows: Suppose there exists an automaton A that accepts $\Sigma^\dagger_{con}$. Set

$$d := \sum_{1 \leq i \leq n} a(ia + i')$$

with $i' := i + 1$ for $1 \leq i < n$ and $n' := 1$, compare Figure 4 with $\alpha := \Im(d)$. α is connected. Thus, A accepts α. For n large enough (i.e., longer than $(|Q| + 1)^2$) there must exist two different occurrences o'_1, o'_2 of labels a on the lower row and o_1, o_2 of their right fathers in the upper row where an accepting computation $\mathcal{C}$ of A reaches in o_1 and o_2 the same state, say s, and in o'_1 and o'_2 a same state, say s'. Now, let β result from re-pointing the arc originally from o_1 to o'_1 now to o'_2 and the arc pointing originally from o_2 to o'_2 now to o'_1. This doesn't introduce

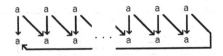

Fig. 4. $\Im(d)$

cycles and the same computation C will still accept β - but β is disconnected
(and still planar).   ∎

However, "bounded" connectivity becomes regular:

Lemma 4.4. *The languages of all connected dags with a fixed or bounded number of roots or leaves, respectively, are regular.*

Ladders of type 1 or 2 and *beams* are dags as presented in figure 5. $D_{1-ladder}$, $D_{2-ladder}$, D_{beam} denote the languages of all type 1 ladders, all type 2 ladders, and all beams, respectively, over a.

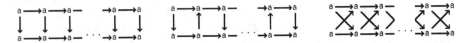

Fig. 5. A type 1 ladder (left), type 2 ladder (midle) and beam (right)

Theorem 3. D_{beam} *and* $D_{2-ladder}$ *are regular,* $D_{1-ladder}$ *is not.*

At a first sight, theorem 3 seems to point to a disadvandage of our concept of dag automata: type 1 ladders and type 2 ladders seem to be so similar that one might think that one type of ladders could result from the other by some "regular transformation" (and automata should preserve "regular transformations"). However, this is not the case. Type 1 and type 2 ladders have very distinct "synchronization properties": An automaton may evaluate the upper row of a type-1-ladder ignoring the evaluation of the lower row, which is impossible for type 2 ladders. The situation is similar for Petri nets: There is a (rather simple) Petri net with $D_{2-ladder}$ as it true-concurrency dag semantics, but no Petri net can possess $D_{1-ladder}$ as its dag semantics, see [12].

One can prove theorem 3 by showing that any dag automaton accepting $D_{1-ladder}$ must also accept some connected but not planar dag (that is not in $D_{1-ladder}$). Thus, $D_{1-ladder}$ can't even be recognized relative to connected dags, i.e., if input dags are restricted to be connected. Both automata α, β of the proof of theorem 2 are planar. Thus, even if all input dags must be planar $\Sigma^{\dagger}_{con}$ is not regular. This implies

Corollary 4.1. *The language* $\Sigma^{\dagger}_{plan}$ *of all planar dags over* Σ *is not regular.* $\Sigma^{\dagger}_{con}$ *is not regular relative to* $\Sigma^{\dagger}_{plan}$ *and* $\Sigma^{\dagger}_{plan}$ *is not regular relative to* $\Sigma^{\dagger}_{con}$.

The *shuffle*

$$D_1 \parallel D_2 := \{\alpha_1 + \alpha_2 | \alpha_i \in D_i\}$$

of two dag languages consists of disjoint unions of one dag from D_1 with one from D_2. $\|^i$ and the big shuffle $\|^*$ are defined as

$$\|^0 D := \{\Im(\theta)\}, \quad \|^{n+1} := (\|^n D) \parallel D, \quad \|^* D := \bigcup_{i \geq 0} \|^n D.$$

It turns out that the big shuffle of even a single dag may be no regular language. Let $\square$ be the beam of length 1, consisting of four nodes and being described by $a(1a + 2a) + a(1 + 2)$. $d := a(a1a + a1)$ describes a dag $\lozenge := \Im(d)$ with also only four nodes and four arcs as $\square$ but only one root and leaf.

Theorem 4. $\|^* \square$ and $\|^* \lozenge$ are nonregular.

This immediately implies:

Lemma 4.5. *The class of regular dag languages is closed under $\|^n$ for any n but not under $\|^*$.*

5   Deterministic DAG Automata

A dag automaton $A = (Q, \Sigma, \delta, \delta_i, \delta_o, \delta_I, \delta_F, I, F)$ is called *deterministic* if $|\delta(s, a)| = 1$ holds for $s \in Q, a \in \Sigma$, $I = \{s_0\}$ for one initial state s_0, there exists a sink state $\perp \in Q$ and $\delta_I(s_0, s_0) = s_0$, $\delta_I(.,.) = \perp$ elsewhere, $\delta_o(s, s) = s$ for some states $s \in Q$ and $\delta_o(.,.) = \perp$ elsewhere. Thus, to ensure a deterministic computation a start configuration for a dag α receives by δ_I the same state s_0 attached to all roots and the same state must be prolonged by δ_o from a father to all sons. As in a parse tree for a context-free derivation, the order of where to apply a transition in a configuration is still free, but $|eval^A(d)| = 1$ will hold. A regular dag language is called *deterministic regular* if it is accepted by some deterministic dag automaton.

If one applies a deterministic (root-to-frontier) dag automata to the reverse α^{rev} of an unranked, unordered tree α it behaves exactly as Courcelle's (frontiers-to-root) $\mathcal{T}_{uu}$-automata applied to α. Tree languages accepted by deterministic root-to-frontier tree automata are a proper subclass of those accepted by deterministic or nondeterministic frontier-to-root automata. Let $\triangleright = \Im(d)$ for $d = a1a + a1$. The trivial language $\{\triangleright\}$ is deterministic regular but $\{\triangleright\} \parallel \{\triangleright\}$ is not, as any deterministic automaton accepting $\triangleright + \triangleright = \Im(a1a + a1 + a2a + a2)$ must also accept $\square = \Im(a(1a + 2a) + a(1 + 2))$. Also, the regular languages $\Sigma^\dagger_{even}$ and $\Sigma^\dagger_{dis}$ are no longer deterministic regular. It is easily seen that the class of deterministic regular dag languages is closed under union, intersection and complement.

When a deterministic dag automaton passes a state from a father node to its sons it cannot react on the possibly different labels of the sons. Thus, deterministic dag automata are *forward blind*. One easily can define with the help of commutative and associative mappings a concept of not forward-blind deterministic root-to-frontier dag automata where the state passed to a son may depend on the state of the father and the multiset of labels of all sons. An *nfb regular* dag language is a regular dag language accepted by a deterministic not forward-blind dag automaton. Table 1 presents some properties of the classes of regular, deterministic regular, nfb regular and semi rational dag languages. A dag language is semi rational if it is the dag semantics of some Petri net, see [13]. In contrast to word languages, dag languages accepted by finite automata must not necessarily be Petri net dag languages, see the last two lines of table 1.

Table 1. Closure Properties, () is a Conjecture

Closed under:	$Reg_{det}^{\dagger}$	$Reg_{nfb}^{\dagger}$	$Reg^{\dagger}$	$SemiRat^{\dagger}$
union	✓	✓	✓	✓
intersection	✓	✓	✓	✓
complement	✓	✓	no	no
reverse	no	no	✓	✓
shuffle	no	no	✓	✓
big shuffle	no	no	no	no
finite sets	no	no	✓	✓
fat embedding of $\mathcal{L}_3$	✓	✓	✓	✓
contains: $\Sigma_{dis}^{\dagger}$	no	no	✓	✓
$D_{r=l}$	no	no	no	✓
$D_{n=}$	no	no	no	✓
can count modulo i the:				
nodes	no	no	✓	✓
roots	no	no	✓	✓
leaves	✓	✓	✓	✓
incoming arcs	✓	✓	✓	(no)
outgoing arcs	no	✓	✓	(no)

6   Comparison to Further Automata Concepts

There are several concepts in the literature of finite automata analyzing graphs or dags with "local conditions". Kaminski and Pinter [11] operate on rooted directed graphs over a double ranked alphabet, and, thus, with a global bound for the in- and out-degree. With their automata $D_{1-ladder}$ is also not recognizable, see Thomas [15]. (One easily may regard $D_{1-ladder}$ as a language over a double ranked alphabet by using different labels for the upper and lower row). Thomas introduces "acceptors" on ranked graphs equivalent to existential monadic second-order logic (EMSL) on those graphs. Those acceptors simulate a tiling of a ranked graph with elementary graphs of a finite set of types plus some nonlocal constraints. $D_{1-ladder}$ becomes now acceptable relative to connected graphs. However, connectedness is expressible in MSL but not in EMSL, and thus not acceptable by those graph acceptors. It is hard to imagine how to generalize Thomas' acceptor concept to unranked graphs and languages with no fixed bound of the in- and out-degree.

It is also known that $D_{1-ladder}$ is acceptable relativ to a class of planar dags (pdags) of Bossut, Dauchet and Warin [3]. They introduce algebraic pdag expressions built from two-sorted letters and operations (iterated) *parallel* and *serial* composition. They can present an automaton that accepts all connected pdag expressions, in contrast to our theorem 2, corollary 4.1 and inexpressibility of connectedness in EMSL. This contradiction is resolved if one notes that their pdag expression cannot describe all planar dags, especially not all planar dags of figure 4 that have been required for violating regularity of connectedness.

Résumé

We have introduced linear expressions that can describe all unranked and unordered abstract dags and finite automata on those unrestricted dags defining a class of regular dag languages. According to theorem 1, these dag automata operate on abstract dags as well as on dag expressions. The closure properties and examples of regular dag languages are just what should be expected from a reasonable concept of nondeterministic and deterministic dag automata. Although our approach is similar to the $\mathcal{T}_{u,u}$ approach of Courcelle we don't have a concept of rationality of dag languages by finite algebras, even not in the deterministic case. The reason is that in our concept of a computation the treatment of sorts l, t and f fits perfectly into the evaluation schema of algebras - such as to replace sa ba $\delta(s,a)$ or $s(f_1 + f_2)$ by $sf_1 + sf_2$. But for synchronization points a global view is involved as we replace si by s only if i occurs exactly once in the overall term. The examples of regular and nonregular dag languages are very similar to those of semi-rational and not semi-rational ones in [13] although the concepts of regularity (acceptance by dag automata) and of semi-rationality (dag semantics of Petri nets) are rather different. This may be a hint that regularity and semi-rationality of dag languages indeed point more to inherent properties of dags than of the chosen concepts of automata and Petri nets.

Acknowledgement

I would like to thank the unknown referees of this and a previous version for their valuable comments.

References

1. Anantharaman, S., Narendran, P., Rusinowitch, M.: Closure properties and decision problems of dag automata. Information Processing Letters 94, 231–240 (2005)
2. Boneva, I., Talbot, J.-M.: Automata and logic for unranked and unordered trees. In: Giesl, J. (ed.) RTA 2005. LNCS, vol. 3467, pp. 500–515. Springer, Heidelberg (2005)

3. Bossut, F., Dauchet, M., Warin, B.: A Kleene theorem for a class of planar acyclic graphs. Theor. Comp. Science Center Report HKUST-TCSC 2001-5 117, 251–265 (1995)
4. Brüggemann-Klein, A., Murata, M., Wood, D.: Regular tree and hedge languages of unranked alphabets. Theor. Comp. Science Center Report HKUST-TCSC 2001 5, 29 (2001)
5. W. Charatonik. Automata on dag representations of finite trees. Technical Report MPI-I-1999-2-001, MPI, Univ. Saarbr$\tilde{A}\frac{1}{4}$cken, 1999.
6. Comon, H., Daucher, M., Gilleron, R., Tison, S., Tommasi, M.: Tree automata techniques and application. Available on the Web from 13ux02.univ-lille.fr in directoty tata (1998)
7. Courcelle, B.: A representation of graphs by algebraic expressions and its use for graph rewriting systems. In: Proc. 3rd Internat. Workshop on Graph-Grammars, pp. 112–132. Springer Verlag, Heidelberg (1988)
8. Courcelle, B.: On recognizable sets and tree automata. In: Aït-Kaci, H., Nivat, M. (eds.) Resolution of Equations in Algebraic Structures, vol. 1, pp. 93–126. Academic Press, London (1989)
9. Fanchon, J., Morin, R.: Regular sets of pomsets with auitoconcurrency. In: Brim, L., Jančar, P., Křetínský, M., Kucera, A. (eds.) CONCUR 2002. LNCS, vol. 2421, pp. 402–417. Springer, Heidelberg (2002)
10. Kamimura, T., Slutzki, G.: Parallel and two-way automata on directed ordered acyclic graphs. Inf. Control 49, 10–51 (1981)
11. Kaminski, M., Pinter, S.: Finite automata on directed graphs. J. Comp. Sys. Sci. 44, 425–446 (1992)
12. Menzel, J.R., Priese, L., Schuth, M.: Some examples of semi-rational dag languages. In: Ibarra, O.H., Dang, Z. (eds.) DLT 2006. LNCS, vol. 4036, pp. 351–362. Springer Verlag, Heidelberg (2006)
13. Priese, L.: Semi-rational sets of dags. In: De Felice, C., Restivo, A. (eds.) DLT 2005. LNCS, vol. 3572, pp. 385–396. Springer Verlag, Heidelberg (2005)
14. Thomas, W.: Finite-state recognizability of graph properties. In: Krob, D. (ed.) Theorie des Automates et Applications, l'Universite de Rouen, France, vol. 172, pp. 147–159 (1992)
15. Thomas, W.: Automata theory on trees and partial orders. In: Bidoit, M., Dauchet, M. (eds.) CAAP 1997, FASE 1997, and TAPSOFT 1997. LNCS, vol. 1214, pp. 20–34. Springer, Heidelberg (1997)

On Almost Periodicity Criteria for Morphic Sequences in Some Particular Cases

Yuri Pritykin*

Moscow State University, Russia
http://lpcs.math.msu.su/~pritykin/
yura@mccme.ru

Abstract. In some particular cases we give criteria for morphic sequences to be almost periodic (=uniformly recurrent). Namely, we deal with fixed points of non-erasing morphisms and with automatic sequences. In both cases a polynomial-time algorithm solving the problem is found. A result more or less supporting the conjecture of decidability of the general problem is given.

1 Introduction

Different problems of decidability in combinatorics on words are always of great interest and difficulty. Here we deal with two main types of symbolic infinite sequences — morphic and almost periodic — and try to understand connections between them. Namely, we are trying to find an algorithmic criterion which given a morphic sequence decides whether it is almost periodic.

Though the main problem still remains open, we propose polynomial-time algorithms solving the problem in two important particular cases: for pure morphic sequences generated by non-erasing morphisms (Section 3) and for automatic sequences (Section 4). In Section 5 we say a few words about connections with monadic logics. In particular, in a curious result of Corollary 4 we give a reason why the main problem may be decidable.

Some attempts to solve the problem were already done. In [3] A. Cobham gives a criterion for automatic sequence to be almost periodic. But even if his criterion gives some effective procedure solving the problem (which is not clear from his result, and he does not care about it at all), this procedure could not be fast. We construct a polynomial-time algorithm solving the problem. In [5] A. Maes deals with pure morphic sequences and finds a criterion for them to belong to a slightly different class of generalized almost periodic sequences (but he calls them almost periodic — see [9] for different definitions). And again, his algorithm does not seem to be polynomial-time.

---

* The work was partially supported by RFBR grants 06-01-00122, 05-01-02803, Kolmogorov grant of Institute of New Technologies, and August Möbius grant of Independent University of Moscow.

2 Preliminaries

Denote the set of natural numbers $\{0, 1, 2, \dots\}$ by $\mathbb{N}$ and the binary alphabet $\{0, 1\}$ by $\mathbb{B}$. Let A be a finite alphabet. We deal with sequences over this alphabet, i. e., mappings $x\colon \mathbb{N} \to A$, and denote the set of these sequences by $A^{\mathbb{N}}$.

Denote by A^* the set of all finite words over A including the empty word Λ. If $i \le j$ are natural, denote by $[i, j]$ the segment of $\mathbb{N}$ with ends in i and j, i. e., the set $\{i, i+1, i+2, \dots, j\}$. Also denote by $x[i, j]$ a subword $x(i)x(i+1)\dots x(j)$ of a sequence x. A segment $[i, j]$ is an occurrence of a word $u \in A^*$ in a sequence x if $x[i, j] = u$. We say that $u \ne \Lambda$ is a factor of x if u occurs in x. A word of the form $x[0, i]$ for some i is called prefix of x, and respectively a sequence of the form $x(i)x(i+1)x(i+2)\dots$ for some i is called suffix of x and is denoted by $x[i, \infty)$. Denote by $|u|$ the length of a word u. The occurrence $u = x[i, j]$ in x is k-aligned if $k|i$.

A sequence x is periodic if for some T we have $x(i) = x(i + T)$ for each $i \in \mathbb{N}$. This T is called a period of x. We denote by $\mathcal{P}$ the class of all periodic sequences. Let us consider an extension of this class.

A sequence x is called *almost periodic*[1] if for every factor u of x there exists a number l such that every factor of x of length l contains at least one occurrence of u (and therefore u occurs in x infinitely many times). Obviously, to show almost periodicity of a sequence it is sufficient to check the mentioned condition only for all prefixes but not for all factors (and even for some increasing sequence of prefixes only). Denote by $\mathcal{AP}$ the class of all almost periodic sequences.

Let A, B be finite alphabets. A mapping $\phi\colon A^* \to B^*$ is called *a morphism* if $\phi(uv) = \phi(u)\phi(v)$ for all $u, v \in A^*$. A morphism is obviously determined by its values on single-letter words. A morphism is *non-erasing* if $|\phi(a)| \ge 1$ for each $a \in A$. A morphism is k-uniform if $|\phi(a)| = k$ for each $a \in A$. A 1-uniform morphism is called a coding. For $x \in A^{\mathbb{N}}$ denote

$$\phi(x) = \phi(x(0))\phi(x(1))\phi(x(2))\dots$$

Further we consider only morphisms of the form $A^* \to A^*$ (but codings are of the form $A \to B$, which in fact does not matter, they can be also of the form $A \to A$ without loss of generality). Let $\phi(s) = su$ for some $s \in A$, $u \in A^*$. Then for all natural $m < n$ the word $\phi^n(s)$ begins with the word $\phi^m(s)$, so $\phi^{\infty}(s) = \lim_{n \to \infty} \phi^n(s) = su\phi(u)\phi^2(u)\phi^3(u)\dots$ is well-defined. If $\forall n \ \phi^n(u) \ne \Lambda$, then $\phi^{\infty}(s)$ is infinite. In this case we say that ϕ is *prolongable* on s. Sequences of the form $h(\phi^{\infty}(s))$ for a coding $h\colon A \to B$ are called *morphic*, of the form $\phi^{\infty}(a)$ are called *pure morphic*.

Notice that there exist almost periodic sequences that are not morphic (in fact, the set of almost periodic sequences has cardinality continuum, while the set of morphic sequences is obviously countable), as well as there exist morphic sequences that are not almost periodic (you will find examples later). Our goal is to determine whether a morphic sequence is almost periodic or not given its constructive definition.

---

[1] It was called *strongly* or *strictly almost periodic* in [7,8].

First of all, observe the following

Lemma 1. *A sequence $\phi^\infty(s)$ is almost periodic iff s occurs in this sequence infinitely many times with bounded distances.*

Proof. In one direction the statement is obviously true by definition.

Suppose now that s occurs in $\phi^\infty(s)$ infinitely many times with bounded distances. Then for every m the word $\phi^m(s)$ also occurs in $\phi^\infty(s)$ infinitely many times with bounded distances. But every word u occurring in $\phi^\infty(s)$ occurs in some prefix $\phi^m(s)$ and thus occurs infinitely many times with bounded distances. □

For a morphism $\phi\colon \{1,\ldots,n\} \to \{1,\ldots,n\}$ we can define a corresponding matrix $M(\phi)$, such that $M(\phi)_{ij}$ is a number of occurrences of symbol i into $\phi(j)$. One can easily check that for each l we have $M(\phi)^l = M(\phi^l)$.

Morphism ϕ is called *primitive* if for some l all the numbers in $M(\phi^l)$ are positive.

Let us construct an oriented graph G corresponding to a morphism. Let its set of vertices be A. In G edges go from $b \in A$ to all the symbols occurring in $\phi(b)$.

For $\phi^\infty(s)$ it can easily be found using the graph corresponding to ϕ which symbols from A really occur in this sequence. Indeed, these symbols form the set of all vertices that can be reached from s. So without loss of generality from now on we assume that all the symbols from A occur in $\phi^\infty(s)$.

A morphism is primitive if and only if its corresponding graph is strongly connected, i. e., there exists an oriented path between every two vertices. This reformulation of the primitiveness notion seems to be more appropriate for computational needs.

By Lemma 1 (and the observation that codings preserve almost periodicity) morphic sequences obtained by primitive morphisms are always almost periodic. Moreover, in the case of increasing morphisms (such that $|\phi(b)| \geqslant 2$ for each b) this sufficient condition is also necessary (and this is a polynomial-time algorithmic criterion). However when we generalize this case even on non-erasing morphisms, it is not enough to consider only the corresponding graph or even the matrix of morphism (which has more information), as it can be seen from the following example.

Let ϕ_1 be as follows: $0 \to 01$, $1 \to 120$, $2 \to 2$, and ϕ_2 be as follows: $0 \to 01$, $1 \to 210$, $2 \to 2$. Then these two morphisms have identical matrices of morphism, but $\phi_1^\infty(0)$ is almost periodic, while $\phi_2^\infty(0)$ is not. Indeed, in $\phi_2^\infty(0)$ there are arbitrary long segments like $222\ldots22$, so $\phi_2^\infty(0) \notin \mathcal{AP}$. There is no such problem in $\phi_1^\infty(0)$. Since 0 occurs in both $\phi_1(0)$ and $\phi_1(1)$, and 22 does not occur in $\phi_1^\infty(0)$, it follows that 0 occurs in $\phi_1^\infty(0)$ with bounded distances. Thus $\phi_1^m(0)$ for every $m \geqslant 0$ occurs in $\phi_1^\infty(0)$ with bounded distances, so $\phi_1^\infty(0) \in \mathcal{AP}$. See Theorem 1 for a general criterion of almost periodicity in the case of fixed points of non-erasing morphisms.

To introduce a bit the notion of almost periodicity, let us formulate an interesting result on this topic. It seems to be first proved in [3], but also follows from the results of [9]. For $x \in A^{\mathbb{N}}$, $y \in B^{\mathbb{N}}$ define $x \times y \in (A \times B)^{\mathbb{N}}$ such that $(x \times y)(i) = \langle x(i), y(i) \rangle$.

Proposition 1. *If x is almost periodic and y is periodic, then $x \times y$ is almost periodic.*

3   Pure Morphic Sequences Generated by Non-erasing Morphisms

Here we consider the case of morphic sequence of the form $\phi^{\infty}(s)$ for non-erasing ϕ. We present an algorithm that determines whether a morphic sequence $\phi^{\infty}(s)$ is almost periodic given an alphabet A, a morphism ϕ and a symbol $s \in A$.

Suppose we have A, ϕ and $s \in A$, such that $|A| = n$, $\max_{b \in A} |\phi(b)| = k$, s begins $\phi(s)$. Remember that we suppose that all the symbols from A appear in $\phi^{\infty}(s)$.

Divide A into two parts. Let I be the set of all symbols $b \in A$ such that $|\phi^m(b)| \to \infty$ as $m \to \infty$. Denote $F = A \setminus I$, it is the set of all symbols b such that $|\phi^m(b)|$ is bounded. Also define $E \subseteq F$ to be the set of all symbols b such that $|\phi(b)| = 1$.

We can find a decomposition $A = I \sqcup F$ in poly(n, k)-time as follows.

Find E. Then find all the cycles in G with all the vertices lying in E. Join all the vertices of all these cycles in a set D. This set is stabilizing: F is the set of all vertices in G such that all infinite paths starting from them stabilize in D. Polynomiality can be checked easily.

Construct "a graph of left tails" L with marked edges. Its set of vertices is I. From each vertex b exactly one edge goes off. To construct this edge, find a representation $\phi(b) = uv$, where $c \in I$, u is the maximal prefix of $\phi(b)$ containing only symbols from F. It follows from the definitions of I and F that u does not coincide with $\phi(b)$, that is why this representation is correct. Then construct in L an edge from b to c and write u on it.

Analogously we construct "a graph of right tails" R. (In this case we consider representations $\phi(b) = vu$ where $u \in F^*$, $c \in I$.)

Now we formulate a general criterion.

Theorem 1. *A sequence $\phi^{\infty}(s)$ is almost periodic iff*

1) G restricted to I is strongly connected;
2) in graphs L and R on each edge of each cycle an empty word Λ is written.

It seems that full and detailed proof of this theorem can only confuse a reader, rather than a proof sketch.

Proof (sketch). By Lemma 1 for almost periodicity it is necessary and sufficient to check whether symbol s occurs infinitely many times with bounded distances.

For every symbol $b \in I$ the symbol s should occur in some $\phi^l(b)$, that is what the 1st part of the criterion says.

Furthermore, in the sequence $\phi^\infty(s)$ all the segments of consecutive symbols from F should be bounded. Indeed, every such segment consists only of symbols from F, but $s \notin F$. That is what the 2nd part of the criterion means, let us explain why.

Consider some $v = buc$ occurring somewhere in $\phi^\infty(a)$, where $b, c \in I$, $u \in F^*$. Every element of sequence of words $v, \phi(v), \phi^2(v), \phi^3(v), \ldots$ occurs in $\phi^\infty(s)$. Somewhere in the middle of $\phi^l(v) = \phi^l(b)\phi^l(u)\phi^l(c)$ a word $\phi^l(u)$ occurs. As l increases, some words from F^* might stick to $\phi^l(u)$ from left or right for these words can come from $\phi^l(b)$ or $\phi^l(c)$. These words exactly correspond to those written on edges of L or R. The 2nd part of the criterion exactly says that this situation can happen only finitely many times, until we get to some cycle in L or R.   □

Let us consider examples with ϕ_1 and ϕ_2 from the end of Section 2. In both cases $I = \{0, 1\}$, $F = \{2\}$. On every edge of R in both cases Λ is written. Almost the same is true for L: the only difference is about the edge going from 1 to 1. In the case of ϕ_1 an empty word is written on this edge, while in the case of ϕ_2 a word 2 is written. That is why $\phi_1^\infty(0)$ is almost periodic, while $\phi_2^\infty(0)$ is not.

Corollary 1. *If for all $b \in A$ we have $|\phi(b)| \geqslant 2$, then $\phi^\infty(s)$ is almost periodic iff ϕ is primitive.*

Proof. Follows from Theorem 1. In that case $A = I$, and on all the edges of L and R the empty word is written.   □

Corollary 2. *There exists a $\mathrm{poly}(n, k)$-algorithm that says whether $\phi^\infty(s)$ is almost periodic.*

Proof. Conditions from Theorem 1 can be checked in polynomial time.   □

It also seems useful to formulate an explicit version of the criterion for the binary case. We do it without any additional assumptions, opposite to the previous.

Corollary 3. *For non-erasing $\phi \colon \mathbb{B} \to \mathbb{B}$ that is prolongable on 0 a sequence $\phi^\infty(0)$ is almost periodic iff one of the following conditions holds:*

1) $\phi(0)$ contains only 0s;
2) $\phi(1)$ contains 0;
3) $\phi(1) = \Lambda$;
4) $\phi(1) = 1$ and $\phi(0) = 0u0$ for some word u.

4 Uniform Morphisms

Now we deal with morphic sequences obtained by uniform morphisms. Again we present a polynomial-time algorithm for solving the problem in this situation.

Suppose we have an alphabet A, a morphism $\phi \colon A^* \to A^*$, a coding $h \colon A \to B$, and $s \in A$, such that $|A| = n$, $|B| \leqslant n$, $\forall b \in A$ $|\phi(b)| = k$, s begins $\phi(s)$. We are interested in whether $h(\phi^\infty(s))$ is almost periodic. Sequences of the form $h(\phi^\infty(s))$ with ϕ being k-uniform are also called k-automatic (see [1]).

4.1 Equivalence Relations and Uniform Morphisms

For each $l \in \mathbb{N}$ define an equivalence relation on A: $b \sim_l c$ iff $h(\phi^l(b)) = h(\phi^l(b))$. We can easily continue this relation on A^*: $u \sim_l v$ iff $h(\phi^l(u)) = h(\phi^l(v))$. In fact, this means $|u| = |v|$ and $u(i) \sim_l v(i)$ for all i, $1 \leqslant i \leqslant |u|$.

Let B_m be the Bell number, i. e., the number of all possible equivalence relations on a finite set with exactly m elements, see [12]. As it follows from this article, we can estimate B_m in the following way.

Lemma 2. $2^m \leqslant B_m \leqslant 2^{Cm \log m}$ *for some constant* C.

Thus the number of all possible relations $\sim_l$ is not greater than $B_n = 2^{O(n \log n)}$. Moreover, the following lemma gives a simple description for the behavior of these relations as l tends to infinity.

Lemma 3. *If* $\sim_r$ *equals* $\sim_s$*, then* $\sim_{r+p}$ *equals* $\sim_{s+p}$ *for all* p.

Proof. Indeed, suppose $\sim_r$ equals $\sim_s$. Then $b \sim_{r+1} c$ iff $\phi(b) \sim_r \phi(c)$ iff $\phi(b) \sim_s \phi(c)$ iff $b \sim_{s+1} c$. So if $\sim_r$ equals $\sim_s$, then $\sim_{r+1}$ equals $\sim_{s+1}$, which implies the lemma statement. □

This lemma means that the sequence $(\sim_l)_{l \in \mathbb{N}}$ turns out to be ultimately periodic with a period and a preperiod both not greater than B_n. Thus we obtain the following

Lemma 4. *For some* $p, q \leqslant B_n$ *we have for all* i *and all* $t > p$ *that* $\sim_t$ *equals* $\sim_{t+iq}$.

4.2 Criterion

Now we are trying to get a criterion which we could check in polynomial time. Notice that the situation is much more difficult than in the pure case because of a coding allowed. In particular, the analogue of Lemma 1 for non-pure case does not hold.

We will move step by step to the appropriate version of the criterion reformulating it several times.

This proposition is quite obvious and follows directly from the definition of almost periodicity since all $h(\phi^m(a))$ are the prefixes of $h(\phi^\infty(a))$.

Proposition 2. *A sequence* $h(\phi^\infty(s))$ *is almost periodic iff for all* m *the word* $h(\phi^m(s))$ *occurs in* $h(\phi^\infty(s))$ *infinitely often with bounded distances.*

And now a bit more complicated version.

Proposition 3. *A sequence* $h(\phi^\infty(s))$ *is almost periodic iff for all* m *the symbols that are* $\sim_m$*-equivalent to* s *occur in* $\phi^\infty(s)$ *infinitely often with bounded distances.*

Proof. $\Leftarrow$. If the distance between two consecutive occurrences in $\phi^\infty(s)$ of symbols that are $\sim_m$-equivalent to s is not greater than t, then the distance between two consecutive occurrences of $h(\phi^m(s))$ in $h(\phi^\infty(s))$ is not greater than tk^m.

$\Rightarrow$. Suppose $h(\phi^\infty(s))$ is almost periodic. Let $y_m = 012\dots(k^m - 2)(k^m - 1)01\dots(k^m - 1)0\dots$ be a periodic sequence with a period k^m. Then by Proposition 1 a sequence $h(\phi^\infty(s)) \times y_m$ is almost periodic, which means that the distances between consecutive k^m-aligned occurrences of $h(\phi^m(s))$ in $h(\phi^\infty(s))$ are bounded. It only remains to notice that if $h(\phi^\infty(s))[ik^m, (i + 1)k^m - 1] = h(\phi^m(s))$, then $\phi^\infty(s)(i) \sim_m s$.   □

Let Y_m be the following statement: symbols that are $\sim_m$-equivalent to s occur in $\phi^\infty(s)$ infinitely often with bounded distances.

Suppose for some T that Y_T is true. This implies that $h(\phi^T(s))$ occurs in $h(\phi^\infty(s))$ with bounded distances. Therefore for all $m \leq T$ a word $h(\phi^m(s))$ occurs in $h(\phi^\infty(s))$ with bounded distances since $h(\phi^m(s))$ is a prefix of $h(\phi^T(s))$. Thus we do not need to check the statements Y_m for all m, but only for all $m \geq T$ for some T.

Furthermore, it follows from Lemma 4, that we are sufficient to check the only one such statement as in the following

Proposition 4. *For all $r \geq B_n$: a sequence $h(\phi^\infty(s))$ is almost periodic iff the symbols that are $\sim_r$-equivalent to s occur in $\phi^\infty(s)$ infinitely often with bounded distances.*

And now the final version of our criterion.

Proposition 5. *For all $r \geq B_n$: a sequence $h(\phi^\infty(s))$ is almost periodic iff for some m the symbols that are $\sim_r$-equivalent to s occur in $\phi^m(b)$ for all $b \in A$.*

Indeed, if the symbols of some set occur with bounded distances, then they occur on each k^m-aligned segment for some sufficiently large m.

4.3 Polynomiality

Now we explain how to check a condition from Proposition 5 in polynomial time. We need to show two things: first, how to choose some $r \geq B_n$ and to find in polynomial time the set of all symbols that are $\sim_r$-equivalent to s (and this is a complicated thing keeping in mind that B_n is exponential), and second, how to check whether for some m the symbols from this set for all $b \in A$ occur in $\phi^m(b)$.

Let us start from the second. Suppose we have found the set H of all the symbols that are $\sim_r$-equivalent to s. For $m \in \mathbb{N}$ let us denote by $P_m^{(b)}$ the set of all the symbols that occur in $\phi^m(b)$. Our aim is to check whether exists m such that for all b we have $P_m^{(b)} \cap H \neq \varnothing$. First of all, notice that if $\forall b\, P_m^{(b)} \cap H \neq \varnothing$, then $\forall b\, P_l^{(b)} \cap H \neq \varnothing$ for all $l \geq m$. Second, notice that the sequence of tuples of sets $((P_m^{(b)})_{b\in \Sigma})_{m=0}^\infty$ is ultimately periodic. Indeed, the sequence $(P_m^{(b)})_{m=0}^\infty$ is obviously ultimately periodic with both period and preperiod not greater than 2^n (recall that n is the size of the alphabet Σ). Thus the period of $((P_m^{(b)})_{b\in \Sigma})_{m=0}^\infty$ is not greater than the least common divisor of that for $(P_m^{(b)})_{m=0}^\infty$, $b \in A$, and the preperiod is not greater than the maximal that of $(P_m^{(b)})_{m=0}^\infty$. So the period is not greater than $(2^n)^n = 2^{n^2}$ and the preperiod is not greater than 2^n. Third, notice that there is a polynomial-time-procedure that given a graph corresponding to some morphism ψ (see Section 2 to recall what is the graph corresponding to a morphism) outputs a graph corresponding to morphism ψ^2. Thus after repeating this procedure $n^2 + 1$ times we obtain a graph by which we can easily find $(P^{(b)}_{2^{n^2}+2^n})_{b \in \Sigma}$, since $2^{n^2+1} > 2^{n^2} + 2^n$.

Similar arguments, even described with more details, are used in deciding our next problem. Here we present a polynomial-time algorithm that finds the set of all symbols that are $\sim_r$-equivalent to s for some $r \geqslant B_n$.

We recursively construct a series of graphs T_i. Let its common set of vertices be the set of all unordered pairs (b, c) such that $b, c \in A$ and $b \neq c$. Thus the number of vertices is $\frac{n(n-1)}{2}$. The set of all vertices connected with (b, c) in the graph T_i we denote by $V_i(b, c)$.

Define a graph T_0. Let $V_0(b, c)$ be the set $\{(\phi(b)(j), \phi(c)(j)) \mid j = 1, \ldots, k, \phi(b)(j) \neq \phi(c)(j)\}$. In other words, $b \sim_{l+1} c$ if and only if $x \sim_l y$ for all $(x, y) \in V_0(b, c)$.

Thus $b \sim_2 c$ if and only if for all $(x, y) \in V_0(b, c)$ for all $(z, t) \in V_0(x, y)$ we have $z \sim_0 t$. For the graph T_1 let $V_1(b, c)$ be the set of all (x, y) such that there is a path of length 2 from (b, c) to (x, y) in T_0. The graph T_1 has the following property: $b \sim_2 c$ if and only if $x \sim_0 y$ for all $(x, y) \in V_1(b, c)$. And even more generally: $b \sim_{l+2} c$ if and only if $x \sim_l y$ for all $(x, y) \in V_1(b, c)$.

Now we can repeat operation made with T_0 to obtain T_1. Namely, in T_2 let $V_2(b, c)$ be the set of all (x, y) such that there is a path of length 2 from (b, c) to (x, y) in T_1. Then we obtain: $b \sim_{l+4} c$ if and only if $x \sim_l y$ for all $(x, y) \in V_2(b, c)$.

It follows from Lemma 2 that $\log_2 B_n \leqslant Cn \log n$. Thus after we repeat our procedure $r = [Cn \log n]$ times, we will obtain the graph T_l such that $b \sim_{2^r} c$ if and only if $x \sim_0 y$ for all $(x, y) \in V_2(b, c)$. Recall that $x \sim_0 y$ means $h(x) = h(y)$, so now we can easily compute the set of symbols that are $\sim_{2^r}$-equivalent to s.

5　Monadic Theories

Combinatorics on words is closely connected with the theory of second order monadic logics. Here we just want to show some examples of these connections. More details can be found, e. g., in [10,11].

We consider monadic logics on $\mathbb{N}$ with the relation "$<$", that is, first-order logics where also unary finite-value function variables and quantifiers over them are allowed. We also suppose that we know some fixed finite-value function $x : \mathbb{N} \to \Sigma$ and can use it in our formulas. Such a theory is denoted by $\mathrm{MT}\langle \mathbb{N}, <, x \rangle$ and is called *monadic theory of* x.

The main question here can be the question of decidability, that is, does there exist an algorithm that given a sentence in a theory says whether this sentence is true of false.

The criterion of decidability for monadic theories of almost periodic sequences can be formulated in terms of some their very natural characteristic, namely, almost periodicity regulator. An almost periodicity regulator of an almost periodic sequence x is a function $f : \mathbb{N} \rightarrow \mathbb{N}$ such that every factor u of x of length n occurs in each factor of x of length $f(n)$. So an almost periodicity regulator somehow regulates how periodic a sequence is. Notice that an almost periodicity regulator of a sequence is not unique: every function greater than regulator is also a regulator.

Theorem 2 (Semenov 1983 [11]). *If x is almost periodic, then* $\mathrm{MT}\langle \mathbb{N}, <, x \rangle$ *is decidable iff x and some its almost periodicity regulator are computable.*

The following result was obtained recently, but uses the technics already used in [10,11].

Theorem 3 (Carton, Thomas 2002 [2]). *If x is morphic, then* $\mathrm{MT}\langle \mathbb{N}, <, x \rangle$ *is decidable.*

A curious result can be implied from two these theorems.

Corollary 4. *If x is both morphic and almost periodic, then some its regulator is computable.*

Proof. Indeed, if x is morphic, then by Theorem 3 the theory $\mathrm{MT}\langle \mathbb{N}, <, x \rangle$ is decidable. Since x is almost periodic, from Theorem 2 it follows that some almost periodicity regulator of x is computable.

Notice that Corollary 4 does not imply the existence of an algorithm that given a morphic sequence computes some almost periodicity regulator of this sequence whenever it is almost periodic (but probably this algorithm can be constructed after deep analyzing the proofs of Theorems 2 and 3 and showing uniformity in a sense). And it also does not imply the decidability of almost periodicity for morphic sequences. This decidability also does not imply Corollary 4.

By the way, Corollary 4 allows us to hope that these algorithms exist. Though the formulation of this statement uses only combinatorics on words, the proof also involves the theory of monadic logics. Of course, it would be interesting to find a simple combinatorial proof of the result.

And the last remark here is that Corollary 4 (and its probable uniform version) seems to be the best progress that we can obtain by this monadic approach. One could try to express in the monadic theory of morphic sequence (which is decidable by Theorem 3) the property of almost periodicity, but it turns out to be impossible.

6 In General Case

We have described two polynomial-time algorithms, but without any precise bound for their working time. Of course, it can be done after deep analyzing of all the previous, but is probably not so interesting.

It is not still known whether the problem of determining almost periodicity of arbitrary morphic sequence is decidable. Corollary 4 somehow supports the conjecture of decidability (but even does not follow from this conjecture!).

Theorem 7.5.1 from [1] allows us to represent an arbitrary morphic sequence $h(\phi^\infty(s))$ as $g(\psi^\infty(b))$ where ψ is non-erasing. So it is sufficient to solve our main problem for $h(\phi^\infty(s))$ with non-erasing ϕ.

It seems that the general problem is tightly connected with a particular case of $h(\phi^\infty(a))$ where $|\phi(b)| \geqslant 2$ for each $b \in A$. There is no strict reduction to this case but solving problem in this case can help to deal with general situation.

The problem of finding an effective periodicity criterion in the case of arbitrary morphic sequences is also of great interest, as well as criteria for variations with periodicity and almost periodicity: ultimate periodicity, generalized almost periodicity, ultimate almost periodicity (see [9] for definitions). If one notion is a particular case of another, it does not mean that corresponding criterion for the first case is more difficult (or less difficult) than for the second.

Acknowledgements

The author is grateful to ⃞An. Muchnik⃞ and A. Semenov for their permanent help in the work, to A. Frid, M. Raskin, K. Saari and to all the participants of Kolmogorov seminar, Moscow [4], for fruitful discussions, and also to anonymous referees for very useful comments.

References

1. Allouche, J.-P., Shallit, J.: Automatic Sequences. Cambridge University Press, Cambridge (2003)
2. Carton, O., Thomas, W.: The Monadic Theory of Morphic Infinite Words and Generalizations. Information and Computation 176, 51–65 (2002)
3. Cobham, A.: Uniform tag sequences. Math. Systems Theory 6, 164–192 (1972)
4. Kolmogorov Seminar: http://lpcs.math.msu.su/kolmogorovseminar/eng/
5. Maes, A.: More on morphisms and almost-periodicity. Theoretical Computer Science 231(2), 205–215 (2000)
6. Morse, M., Hedlund, G.A.: Symbolic dynamics. American Journal of Mathematics 60, 815–866 (1938)
7. Muchnik, An., Semenov, M., Ushakov, M.: Almost periodic sequences. Theoretical Computer Science 304, 1–33 (2003)
8. Pritykin, Yu.L.: Finite-Automaton Transformations of Strictly Almost-Periodic Sequences. Mathematical Notes 80(5), 710–714 (2006)
9. Pritykin, Yu.: Almost Periodicity, Finite Automata Mappings and Related Effectiveness Issues. In: Proceedings of WoWA'06, St. Petersburg, Russia (satellite to CSR'06). Izvestia VUZov. Mathematics (To appear, 2007)
10. Semenov, A.L.: On certain extensions of the arithmetic of addition of natural numbers. Math. of USSR, Izvestia 15, 401–418 (1980)
11. Semenov, A.L.: Logical theories of one-place functions on the set of natural numbers. Math. of USSR, Izvestia 22, 587–618 (1983)
12. Weisstein, Eric W.: Weisstein. Bell Number. From MathWorld — A Wolfram Web Resource, http://mathworld.wolfram.com/BellNumber.html

A Local Balance Property of Episturmian Words

Gwénaël Richomme

UPJV, LaRIA,
33, Rue Saint Leu
80039 Amiens cedex 01, France
gwenael.richomme@u-picardie.fr
http://www.laria.u-picardie.fr/~richomme/

Abstract. We prove that episturmian words and Arnoux-Rauzy sequences can be characterized using a local balance property. We also give a new characterization of epistandard words.

Keywords: Arnoux-Rauzy sequences, episturmian words, balance property.

1 Introduction

M. Morse and G.A. Hedlund [19] were the first to study in depth a family of words called Sturmian words. Now a large literature exists on these words for which many fascinating characterizations have been found (see for instance [1,3,20]).

Sturmian words are defined over a binary alphabet. From their various characteristic properties, some generalizations of Sturmian words have emerged over larger alphabets. One of them, the so-called Arnoux-Rauzy sequences, is based on the notion of complexity of a word and is interesting by its geometrical, arithmetic, ergodic and combinatorial aspects (see for instance [20]).

One of the first properties of Sturmian words stated by M. Morse and G.A. Hedlund [19] is the balance property: any infinite word w over the alphabet $\{a, b\}$ is Sturmian if and only if it is non-ultimately periodic and balanced, that is the number of occurrences of the letter a differs in two factors of same length of w by at most one. Generalizations of balanced words were studied for instance by P. Hubert [13] (see also [23] for a survey of this property). J. Justin and L. Vuillon have stated a non-characteristic kind of the balance property [15] for Arnoux-Rauzy sequences. Although it was first conjectured that the Arnoux-Rauzy sequences are balanced [9], J. Cassaigne, S. Ferenczi and L.Q. Zamboni have proved that this does not necessarily hold [6].

In 1973, E.M. Coven and G.A. Hedlund [7] stated that a word w over $\{a, b\}$ is not balanced if and only if there exists a palindrome t such that ata and btb are both factors of w. This could be seen as a local balance property of Sturmian words since to check the balance property we do not have to compare all factors of the same length but only factors on the sets AtA for t being a factor of w. The previous property can be rephrased as follows: an infinite word w over the alphabet $A = \{a, b\}$ is Sturmian if and only if it is non-ultimately periodic and

T. Harju, J. Karhumäki, and A. Lepistö (Eds.): DLT 2007, LNCS 4588, pp. 371–381, 2007.
© Springer-Verlag Berlin Heidelberg 2007

for any factor t of w, the set of factors belonging to AtA is a subset of $atA \cup Ata$ or a subset of $btA \cup Atb$. In Section 3, we show that this result can be generalized to Arnoux-Rauzy sequences.

Actually our result concerns a larger family of infinite words presented in Section 2. Based on ideas of A. de Luca [8], Episturmian words were proposed by X. Droubay, J. Justin and G. Pirillo [9] as a generalization of Sturmian words. They have observed that Arnoux-Rauzy words are special episturmian words which they called strict episturmian words. In the binary case episturmian words are the Sturmian words and the balanced periodic infinite words. Let us note that the remaining balanced words, namely the skew ones, have recently been generalized [11,12].

In [9], episturmian words are defined as an extension to standard episturmian words (Here we will call *epistandard* these standard episturmian words) previously introduced as a generalization of standard Sturmian words. In Section 4, we generalize to epistandard words a characterization of standard words proving a converse of a theorem in [14] and stating that an infinite word w is epistandard if and only if there exists at least two letters such that aw and bw are both episturmian. Interested readers can also consult [12] and its references for other characterizations of episturmian words using left extension in the context of an ordered alphabet.

Our last section comes back to the generalization of the local balance property introduced by E.M. Coven and G.A. Hedlund. One another way to rephrase it is: an infinite word w over the alphabet $A = \{a, b\}$ is Sturmian if and only if it is non-ultimately periodic and for any factor t of w, the set of factors belonging to AtA is balanced. This yields a new family of words on which we give partial results.

2   Episturmian and Epistandard Words

Even if we assume the reader is familiar with combinatorics on words (see, e.g., [18]), we specify our notation. Given an alphabet A (a finite non-empty set of letters), A^* is the set of finite words over A including the empty word ε. The length of a word w is denoted by $|w|$ and the number of occurrences of a letter a in w is denoted by $|w|_a$. The *mirror image* of a finite word $w = w_1 \ldots w_n$ ($w_i \in A$, for $i = 1, \ldots, n$) is the word $w_n \ldots w_1$ (the mirror image of ε is ε itself). A word equals to its mirror image is a *palindrome*. A word u is a *factor* of w if there exist words p and s such that $w = pus$. If $p = \varepsilon$ (resp. $s = \varepsilon$), u is a *prefix* (resp. *suffix*) of w. A word u is a *left special* (resp. *right special*) factor of w if there exist (at least) two different letters a and b such that au and bu (resp. ua and ub) are factors of w. A *bispecial factor* is any word which is both a left and a right special factor (see, e.g., [5] for more informations on special factors). The set of factors of a word w will be denoted $Fact(w)$.

Most of previous notions can be extended in a natural way to any infinite words. Moreover any *ultimately periodic* infinite word can be written uv^ω for two finite words u, v ($v \neq \varepsilon$): it is then the infinite word obtained concatenating infinitely often v to u. If $u = \varepsilon$, the word is said *periodic*.

A word w is *episturmian* if and only if its set of factors is closed by mirror image and w contains at most one left (or equivalently right) special factor of each length. A word w is *epistandard Sturmian* or *epistandard*, if w is episturmian and all its left special factors are prefixes of w. Let us note that, in [9], epistandard words were introduced by several equivalent ways, and then episturmian words were defined as words having same set of factors as an epistandard one.

The two theorems below recall a very useful property of episturmian words which is the possibility to decompose infinitely an episturmian word using some morphisms. This property already seen for Arnoux-Rauzy sequences in [2] is related to the notion of S-adic dynamical system (see, e.g., [20] for more details). This property could be useful to get information on the structure of episturmian words (see for instance [4,16,17,22] for some uses in the binary cases).

Given an alphabet A, a *morphism* f on A is a mapping from A^* to A^* such that $f(uv) = f(u)f(v)$ for any words u, v over A. A morphism on A is entirely defined by the images of elements of A.

Episturmian morphisms studied in [14,21] are the morphisms defined by composition of the permutation morphisms and the morphisms L_a and R_a defined, for a being a letter, by

$$L_a \begin{cases} a \mapsto a \\ b \mapsto ab, \text{ if } b \neq a, \end{cases} \qquad R_a \begin{cases} a \mapsto a \\ b \mapsto ba, \text{ if } b \neq a. \end{cases}$$

Theorem 1. [14] *An infinite word w is epistandard if and only if there exist an infinite sequence of infinite words $(w^{(n)})_{n \geq 0}$ and an infinite sequence of letters $(x_n)_{n \geq 1}$ such that $w^{(0)} = w$ and for all $n \geq 1$, $w^{(n-1)} = L_{x_n}(w^{(n)})$.*

It is worth noting that any episturmian word is *recurrent*, that is, each factor of w occurs infinitely often. An infinite word w is recurrent if and only if each factor of w occurs at least twice. Equivalently each factor of w occurs at a non-prefix position. Thus an infinite word w over an alphabet A is recurrent if and only if for each of its factors u the set $AuA \cap Fact(w)$ (or simply $Au \cap Fact(w)$) is not empty.

Theorem 2. [14] *An infinite word w is episturmian if and only if there exist an infinite sequence of recurrent infinite words $(w^{(n)})_{n \geq 0}$ and an infinite sequence of letters $(x_n)_{n \geq 1}$ such that $w^{(0)} = w$ and for all $n \geq 0$, $w^{(n-1)} = L_{x_n}(w^{(n)})$ or $w^{(n-1)} = R_{x_n}(w^{(n)})$.*

Moreover, w has the same set of factors as the epistandard word directed by $(x_n)_{n \geq 1}$.

The infinite sequence $(x_n)_{n \geq 1}$ which appears in the two previous theorem is called the *directive word* of w and is denoted $\Delta(w)$: Actually in terms of [14], it is the directive word of the epistandard word having the same set of factors as w. Each episturmian word has a unique directive word.

We denote as in [14] $Ult(w)$ the set of letters occurring infinitely often in w. For B a subset of the alphabet, we introduce a new definition: we call *ultimately B-strict episturmian* any episturmian word w for which $Ult(\Delta(w)) = B$. Of course this notion is related to the notion of B-*strict episturmian* word (see [14, def. 2.3]) which is an ultimately B-strict episturmian word whose alphabet (the letters occurring in w) is exactly B: B-strict episturmian words are also the Arnoux-Rauzy sequences over B.

As shown in [9], there is a close relation between the directive word of an episturmian word and its special words. Corollary 1 below will show it again for ultimately strict episturmian words.

Let w be an episturmian word and $\Delta(w) = (x_n)_{n \geq 1}$ its directive word. With notations of Theorem 2, for $n \geq 1$, we denote $u_{n,w}$ (or simply u_n) the word:

$$u_{n,w} = L_{x_1}(L_{x_2}(\ldots(L_{x_{n-1}}(\varepsilon)x_{n-1})\ldots)x_2)x_1$$

When $n = 1$, $u_{n,w} = \varepsilon$. These words play an important role in the initial definition of episturmian word by palindromic closure (see [14, Sec. 2]). In particular, each u_n is a palindrom (see for instance [14, Lem. 2.5]). One can also observe that, if $Ult(\Delta(w))$ contains at least two letters, then each u_n is a bispecial factor of w. Indeed for $n \geq 1$, u_n is a prefix of the epistandard word s directed by $\Delta(w)$ and so, by definition of an epistandard word, it is a left special factor of s and so of w by Theorem 2. Since the set of factors of w is closed by mirror image and since u_n is a palindrom, u_n is a right special factor of w. Conversely let us observe that any bispecial factor of an episturmian word is a palindrom. Indeed if u is a bispecial factor, then u and its mirror image $\tilde{u}$ are left special factors of an infinite word containing at most one left special word of length $|u|$. It follows the construction of an epistandard word w by palindromic closure [9], that the the words $u_{n,w}$ are the only palindroms prefixes of w. From what precedes, we deduce the following fact that does not seem to have been already quoted in the literature:

Remark 1. For an episturmian word w with the directive word $(x_n)_{n \geq 1}$, a factor u is bispecial if and only if $u = u_{n,w}$ for an integer $n \geq 1$.

Another result involving the palindroms u_n is:

Theorem 3. [9, Th. 6] *Let s be an epistandard word over the alphabet A with the directive word $\Delta(s) = (x_n)_{n \geq 1}$. For $n \geq 1$ and $x \in A$, $u_{n,s}x$ (or equivalently $xu_{n,s}$) is a factor of s if and only if x belongs to $\{x_i \mid i \geq n\}$.*

By Theorem 2, an episturmian word w with a directive word Δ has the same set of factors as the epistandard word with the directive word Δ. Hence the previous theorem is still valid for any episturmian word, and we can deduce:

Corollary 1. *Let w be an episturmian word over an alphabet A and let $B \subseteq A$ be a set containing at least two different letters. The word w is an ultimately B-strict episturmian word if and only if for an integer n_0, each left special factor with $|u| \geq n_0$ verifies $Au \cap Fact(w) = Bu$.*

Moreover for each left special factors with $|u| < n_0$, $Bu \subseteq Fact(w)$.

The restriction on the cardinality of B (≥ 2) will be used in all the rest of the paper. It is needed to have special factors of arbitrary length.

3 A New Characterization of Episturmian Words

Now we give our first main result presented in the introduction as a kind of local characteristic balance property of episturmian words.

Theorem 4. *For a recurrent infinite word w, the following assertions are equivalent:*

1. *w is episturmian;*
2. *for each factor u of w, a letter a exists such that $AuA \cap Fact(w) \subseteq auA \cup Aua$;*
3. *for each palindromic factor u of w,*
   a letter a exists such that $AuA \cap Fact(w) \subseteq auA \cup Aua$.

In the previous theorem, the letter a and the cardinality of the set AuA depends on u. This is shown for instance by the Fibonacci word (abaababaabaa...), the epistandard word having $(ab)^\omega$ as the directive word, for which $A\varepsilon A \cap Fact(w) = \{aa, ab, ba\}$, $AaA \cap Fact(w) = \{aab, baa\}$, $AbA \cap Fact(w) = \{aba\}$, $AaaA \cap Fact(w) = \{baab\}$, ...

Proof of Theorem 4.
Proof of $1 \Rightarrow 2$. Assume w is episturmian. Since the result deals only with factors of w, and since by Theorem 2 an episturmian word has the same set of factors as an epistandard word, without loss of generality we can assume that w is epistandard. Let u be a factor of w. Property 2 is immediate if u is not a bispecial factor of w. If u is bispecial in w, by Remark 1, an integer $n \geq 1$ exists such that $u = u_{n,w}$. Let $\Delta = (x_i)_{i \geq 1}$ be the directive word of w, let s (resp. t) be the epistandard word with $(x_i)_{i \geq n}$ (resp. $(x_i)_{i \geq n+1}$) as the directive word and let $a = x_n$. Letters occurring in t are exactly the letters of the set $B = \{x_i \mid i \geq n+1\}$. Since $s = L_{x_n}(t)$, the factors of length 2 in s are the words ab and ba with $b \in B$. By definition of Δ and $u_{n,w}$, $w = L_{x_1}(L_{x_2}(\ldots L_{x_{n-1}}(s)\ldots))$ and $u_{n,w} = L_{x_1}(L_{x_2}(\ldots (L_{x_{n-1}}(\varepsilon)x_{n-1})\ldots)x_2)x_1$. Hence by an easy induction on n, we deduce $AuA \cap Fact(w) = auB \cup Bua \subseteq auA \cup Aua$.

Proof of $2 \Rightarrow 1$. Assume that, for any factor u of w, a letter a exists such that $AuA \cap Fact(w) \subseteq auA \cup Aua$. In particular, considering the empty word, we deduce that $AA \cap Fact(w) \subseteq aA \cup Aa$ for a letter a. Hence, for an infinite word x, $w = L_a(y)$ if w starts with a and $w = R_a(y)$ otherwise.

Let us prove that for each factor v of y, $AvA \cap Fact(w) \subseteq bvA \cup Avb$ for a letter b. We consider $w = L_a(y)$ (resp. $w = R_a(y)$). Let v be a factor of y and let $u = L_a(v)a$ (resp. $u = aR_a(v)$). We observe that for letters c, d, the words cud is a factor of w if and only if cvd is a factor of y. By hypothesis there exists a letter b such that $AuA \cap Fact(w) \subseteq buA \cup Aub$. Hence $AvA \cap Fact(w) \subseteq bvA \cup Avb$.

Letting $x_1 = a$ and iterating infinitely the previous step, we get an infinite sequence of letters $(x_i)_{i \geq 1}$ and an infinite sequence of words $(w^{(i)})_{i \geq 0}$ such that $w^{(0)} = w$ and for all $i \geq 1$, $w^{(i-1)} = L_{x_i}(w^{(i)})$ or $w^{(i-1)} = R_{x_i}(w^{(i)})$. Due to the fact that w is recurrent, each word $w^{(i)}$ is also recurrent. By Theorem 2, the word w is episturmian.

The proof of $1 \Leftrightarrow 3$ is similar to the proof of $1 \Leftrightarrow 2$. Actually, $1 \Rightarrow 3$ is a particular case of $1 \Rightarrow 2$. When proving $3 \Rightarrow 1$, we need to prove in the inductive step that u is a palindrome if and only if v is a palindrome. This is stated by Lemma 2. 5 in [14] : *a word u is a palindrome if and only the word $L_a(u)a = aR_a(u)$ is a palindrome.* □

We end this section with few remarks concerning results that can be proved similarly.

Remark 2. Since an infinite word w over an alphabet A is recurrent if and only if for each factor of w the set $AuA \cap Fact(w)$ is not empty, we have: an infinite word is episturmian if and only if for each (resp. *palindromic*) factor u of w, $AuA \cap Fact(w)$ is not empty and a letter a exists such that $AuA \cap Fact(w) \subseteq auA \cup Aua$.

Remark 3. We have already said that Arnoux-Rauzy sequences over an alphabet A are exactly the (ultimately) A-strict episturmian word. One can ask for a characterization of these words in a way quite similar to Theorem 4. Corollary 1 can fulfill this purpose. But the proof of Theorem 4 can also be easily reworked to state : *an episturmian word w over an alphabet A is an ultimately B-strict episturmian word with $B \subseteq A$ if and only if for all $n \geq 0$, there exists a (resp. palindromic) word u of length at least n and a letter a such that $AuA \cap Fact(w) = auB \cup Bua$.*

Remark 4. Another adaptation of the proof of Theorem 4 concerns finite words: *a finite word w is a factor of an infinite episturmian word if and only if for each factor u of w, a letter a exists such that $AuA \cap Fact(w) \subseteq auA \cup Aua$.* We let the reader verify this result. The main difficulty of the proof is that in the "if part", we do not have necessarily $w = L_a(y)$ or $w = R_a(y)$. But we have one of the four following cases depending on the fact that w ends or not with a: $w = L_a(y)$, $w = aL_a(y)$, or $wa = L_a(y)$ or $wa = aL_a(y)$. Except in small cases, we have $|y| < |w|$ and the technique of the proof of Theorem 4 can be applied.

4   A Characterization of Epistandard Words

Let us note that for any episturmian word w, there exists at least one letter a such that aw is also episturmian. Indeed, since any episturmian word is recurrent, for any prefix p of w, there exists a letter a_p such that $a_p p$ is a factor of w. We work with a finite alphabet hence an infinity of letters a_p are mutually equal: there exists a letter a such that ap is a factor of p for an infinity of prefixes (and so for all prefixes) of w. The word aw has the same set of factors as w: it is episturmian.

In restriction to epistandard words, a more precise result is already know:

Theorem 5. [14, Th. 3.17] *If a word s is epistandard, then for each letter a in $Ult(\Delta(s))$, as is episturmian.*

As far as we know the converse of this result has already been stated only in the Sturmian case (see [3, Prop. 2.1.22]): *For every Sturmian word w over $\{a, b\}$, w is standard episturmian if and only if aw and bw are both Sturmian.* We generalize here this result, proving a converse to Theorem 5 (when $Ult(\Delta(s))$ contains at least two elements).

Proposition 1. *A non-periodic word w is epistandard if and only if, for (at least) two different letters a and b, aw and bw are episturmian.*

Proof. Let w be a non-periodic epistandard word w. By [9, Th. 3], we know that $Ult(\Delta(w))$ contains at least two different letters, say a and b. By Theorem 5, aw and bw are episturmian.

Assume now that for two different letters a and b, aw and bw are episturmian. Since aw (and also bw) is recurrent, w has the same set of factors as aw and so w is episturmian. Moreover each prefix p is left special (since ap and bp are factors of w). Since any episturmian word has at most one left special factor for each length, the left special factors of w are its prefixes: w is epistandard. □

Let us give a more precise result:

Theorem 6. *Let w be an infinite word over the alphabet A and assume $B \subseteq A$ contains at least two different letters. The two following assertions are equivalent:*

1. *The word w is ultimately B-strict epistandard;*
2. *For each letter a in A, aw is episturmian if and only if a belongs to B.*

Proof. Assume first that w is B-strict epistandard, that is, $Ult(\Delta(w)) = B$. By Theorem 5, for each letter a in B, aw is episturmian. For any integer $n \geq 0$, the word $u_{n,w}$ is a prefix of w. If a does not belong to B, by Theorem 3, for at least one integer $n \geq 0$, $au_{n,w}$ is not a factor of w. Thus the word aw is not recurrent and so it is not episturmian. Hence if w is B-strict epistandard, for each letter a in A, aw is episturmian if and only if a belongs to B.

Assume now that for each letter a in A, aw is episturmian if and only if a belongs to B. Since B contains at least two letters, by Proposition 1, w is epistandard. As a consequence of Theorem 3, we can deduce $Ult(\Delta(w)) = B$. □

5   A New Family of Words

In this section, we consider recurrent infinite words w over an alphabet A having the following property:

Property $\mathcal{P}$: for any word u over A, the set of factors of w belonging to AuA is balanced, that is, for any word u and for any letters a, b, c, d, if aub and cud are factors of w then $\{a, b\} \cap \{c, d\} \neq \emptyset$.

Any word verifying Assertion 2 in Theorem 4 also verifies Property $\mathcal{P}$. As shown by the word $(abc)^\omega$, the converse does not hold. In other words, any episturmian word verifies Property $\mathcal{P}$, but this is not a characteristic property (except in the binary case for which it is immediate that a word w verifies Property $\mathcal{P}$ if and only if for all words u, aua or bub is not a factor of w).

We prove:

Proposition 2. *A recurrent word w over an alphabet A verifies property $\mathcal{P}$ if and only if one of the two following assertions holds:*

1. *w is episturmian;*
2. *there exist three different letters a, b, c in A, a word w' over $\{a, b, c\}$ and an episturmian morphism on $\{a, b, c\}$ such that $w = f(w')$, w' verifies Property $\mathcal{P}$ and the three words ab, bc and ca are factors of w'.*

This proposition is a consequence of the next two lemmas.

Lemma 1. *If a recurrent infinite word w verifies property $\mathcal{P}$, then one of the two following assertions holds:*

1. *$w = L_\alpha(w')$ or $w = R_\alpha(w')$ for a letter α and a recurrent infinite word w';*
2. *there exist three different letters a, b, c such that $w \in \{a, b, c\}^\omega$ and the three words ab, bc and ca are factors of w.*

Proof. We first observe that if $AA \cap Fact(w) \subseteq \alpha A \cup A\alpha$ then (as in the proof of Theorem 6) $w = L_\alpha(w')$ or $w = R_\alpha(w')$, for a letter α and a recurrent infinite word w'.

We assume from now on that $AA \cap Fact(w) \not\subseteq \alpha A \cup A\alpha$.

For any letter α in A, $\alpha\alpha$ is not a factor of w. Indeed if such a word is a factor of w, then, for any factor $\beta\gamma$ with β and γ letters, by Property $\mathcal{P}$, $\beta = \alpha$ or $\gamma = \alpha$, that is $AA \cap Fact(w) \subseteq \alpha A \cup A\alpha$.

The alphabet A contains at least three letters. Indeed if A contains at most two letters a and b, then Property $\mathcal{P}$ implies that aa and bb are not simultaneously factors of w, and so we have $AA \cap Fact(w) \subseteq aA \cup Aa$ or $AA \cap Fact(w) \subseteq bA \cup Ab$.

Let us prove that A contains exactly three letters. Assume by contradiction that A contains at least four letters. Let a (resp. b) be the first (resp. the second) letter of w. Since aa is not a factor of w, $a \neq b$. At least two other letters c and d occur in w ($c, d \notin \{a, b\}$, $c \neq d$). By Property $\mathcal{P}$, each occurrence of c is preceded by a or by b. Assume that ac occurs in w. Since ab also occurs, for any letter α not in $\{a, b, c\}$, each occurrence of α is preceded and followed by the letter a. But $AA \cap Fact(w) \not\subseteq aA \cup Aa$. Hence bc or cb occurs in w. But then the factor ad contradicts Property $\mathcal{P}$. Assume now that bc occurs in w. Since ab also occurs, for any letter α not in $\{a, b, c\}$, each occurrence of α is preceded and followed by b. But $AA \cap Fact(w) \not\subseteq bA \cup Ab$. Hence ac or ca occurs in w. But then the factor db contradicts Property $\mathcal{P}$.

Until now we have proved that w is written on a three-letter alphabet and contains no word $\alpha\alpha$ with α a letter. Assume that, for two letters a and b, ab is a factor of w but not ba. Then for an integer $n \geq 1$, $a(bc)^n a$ (let recall that aa, bb, cc and ba are not factors of w), and so ab, bc and ca are factors of w. Now if, for all letters α and β, $\alpha\beta$ and $\beta\alpha$ are factors of w then denoting by a, b and c the letters occurring in w, once again ab, bc and ca are factors of w. □

Lemma 2. *Let α be a letter, w and w' be recurrent words such that $w = L_\alpha(w')$ or $w = R_\alpha(w')$. The word w verifies Property $\mathcal{P}$ if and only if w' verifies Property $\mathcal{P}$.*

Proof. We first assume $w = L_\alpha(w')$.

Assume that w does not verify Property $\mathcal{P}$: aub and cud are factors of w for some letters a, b, c, d and a word u such that $\{a, b\} \cap \{c, d\} = \emptyset$. At least one of the two letters a and b is different from α and at least one of the two letters c and d is different from α. Since $w = L_\alpha(w')$, we deduce that $u \neq \varepsilon$, and that u begins and ends with α: $u = L_\alpha(v)\alpha$ for a word v. Thus $aub = aL_\alpha(v)\alpha b$ and $cud = cL_\alpha(v)\alpha d$. We observe that if $a \neq \alpha$ (resp. $c \neq \alpha$), $\alpha a L_\alpha(v)\alpha b$ (resp. $\alpha c L_\alpha(v)\alpha d$) is a factor of w. Thus we can deduce that avb and cvd are factors of w' (even if one of the letters a, b, c, d is α): the word w' does not verify Property $\mathcal{P}$.

Assume conversely that the word w' does not verify Property $\mathcal{P}$: aub and cud are factors of w' for some letters a, b, c, d and a word u such that $\{a, b\} \cap \{c, d\} = \emptyset$. The word $aL_\alpha(u)\alpha b$ is a factor of w (if $b = \alpha$, this is still true since we work with infinite words and so in this case $au\alpha b'$ is a factor of w for a letter b'). Similarly $cL_\alpha(u)\alpha d$ is a factor of w: the word w does not verify Property $\mathcal{P}$.

The proof when $w = R_\alpha(w')$ is similar. Note that the fact that w' is recurrent is needed for the last part of the proof to know when $a = \alpha$, that $a'\alpha ub$ is a factor of w' for a letter a'.   □

Proof of Proposition 2. Assume w is a recurrent word that verifies Property $\mathcal{P}$ but that does not verify Assertion 2 of Lemma 1. Then $w = L_\alpha(w')$ or $w = R_\alpha(w')$, with w' a recurrent word. By Lemma 2, w' verifies Property $\mathcal{P}$.

Thus using Lemmas 1 and 2, we can prove by induction that, for any integer $n \geq 0$, one of the two following assertions holds:

- there exist recurrent infinite words $w^{(0)} = w$, $w^{(1)}, \ldots w^{(n)}$, and letters $a_1, \ldots, a_n$ such that for each $1 \leq p \leq n$, $w^{(p-1)} = L_{a_p}(w^{(p)})$ or $w^{(p-1)} = R_{a_p}(w^{(p)})$, and w^n verifies property $\mathcal{P}$;
- for an integer $m \leq n$, there exist recurrent infinite words $w^{(0)} = w$, $w^{(1)}, \ldots w^{(m)}$, and letters $a_1, \ldots, a_m$ such that for each $1 \leq p \leq m$, $w^{(p-1)} = L_{a_p}(w^{(p)})$ or $w^{(p-1)} = R_{a_p}(w^{(p)})$, and $w^{(m)}$ verifies Assertion 2 of Lemma 1.

Hence the proposition is a consequence of Theorem 2.   □

6   Conclusion

The reader has certainly noticed that words verifying Property $\mathcal{P}$ are not completely characterized. For this, one should have to better know ternary recurrent words verifying Property $\mathcal{P}$ and containing the words ab, bc and ca as factors.

Let us give examples of such words. One can immediately verify that if ab, bc and ca are the only words of length 2 that are factors of a word w, then w is $(abc)^\omega$, $(bca)^\omega$ or $(cab)^\omega$. When a recurrent word w verifying property $\mathcal{P}$ has exactly the words ab, bc, ca and ba as factors of length 2, one can see that

w is a suffix of a word $f(w')$ where w' is a Sturmian word over $\{a, b\}$ and f is the morphism defined by $f(a) = (ab)^n c$ and $f(b) = (ab)^{n+1} c$ for an integer $n \geq 1$. When f is replaced by one of the following morphisms g_1 or g_2, we can get other examples of ternary words verifying Property $\mathcal{P}$ (and containing exactly 5 factors of length 2 with amongst them ab, bc and ca) : $g_1(a) = (ab)^n c$, $g_1(b) = (ab)^n cb$, $g_2(a) = (ab)^n c$, $g_2(b) = (ab)^{n+1} cb$. Our final example is the periodic word $(abcabacbabcb)^\omega$ which verifies Property $\mathcal{P}$ and contains as factors all words of length 2 except aa, bb, cc: this word could be seen as the morphic image of a^ω by the morphism that maps a onto $abcabacbabcb$.

All these examples lead to the question: Are all ternary recurrent words verifying Property $\mathcal{P}$ and containing ab, bc and ca as factors are suffixes of a word $f(w')$ with w' a recurrent balanced word (that is a Sturmian word or a periodic balanced word) and with f a morphism? If it is true, which are the possible values for f?

We end with another question about the set of finite words that are not factors of episturmian words. In [10] (see also [3, Pb. 2.1.4, p. 102]), it is proved: *a word w is not balanced (and so not a factor of a Sturmian word) if and only if it can be written $w = xauaybübz$ for some words u, x, y, z and distinct letters a and b* ($\tilde{u}$ is the mirror image of u). This result was then used to state that the set of non-balanced word is context-free. Is it still true for words that are not factors of episturmian words over a (at least) ternary alphabet? Can Theorem 4 be useful to prove the context-freness or the non-context-freeness of this set of words?

Acknowledgements. The author would like to thanks J.-P. Allouche for his questions that have initiated the present work. Many thanks also to the referees: one provided a lot of helpful remarks, two others found an error in an attempt to state the non-context-freeness of the set of words that are not factors of episturmian words.

References

1. Allouche, J.-P., Shallit, J.: Automatic sequences. Cambridge University Press, Cambridge (2003)
2. Arnoux, P., Rauzy, G.: Représentation géométrique de suites de complexités $2n+1$. Bull. Soc. Math. France 119, 199–215 (1991)
3. Berstel, J., Séébold, P.: Sturmian words. In: Lothaire, M. (ed.) Algebraic Combinatorics on Words, vol. 90, ch. 2, Cambridge Mathematical Library (2002)
4. Berthé, V., Holton, C., Zamboni, L.Q.: Initial powers of sturmian sequences. Acta. arithmetica 122, 315–347 (2006)
5. Cassaigne, J.: Complexité et facteurs spéciaux. Belg. Bull. Math. Soc. 4, 67–88 (1997)
6. Cassaigne, J., Ferenczi, S., Zamboni, L.Q.: Imbalances in Arnoux-Rauzy sequences. Ann. Inst. Fourier, Grenoble 50(4), 1265–1276 (2000)
7. Coven, E.M., Hedlund, G.A.: Sequences with minimal block growth. Math. Syst. Th. 7, 138–153 (1973)
8. de Luca, A.: On standard Sturmian morphisms. Theoretical Computer Science 178, 205–224 (1997)

9. Droubay, X., Justin, J., Pirillo, G.: Episturmian words and some constructions of de Luca and Rauzy. Theoretical Computer Science 255, 539–553 (2001)
10. Dulucq, S., Gouyou-Beauchamps, D.: Sur les facteurs des suites de Sturm. Theoretical Computer Science 71, 381–400 (1990)
11. Glen, A.: A characterization of fine words over a finite alphabet. In: International School and Conference on Combinatorics, Automata and Number Theory (Cant'06), p. 9. Université de Liège, Belgium (2006)
12. Glen, A., Justin, J., Pirillo, G.: Characterizations of finite and infinite episturmian words via lexicographic orderings (submitted, 2006)
13. Hubert, P.: Suites équilibrées. Theoretical Computer Science 242, 91–108 (2000)
14. Justin, J., Pirillo, G.: Episturmian words and episturmian morphisms. Theoretical Computer Science 276(1-2), 281–313 (2002)
15. Justin, J., Vuillon, L.: Return words in Sturmian and episturmian words. RAIRO Theoret. Infor. Appl. 34, 343–356 (2000)
16. Levé, F., Richomme, G.: Quasiperiodic infinite words: some answers. Bull. Europ. Assoc. Theoret. Comput. Sci. 84, 128–238 (2004)
17. Levé, F., Richomme, G.: Quasiperiodic Sturmian words and morphisms. Theoretical Computer Science 372(1), 15–25 (2007)
18. Lothaire, M.: Combinatorics on words. Encyclopedia of Mathematics and its Applications, vol. 17. Addison-Wesley, London (1997) Reprinted in the Cambridge Mathematical Library, Cambridge University Press, UK, 1997
19. Morse, M., Hedlund, G.A.: Symbolic Dynamics II: Sturmian trajectories. Amer. J. Math. 61, 1–42 (1940)
20. Pytheas Fogg, N.: Substitutions in Dynamics, Arithmetics and Combinatorics. In: Berthé, V., Ferenczi, S., Mauduit, C., Siegel, A. (eds.) Frontiers of Combining Systems. LNCS, vol. 1794, Springer, Heidelberg (2000)
21. Richomme, G.: Conjugacy and episturmian morphisms. Theoretical Computer Science 302, 1–34 (2003)
22. Richomme, G.: Conjugacy of morphisms and Lyndon decomposition of standard Sturmian words. Theoretical Computer Science (Words'05 special number) (to appear)
23. Vuillon, L.: Balanced words. Bull. Belg. Math.Soc. 10(5), 787–805 (2003)

Suffix Automata and Standard Sturmian Words

Marinella Sciortino[1] and Luca Q. Zamboni[2]

[1] Dipartimento di Matematica ed Applicazioni, University of Palermo,
Via Archirafi 34 - 90123, Palermo, Italy
mari@math.unipa.it

[2] Department of Mathematics, PO Box 311430, University of North Texas,
Denton, TX 76203-1430, USA
luca@unt.edu

Abstract. Blumer *et al.* showed (cf. [3,2]) that the suffix automaton of a word w must have at least $|w| + 1$ states and at most $2|w| - 1$ states. In this paper we characterize the language L of all binary words w whose minimal suffix automaton $\mathcal{S}(w)$ has exactly $|w| + 1$ states; they are precisely all prefixes of standard Sturmian words. In particular, we give an explicit construction of suffix automaton of words that are palindromic prefixes of standard words. Moreover, we establish a necessary and sufficient condition on $\mathcal{S}(w)$ which ensures that if $w \in L$ and $a \in \{0, 1\}$ then $wa \in L$. By using such a condition, we show how to construct the automaton $\mathcal{S}(wa)$ from $\mathcal{S}(w)$. More generally, we provide a simple construction that by starting from an automaton recognizing all suffixes of a word w over a finite alphabet A, allows to obtain an automaton that recognizes the suffixes of $wa, a \in A$.

1   Introduction

Several structures are used to store the suffixes of a text and are designed to give a fast access to all factors of the text itself. For this reason such structures have a lot of applications in text processing. Suffix tries provide a representation of all the suffixes of a word by an ordinary tree. It has the advantage of being simple but can lead to a memory size that is quadratic in the length of the considered word. Suffix trees are compact representations of suffix tries and have been first introduced by Weiner [17], but the most practical algorithms are by McCreight [13] and Ukkonen [16]. The total size of a suffix tree is linear in the length of the considered word. Automata are alternative data structure to recognize all suffixes of a word. Whereas trees put together common prefixes of all suffixes of a word, automata gather also the common suffixes. The suffix automaton of a word is obtained by the minimization (related to automata) of the suffix trie of the word itself. Suffix automata and suffix trees have similar applications to the implementation of indexes (inverted files), to pattern matching, and to data compression.

The linear size of suffix automata has been first noticed by Blumer *et al.* in [3,2] where a linear algorithm on a fixed alphabet is given. In particular they showed that the suffix automaton of a word w must have at least $|w| + 1$ states and at most $2|w| - 1$ states. Moreover, in [5] Crochemore proved the minimality of such a structure as an automaton and showed that the factor automaton of a word (i.e. the minimal deterministic automaton recognizing all the factors of the word) could be build within the same

T. Harju, J. Karhumäki, and A. Lepistö (Eds.): DLT 2007, LNCS 4588, pp. 382–398, 2007.
© Springer-Verlag Berlin Heidelberg 2007

complexity. Upper and lower bounds on the number of states of a factor automaton are also given in [4].

We focus on the suffix automata of a word and the language of the binary words whose suffix automaton has exactly $|w| + 1$ states.

Let A be a non-empty finite alphabet. Given a word $w \in A^*$, we denote by $\mathcal{S}(w)$ the suffix automaton of w, i.e. the minimal deterministic (non necessarily complete) automaton which recognizes the finite set of suffixes of w (see [6]). The set of states and the set of edges of $\mathcal{S}(w)$ are denoted by Q_w and E_w, respectively.

Let $w = w_1 w_2 \ldots w_n \in A^+$; then the suffix automaton of w must always contain (as a sub-automaton) the automaton described in Figure 1 where the state X_0 is initial.

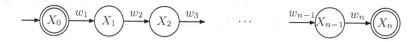

Fig. 1. Sub-automaton of $\mathcal{S}(w)$

Moreover X_0 and X_n are always terminal states, corresponding to the empty word and the entire word w, respectively. Typically other states X_i will also be terminal states, but if w is unbordered, i.e. no proper non-empty prefix of w is also a suffix of w, then X_0 and X_n are the only terminal states.

In general, the suffix automaton of a word $w = w_1 w_2 \ldots w_n$ will contain other states in addition to $X_0, X_1, \ldots X_n$.

Example 1. Consider the factor $w = 10100100101$ of the infinite Fibonacci word. The suffix automaton $\mathcal{S}(w)$ has 18 states as shown in Figure 2:

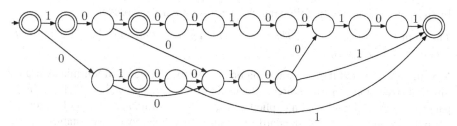

Fig. 2. Suffix automaton of 10100100101

As reported in [7], the following two propositions provide tight lower and upper bounds on the size of the suffix automaton both in terms of the number of states and number of edges. In particular one can show that the suffix automaton of a word w can have as many as $2|w| - 1$ states.

Proposition 1. *Let A be a finite alphabet and w a word in A^*. If $|w| = 0$ then $\#(Q_w) = 1$; and if $|w| = 1$ then $\#(Q_w) = 2$. If $|w| \geq 2$ then $|w| + 1 \leq \#(Q_w) \leq 2|w| - 1$ and the upper bound is reached when w has the form $ab^{|w|-1}$ for distinct letters a and b.*

Proposition 2. *Let A be a finite alphabet and w a word in A^*. If $|w| = 0$, $\#(E_w) = 0$; and if $|w| = 1$, $\#(E_w) = 1$. If $|w| = 2$, $2 \leq \#(E_w) \leq 3$ and if $|w| \geq 3$ then $|w| \leq \#(E_w) \leq 3|w| - 4$ and the upper bound is reached when w has the form $ab^{|w|-2}c$ for three distinct letters a, b and c.*

In this paper we deal with the lower bound on the number of states in the case of binary alphabets. More precisely, let us denote by L the following language

$$L = \{w \in \{0,1\}^* \mid \#(Q_w) = |w| + 1\}.$$

Note that if $w \in L$, then the only states of $S(w)$ are $X_0, X_1, \ldots, X_n$. Also, unless $w = a^n$, $a \in A = \{0,1\}$, the suffix automaton $S(w)$ will contain additional edges of the form $X_i \xrightarrow{a} X_j$, where $j - i > 1$. We call such an edge a *bypass edge* while edges of the form $X_i \xrightarrow{a} X_{i+1}$ will be called *direct edges*.

Example 2. Let $w = 00100010010$. Then the suffix automaton $S(w)$ has 4 terminal states and 4 bypass edges as described in Figure 3.

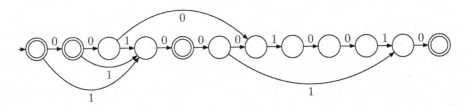

Fig. 3. Suffix automaton of 00100010010

In this paper we obtain the following characterization of the words in L :

Theorem 1. *Let $w \in \{0,1\}^*$. The word w belongs to L if and only if w is a prefix of a standard Sturmian word.*

Note that the fact that the suffix automaton of a prefix of a standard Sturmian word has minimal size could be proved directly by using some results proved in [14,4,10]. In this paper we introduce and use an explicit construction of suffix automaton of words that are palindromic prefixes of standard Sturmian words. Moreover, we establish a necessary and sufficient condition (name *extendability condition*) on the suffix automaton $S(w)$ of a word w, which assures that a word in L is extendable in L. Such a condition also allows a simple construction of the suffix automaton of wa ($a \in A$) by starting from $S(w)$. Many of the ideas involved in the proof of such a theorem apply more generally to the case of words on larger alphabets A.

In Section 2 we prove that the language L is closed under prefixes. Actually, given a *trim* automaton A recognizing all suffixes of a word v, we construct a new automaton $P(A)$, called the *pruning* of A, that recognizes all suffixes of the $|v| - 1$ length prefix of v. Moreover we introduce the extendability condition above mentioned. Finally, by starting from an automaton A recognizing all suffixes of a word w, we show how to construct the automaton $E_a(A)$ that recognizes the suffixes of wa, $a \in A$.

In Section 3, we give an explicit construction of suffix automaton of words that are palindromic prefixes of standard sturmian words. We show that such words belong to L. Section 4 contains the proof of the main result of the paper.

2   Closure Under Prefixes and Extendability Condition

Such a section is devoted to describe some properties of words in the language L defined in previous section. First of all we prove that the language L is closed under prefixes. Actually we define a construction, called *pruning* of an automaton, that allow to build the suffix automaton of a word $w \in A^*$ starting from the suffix automaton of wa, $a \in A$. Moreover we deal with the problem of extending a word in L to a word also belonging to L. More generally, we introduce a construction that, by starting from an automaton $\mathcal{A}$ recognizing all suffixes of a word w over a finite alphabet A, allow to obtain the automaton $\mathcal{E}_a(\mathcal{A})$ that recognizes the suffixes of wa, $a \in A$. Such a construction is based on the notion of extendability condition that we establish for a generic deterministic finite automaton recognizing a finite set of words in A^+. In case of a binary word w of L, the extendability condition on $\mathcal{S}(w)$ represents a necessary and sufficient condition that ensures that $wa \in L$, where $a \in A$.

Definition 1. *Let $\mathcal{A}$ be a deterministic automaton, where X_0 denotes its initial state. We say that $\mathcal{A}$ is* trim *if each of the following holds:*

- *For each state X in $\mathcal{A}$, there is at least one path from X_0 to X, i.e. each state is accessible from the initial state.*
- *For each state X in $\mathcal{A}$ there exists a path starting from X and terminating at a final state, i.e. each state is coaccessible from a final state.*

Let us denote by $F_{\mathcal{A}}$ be the set of all terminal states having no outgoing edges. A priori this set may be empty.

Let $w \in A^+$ and $\mathcal{B}$ be an automaton which recognizes the set of all suffixes of w.

Thus if $\mathcal{B}$ is trim, it follows that i) $F_{\mathcal{B}} \neq \emptyset$, and ii) for each state Y of $\mathcal{B}$, any two edges pointing into Y have the same label. From here on, all automata recognizing all suffixes of any word are assumed to be trim.

Let $\mathcal{A}$ be a trim deterministic finite automaton. We denote by $Q_{\mathcal{A}}$ the set of the states of $\mathcal{A}$. We define a new automaton $\mathcal{P}(\mathcal{A})$, called the *pruning* of $\mathcal{A}$, as follows.

i) The set of states $Q_{\mathcal{P}(\mathcal{A})}$ of $\mathcal{P}(\mathcal{A})$ is equal to $Q_{\mathcal{A}} - F_{\mathcal{A}}$. The initial state of $\mathcal{P}(\mathcal{A})$ is equal to the initial state of $\mathcal{A}$.

ii) For $X, X' \in Q_{\mathcal{P}(\mathcal{A})}$, in $\mathcal{P}(\mathcal{A})$ there is an edge labeled b ($b \in A$) directed from X to X' if and only if in $\mathcal{A}$ there exists an edge labeled b from X to X'.

iii) A state $X \in Q_{\mathcal{P}(\mathcal{A})}$ is a terminal state of $\mathcal{P}(\mathcal{A})$ if and only if in $\mathcal{A}$ there exists an edge labeled a from X to Y and Y is a terminal state in $\mathcal{A}$.

In short, $\mathcal{P}(\mathcal{A})$ is obtained from $\mathcal{A}$ by simply deleting all states $X \in F_{\mathcal{A}}$ together with all edges pointing into X, and by changing the choice of terminal states according to the rule: X is a terminal state of $\mathcal{P}(\mathcal{A})$ if and only if in $\mathcal{A}$ there is an edge labeled a from X to a terminal state Y of $\mathcal{A}$. We note that $\#Q_{\mathcal{P}(\mathcal{A})} < \#Q_{\mathcal{A}}$.

Note that if $\mathcal{A}$ is trim, it follows that i) $F_\mathcal{A} \neq \emptyset$ and ii) for each state Y of $\mathcal{A}$, any two edges pointing into Y have the same label. Moreover, we assume that all automata recognizing all suffixes of any word are trim.

Proposition 3. *Let $\mathcal{A}$ be an automaton which recognizes the suffixes of wa, where $w \in A^+$ and $a \in A$. Then $\mathcal{P}(\mathcal{A})$ defined above recognizes the suffixes of w. Moreover $\mathcal{P}(\mathcal{A})$ is trim.*

Proof. Let $x_1 x_2 \ldots x_k$ denote a path in $\mathcal{P}(\mathcal{A})$ starting from the initial state X_0 and ending at some terminal state X in $\mathcal{P}(\mathcal{A})$. Then, there is an edge labeled a in $\mathcal{A}$ from X to some terminal state Y. Thus $x_1 x_2 \ldots x_k a$ defines a successful path in $\mathcal{A}$ and hence $x_1 x_2 \ldots x_k a$ is a suffix of wa. It follows that $x_1 x_2 \ldots x_k$ is a suffix of w.

Conversely, suppose $x_1 x_2 \ldots x_k$ is a suffix of w. Then, $x_1 x_2 \ldots x_k a$ is a suffix of wa and hence defines a path in $\mathcal{A}$ from X_0 to some terminal state Y of $\mathcal{A}$. Let X denote the final state in $\mathcal{A}$ corresponding to the path $x_1 x_2 \ldots x_k$. Then by (iii), X is a terminal state of $\mathcal{P}(\mathcal{A})$. Hence $x_1 x_2 \ldots x_k$ defines a successful path in $\mathcal{P}(\mathcal{A})$.

We can note also that, since $\mathcal{A}$ is trim then $\mathcal{P}(\mathcal{A})$ is trim, too.   □

Corollary 1. *Let $\mathcal{A}$ be an automaton recognizing the suffixes of wa. Then there is an automaton $\mathcal{B}$ recognizing the suffixes of w with $\#Q_\mathcal{B} \leq \#Q_\mathcal{A} - 1$.*

Proof. We can simply take $\mathcal{B} = \mathcal{P}(\mathcal{A})$. In fact, it follows from Proposition 3 that $\mathcal{P}(\mathcal{A})$ recognizes the set of suffixes of w, and $\#Q_{\mathcal{P}(\mathcal{A})} < \#Q_\mathcal{A}$.   □

Example 3. Note that even if $\mathcal{A}$ is minimal, $\mathcal{P}(\mathcal{A})$ need not be minimal. For instance, consider the minimal suffix automaton of the word 100 shown below in Figure 4.

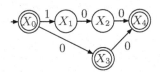

Fig. 4. Suffix automaton of 100

Then $\mathcal{P}(\mathcal{A})$, shown in Figure 5, recognizes the suffixes of 10 but is not the minimal such automaton. In fact the minimal automaton is shown in Figure 6.

As a consequence of Corollary 1 we deduce the following closure property of the words of L.

Corollary 2. *Let $w \in A^+$. If $\mathcal{S}(w)$ has $|w| + 1$ many states, then for each prefix u of w, $\mathcal{S}(u)$ has $|u| + 1$ many states. In particular, if w is in L, then so is every prefix of w.*

Proof. Let $w = w_1 w_2 \ldots w_n$, and suppose that $\mathcal{S}(w)$ has $n + 1$ states. Set $w' = w_1 w_2 \ldots w_{n-1}$ and $a = w_n$, so that $w = w'a$. It suffices to show that $\mathcal{S}(w')$ has n states. By Corollary 1, there is an automaton $\mathcal{B}$ having at most n states which recognizes the suffixes of w'. But by Proposition 1, $\mathcal{B}$ must have at least $n = |w'| + 1$ states. Whence, $\mathcal{B}$ has exactly n states.   □

As a consequence of Proposition 3, if $w = w'a \in L$, then $\mathcal{P}(\mathcal{S}(w)) = \mathcal{S}(w')$.

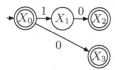

Fig. 5. Automaton recognizing the suffixes of 10

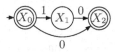

Fig. 6. Minimal automaton recognizing the suffixes of 10

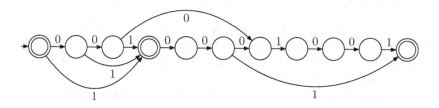

Fig. 7. Suffix automaton of 0010001001

Example 4. We can use the 'pruning' method described in this section to construct the suffix automaton (shown in Figure 7) of the word $w' = 0010001001$, prefix of the word w used in the Example 2.

Let us introduce now the extendability condition that we will use to extend the words in L to words belonging to L.

Let $\mathcal{M}$ be a deterministic finite automaton recognizing a finite set of words in A^+. Let $Q_{\mathcal{M}}$ denote the set of states in $\mathcal{M}$.

Definition 2. *We say the automaton $\mathcal{M}$ satisfies the* extendability condition *for the letter $a \in A$ if for each state Y in $\mathcal{M}$ pointed by edges labeled a from states X_i and X_j, then X_i is a terminal state of $\mathcal{M}$ if and only if X_j is a terminal state of $\mathcal{M}$.*

Let us suppose $\mathcal{M}$ is trim. By definition of trim automaton the following two conditions hold:

- For each state X of $\mathcal{M}$ having no outgoing edges, X is a terminal state.
- $F_{\mathcal{M}} \neq \emptyset$.

Recall that $F_{\mathcal{M}}$ was defined as the set of all terminal states of $\mathcal{M}$ having no outgoing edges.

As $\mathcal{M}$ is trim, associated to $\mathcal{M}$ is a new automaton $\mathcal{E}_a(\mathcal{M})$, the *extension of $\mathcal{M}$ with respect to* a, defined as follows.

We distinguish two cases:

1. $\mathcal{M}$ satisfies the extendability condition for the letter a:
   i) starting with $\mathcal{M}$ we adjoin a new state X^*.
   ii) for each terminal state X in $\mathcal{M}$, if X has no outgoing edge labeled a, then we adjoin an edge labeled a from X to X^*.
   iii) a state Y in $\mathcal{E}_a(\mathcal{M})$ is terminal if and only if there exists in $\mathcal{E}_a(\mathcal{M})$ an edge labeled a from X to Y and X is a terminal state in $\mathcal{M}$.
2. $\mathcal{M}$ does not satisfy the extendability condition for the letter a. This means that there exists at least one state Y, and two states X_i and X_j and edges labeled a from X_i to Y and X_j to Y and where X_i is terminal and X_j is not terminal. In this case, we first we construct an intermediate automaton $\mathcal{M}'$ as follows.
   For each state Y violating the extendability condition for a we do the following:
   a. we adjoin to $\mathcal{M}$ a state Y' that is terminal in $\mathcal{M}'$ if and only Y is terminal in $\mathcal{M}$;
   b. if in $\mathcal{M}$ there is an edge labeled a from a terminal state X to Y then in $\mathcal{M}'$ we re-direct that edge to an edge labeled a from X to Y';
   c. for each edge labeled $b \in A$ from Y to Z in $\mathcal{M}$, in $\mathcal{M}'$ we adjoin an edge labeled b from Y' to Z.
   It is now easy to see that the automaton $\mathcal{M}'$ satisfies the extendability condition for a, so we construct $\mathcal{E}_a(\mathcal{M})$ by applying item (1) to the automaton $\mathcal{M}'$.

Remark 1. We note that the intermediate automaton $\mathcal{M}'$ is never minimal as the two states Y and Y' are indistinguishable.

The following proposition shows that, by starting from an automaton recognizing all suffixes of a word w, the above construction allows to build the automaton that recognizes all suffixes of wa, $a \in A$. Recall that we suppose that all automata recognizing all suffixes of any word are trim.

Proposition 4. *Let $\mathcal{A}$ be an automaton which recognizes the suffixes of a word $w \in A^*$. Then $\mathcal{E}_a(\mathcal{A})$ recognizes the suffixes of wa. Moreover $\mathcal{E}_a(\mathcal{A})$ is trim, too.*

Proof. First let us suppose that $\mathcal{A}$ satisfies the extendability condition for a. If $x_1 x_2 \ldots x_k a$ is a suffix of wa then $x_1 x_2 \ldots x_k$ is a suffix of w. Hence $x_1 \ldots x_k$ defines a path in $\mathcal{A}$ from the initial state X_0 to some terminal state X. By construction, in $\mathcal{E}_a(\mathcal{A})$ there is an edge labeled a from X to some terminal state Y. Hence $x_1 x_2 \ldots x_k a$ defines a path in $\mathcal{E}_a(\mathcal{A})$ from X_0 to some terminal state Y. Conversely, suppose $x_1 x_2 \ldots x_k x$ defines a path in $\mathcal{E}_a(\mathcal{A})$ from X_0 to some terminal state Y. Since all edges terminating at Y are labeled a, it follows that $x = a$. By construction, $x_1 x_2 \ldots x_k$ defines a path in $\mathcal{A}$ from X_0 to some state X; since Y is a terminal state of $\mathcal{E}_a(\mathcal{A})$, the extendability condition guarantees that X is a terminal state of $\mathcal{A}$. Hence $x_1 x_2 \ldots x_k$ is a suffix of w, whence $x_1 x_2 \ldots x_k a$ is a suffix of wa.

Let us suppose now that $Y_1, Y_2, \ldots, Y_k$ are the states violating the extendability condition for a. We claim that the intermediate automaton $\mathcal{A}'$ defined above recognizes the suffixes of w. Let $Y_1', Y_2', \ldots, Y_k'$ be the states adjoined by applying the item (2).a. It is easy to see that for $i = 1, \ldots, k$ Y_i and Y_i' are indistinguishable. In fact for each

$z \in A^*$, there exists a path z from Y_i to some terminal state Z if and only if there exists the path z from Y_i' to Z. We claim that v defines a successful path in $\mathcal{A}$ if and only if it defines a successful path in $\mathcal{A}'$. The case in which the path does not go through any state Y_i is trivial. Otherwise we write $v = uz$ such that u is a path from X_0 to Y_j and z is a path from Y_j to a terminal state in $\mathcal{A}$. By construction there is a path labeled u in $\mathcal{A}'$ from X_0 to Y_i or from X_0 to Y_i'. So, v is recognized by $\mathcal{A}$ if and only if v is recognized from $\mathcal{A}'$. Since $\mathcal{A}'$ satisfies the extendability condition for a, by using the first part of this proof the result now follows.

Moreover, since $\mathcal{A}$ is trim it follows by construction that each state of $\mathcal{E}_a(\mathcal{A})$ is both accessible and coaccessible. So, $\mathcal{E}_a(\mathcal{A})$ is trim, too. $\qquad\square$

Corollary 3. *Let $\mathcal{A}$ be an automaton recognizing the suffixes of w. If $\mathcal{A}$ satisfies the extendability condition for a, then there is an automaton $\mathcal{B}$ recognizing the suffixes of wa with $\#Q_\mathcal{B} = \#Q_\mathcal{A} + 1$.*

Proof. From Proposition 4 it follows that $\mathcal{E}_a(\mathcal{A})$ is the required automaton $\mathcal{B}$. In fact $\mathcal{E}_a(\mathcal{A})$ recognizes all suffixes of wa and it has one more state than $\mathcal{A}$. $\qquad\square$

Corollary 4. *Let $\mathcal{A}$ be an automaton recognizing the suffixes of w. If $\mathcal{A}$ does not satisfy the extendability condition for a, then there is an automaton $\mathcal{B}$ recognizing the suffixes of wa with $\#Q_\mathcal{B} = \#Q_\mathcal{A} + k + 1$, where k is the number of states that violate the extendability condition for a.*

Proof. From Proposition 4 it follows that $\mathcal{E}_a(\mathcal{A})$ is the required automaton $\mathcal{B}$. In fact $\mathcal{A}'$ is obtained by adding k states. Then $\mathcal{E}_a(\mathcal{A}')$ has one more state than $\mathcal{A}'$. $\qquad\square$

Proposition 5. *Let $w \in A^*$ and let $\mathcal{S}(w)$ be the suffix automaton recognizing the suffixes of w. Then for each $a \in A$, there is at most one state in $\mathcal{S}(w)$ which does not satisfy the extendability condition for a.*

Proof. Let $a \in A$, and suppose to the contrary that there are at least two distinct states Z_1 and Z_2 in $\mathcal{S}(w)$ which violate the extendability condition for a. Thus there exist states X, Y, X', Y' in $\mathcal{A}$ with X and Y terminal, X' and Y' not terminal, and edges $X \xrightarrow{a} Z_1$, $X' \xrightarrow{a} Z_1$, $Y \xrightarrow{a} Z_2$, $Y' \xrightarrow{a} Z_2$. Let u and u' be paths from the initial state X_0 to X and X' respectively, and v and v' be paths from X_0 to Y and Y' respectively. Since both X and Y are terminal it follows that both u and v are suffixes of w. We can suppose without loss of generality that u is a proper suffix of v. We also know that u' and v' are not suffixes of w since neither X' nor Y' are terminal states. So we have:

1. u and v are suffixes of w, while u' and v' are not.
2. u is a proper suffix of v, hence every occurrence in w of v is an occurrence of u.
3. $ua, u'a, va, v'a$ are all factors of w.
4. Every occurrence of ua in w is an occurence of $u'a$ in w, and every occurrence of va in w is an occurrence of $v'a$ in w.
5. There is an occurrence of ua in w which is not an occurrence of va in w.

Item (5) is a consequence of the fact that $\mathcal{S}(w)$ is a minimal automaton, for otherwise the states Z_1 and Z_2 would be identified. Now consider an occurrence of va in w. Then since this is also an occurrence of ua in w, we have that either $u'a$ is a suffix of va or va a suffix of $u'a$. But the first case implies that u' is a suffix of v, and hence u' is a suffix of w, contradicting 1. In the second case we would have that every occurrence of ua in w is an occurrence of va in w, contradicting 5. This contradiction implies the desired result.   □

Proposition 6. *Let $w \in A^*$ and $a \in A$. Then*

$$\mathcal{E}_a(\mathcal{S}(w)) = \mathcal{S}(wa).$$

Proof. By Proposition 4, $\mathcal{E}_a(\mathcal{S}(w))$ recognizes the suffixes of wa. It remains to prove that $\mathcal{E}_a(\mathcal{S}(w))$ is minimal. Thus we must show that any two distinct states P and Q of $\mathcal{E}_a(\mathcal{S}(w))$ are distinguishable from one another. If there exists a $z \in A^*$ defined at one of P and Q but not the other, then clearly P and Q are distinguishable. Thus we can assume that for each $z \in A^*$, the path labeled z is defined at P if and only if it is defined at Q. In this case, we must show that there exists a $z \in A^*$ such that the path starting at P labeled z ends at a terminal state if and only if the path starting at Q labeled z does not end at a terminal state.

First suppose $\mathcal{S}(w)$ satisfies the extendability condition for a. In this case both P and Q are distinguishable states of $\mathcal{S}(w)$. Hence without loss of generality, there exists a $z' \in A^*$ such that the path labeled z' starting at P ends at a terminal state of $\mathcal{S}(w)$ while the path labeled z' starting at Q does not end at a terminal state of $\mathcal{S}(w)$. In this case, in $\mathcal{E}_a(\mathcal{S}(w))$, the path $z = z'a$ starting at P ends at a terminal state of $\mathcal{E}_a(\mathcal{S}(w))$ while the path labeled z starting at Q does not end at a terminal state of $\mathcal{E}_a(\mathcal{S}(w))$. Thus P and Q are distinguishable in $\mathcal{E}_a(\mathcal{S}(w))$.

Next suppose that $\mathcal{S}(w)$ does not satisfy the extendability condition for a. So we consider the states P and Q of the intermediate automaton $\mathcal{S}(w)'$. Again, if P and Q are distinguishable states of $\mathcal{S}(w)'$, then the above argument shows that they are distinguishable states of $\mathcal{E}_a(\mathcal{S}(w))$. So it remains to consider the case in which $P = Y$ and $Q = Y'$ where Y and Y' are the two indistinguishable states of $\mathcal{S}(w)'$. But in this case, Y' is a terminal state of $\mathcal{E}_a(\mathcal{S}(w))$ while Y is not. Hence again P and Q are distinguishable in $\mathcal{E}_a(\mathcal{S}(w))$.   □

As a consequence we recover the result reported in Proposition 1 on the upper bound on the number of states of $\mathcal{S}(w)$, the lower bound being obvious.

Corollary 5. *Let $w \in A^*$ and $a \in A$. Then $\#(Q_{wa}) \leq \#(Q_w) + 2$. Moreover $\#(Q_{wa}) = \#(Q_w) + 1$ if and only if $\mathcal{S}(w)$ satisfies the extendability condition for the letter a. Hence if $|w| \geq 2$, we have $|w| + 1 \leq \#(Q_w) \leq 2|w| - 1$.*

Now we consider the language L of all binary words such that $\mathcal{S}(w)$ has $|w| + 1$ states. The following lemma establishes a close relation between the extendability of the words in L and the extendability condition for the automaton $\mathcal{S}(w)$.

Lemma 1. *Suppose $w \in L$ and $a \in \{0, 1\}$. Then $wa \in L$ if and only if $\mathcal{S}(w)$ satisfies the extendability condition for the letter a.*

Proof. Suppose $wa \in L$; then $\mathcal{S}(w)$ is obtained from $\mathcal{S}(wa)$ in the way above described. Suppose in $\mathcal{S}(w)$ there exist edges labeled a from X_i and X_j to Y. Then X_i is a terminal state of $\mathcal{S}(w)$ if and only if Y is a terminal state of $\mathcal{S}(wa)$. Similarly for X_j. Hence X_i is a terminal state of $\mathcal{S}(w)$ if and only if X_j is a terminal state of $\mathcal{S}(w)$. Thus $\mathcal{S}(w)$ satisfies the extendability condition for the letter a.

Conversely, suppose that the extendability condition holds. Since $w \in L$, the automaton $\mathcal{S}(w)$ has $|w| + 1$ states. From the previous proposition it follows that $\mathcal{S}(wa)$ has one state nore than $\mathcal{S}(w)$, so it has $|wa| + 1$ states. Hence $wa \in L$.   □

Example 5. Let w be any word in L beginning in 001 and ending in 10. Then $\mathcal{S}(w)$ must begin in

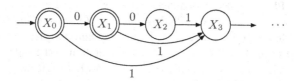

X_1 is a terminal state since w ends in 0 but X_2 is not terminal since 00 is not a suffix of w. Yet both X_1 and X_2 have an edge labeled 1 to X_3. Thus $\mathcal{S}(w)$ does not satisfy the extendability condition for 1. Hence $w1 \notin L$.

3   Suffix Automaton of Words in PER

In this section we give an explicit construction of suffix automaton of words that are palindromic prefixes of standard Sturmian words.

Sturmian words were introduced in the forties by Morse and Hedlund (cf. [15]). They are defined as the infinite binary sequences having exactly $n + 1$ distinct factors of length n. Their numerous properties have lead to a great development of many fields of research. Such a versatility explains also the existence of many equivalent definitions (see [1]).

In particular, a Sturmian word can be defined by considering the intersections with a squared-lattice of a semiline having a slope which is an irrational number. A vertical intersection is denoted by the letter 0, an horizontal intersection by 1 and the intersection with a corner by 01 or 10. If the semiline starts from the origin the corresponding Sturmian words is called characteristic. Since the language of factors of a Sturmian word depends only on the slope of the corresponding semiline, characteristic words capture most of the properties of Sturmian words.

Characteristic Sturmian words can be constructed by a family of finite words called *standard Sturmian words* (or simpler *standard words*), in the sense that every characteristic word is the limit of a sequence of standard words (cf. [1]).

Let $d_1, d_2 \ldots d_n, \ldots$, be a sequence of natural integers, with $d_1 \geq 0$ and $d_i > 0$ for $i = 2, \ldots, n, \ldots$. Consider the following sequence of words $\{s_n\}_{n \geq 0}$: $s_0 = 1$, $s_1 = 0$, and $s_{n+1} = s_n^{d_n} s_{n-1}$ for $n \geq 1$.

Each finite word s_n in the sequence is called a *standard word*. It is univocally determined by the finite sequence $(d_0, d_1, \ldots, d_{n-1})$. We denote by STAND the language of all standard words.

Standard words have several characterizations (cf. [12,8] and references therein). A characterization, that we use from here on, is based on the notion of periodicity of words and it is closely related to Fine and Wilf's theorem (cf. [11]). Let $w \in A^*$ and $\Pi(w)$ be the set of all periods of w. The set PER is defined as all words w having two periods $p, q \in \Pi(w)$ which are coprime and such that $|w| = p + q - 2$. Thus, a word w belongs to PER if it is a power of a single letter or if it is a word of maximal length for which the theorem of Fine and Wilf does not apply. In [9] it is proved that

$$\text{STAND} = A \cup \text{PER}\{01, 10\}.$$

The words in PER are also called *central words*. Remind that (cf. [9,1] and references therein) *a word w belongs to* PER *if and only if w is a palindromic prefix of a standard word*. Moreover it is proved in [9] that *a palindromic prefix of a word belonging to* PER, *is also in* PER. Finally, in [4] a computation of the index of the Nerode equivalence of the language of all factors of a word in PER is given.

Let $w = w_1 w_2 \ldots w_n \in$ PER, i.e. w is a palindromic prefix of length n of a standard word. In this section we show that $w \in L$ and give an explicit description of $\mathcal{S}(w)$.

More precisely, we will prove that $\mathcal{S}(w)$ is constructed as follows: start with the skeletal sub-automaton of $n + 1$ states $X_0, X_1, \ldots, X_n$ and direct edges labeled w_i from X_{i-1} to X_i as shown below.

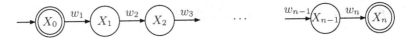

1. We modify some of the labels in the skeletal sub-automaton as follows: change the label w_i to $\widehat{w_i}$ if and only if either $i = 1$ or $w_1 \ldots w_{i-1}$ is a palindrome. Otherwise leave the label w_i the same. For each $a \in \{0, 1\}$, we say the state X_i is of *type* a if and only if the direct edge coming out of X_i is $\hat{a}$, and is of *co-type* a if and only if the direct edge coming into X_i is $\hat{a}$.
2. For each $a \in \{0, 1\}$, we adjoin a bypass edge labeled a from each state of type $1 - a$ to the next state of co-type a (if such state exists).
3. Finally the state X_i is a terminal state if and only if $i = n$ or if X_i is a state of type a for some $a \in \{0, 1\}$.

Example 6. For example, consider the palindromic prefix $w = 001000100$ of the Fibonacci word. We modify the edges labels of skeletal sub-automaton according to (1) to obtain the automaton in Figure 8.

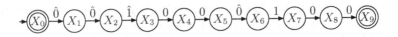

Fig. 8. Skeletal sub-automaton of $\mathcal{S}(001000100)$

Finally we add the bypass edges according to (2) and assign the terminal states according to (3) to obtain the automaton in Figure 9.

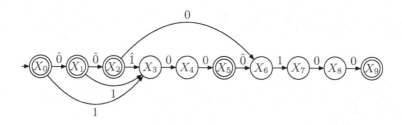

Fig. 9. The final automaton $\mathcal{S}(001000100)$

With the following proposition we show that the above construction generates the suffix automaton $\mathcal{S}(w)$ where of course it is understood that any label of the form $\hat{a}$ is replaced with the label a.

Proposition 7. *Let* $w = w_1 w_2 \ldots w_n$ *be a word in* PER. *Then the automaton defined according to items (1), (2) and (3) above defines the suffix automaton* $S(w)$. *In particular, we have that* $w \in L$.

Proof. We proceed by induction on $|w|$. If $|w| = 1$, so $w = a$, $a \in \{0, 1\}$. Then we obtain the automaton in Fig. 10 as required. Note that X_0 is of type a while X_1 is of co-type a. More generally if $w = a^k$ then the above construction gives rise to the automaton shown in Figure 11. It is easy to see that this automaton is in fact the suffix automaton of a^k.

Thus, we can assume that $|w| > 1$ and w contains both 0 and 1. We write $w = ubv$, $b \in \{0, 1\}$, where u is the longest palindromic proper prefix of w. If b does not occur in u then w is of the form $w = a^k b a^k$ in which case our construction yields the automaton shown in Figure 12.

Finally suppose b occurs in u. Since u is a palindromic prefix, u can be factorized as $u = zbv$, where z is the longest palindromic proper prefix followed by b. We write $z = z_1 z_2 \ldots z_k$ and $v = v_1 v_2 \ldots v_l$. By induction hypothesis (since u is a palindromic prefix of word in PER of length smaller than $|w|$), $\mathcal{S}(u)$ has the required structure shown in Figure 13. The state X_k is the last state of type b. Moreover, if a state X_j between X_k and X_{k+l+1} is a terminal state, then such X_j is of type $1 - b$, and in this case there would be a bypass edge from X_k to X_j. Note that any suffix of u of length greater than k is represented by a path in $\mathcal{S}(u)$ beginning at X_0 and ending at some terminal state X_j for some $j > k$. Thus such X_j is either X_{k+l+1} or a state of type $1 - b$.

Our automaton construction (applied to $w = ubv$) may be obtained by extending $\mathcal{S}(u)$ as follows: adjoin to $\mathcal{S}(u)$ the automaton shown in Figure 14 by a direct edge labeled $\hat{b}$ from X_{k+l+1} to Y_0.

$$\overset{X_0}{\bigcirc} \overset{\hat{a}}{\longrightarrow} \overset{X_1}{\bigcirc}$$

Fig. 10. Suffix automaton of $w = a$

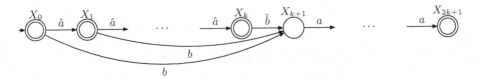

Fig. 11. Suffix automaton of a^k

Fig. 12. Suffix automaton of $w = a^k b a^k$

Also put an edge labeled b from any state X_j ($j > k$) of type $1 - b$ to Y_0. We claim that the resulting automaton (call it $\mathcal{A}$) defines $\mathcal{S}(w)$. In fact, if s is a suffix of w of length less or equal to $|u|$, then s is a suffix of u, and hence is represented by a path in $\mathcal{A}$ starting at X_0 and terminating at some terminal state X_i for some $1 \leq i \leq k+l+1$; and, conversely, any such path defines a suffix of u and hence a suffix of w of length smaller than or equal to $|u|$.

If s is a suffix of w of length greater than $|u|$, then we can write $s = xbv$ where x is a suffix of u of length greater than $|z|$. Thus x is represented by a path in $\mathcal{A}$ beginning at X_0 and terminating at some terminal state X_j for some $k < j \leq k+l+1$. But in our construction of $\mathcal{A}$, there is an edge labeled b from X_j to Y_0. Thus s is represented by a path in $\mathcal{A}$ from X_0 to Y_l. Conversely, any path σ from X_0 to Y_l can be factorized as $\sigma = \sigma_1 b v_1 \ldots v_l$, where σ_1 is a path from X_0 to a terminal state X_j ($k < j \leq k+l+1$). Hence σ_1 corresponds to a suffix of u and hence σ to a suffix of w. $\qquad\square$

4   Main Result

In this section we give some characterizations of the prefixes of standard words. In particular, in Theorem 1 we prove that the binary words whose suffix automaton has minimal number of states are exactly the prefixes of standard words.

Theorem 1. *A word w belongs to L if and only if w is a prefix of a standard word.*

Proof. First, if w is a prefix of a standard word, then w is a prefix of a palindrome u which is also a prefix of a standard word. In the previous section we saw that $u \in L$. Since w is a prefix of u, by Corollary 2 it follows that $w \in L$.

We now show the converse; let us suppose $w \in L$. We proceed by induction on $|w|$. If $|w| = 1$ then w is a prefix of a standard word. Next suppose $w = ua \in L$ for some $a \in \{0, 1\}$. Since L is closed under prefixes, it follows that $u \in L$. Since $|u| < |w|$ we can apply our induction hypothesis to u and conclude that u is a prefix of a standard word. If u is a palindrome, then both $u0$ and $u1$ are prefixes of a standard word (cf. [9]), hence w is a prefix of a standard word. So, suppose u is not a palindrome. In this case there exists a unique $a' \in \{0, 1\}$ for which ua' is a prefix of a standard word. So,

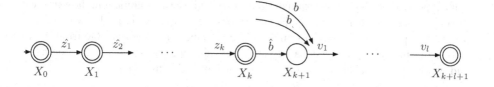

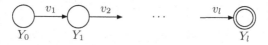

Fig. 13. Suffix automaton of $u = zbv$ (z_k as well as some of the v_i may be hatted)

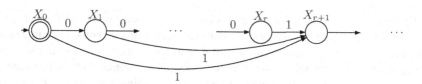

Fig. 14. Sub-automaton of $\mathcal{S}(v_1 v_2 \ldots v_l)$

$ua' \in L$. We want to show $a = a'$. To do this we will show that $ub \notin L$ for $b \neq a'$. Let u^+ denote the right palindromic closure of u, i.e. u^+ is the shortest palindrome having u as a prefix. Then u^+ is a palindromic prefix of a standard word and so $\mathcal{S}(u^+)$ has the structure given in the previous section. $\mathcal{S}(u)$ can be viewed as a sub-automaton of $\mathcal{S}(u^+)$ except that a terminal state of $\mathcal{S}(u)$ may not be a terminal state of $\mathcal{S}(u^+)$. Recall that the terminal states in $\mathcal{S}(u^+)$ are all states of type 0, all states of type 1 and the final state.

Without loss of generalities we can assume that u begins in $0^r 1$ (for some $r \geq 1$). So, $\mathcal{S}(u)$ begins in the automaton shown in Figure 15.

Fig. 15. Suffix automaton of $u = 0^r 1 \ldots$

If for some X_i (with $1 \leq i < r$) is not a terminal state of $\mathcal{S}(u)$, then neither is X_r, in other words, if 0^i is not a suffix of u for some $1 \leq i \leq r$, then neither is 0^r. But then $\mathcal{S}(u)$ does not satisfy the extendability condition for 1 since both X_0 and X_r have an edge labeled 1 to X_{r+1}, but X_0 is a terminal state while X_r is not. Hence $u1 \notin L$. Note that we can deduce that if $w = u0$ then either $w = 0^k$ or w must begin in $0^r 1$. In this case $u1 \notin L$. Similarly, if $w = u1$, then either $w = 1^k$ or w must begin in $1^r 0$. In this case $u0 \notin L$. Hence w is a prefix of a standard word.

So suppose $X_0, X_1, \ldots, X_r$ are all terminal states of $\mathcal{S}(u)$. We claim that there exists a letter $b \in \{0, 1\}$ and a state X_k of type b in $\mathcal{S}(u)$ which is not a terminal state and such that for every other state X_i for $i < k$, if X_i is either of type 0 or of type 1, then X_i is a terminal state of $\mathcal{S}(u)$. In other words, X_k is the first state in $\mathcal{S}(u)$ which is of some type and which is not a terminal state. If we believe for a moment in the existence of such a state X_k, then $\mathcal{S}(u)$ has the sub-automaton shown in Figure 16, where X_j is a state of type $1 - b$ which is a terminal state, and there is a bypass edge from X_j to X_{k+1}. But then $\mathcal{S}(u)$ does not satisfy the extendability condition for b since both X_j and X_k have edges labeled b to X_{k+1} but X_j is terminal while X_k is not.

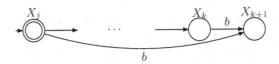

Fig. 16. Suffix automaton of $u = 0^r 1 \ldots$

So, it remains to show the existence of this first state X_k which is not terminal but which is either of type 0 or of type 1. We can write $u = u'cv$, with $c \in \{0, 1\}$ and where u' is the longest palindromic prefix of u. Then $u^+ = (u'c)^+$ so $|cv| \le |u'|$. Writing $u' = u_1 u_2 \ldots u_t$ we have that $\mathcal{S}(u)$ must begin in the automaton described in Figure 17.

Fig. 17. First states in $\mathcal{S}(u)$

We claim that X_t (which is of type c) is not a terminal state of $\mathcal{S}(u)$. If it were, then $u' = u_1 u_2 \ldots u_t$ would be a suffix of u. Thus the palindrome u' would be both a prefix and a suffix of u and two occurrences of u' in u would overlap. This would imply that u is a palindrome (a contradiction). This shows that $\mathcal{S}(u)$ contains at least one state X_i (for some $i > r$) which is of type 0 or 1 and is not a terminal state. Hence there exists at least such i. □

Let w be a word on a finite alphabet A, and $\mathcal{S}(w)$ the suffix automaton of w. The following remark points out some properties of the suffix automaton of any word w (cf. [7]).

Remark 2. Let X be any state of $\mathcal{S}(w)$, and u and v be any two paths in $\mathcal{S}(w)$ beginning at the initial state X_0 and terminating at X. Then, by minimality of $\mathcal{S}(w)$ we have that (i) either u is a suffix of v or v a suffix of u. (ii) Every occurrence of u in w is an occurrence of v in w and vice versa.

Recall that a factor u of a finite word w is called *left special* if there exist two distinct letters x and y such that both xu and yu are factors of w. From the above remark one easily deduces the following:

Proposition 8. *Let w a word in A^* and $\mathcal{S}(w)$ its suffix automaton.*

a) *The number of states of $\mathcal{S}(w)$ is $|w| + 1$ if and only if any left special factors of w is a prefix of w, i.e., if au and bu are each factors of w with $a \neq b$, then u is a prefix of w.*

b) *For each factor u of w, let $p_w(u)$ denote the shortest prefix of w ending in u. Then $\mathcal{S}(w)$ has $|w| + 1$ states if and only if for each factor u of w, we have that each prefix of w ending in u also ends in $p_w(u)$, i.e., every occurrence of u is an occurrence of $p_w(u)$.*

Thus in case w is a binary word, by Theorem 1 and Proposition 8, we have the following characterization of prefixes of standard words:

Corollary 6. *Let w be a binary word. The following statements are equivalent:*

1. *w is a prefix of a standard word;*
2. *every left special factor of w is a prefix of w*
3. *for every factor u of w, every occurrence of u in w is an occurrence of $p_w(u)$.*

Many of the ideas involved in the proof of the main theorem and in this paper apply more generally to the case of words on larger alphabets A. Further research will be devoted to deeply investigate the bounds on the size of suffix automaton of words over a finite alphabet A. Similar results can be obtained for the factor automaton of a word.

References

1. Berstel, J., Séébold, P.: Sturmian words. In: Lothaire, M. (ed.) Algebraic Combinatorics on Words. Encyclopedia of Mathematics and its Applications, vol. 90, ch. 2, pp. 45–110. Cambridge University Press, Cambridge (2002)
2. Blumer, A., Blumer, J., Ehrenfeucht, A., Haussler, D., Chen, M.T., Seiferas, J.: The smallest automaton recognizing the subword of a text. Theoret. Comput. Sci. (40), 31–55 (1985)
3. Blumer, A., Blumer, J., Ehrenfeucht, A., Haussler, D., McConnell, R.: Linear size finite automata for the set of all subwords of a word: An outline of results. Bull. Eur. Assoc. Theoret. Comput. Sci. (21), 12–20 (1983)
4. Carpi, A., de Luca, A.: Words and special factors. Theoret. Comput. Sci. (259), 145–182 (2001)
5. Crochemore, M.: Transducers and repetions. Theoret. Compu. Sci. 45, 63–86 (1986)
6. Crochemore, M.: Structures for indexes. In: Lothaire, M. (ed.) Applied Combinatorics on Words, ch. 2, pp. 106–163. Cambridge University Press, Cambridge (2005)
7. Crochemore, M., Hancart, C.: Automata for matching patterns. In: Rozenberg, G., Salomaa, A. (eds.) Handbook of Formal Languages. Linear Modeling: Background and Application, vol. 2, ch. 9, pp. 399–462. Springer, Heidelberg (1997)
8. de Luca, A.: Combinatorics of standard sturmian words. In: Mycielski, J., Rozenberg, G., Salomaa, A. (eds.) Structures in Logic and Computer Science. LNCS, vol. 1261, pp. 249–267. Springer, Heidelberg (1997)
9. de Luca, A., Mignosi, F.: Some combinatorial properties of sturmian words. Theoret. Comput. Sci. (136), 361–385 (1994)
10. Epifanio, C., Mignosi, F., Shallit, J., Venturini, I.: Sturmian graphs and a conjecture of moser. In: Calude, C.S., Calude, E., Dinneen, M.J. (eds.) DLT 2004. LNCS, vol. 3340, pp. 175–187. Springer, Heidelberg (2004)

11. Fine, N.J., Wilf, H.S.: Uniqueness theorem for periodic functions. Proc. Am. Mathematical Society (16), 109–114 (1965)
12. Mantaci, S., Restivo, A., Sciortino, M.: Burrows-Wheeler transform and Sturmian words. Informat. Proc. Lett. 86, 241–246 (2003)
13. McCreight, E.M.: A space-economical suffix tree construction algorithm. Journal of the ACM 23(2), 262–272 (1976)
14. Mignosi, F.: Infinite words with linear subword complexity. Theoret. Comput. Sci. (65), 221–242 (1989)
15. Morse, M., Hedlund, G.A.: Symbolic dynamics II: Sturmian trajectories. Amer. J. Math. (62), 1–42 (1940)
16. Ukkonen, E.: Constructing suffix trees online in linear time. In: Proceedings of the IFIP 12th World Computer Congress, pp. 484–492, (1992)
17. Weiner, P.: Linear pattern matching algorithm. In: Proceedings of the 14th Annual IEEE Symposium on Switching and Automata Theory.

Fine Hierarchy of Regular Aperiodic ω-Languages

Victor L. Selivanov[*]

A.P.Ershov Institute of Informatics Systems
Siberian Division Russian Academy of Sciences
and
Theoretische Informatik, Universität Würzburg
selivanov@informatik.uni-wuerzburg.de

Abstract. We develop a theory of regular aperiodic ω-languages in parallel with the theory around the Wagner hierarchy. In particular, we characterize the Wadge degrees of regular aperiodic ω-languages, find an effective version of the Wadge reducibility adequate for this class of languages and prove "aperiodic analogs" of the Büchi-Landweber determinacy theorem and of Landweber's description of regular open and regular G_δ sets.

Keywords: Automaton, acceptor, transducer, regular aperiodic ω-language, Wagner hierarchy, reducibility.

1 Introduction

This paper is devoted to the theory of infinite behavior of computing devices that is of primary importance for theoretical and practical computer science. More exactly, we consider topological aspects of this theory in the simplest case of finite automata.

A series of papers in this direction culminated with the paper [Wag79] giving in a sense the finest possible topological classification of regular ω-languages (i.e., of subsets of X^ω for a finite alphabet X recognized by finite automata) known as the Wagner hierarchy. In [Se94a, Se95, Se98] the Wagner hierarchy of regular ω-languages was related to the Wadge hierarchy and to the author's fine hierarchy [Se95a]. This provided new proofs of results in [Wag79] and yielded some new results on the Wagner hierarchy. See also an alternative approach [CP97, CP99, DR06].

Later some results from [Wag79, Se98] were extended to more complicated computing devices. In particular, the Wadge degrees of deterministic context-free ω-languages, of ω-languages recognized by deterministic Turing machines, and of infinite tree languages recognized by deterministic tree automata were

---

[*] Supported by the Alexander von Humboldt Foundation, by DFG Mercator program, and by DFG-RFBR Grant 06-01-04002.

T. Harju, J. Karhumäki, and A. Lepistö (Eds.): DLT 2007, LNCS 4588, pp. 399–410, 2007.

determined respectively in [D03, Se03, M06]. Note that in all these three cases some important properties of the Wagner hierarchy are either false or still open.

In this paper, we develop a complete analog of the theory from [Wag79, Se98] for the class $\mathcal{A}$ of regular aperiodic ω-languages. The class $\mathcal{A}$ is certainly the most important subclass of $\mathcal{R}$ which has several remarkable characterizations and is essential in the field of specification and verification of finite-state systems. To explain our results, let us recall some results on the Wagner hierarchy in more details. In [Wag79] the following results (among others) were established:

1) The structure $(\mathcal{R}; \leq_{CA})$ of regular set under reducibility by continuous functions is almost well-ordered with order type ω^ω, i.e. there are $A_\alpha \in \mathcal{R}$, $\alpha < \omega^\omega$, such that $A_\alpha <_{CA} A_\alpha \oplus \overline{A}_\alpha <_{CA} A_\beta$ for $\alpha < \beta < \omega^\omega$ and any regular set is CA-equivalent to one of sets $A_\alpha, \overline{A}_\alpha, A_\alpha \oplus \overline{A}_\alpha (\alpha < \omega^\omega)$.

2) CA-reducibility coincides on $\mathcal{R}$ with DA-reducibility, i.e. the reducibility by functions computed by deterministic asynchronous finite transducers, and $\mathcal{R}$ is closed under DA-reducibility.

3) Any level $\mathcal{R}_\alpha = \{C \mid C \leq_{DA} A_\alpha\}$ of the Wagner hierarchy is decidable.

In [Se98] the following additional facts (among others) about the Wagner hierarchy were established:

4) Any class $\mathcal{R}_\alpha$ has a natural set-theoretic description in terms of classes $\mathcal{L}_0$ of regular open sets and $\mathcal{L}_1$ of regula F_σ sets. In particular, there is a Boolean term $t(x_1, \ldots, x_n, y_1, \ldots, y_n)$ with $\mathcal{R}_\alpha = t(\mathcal{L}_0, \mathcal{L}_1)$, where $t(\mathcal{L}_0, \mathcal{L}_1)$ is the set of values of t when $x_1, \ldots, x_n$ range over $\mathcal{L}_0$ and $y_1, \ldots, y_n$ range over $\mathcal{L}_1$.

5) For every term t as above, the set $t(\mathcal{L}_0, \mathcal{L}_1)$ coincides with one of classes $\mathcal{R}_\alpha$ or their duals.

6) If a regular set R is represented as $R = t(B_1, \ldots, B_n, C_1, \ldots, C_n)$ for some term t as above, open sets $B_1, \ldots, B_n$ and F_σ-sets $C_1, \ldots, C_n$, then $R = t(B_1', \ldots, B_n', C_1', \ldots, C_n')$ for some $B_1', \ldots, B_n' \in \mathcal{L}_0$ and $C_1', \ldots, C_n' \in \mathcal{L}_1$.

In this paper we show that the sets A_α from 1) may be chosen from $\mathcal{A}$, and thus the Wadge degrees (as well as DA-degrees) of sets in $\mathcal{A}$ are the same as those of sets in $\mathcal{R}$ (this result is not hard and, probably, may be obtained by a careful analysis of the corresponding automata in [Se98]; our approach here is different). We will find a reducibility $\leq_{AA}$ related to the class $\mathcal{A}$ in exactly the same way as DA-reducibility is related to $\mathcal{R}$. Thus, we obtain analogs of 1)—3). We also show that classes $\mathcal{A}_\alpha = \{C \mid C \leq_{AA} A_\alpha\}$ of the fine hierarchy of regular aperiodic languages have properties 4)—6), if we take the classes $\mathcal{K}_0$ of regular aperiodic open and $\mathcal{K}_1$ of regular aperiodic F_σ-sets in place of the classes $\mathcal{L}_0$ and $\mathcal{L}_1$, respectively. We obtain also some facts of independent interest, e.g. "aperiodic analogs" of the Büchi-Landweber determinacy theorem and of Landweber's description of regular open and regular G_δ sets (cf. the algebraic approach in [C00]). Note that these results actually subsume the corresponding results about regular sets, hence this paper formally subsumes many results from [Wag79, Se98]. Note that several proofs in this paper are not straightforward generalizations of those in [Wag79, Se98] because this time we have to take care that corresponding automata are aperiodic (or counter-free) which is often not obvious.

The rest of the paper is organized as follows. In Section 2 we collect notation and known facts we will rely upon. In Section 3 we establish some necessary facts about aperiodic automata (known also as counter-free automata). In Section 4 we obtain the "aperiodic analog" of the Büchi-Landweber determinacy theorem. In Section 5 we establish some facts about regular aperiodic sets in the Borel hierarchy. Section 6 deals with the fine hierarchy of regular aperiodic sets, while Section 7 — with reducibilities on such sets.

Because of space bounds, we omit all the proofs (most of which are short) in this conference paper. They may be found in the full version [Se06].

2 Notation and Reminder

For a set S, $P(S)$ is the class of subsets of S. For a class $\mathcal{C} \subseteq P(S)$, $\check{\mathcal{C}}$ is the dual class $\{\overline{C} \mid C \in \mathcal{C}\}$ and $B(\mathcal{C})$ is the Boolean closure of $\mathcal{C}$.

Fix a finite alphabet X containing more than one symbol (for simplicity we may assume that $X = \{x \mid x < k\}$ for a natural number $k > 1$, so $0, 1 \in X$). Let X^* and X^ω denote respectively the sets of all words and of all ω-words (i.e. sequences $\alpha : \omega \to X$) over X. The empty word is denoted by ε. Let $X^+ = X^* \setminus \{\varepsilon\}$ and $X^{\leq \omega} = X^* \cup X^\omega$. For $n < \omega$, let X^n be the set of words of length n. Sets of words $X^{\leq n}$ and $X^{>n}$ are defined in the same way. For $X = \{0, 1\}$ we write 2^* in place of X^*, 2^ω in place of X^ω and so on.

We use some almost standard notation concerning words and ω-words, so we are not too casual in reminding it here. For $w \in X^*$ and $\xi \in X^{\leq \omega}$, $w \sqsubseteq \xi$ means that w is a substring of ξ, $w \cdot \xi = w\xi$ denote the concatenation, $l = |w|$ is the length of $w = w(0) \cdots w(l-1)$. For $w \in X^*, W \subseteq X^*$ and $A \subseteq X^{\leq \omega}$, let $w \cdot A = \{w\xi \mid \xi \in A\}$ and $W \cdot A = \{w\xi \mid w \in W, \xi \in A\}$. For $k, l < \omega$ and $\xi \in X^{\leq \omega}$, let $\xi[k, l) = \xi(k) \cdots \xi(l-1)$ and $\xi[k] = \xi[0, k)$. For $u \in X^*$ and $n < \omega$, u^n denote the concatenation of n copies of the word u. Our notation does not distinguish a word of length 1 and the corresponding letter.

Note that usually we work with the fixed alphabet X but sometimes we are forced to consider several alphabets simultaneously; in this case we denote the alphabets by Y, Z, possibly with indices, and include the alphabets in the corresponding notation. The "fixed-alphabet mode" is the default one.

The set X^ω carries the Cantor topology with the open sets $W \cdot X^\omega$, where $W \subseteq X^*$. Continuous functions in this topology are called also CA-functions. A CS-function is a function $f : X^\omega \to Y^\omega$ satisfying $f(\xi)(n) = \phi(\xi[n+1])$ for some $\phi : X^+ \to Y$. A delayed CS-function is a function $f : X^\omega \to Y^\omega$ satisfying $f(\xi)(n) = \phi(\xi[n])$ for some $\phi : X^* \to Y$. Every delayed CS-function is a CS-function, and every CS-function is a CA-function. In descriptive set theory CS-functions are known as Lipschitz functions. All three classes of functions are closed under composition.

Let $\mathcal{B}$ denote the class of Borel subsets of X^ω, i.e. the least class containing the open sets and closed under complement and countable union. Borel sets are organized in the Borel hierarchy the lowest levels of which are as follows: G and F are the classes of open and closed sets, respectively; G_δ (F_σ) is the class of countable intersections (unions) of open (resp. closed) sets; $G_{\delta\sigma}$ ($F_{\sigma\delta}$) is the class of countable unions (intersections) of G_δ- (resp. of F_σ-) sets, and so on. In the modern notation of hierarchy theory, $\Sigma_1^0 = G$, $\Sigma_2^0 = F_\sigma$, $\Sigma_3^0 = G_{\delta\sigma}$, $\Sigma_4^0 = F_{\sigma\delta\sigma}$ and so on, Π_n^0 is the dual class for Σ_n^0, and $\Delta_n^0 = \Sigma_n^0 \cap \Pi_n^0$. For any $n > 0$, the class Σ_n^0 contains $\emptyset, X^\omega$ and is closed under countable unions and finite intersections, while the class Δ_n^0 is a Boolean algebra. For any $n > 0$, $\Sigma_n^0 \cup \Pi_n^0 \subseteq \Delta_{n+1}^0$, and $\Sigma_n^0 \not\subseteq \Pi_n^0$.

By *automaton* (over X) we mean a triple $\mathcal{M} = (Q, X, f)$ consisting of a finite non-empty set Q of states, an input alphabet X and a transition function $f : Q \times X \to Q$. The transition function is naturally extended to the function $f : Q \times X^* \to Q$ defined by induction $f(q, \varepsilon) = q$ and $f(q, u \cdot x) = f(f(q, u), x)$, where $u \in X^*$ and $x \in X$. Similarly, we may define the function $f : Q \times X^\omega \to Q^\omega$ by $f(q, \xi)(n) = f(q, \xi[n])$. Note that in this paper we consider only deterministic finite automata. A *partial automaton* is defined in the same way, only now $f : Q \times X \to Q$ is a partial function.

Automata equipped with appropriate additional structures are used as acceptors (devises accepting words or ω-words) and transducers (devices computing functions on words or ω-words). A *word acceptor* is a triple $(\mathcal{M}, i, F)$ consisting of an automaton $\mathcal{M}$, an initial state i of $\mathcal{M}$ and a set of final states $F \subseteq Q$. Such an acceptor recognizes the language $L(\mathcal{M}, i, F) = \{u \in X^* \mid f(i, u) \in F\}$. Languages recognized by such acceptors are called *regular*. The languages recognized by partial automata are defined in the same way and coincide with the regular languages.

For the case of ω-words, there are several notions of acceptors of which we will use only three. A *Büchi acceptor* has the form $(\mathcal{M}, i, F)$ as above and recognizes the set $L_\omega(\mathcal{M}, i, F) = \{\xi \in X^* \mid In(f(i, \xi)) \cap F \neq \emptyset\}$, where $In(f(i, \xi))$ is the set of states which occur infinitely often in the sequence $f(i, \xi) \in Q^\omega$. A *Muller acceptor* has the form $(\mathcal{M}, i, \mathcal{F})$, where $\mathcal{M}, i$ are as above and $\mathcal{F} \subseteq P(Q)$; it recognizes the set $L_\omega(\mathcal{M}, i, \mathcal{F}) = \{\xi \in X^\omega \mid In(f(i, \xi)) \in \mathcal{F}\}$. A *Mostowski acceptor* (known also as Rabin chain acceptor or parity acceptor) has the form $(\mathcal{M}, i, \Omega)$, where $\mathcal{M}, i$ are as above and $\Omega = (E_1, F_1, \ldots, E_n, F_n)$ for some $E_1 \subseteq F_1 \subseteq \cdots \subseteq E_n \subseteq F_n \subseteq Q$; it recognizes the set $L_\omega(\mathcal{M}, i, \Omega) = \{\xi \in X^\omega \mid \exists k(In(f(i, \xi)) \cap E_k = \emptyset \wedge In(f(i, \xi)) \cap F_k \neq \emptyset)\}$. It is well known that Muller and Mostowski acceptors recognize the same ω-languages; these are called *regular ω-languages* or just regular sets. The class $\mathcal{R}$ of all regular ω-languages is a proper subclass of $B(\Sigma_2^0)$ that in turn is a proper subclass of Δ_3^0. Büchi acceptors recognize a smaller class of sets, namely exactly the regular Π_2^0-sets.

A *synchronous transducer* (over X, Y) is a tuple $\mathcal{T} = (Q, X, Y, f, g, i)$, also written as $\mathcal{T} = (\mathcal{M}, Y, g, i)$, consisting of an automaton $\mathcal{M}$ as above, an initial state i and an output function $g : Q \times X \to Y$. The output function is naturally extended to a function $g : Q \times X^* \to Y^*$ and to a function $g : Q \times X^\omega \to Y^\omega$ denoted by the same letter. The transducer $\mathcal{T}$ *computes* the function $g_\mathcal{T} : X^\omega \to Y^\omega$ defined by $g_\mathcal{T}(\xi) = g(i, \xi)$. If the output function is of the form $g : Q \to Y$ (i.e., does not really depend on the second argument), then $\mathcal{T}$ is called a *delayed synchronous transducer*. An *asynchronous transducer* (over alphabets

X, Y) is defined as a synchronous transducer with only one exception: this time the output function g maps $Q \times X$ into Y^*. As a result, the value $g(q, \xi)$ is in $Y^{\leq \omega}$, and the function g_T maps X^ω into $Y^{\leq \omega}$. Usually we consider the case when g_T maps X^ω into Y^ω.

Functions computed by synchronous (delayed synchronous, asynchronous) transducers are called DS-functions (respectively delayed DS-functions and DA-functions). As is well known, all tree classes of functions are closed under composition, and every DS-function (delayed DS-function, DA-function) is a CS-function (delayed CS-function, CA-function).

We will study several reducibilities on subsets of X^ω. For $A, B \subseteq X^\omega$, A is said to be CA-*reducible* to B (in symbols $A \leq_{CA} B$), if $A = g^{-1}(B)$ for some CA-function $g : X^\omega \to X^\omega$. The relations $\leq_{DA}$, $\leq_{CS}$ and $\leq_{DS}$ on $P(X^\omega)$ are defined in the same way but using the other three classes of functions. The introduced relations on $P(X^\omega)$ are preorderings. The CA-reducibility is widely known as Wadge reducibility, and CS-reducibility as Lipschitz reducibility. The other two reducibilities are effective automatic versions of these. By $\equiv_{CA}$ we denote the induced equivalence relation which gives rise to the corresponding quotient partial ordering. Following a well established jargon, we call this ordering the structure of CA-degrees. The same applies to the other reducibilities (and to reducibilities to be introduced later). In the "alphabet-dependent mode", we say that $A \subseteq X^\omega$ is CA-reducible to $B \subseteq Y^\omega$, if $A = g^{-1}(B)$ for some CA-function $g : X^\omega \to Y^\omega$. Sometimes such variations are also of use.

The operation $A \oplus B = \{0 \cdot \alpha, i \cdot \beta \mid 0 < i < k, \alpha \in A, \beta \in B\}$ on subsets of X^ω, $X = \{0, \ldots, k-1\}$, induces the operation of least upper bound in the structures of degrees under all four reducibilities introduced above. Any level of the Borel hierarchy is closed under CA-reducibility (and thus under all four reducibilities) in the sense that every set reducible to a set in the level is itself in that level. The class $\mathcal{R}$ is closed under DA- and DS-reducibilities but is not closed under CA- and CS-reducibilities. Every Σ-level $\mathcal{C}$ (and also every Π-level) of the Borel hierarchy has a CA-complete set C which means that $\mathcal{C} = \{A \mid A \leq_{CA} C\}$.

More detailed information related to the introduced notions may be found in many sources including [Sta97, Th90, Th96, TB70, PP04].

3   Aperiodic Acceptors and Transducers

Here we formulate some facts on the regular aperiodic sets, the main object of this paper, and on the closely related aperiodic automata (known also as counter-free automata).

Aperiodic languages were characterized in several ways, in particular as: languages defined by extended regular star-free expressions; languages of words described by first-order sentences of a natural signature; languages of words satisfying a formula of linear time temporal logic; languages recognized by aperiodic acceptors. Similar characterizations exist also for regular aperiodic ω-languages (see e.g. [Th79, Th90, Th96] and references therein). It is well-known (and follows from the mentioned characterizations) that classes of regular aperiodic languages and ω-languages are closed under Boolean operations. For our paper, the characterization in terms of aperiodic acceptors is the most relevant. Let us recall the corresponding definition from [MP71].

Definition 1. *An automaton* $\mathcal{M} = (Q, X, f)$ *is aperiodic if for all* $q \in Q$, $u \in X^+$ *and* $n > 0$ *the equality* $f(q, u^n) = q$ *implies* $f(q, u) = q$. *This is clearly equivalent to say that for all* $q \in Q$ *and* $u \in X^+$ *there is* $m < \omega$ *with* $f(q, u^{m+1}) = f(q, u^m)$. *An acceptor (or a transducer) is aperiodic if so is the corresponding automaton.*

A basic fact proved in [MP71] states that a regular language $A \subseteq X^*$ is aperiodic iff it is recognized by an aperiodic acceptor. It is known and easy to check that the same is true for ω-languages and aperiodic Muller acceptors.

Let us establish some results on functions g_T computed by aperiodic transducers T (see Section 2).

Definition 2. *A function* $h : X^\omega \to Y^{\leq \omega}$ *is called an AA-function (an AS-function) if it is computed by an aperiodic asynchronous (respectively, aperiodic synchronous) transducer* T *over* X, Y, *i.e.* $h = g_T$.

We have the following natural closure property of the introduced classes of functions.

Proposition 1. *The classes of AA-functions and of AS-functions are closed under composition.*

We say that a set $A \subseteq X^\omega$ is AA-reducible (AS-reducible) to a set $B \subseteq Y^\omega$, in symbols $A \leq_{AA} B$ ($A \leq_{AS} B$) if $A = g^{-1}(B)$ for some AA-function (respectively, AS-function) $g : X^\omega \to Y^\omega$. For $X = Y$ we obtain relations $\leq_{AA}$ and $\leq_{AS}$ on $P(X^\omega)$ called AA- and AS-reducibilities.

Corollary 1. *The relations* $\leq_{AA}$ *and* $\leq_{AS}$ *on* $P(X^\omega)$ *are preorders. The corresponding quotient partial orderings of AA- and AS-degrees are upper semilattices under operation induced by the operations* $\oplus$ *from Section 2.*

Now we relate AA-transducers to regular aperiodic sets. We say that a set is *Büchi aperiodic* if it is recognized by an aperiodic Büchi acceptor.

Proposition 2. *The classes of Büchi aperiodic and of regular aperiodic* ω-*languages are closed under preimages of AA-functions and hence also under AA- and AS-reducibilities.*

4   Aperiodic Determinacy

Here we establish an "aperiodic version" of the Büchi-Landweber regular determinacy theorem. This may be of independent interest and is also an important technical tool to prove some results below.

We start with recalling some relevant information on Gale-Stewart games. Relate to any set $A \subseteq (X \times Y)^\omega$ the Gale-Stewart game $G(A)$ played by two opponents 0 and 1 as follows. Player 0 chooses a letter $x_0 \in X$, then player 1 chooses a letter $y_0 \in Y$, then 0 chooses $x_1 \in X$, then 1 chooses $y_1 \in Y$ and so on. Each player knows all the previous moves. After ω moves, 0 has constructed a word $\xi = x_0 x_1 \cdots \in X^\omega$ while 1 has constructed a word $\eta = y_0 y_1 \cdots \in Y^\omega$. Player 1 wins this particular play if $\xi \times \eta \in A$, otherwise 0 wins.

A *strategy for player* 1 (0) in the game $G(A)$ is a function $h : X^+ \to Y$ (respectively, $h : Y^* \to X$) that prompts the 1's move (respectively, the 0's move) for any finite string of the opponent's previous moves. It is clear that strategies for 1 (for 0) are in a bijective correspondence with CS-functions $h : X^\omega \to Y^\omega$ (respectively, with delayed CS-functions $h : Y^\omega \to X^\omega$); we identify strategies with the corresponding CS-functions.

A strategy h for player 1 (0) in the game $G(A)$ is *winning* if the player wins each play when following the strategy, i.e. if $\xi \times h(\xi) \in A$ for all $\xi \in X^\omega$ (respectively, $h(\eta) \times \eta \in \overline{A}$ for all $\eta \in Y^\omega$). A set $A \subseteq (X \times Y)^\omega$ is *determined* if one of the players has a winning strategy in $G(A)$. It is interesting and useful to know which sets are determined and, in case of determinacy, how complicated it is to find the winner and how complicated is his winning strategy.

One of the best results of descriptive set theory is the Martin determinacy theorem (see e.g. [Ke94]) stating that any Borel set is determined. Note that, since any regular set is Borel, this implies determinacy of regular sets. One of the best results of automata theory is the Büchi-Landweber regular determinacy theorem stating that for any regular set A the winner in $G(A)$ may be computed effectively, he has a winning strategy which is a DS-function, and the strategy is also computed effectively.

Now we are ready to establish the aperiodic version of the Büchi-Landweber theorem.

Theorem 1. *For any regular aperiodic set $A \subseteq (X \times Y)^\omega$ the winner of the game $G(A)$ may be computed effectively, he has a winning strategy which is an AS-function, and the strategy is also computed effectively.*

Corollary 2

(i) Let $A \subseteq (X \times Y)^\omega$ be regular aperiodic and player i has a winning CS-strategy in the game $G(A)$. Then i has also a winning AS-strategy.

(ii) Let $B \subseteq X^\omega$ and $C \subseteq Y^\omega$ be regular aperiodic and let $h : X^\omega \to Y^\omega$ be a CS-function satisfying $B = h^{-1}(C)$. Then $B = g^{-1}(C)$ for some AS-function $g : X^\omega \to Y^\omega$.

(iii) Let $B, C \subseteq X^\omega$ be regular aperiodic. Then $B \leq_{AS} C$ or $\overline{C} \leq_{AS} B$.

Define subsets K_0, K_1 of 2^ω which play an essential role in further considerations by $K_0 = 0^*1(0 \cup 1)^\omega$ and $K_1 = (0 \cup 1)^*0^\omega$.

Corollary 3

(i) The sets K_0 and $\overline{K}_1$ are Büchi aperiodic.

(ii) $K_0 \in \mathbf{\Sigma}_1^0 \setminus \mathbf{\Pi}_1^0$ and $K_1 \in \mathbf{\Sigma}_2^0 \setminus \mathbf{\Pi}_2^0$.

(iii) Any $\mathbf{\Sigma}_1^0$-set ($\mathbf{\Sigma}_2^0$-set) $B \subseteq X^\omega$ is CS-reducible to K_0 (respectively, to K_1).

(iv) *Any regular Σ_1^0-set (regular Σ_2^0-set) $B \subseteq X^\omega$ is DS-reducible to K_0 (respectively, to K_1).*

(v) *Any regular aperiodic Σ_1^0-set (regular aperiodic Σ_2^0-set) $B \subseteq X^\omega$ is AS-reducible to K_0 (respectively, to K_1).*

5   Aperiodic Σ_n^0-Sets

Here we establish some facts on regular aperiodic sets in the Borel hierarchy. Our first result is an "aperiodic version" of Landweber's characterizations of regular Σ_n^0-sets (see e.g. [Th90]).

Theorem 2

(i) *A regular aperiodic set $A \subseteq X^\omega$ is Σ_1^0 iff $A = W \cdot X^\omega$ for some regular aperiodic set $W \subseteq X^*$.*

(ii) *A regular aperiodic set $A \subseteq X^\omega$ is Π_2^0 iff A is Büchi aperiodic.*

Recall that a class $\mathcal{C}$ has the *reduction property*, if for all $C_0, C_1 \in \mathcal{C}$ there are disjoint $C_0', C_1' \in \mathcal{C}$ such that $C_i' \subseteq C_i$ for both $i < 2$ and $C_0 \cup C_1 = C_0' \cup C_1'$; such a pair (C_0', C_1') is called a *reduct* of (C_0, C_1). A class $\mathcal{C}$ has the *separation property*, if all disjoint sets $C_0, C_1 \in \mathcal{C}$ are separable by a set $B \in \mathcal{C} \cap \check{\mathcal{C}}$ (i.e. $C_0 \subseteq B \subseteq \overline{C_1}$). It is well known that if a class $\mathcal{C}$ has the reduction property then the dual class $\check{\mathcal{C}}$ has the separation property.

As is well known [Ke94], any level Σ_n^0 of the Borel hierarchy has the reduction property. In [Se98] it was shown that classes $\mathcal{L}_n = \mathcal{R} \cap \Sigma_{n+1}^0$, $n < 2$, have the reduction property (this trivially holds also for $n \geq 2$, because in this case we have $\mathcal{L}_n = \mathcal{R}$ and this class is closed under the Boolean operations). The next result is an "aperiodic version" of the last fact. Let $\mathcal{K}_n = \mathcal{A} \cap \Sigma_{n+1}^0$, where $\mathcal{A}$ is the class of regular aperiodic ω-languages. Since again $\mathcal{K}_n = \mathcal{A}$ is a Boolean algebra for $n \geq 2$, these classes trivially have the reduction property.

Theorem 3. *The classes $\mathcal{K}_0$ and $\mathcal{K}_1$ have the reduction property.*

Corollary 4. *The classes $\check{\mathcal{K}}_0$ and $\check{\mathcal{K}}_1$ have the separation property.*

Next we establish the "aperiodic version" of a theorem due to L. Staiger and K. Wagner [SW74].

Theorem 4. *Every regular aperiodic Δ_2^0-set is a Boolean combination of open regular aperiodic sets. In symbols, $\mathcal{K}_1 \cap \check{\mathcal{K}}_1 = B(\mathcal{K}_0)$.*

We conclude this section with a corollary which is crucial for the subsequent sections. Recall that *a base* (in a set S) is a sequence $\{L_n\}_{n<\omega}$ of sublattices of $(P(S); \cup, \cap, 0, 1)$ satisfying $L_n \cup \check{L}_n \subseteq L_{n+1}$. A base L is *reducible* if every L_n has the reduction property. A base L is *interpolable* if for all $n < \omega$ any two disjoint elements $A, B \in \check{L}_{n+1}$ are separable by a Boolean combination of elements of L_n.

In [Se98] we have shown that the base $\{\mathcal{L}_n\}_{n<\omega}$ is reducible and interpolable. From results of this section we obtain

Corollary 5. *The base $\{\mathcal{K}_n\}_{n<\omega}$ is reducible and interpolable.*

6   Fine Hierarchy

Here we describe some basic properties of the fine hierarchy of regular aperiodic ω-languages which is just the fine hierarchy over the base $\mathcal{K} = \{\mathcal{K}_n\}_{n<\omega}$. For background on the fine hierarchy see e.g. [Se95a, Se98]. Results of this section are particular cases of the corresponding general facts about the fine hierarchy [Se94, Se95].

In our definition of the fine hierarchy we use an operation $bisep$ that relates to classes $\mathcal{C}, \mathcal{D}_0, \mathcal{D}_1$ and $\mathcal{D}_2$ of ω-languages the class $bisep(\mathcal{C}, \mathcal{D}_0, \mathcal{D}_1, \mathcal{D}_2)$ of all sets $\{C_0 \cap D_0\} \cup (C_1 \cap D_1) \cup (\overline{C}_0 \cap \overline{C}_1 \cap D_2)$ where $C_i \in \mathcal{C}$, $D_j \in \mathcal{D}_j$ and $C_0 \cap C_1 = 0$.

Definition 3. *The fine hierarchy over* $\mathcal{K}$ *is the sequence* $\{\mathcal{A}_\alpha\}_{\alpha<\omega^\omega}$ *defined by induction on* $\alpha < \omega^\omega$ *as follows:*
   $\mathcal{A}_n = D_n(\mathcal{K}_0)$ *is the n-the level of the difference hierarchy over* $\mathcal{K}_0$ *for* $n < \omega$;
   $\mathcal{A}_{\omega^n} = D_n(\mathcal{K}_1)$ *for* $0 < n < \omega$;
   $\mathcal{A}_{\beta+\omega^n} = bisep(\mathcal{K}_0, \mathcal{A}_\beta, \check{\mathcal{A}}_\beta, \mathcal{A}_{\omega^n})$ *for* $0 < n < \omega$ *and* β *of the form* $\beta = \omega^n \cdot \beta_1$ *for some* $\beta_1, 0 < \beta_1 < \omega^\omega$;
   $\mathcal{A}_{\beta+1} = bisep(\mathcal{K}_0, \mathcal{A}_\beta, \check{\mathcal{A}}_\beta, \mathcal{A}_0)$ *for* $\omega \le \beta < \omega^\omega$.

The definition is correct since every non-zero ordinal $\alpha < \omega^\omega$ is uniquely representable in the form $\alpha = \omega^{n_0} + \cdots + \omega^{n_k}$ for a finite sequence $n_0 \ge \cdots \ge n_k$ of ordinals $< \omega$; applying it we subsequently get $\mathcal{A}_{\omega^{n_0}}, \mathcal{A}_{\omega^{n_0}+\omega^{n_1}}, \ldots, \mathcal{A}_\alpha$. Note that the definition applies to any base in place of the base $\mathcal{K}$ above. For this paper, the fine hierarchy $\{\mathcal{S}_\alpha\}$ over the base $\{\mathbf{\Sigma}_{n+1}^0\}$ and the fine hierarchy $\{\mathcal{R}_\alpha\}$ over the base $\mathcal{L} = \{\mathcal{L}_n\}_{n<\omega}$ are also relevant. The hierarchy $\{\mathcal{R}_\alpha\}$ coincides with the Wagner hierarchy [Se98]. Note also that our definition here slightly differs from (but is equivalent to) the definition of the fine hierarchy in [Se98].

Proof of the next result uses heavily the fact from the previous section that the base $\mathcal{K}$ is reducible and interpolable.

Theorem 5

   (i) *For all* $\alpha < \beta < \omega^\omega$, $\mathcal{A}_\alpha \cup \check{\mathcal{A}}_\alpha \subseteq \mathcal{A}_\beta$.
   (ii) $\bigcup_{\alpha<\omega^\omega} \mathcal{A}_\alpha = \mathcal{A}$.
   (iii) *For any limit ordinal* $\lambda < \omega^\omega$, $\mathcal{A}_\lambda \cap \check{\mathcal{A}}_\lambda = \bigcup_{\alpha<\lambda} \mathcal{A}_\alpha$.
   (iv) *For any* $\alpha < \omega^\omega$, $\mathcal{A}_{\alpha+1} \cap \check{\mathcal{A}}_{\alpha+1}$ *coincides with the class of sets* $(A \cap C) \cup (B \cap \overline{C})$ *where* $A \in \mathcal{A}_\alpha$, $B \in \check{\mathcal{A}}_\alpha$ *and* $C \in \mathbf{\Delta}_1^0 = \mathcal{K}_0 \cap \check{\mathcal{K}}_0$.
   (v) *For any* $\alpha < \omega^\omega$, $\check{\mathcal{A}}_\alpha$ *has the separation property.*

Next we give an alternative description of the fine hierarchy $\{\mathcal{A}_\alpha\}$. Recall that a *typed boolean term* is a term of signature $\{\cup, \cap, ^-, 0, 1\}$ with variables $v_n^0, v_n^1 (n < \omega)$. Variables v_n^0 are of type 0 while variables v_n^1 are of type 1. For a typed boolean term t and a base L, let $t(L_0, L_1)$ be the class of values of t when variables $v_n^i (n < \omega)$ of type i range over L_i, $i < 2$.

The next result shows the close relation of classes $t(\mathcal{K}_0, \mathcal{K}_1)$ to the fine hierarchy $\{\mathcal{A}_\alpha\}$. The result follows from the reducibility of the base $\mathcal{K}$. Similar facts for the bases $\{\mathbf{\Sigma}_{n+1}^0\}$ and $\mathcal{L}$ were established in [Se95a, Se98].

Theorem 6. *For every $\alpha < \omega^\omega$ one can effectively find a typed boolean term $t = t_\alpha$ such that $\mathcal{A}_\alpha = t(\mathcal{K}_0, \mathcal{K}_1)$. Conversely, for every typed boolean term t the class $t(\mathcal{K}_0, \mathcal{K}_1)$ coincides with one of classes $\mathcal{A}_\alpha, \check{\mathcal{A}}_\alpha$ for some $\alpha < \omega^\omega$, and the class is effectively computable from t.*

7 Reducibilities on $\mathcal{A}$

Here we establish the non-collapse property of the fine hierarchy $\{\mathcal{A}_\alpha\}_{\alpha<\omega^\omega}$ from the previous section and describe the structures of AA- and AS-degrees of regular aperiodic sets. First we will show that for every $\alpha < \omega^\omega$ the class $\mathcal{A}_\alpha$ has an AA-complete set.

Theorem 7. *For every $\alpha < \omega^\omega$ the class $\mathcal{A}_\alpha$ has an AA-complete set A_α.*

Corollary 6

(i) *The set A_α is DA-complete in $\mathcal{R}_\alpha$ and CA-complete in $\mathcal{S}_\alpha$.*
(ii) *The set $A_\alpha \oplus \bar{A}_\alpha$ is AA-complete in $(\mathcal{A}_{\alpha+1} \cap \check{\mathcal{A}}_{\alpha+1}) \setminus (\mathcal{A}_\alpha \cup \check{\mathcal{A}}_\alpha)$ and DA-complete in $(\mathcal{R}_{\alpha+1} \cap \check{\mathcal{R}}_{\alpha+1}) \setminus (\mathcal{R}_\alpha \cup \check{\mathcal{R}}_\alpha)$.*

The non-collapse property is also an easy corollary.

Corollary 7. *The hierarchy $\{\mathcal{A}_\alpha\}$ does not collapse, i.e. $\mathcal{A}_\alpha \not\subseteq \check{\mathcal{A}}_\alpha$ for all $\alpha<\omega^\omega$.*

The relation of the hierarchy $\{\mathcal{A}_\alpha\}$ to AA-reducibility is even tighter than Theorem 7 and Corollary 6 suggest.

Theorem 8. *For every $\alpha < \omega^\omega$, $\mathcal{A}_\alpha \setminus \check{\mathcal{A}}_\alpha = \{C \mid C \equiv_{AA} A_\alpha\}$ and $(\mathcal{R}_{\alpha+1} \cap \check{\mathcal{R}}_{\alpha+1}) \setminus (\mathcal{R}_\alpha \cup \check{\mathcal{R}}_\alpha) = \{C \mid C \equiv_{AA} A_\alpha \oplus \bar{A}_\alpha\}$.*

Let us summarize some facts on AA-degrees of regular aperiodic sets.

Corollary 8

(i) *For any $\alpha < \omega^\omega$, $A_\alpha \not\leq_{AA} \bar{A}_\alpha$.*
(ii) *For all $\alpha < \beta < \omega^\omega$, $A_\alpha \oplus \bar{A}_\alpha <_{AA} A_\beta$.*
(iii) *Any regular aperiodic ω-language is AA-equivalent to exactly one of the sets $A_\alpha, \bar{A}_\alpha, A_\alpha \oplus \bar{A}_\alpha$ $(\alpha < \omega^\omega)$.*
(iv) *The relations $\leq_{AA}, \leq_{DA}$ and $\leq_{CA}$ coincide on $\mathcal{A}$.*

Another corollary states interesting relationships between hierarchies $\{\mathcal{A}_\alpha\}$ and $\{\mathcal{S}_\alpha\}$ parallel to those between hierarchies $\{\mathcal{R}_\alpha\}$ and $\{\mathcal{S}_\alpha\}$ established in [Se98].

Corollary 9

(i) *For any $\alpha < \omega^\omega$, $\mathcal{A}_\alpha = \mathcal{A} \cap \mathcal{S}_\alpha = \mathcal{A} \cap \mathcal{R}_\alpha$, hence any level $\mathcal{A}_\alpha$ is decidable.*
(ii) *Let $t = t(x_1, \ldots, x_n, y_1, \ldots, y_n)$ be a typed boolean term, where x_i are variables of type 0 and y_i are variables of type 1. Let A be a regular aperiodic set such that $A = t(B_1, \ldots, B_n, C_1, \ldots, C_n)$ for some $B_1, \ldots, B_n \in \Sigma_1^0$ and $C_1, \ldots, C_n \in \Sigma_2^0$. Then $A = t(B_1', \ldots, B_n', C_1', \ldots, C_n')$ for some $B_1', \ldots, B_n' \in \mathcal{K}_0$ and $C_1', \ldots, C_n' \in \mathcal{K}_1$.*

We conclude with a characterization of the structure $(\mathcal{A}; \leq_{AS})$ similar to the characterization of the structure $(\mathcal{R}; \leq_{DS})$ in [Wag79, Se98]. To this end, we define for all $\alpha < \omega^\omega$ and $n < \omega$ the regular aperiodic set $A_\alpha^n = 0^{n+1} \cdot A_\alpha \cup (\cup\{u \cdot \overline{A}_\alpha \mid u \in X^{k+1}, u \neq 0^{n+1}\})$.

Theorem 9. *Let $\alpha < \omega^\omega$, $n < \omega$ and $C \in \mathcal{A}$.*

(i) $C \leq_{AA} A_\alpha$ iff $C \leq_{AS} A_\alpha$.
(ii) $\mathcal{A}_\alpha \setminus \check{\mathcal{A}}_\alpha = \{C \mid C \equiv_{AS} A_\alpha\}$.
(iii) $A_\alpha^n \equiv_{AA} A_\alpha \oplus \overline{A}_\alpha$, $A_\alpha^n <_{AS} A_\alpha^{n+1}$ and $A_\alpha^n \equiv_{AS} \overline{A}_\alpha^n$.
(iv) $C_\alpha \equiv_{AA} A_\alpha \oplus \overline{A}_\alpha$ iff $C_\alpha \equiv_{AS} A_\alpha^k$ for a unique $k < \omega$.
(v) Analogs of (i)—(iv) hold true for $\mathcal{R}$ in place of $\mathcal{A}$ and DS-reducibility in place of AS-reducibility.

Let us summarize some facts established above.

Corollary 10

(i) For all $\alpha < \beta < \omega^\omega$ and $n < \omega$, $A_\alpha, \overline{A}_\alpha <_{AS} A_\alpha^n <_{AS} A_\alpha^{n+1} <_{AS} A_\beta, \overline{A}_\beta$.
(ii) Any regular aperiodic ω-language is AS-equivalent to exactly one of the sets $A_\alpha, \overline{A}_\alpha, A_\alpha^n$ $(\alpha < \omega^\omega, n < \omega)$.
(ii) The relations $\leq_{AS}$, $\leq_{DS}$ and $\leq_{CS}$ coincide on $\mathcal{A}$.
(iv) For every $R \in \mathcal{R}$ there is $A \in \mathcal{A}$ with $R \equiv_{DS} A$ (and hence $R \equiv_{CS} A$). In particular, the Lipschitz degrees of regular aperiodic sets coincide with the Lipschitz degrees of regular sets.

Acknowledgement. The research reported in this paper was initiated during a visit of the author to RWTH Aachen in 1999. Some results were obtained during that visit and were reported at several seminars and conferences. But only recently I was able to answer remaining open questions and produce a reasonably closed text. I am grateful to Wolfgang Thomas for organizing the visit to Aachen, to Klaus Wagner for hosting my Mercator visiting professorship at the University of Würzburg and to both of them for useful discussions and bibliographical hints.

References

[BL69]   Büchi, J.R., Landweber, L.H.: Solving sequential conditions by finite-state strategies. Trans. Amer. Math. Soc. 138, 295–311 (1969)

[C00]   Carton, O.: Wreath product and infinite words. J. Pure and Applied Algebra 153, 129–150 (2000)

[CP97]   Carton, O., Perrin, D.: Chains and superchains for ω-rational sets, automata and semigroups. International Journal of Algebra and Computation 7(7), 673–695 (1997)

[CP99]   Carton, O., Perrin, D.: The Wagner hierarchy of ω-rational sets. International Journal of Algebra and Computation 9(7), 673–695 (1999)

[D03]   Duparc, J.: A hierarchy of deterministic context-free ω-languages. Theoretical Computer Science 290(3), 1253–1300 (2003)

[DR06]   Duparc, J., Riss, M.: The missing link for ω-rational sets, automata, and semigroups. International Journal of Algebra and Computation 16(1), 161–185 (2006)

[Ke94]   Kechris, A.S.: Classical Descriptive Set Theory. Springer, Berlin Heidelberg (1994)

[MP71]   McNaughton, R., Papert, S.: Counter-free automata. MIT Press, Cambridge, Massachusets (1971)

[M06]   Murlak, F.: The Wadge hierarchy of deterministic tree languages. In: Bugliesi, M., Preneel, B., Sassone, V., Wegener, I. (eds.) ICALP 2006. LNCS, vol. 4052, pp. 408–419. Springer, Heidelberg (2006)

[PP04]   Perrin, D., Pin, J.-E.: Infinite Words. Pure and Applied Mathematics, vol. 141. Elsevier, Amsterdam (2004)

[Se94]   Selivanov, V.L.: Two refinements of the polynomial hierarchy. In: Enjalbert, P., Mayr, E.W., Wagner, K.W. (eds.) STACS 94. LNCS, vol. 775, pp. 439–448. Springer, Heidelberg (1994)

[Se94a]   Selivanov, V.L.: Fine hierarchy of regular ω-languages. Preprint N 14, the University of Heidelberg, Chair of Mathematical Logic, p. 13 (1994)

[Se95]   Selivanov, V.L.: Fine hierarchy of regular ω-languages. In: Mosses, P.D., Schwartzbach, M.I., Nielsen, M. (eds.) CAAP 1995, FASE 1995, and TAPSOFT 1995. LNCS, vol. 915, pp. 277–287. Springer, Heidelberg (1995)

[Se95a]   Selivanov, V.L.: Fine hierarchies and Boolean terms. Journal of Symbolic Logic 60, 289–317 (1995)

[Se98]   Selivanov, V.L.: Fine hierarchy of regular ω-languages. Theoretical Computer Science 191, 37–59 (1998)

[Se03]   Selivanov, V.L.: Wadge degrees of ω-languages of deterministic Turing machines. Theoretical Informatics and Applications 37, 67–83 (2003)

[Se06]   Selivanov, V.L.: Fine hierarchy of regular aperiodic ω-languages. Technical report No 390, Institute of Informatics, University of Würzburg, p. 21 (2006)

[Sta97]   Staiger, L.: ω-Languages. In: Handbook of Formal Languages, vol. 3, pp. 339–387. Springer, Berlin Heidelberg (1997)

[SW74]   Staiger, L., Wagner, K.: Automatentheoretische und automatenfreie Characterisierungen topologischer klassen regulärer Folgenmengen. Elektron.Inf. verarb. Kybern. EIK 10(7), 379–392 (1974)

[Th79]   Thomas, W.: Star-free regular sets of ω-sequences. Information and Control 42, 148–156 (1979)

[Th90]   Thomas, W.: Automata on infinite objects. Handbook of Theoretical Computer Science B, 133–191 (1990)

[Th96]   Thomas, W.: Languages, automata and logic. Handbook of Formal Language theory B, 133–191 (1996)

[TB70]   Trakhtenbrot, B.A., Barzdin, J.M.: Finite automata. Behaviour and Synthesis (Russian, English translation 1973), Mir, Moscow. North Holland, Amsterdam (1970)

[Wag79]   Wagner, K.: On ω-regular sets. Information and Control 43, 123–177 (1979)

On Transition Minimality of Bideterministic Automata

Hellis Tamm[*]

Institute of Cybernetics, Akadeemia tee 21, 12618 Tallinn, Estonia
hellis@cs.ioc.ee

Abstract. Bideterministic automata are deterministic automata with the property of their reversal automata also being deterministic. Bideterministic automata have previously been shown to be unique (up to an isomorphism) minimal NFAs with respect to the number of states. In this paper, we show that in addition to state minimality, bideterministic automata are also transition-minimal NFAs. However, as this transition minimality is not necessarily unique, we also present the necessary and sufficient conditions for a bideterministic automaton to be uniquely transition-minimal among NFAs. Furthermore, we show that bideterministic automata are transition-minimal ϵ-NFAs.

1 Introduction

In automata theory, problems related to descriptional complexity issues have been of interest for decades and a lot of research has been done on the subject since the fifties. Many new results on state complexity were obtained in the last ten years and there are still open problems in this research area [1,2]. For example, it is long and well known that the *deterministic* state complexity, that is, the number of states of the minimal deterministic finite automaton (DFA) for a given language, can be exponentially larger than the *nondeterministic* state complexity – the number of states in a minimal nondeterministic finite automaton (NFA) [3]. However, it is also well known that the DFA minimization can be done efficiently whereas the NFA minimization problem is PSPACE-complete [4].

But obviously, there are many cases where the maximal blow-up of size when converting an NFA to DFA does not occur. Recently, some sufficient conditions have been identified which imply that the deterministic and nondeterministic state complexities are the same [5]. Especially, it was shown in [5] that every *bideterministic* automaton – which is any deterministic automaton such that its reversal automaton is also deterministic – is the unique state-minimal NFA accepting its language. Since a bideterministic automaton was known to be the minimal DFA [6,7], it was thus established that for any language accepted by a bideterministic automaton, the deterministic and nondeterministic state complexities coincide. Later, a larger class of automata (that includes bideterministic

---

[*] Supported by EU structural funds (RAK INNOVE project nr. 1.0101-0275) and the Estonian Science Foundation grant 6940.

T. Harju, J. Karhumäki, and A. Lepistö (Eds.): DLT 2007, LNCS 4588, pp. 411–421, 2007.

automata) was shown to have the property of having a unique minimal NFA [8] but in that class, the deterministic and nondeterministic state complexities are not necessarily equal.

While the state-minimal DFA is also minimal with respect to the number of transitions, this is not necessarily the case with NFAs. Vice versa, even allowing one more state in an NFA can produce a considerable reduction in the number of transitions. By [9], there are languages $L(n)$ such that the number of transitions needed by a state-minimal NFA for $L(n)$ is $\Omega(n^2)$ but if an NFA for $L(n)$ can have one more state then it needs only $O(n)$ transitions. Therefore, the number of transitions may be even a better measure for the size of an NFA than the number of states. Recently, lower bounds for transition complexity in terms of nondeterministic state complexity have been studied in [9,10].

It is well known that ϵ-NFAs can be more compact with respect to the number of transitions than NFAs. By [11], there exist regular languages that can be accepted by ϵ-NFAs with $O(n \log_2 n)$ transitions but every NFA for these languages needs at least $\Omega(n^2)$ transitions. This means that an ϵ-NFA may have almost quadratically less transitions than an NFA.

In this paper, we present some transition complexity results for bideterministic automata, in addition to the earlier state complexity result. First, we show that a bideterministic automaton is a transition-minimal NFA. However, as this transition minimality is not necessarily unique, we also present the necessary and sufficient conditions for a bideterministic automaton to be uniquely transition-minimal among NFAs. And second, moreover, we show that a bideterministic automaton is a transition-minimal ϵ-NFA. The first result can be derived using a canonical automaton, called *universal automaton* of a regular language. This automaton has been studied, for example, in [12,13]. The second, more general result, is obtained by applying the theory for finding transition-minimal ϵ-NFAs developed by S. John [14,15], to bideterministic automata.

Bideterministic automata or bideterministic languages have been considered, for example, in the context of machine learning [6], as a special case of reversible automata and languages [7], and in coding theory [16]. They have also been considered in the study of the star height problem in [17,18] where it was shown that the star height of a bideterministic language equals to the loop complexity of the corresponding bideterministic automaton. This is an interesting result, considering the fact that the star height of a regular language is not necessarily equal to the loop complexity of the minimal DFA.

The rest of the paper is organized as follows. After giving some basic definitions of automata in Section 2, we will present the universal automaton of a regular language in Section 3 and use it to show the transition minimality of bideterministic automata among NFAs. In the same section, we will also give the necessary and sufficient conditions for a bideterministic automaton to be a unique transition-minimal NFA accepting the given language. In Section 4 we will present the main ideas of the theory of finding transition-minimal ϵ-NFAs of [14,15], and by applying that theory in Section 5 we will show that bideterministic automata are transition-minimal among ϵ-NFAs.

2 Definitions

A nondeterministic finite automaton with ϵ-transitions (ϵ-NFA) A is presented by $A = (Q, \Sigma, E, I, F)$ where Q is a finite set of states, Σ is an input alphabet, $E \subseteq Q \times (\Sigma \cup \{\epsilon\}) \times Q$ is a set of transitions with ϵ being the empty string, $I \subseteq Q$ is a set of initial states and $F \subseteq Q$ is a set of final states. Let $p, q \in Q$, $a \in \Sigma$ and $c \in \Sigma \cup \{\epsilon\}$. Given a transition $t = (p, c, q) \in E$, we say that t leaves p and enters q. We denote by $indegree(p)$ the number of transitions that enter p, and by $outdegree(p)$ the number of transitions leaving p. Let $p \cdot \epsilon$ denote the ϵ-closure of p, that is, a subset of Q consisting of p and all such states which can be reached from p by a path consisting of ϵ-transitions only. Let $p \cdot a$ denote the set $\{q \in Q \mid$ there are $p', q' \in Q$ such that $p' \in p \cdot \epsilon$, $(p', a, q') \in E$ and $q \in q' \cdot \epsilon\}$. We extend this definition in the following way: for all $P \subseteq Q$ and $x \in \Sigma^*$, $P \cdot c = \bigcup_{p \in P} p \cdot c$, $P \cdot ax = (P \cdot a) \cdot x$.

A special case of an ϵ-NFA is a nondeterministic finite automaton (NFA) if there are no ϵ-transitions in the automaton. In turn, a special case of an NFA is a deterministic finite automaton (DFA) which has a unique initial state and which does not have any pair of transitions (p, a, q) and (p, a, r) such that $q \neq r$, for any $p \in Q$ and $a \in \Sigma$. The *reversal* of an automaton A is the automaton $A^R = (Q, \Sigma, E^R, F, I)$ where for each $p, q \in Q$ and $c \in \Sigma \cup \{\epsilon\}$, $(p, c, q) \in E^R$ if and only if $(q, c, p) \in E$. An automaton A is called *bideterministic* if both A and its reversal automaton A^R are deterministic.

A string $x \in \Sigma^*$ is *accepted* by A if and only if $I \cdot x \cap F \neq \emptyset$. The set of all strings accepted by A is the *language* of A denoted by $L(A)$. Let $q \in Q$. The set $L_L(A, q) = \{x \in \Sigma^* \mid q \in I \cdot x\}$ is the *left language* of q and the set $L_R(A, q) = \{x \in \Sigma^* \mid q \cdot x \cap F \neq \emptyset\}$ is the *right language* of q.

A state q of A is *useful* if it is on some path from an initial state to a final state of A. An automaton A is *trim* if all of its states are useful. A language accepted by a bideterministic automaton is a *bideterministic language*. Two automata are equivalent if they accept the same language.

Given an NFA A, using the well-known operation of the *subset construction*, we obtain an equivalent deterministic automaton $D(A)$ [19]. This operation is also called *determinization*. The automaton $D(A)$ is trim.

Let $A = (Q, \Sigma, E, I, F)$ and $A' = (Q', \Sigma, E', I', F')$ be two NFAs. Then a mapping μ from Q into Q' is a *morphism* of automata if and only if $p \in I$ implies $p\mu \in I'$, $p \in F$ implies $p\mu \in F'$, and $(p, a, q) \in E$ implies $(p\mu, a, q\mu) \in E'$ for all $p, q \in Q$ and $a \in \Sigma$.

3 Bideterministic Automata: Universal and Minimal

Applying the theory of state minimization of NFAs developed by Kameda and Weiner [20], it was shown in [5] that any bideterministic automaton has the minimum number of states among all NFAs accepting the same language, and moreover, it is the only minimal NFA for the given language. By applying the same theory, it can be proven that a bideterministic automaton is a minimal NFA with respect to the number of transitions also [21, Theorem 3.6].

However, there are other means to show the state minimality as well as transition minimality of bideterministic automata. Polak [22] showed how the unique state minimality of bideterministic automata among NFAs can be obtained by using the notion of the universal automaton of the given language. In this section, we will demonstrate how the universal automaton can be used to obtain the transition minimality result as well.

A universal automaton is a canonical automaton of a given regular language. Its properties have been studied, for example, in [12,13,23,22]. In the following, we will give the definition and some basic known properties of this automaton.

Let Σ be a finite alphabet and let $L \subseteq \Sigma^*$. Let ϵ denote the empty string.

Definition 1. *A* factorization *of L is a maximal couple (with respect to the inclusion) of languages (U, V) such that $UV \subseteq L$. The* universal automaton *of L is $U_L = (Q, \Sigma, E, I, F)$ where Q is the set of factorizations of L and $I = \{(U, V) \in Q \mid \epsilon \in U\}$, $F = \{(U, V) \in Q \mid U \subseteq L\}$, $E = \{((U, V), a, (U', V')) \in Q \times a \times Q \mid Ua \subseteq U'\}$.*

The following two propositions state the basic known properties of the universal automaton:

Proposition 1. *The universal automaton of the language L is a finite automaton that accepts L.*

Proposition 2. *Let A be a trim automaton that accepts L. Then there exists an automaton morphism from A into U_L. In particular, U_L contains as a subautomaton every state-minimal NFA accepting L.*

The following proposition by Lombardy [23] gives an effective method for constructing the universal automaton from the minimal DFA of the given language.

Proposition 3 (Lombardy [23, Proposition 6]). *Let $A = (Q, \Sigma, E, \{q_0\}, F)$ be the minimal DFA accepting L and let P be the set of states of the automaton $D(A^R)$. Let $P_\cap$ be the closure of P under intersection, without the empty set: if $X, Y \in P_\cap$ and $X \cap Y \neq \emptyset$ then $X \cap Y \in P_\cap$. Then, the universal automaton U_L is isomorphic to $(P_\cap, \Sigma, H, I, J)$ where $H = \{(X, a, Y) \in P_\cap \times \Sigma \times P_\cap \mid X \cdot a \subseteq Y$ and for all $p \in X, p \cdot a \neq \emptyset\}$, $I = \{X \in P_\cap \mid q_0 \in X\}$, and $J = \{X \in P_\cap \mid X \subseteq F\}$.*

Now, let us apply the method of Proposition 3 to construct the universal automaton of a bideterministic language L. Let $A = (Q, \Sigma, E, \{q_0\}, \{q_f\})$ be a trim bideterministic automaton. It is known that A is the minimal DFA. Since the reversal automaton of A is deterministic, $D(A^R) = A^R$ and the set P as well as $P_\cap$ of Proposition 3 consist of all sets $\{q\}$ such that $q \in Q$. Since it is easy to see that also the transition relation H of U_L is equal to E, $I = \{q_0\}$, and $J = \{q_f\}$, it is concluded that U_L is isomorphic to A. Thus, the following proposition holds:

Proposition 4. *The universal automaton of a bideterministic language is isomorphic to the corresponding bideterministic automaton.*

By using algebraic considerations, the fact given by Proposition 4 about a bideterministic automaton being the universal automaton of its language has been observed by Polak [22].

By applying the theory of state minimization of NFAs developed by Kameda and Weiner [20], it has been proved in [5] that any bideterministic automaton is the unique state-minimal NFA accepting its language. Later, based on algebraic observations about the universal automaton, the same result was obtained by Polak [22]. Indeed, based on Propositions 2 and 4, this result can easily be obtained since a bideterministic automaton is a minimal DFA, and for any minimal DFA, no strict subautomaton of it can accept the same language. Thus, a bideterministic automaton is the only state-minimal NFA for the given language (up to an isomorphism).

Let $A = (Q, \Sigma, E, \{q_0\}, \{q_f\})$ be a trim bideterministic automaton and let $A' = (Q', \Sigma, E', I', F')$ be another trim automaton (non-isomorphic to A) accepting the same language. From above we know that A' has more states than A does. Since $A = U_{L(A)}$ (Proposition 4), then by Proposition 2 there exists an automaton morphism μ from A' into A. In fact, we will show that μ defines an automaton transformation from A' to A. To see this, we present the following two propositions:

Proposition 5. μ *is surjective.*

Proof. Since A is a state-minimal NFA then for each state q of A there exists at least one state q' of A' such that $q'\mu = q$. Thus, μ is surjective. □

Proposition 6. *There is a transition (p, a, q) of A if and only if there is a transition (p', a, q') of A' such that $p'\mu = p$ and $q'\mu = q$.*

Proof. The "if" part follows immediately from the definition of automaton morphism. Indeed, if there is a transition (p', a, q') of A' such that $p'\mu = p$ and $q'\mu = q$ for some $p, q \in Q$ and $a \in \Sigma$ then by the definition of automaton morphism, (p, a, q) is a transition of A.

Now, the "only-if" part. Let us suppose that (p, a, q) is a transition of A but there is no transition (p', a, q') of A' such that $p'\mu = p$ and $q'\mu = q$. Let $B = (Q, \Sigma, E \setminus \{(p, a, q)\}, \{q_0\}, \{q_f\})$ be a subautomaton of A (without the transition (p, a, q)). It is clear that μ is an automaton morphism from A' into B. Let us recall that for any automaton morphism ν from some automaton X into an automaton Y, for any state r of X, it holds that $L_L(X, r) \subseteq L_L(Y, r\nu)$ and $L_R(X, r) \subseteq L_R(Y, r\nu)$ [12,23], and thus $L(X) \subseteq L(Y)$. Therefore, $L(A') \subseteq L(B)$. Since $L(A) = L(A')$, we also get $L(A) \subseteq L(B)$. But, since A is the unique minimal DFA and B has less transitions than A, it must be that $L(B) \subset L(A)$. We have obtained a contradiction. □

Now, based on Propositions 5 and 6, it is not difficult to see that μ defines an automaton transformation from A' to A. Let $Q = \{q_0, \ldots, q_{n-1}\}$. Since μ is surjective, we can form a partition $\Pi = \{Q'_0, \ldots, Q'_{n-1}\}$ of Q' into $n = |Q|$ disjoint non-empty subsets so that for every $q' \in Q'$ and $i \in \{0, \ldots, n-1\}$,

$q' \in Q'_i$ if and only if $q'\mu = q_i$. Using the partition Π, A' can be transformed into an equivalent automaton A'' in the following obvious way: for every $i \in \{0, \ldots, n-1\}$, all states in Q'_i are merged into a single state q''_i of A'' so that all incoming and outgoing transitions of each $q'_i \in Q'_i$ will respectively become the incoming and outgoing transitions of q''_i (with the elimination of duplicate transitions). By Propositions 5 and 6, it is clear that A'' is isomorphic to A. Also, it is clear that the number of transitions of A'' is no more than the number of transitions of A'. Thus, the following proposition holds:

Proposition 7. *Any bideterministic automaton is a transition-minimal NFA.*

However, differently from the state minimality, a bideterministic automaton is not necessarily the only transition-minimal NFA for the corresponding language. A simple example of this kind of language would be $L = \{a, b\}$. The bideterministic automaton accepting L is $A = (\{q_0, q_1\}, \{a, b\}, \{(q_0, a, q_1), (q_0, b, q_1)\}, \{q_0\}, \{q_1\})$ which has two states and two transitions. Another automaton with three states and two transitions accepting the same language is, for example, $A' = (\{q_0, q_1, q_2\}, \{a, b\}, \{(q_0, a, q_1), (q_0, b, q_2)\}, \{q_0\}, \{q_1, q_2\})$.

Next, we are interested in finding the necessary and sufficient conditions for a bideterministic automaton to be uniquely transition-minimal among NFAs accepting the same language. We will prove the following theorem:

Theorem 1. *A trim bideterministic automaton $A = (Q, \Sigma, E, \{q_0\}, \{q_f\})$ is a unique transition-minimal NFA if and only if the following three conditions hold:*

(i) $q_0 \neq q_f$,
(ii) $indegree(q_0) > 0$ or $outdegree(q_0) = 1$,
(iii) $indegree(q_f) = 1$ or $outdegree(q_f) > 0$.

Proof. To prove the necessity part, let us assume that A is a unique transition-minimal NFA and suppose first that (i) does not hold true, that is, $q_0 = q_f$. Let $p \notin Q$ be some new state. Then it is easy to see that there is another automaton $A' = (Q \cup \{p\}, \Sigma, E, \{q_0, p\}, \{q_0, p\})$ with the same number of transitions which accepts the same language. Thus, the condition (i) must necessarily hold true for a unique transition-minimal NFA. Second, let us suppose that (ii) does not hold. Then the initial state q_0 of A has no in-transitions but, since $q_0 \neq q_f$, it must have at least two out-transitions. Now, consider another automaton A' which is obtained from A by splitting q_0 into, for example, as many states as there are out-transitions from q_0 so that each of these states is initial and has only one out-transition. Clearly, A' is equivalent to A, with the same number of transitions. We have obtained a contradiction to our assumption that A is uniquely transition-minimal. Similarly, a supposition that (iii) does not hold allows us to build an equivalent automaton that is non-isomorphic to A but has the same number of transitions.

Now, the sufficiency direction. Let us assume that the conditions (i), (ii) and (iii) hold. Suppose that there is another automaton $A' = (Q', \Sigma, E', I', F')$, different from A, accepting the same language, with the same number of transitions. Let $Q = \{q_0, \ldots, q_{n-1}\}$ and let $\Pi = \{Q'_0, \ldots, Q'_{n-1}\}$ be the partition of

Q' as described above. From Proposition 6 and $|E| = |E'|$, it follows that for any transition (q_i, a, q_j) of A where $i, j \in \{0, \ldots, n-1\}$ and $a \in \Sigma$, there is exactly one transition (q'_i, a, q'_j) of A' such that $q'_i \in Q'_i$ and $q'_j \in Q'_j$. This implies that for $i = 0, \ldots, n-1$, the equations $indegree(q_i) = \Sigma_{q' \in Q'_i} indegree(q')$ and $outdegree(q_i) = \Sigma_{q' \in Q'_i} outdegree(q')$ must hold. Since $|Q'| > |Q|$, there has to be some $k \in \{0, \ldots, n-1\}$ such that $Q'_k = \{q'_{k_1}, \ldots, q'_{k_m}\}$, $m \geq 2$. Suppose $indegree(q_k) \geq 1$ and $outdegree(q_k) \geq 1$. Then there is a pair of transitions of A consisting of some in-transition (q_h, a, q_k) of q_k and some out-transition (q_k, b, q_l) of q_k, with $h, l \in \{0, \ldots, n-1\}$ and $a, b \in \Sigma$, such that the corresponding transitions of A' are (q'_h, a, q'_{k_1}) and (q'_{k_2}, b, q'_l) where $q'_h \in Q'_h$ and $q'_l \in Q'_l$. Then, there is a word $w \in L(A)$ with its accepting computation passing through the transition (q_h, a, q_k) followed by (q_k, b, q_l) such that no computation of A' which goes through (q'_h, a, q'_{k_1}) can take (q'_{k_2}, b, q'_l) as the next transition. Since it can be seen that A as well as A' are unambiguous, it is concluded that $w \notin L(A')$ which leads to a contradiction. So, it must be that $indegree(q_k) = 0$ or $outdegree(q_k) = 0$. This means that q_k must be either the initial or the final state of A because otherwise q_k cannot be useful. If $indegree(q_k) = 0$ then $q_k = q_0$ and, since $|Q'_k| > 1$, we conclude that $outdegree(q_0) > 1$ which is a contradiction to (ii). Similarly, if $outdegree(q_k) = 0$ then $q_k = q_f$ and from $|Q'_k| > 1$ we get that $indegree(q_f) > 1$. But, this is a contradiction to (iii).

Thus, we have proved that the conditions (i), (ii) and (iii) are necessary and, as together, also sufficient conditions for a bideterministic automaton to be uniquely (up to an isomorphism) transition-minimal among NFAs accepting the same language.   □

4   Transition-Minimal Unambiguous ϵ-NFA

S. John [15,14] has developed a theory to reduce the number of transitions of ϵ-NFAs. In the following we present the main ideas from this theory that we need to prove our result in Section 5.

Let A be an ϵ-NFA (Q, Σ, E, I, F) where the transition relation E is partitioned into subrelations E_Σ and E_ϵ such that $E_\Sigma = \{(p, a, q) \mid (p, a, q) \in E, a \in \Sigma\}$ and $E_\epsilon = \{(p, \epsilon, q) \mid (p, \epsilon, q) \in E\}$. Let $t_0 \notin E$ be a new special transition and let $E_0 = E_\Sigma \cup \{t_0\}$. Let the source and target states of a transition t be denoted as $source(t)$ and $target(t)$. The follow-relation $\longrightarrow$ is defined on $E_0 \times E_0$ as follows:

Definition 2 (John [15, Definition 3]). *For $s, t \in E_\Sigma$:*

$$s \longrightarrow t \; :\Leftrightarrow \; target(s) \; E_\epsilon^* \; source(t)$$
$$t_0 \longrightarrow t \; :\Leftrightarrow \; there\ is\ an\ initial\ state\ q \in I\ with\ q\ E_\epsilon^*\ source(t)$$
$$s \longrightarrow t_0 \; :\Leftrightarrow \; there\ is\ a\ final\ state\ q \in F\ with\ target(s)\ E_\epsilon^*\ q$$

A path $\eta \in E_\Sigma^*$ is a sequence $\eta = \eta_1 \cdots \eta_m$ with $m \geq 0$ of transitions $\eta_i \in E_\Sigma$ connected by the follow-relation. The transitions η_i are labeled by $l(\eta_i) \in \Sigma$. Let $l(t_0) = l(\epsilon) = \epsilon$. Then, the string yielded by the path η is defined to be $l(\eta) = l(\eta_1) \cdots l(\eta_m)$.

Definition 3 (John [15, Definition 5]). *Let A be an ϵ-NFA. Then $L(A) = \{w \in \Sigma^* \mid$ there is a path $\eta \in E_{\Sigma}^*$ with $l(t_0\eta t_0) = w\}$. The automaton A is unambiguous if and only if for each $w \in L(A)$ there is exactly one path η with $l(t_0\eta t_0) = w$.*

Definition 4 (John [15]). *Let $t \in E_{\Sigma}$ with $l(t) = a$. The future of t is the set $\varphi(t) = \{w \in \Sigma^* \mid$ there is a path η with $l(\eta t_0) = w$ and $\eta_1 = t\}$. The past of t is the set $\pi(t) = \{w \in \Sigma^* \mid$ there is a path η with $l(t_0\eta) = w$ and $\eta_{|w|} = t\}$. Also, the strict future of t is the set $\hat{\varphi}(t) = \{w \in \Sigma^* \mid aw \in \varphi(t)\}$ and the strict past of t is the set $\hat{\pi}(t) = \{w \in \Sigma^* \mid wa \in \pi(t)\}$.*

Lemma 1 (John [15, Lemma 1]). *Let $v, w \in \Sigma^*$ and $a \in \Sigma$. Then $vaw \in L(A)$ if and only if there exists a transition t such that $va \in \pi(t)$ and $aw \in \varphi(t)$.*

Proposition 8 (John [15, Proposition 3]). $L(A) = \bigcup_{t \in E_{\Sigma}} \hat{\pi}(t)l(t)\hat{\varphi}(t)$.

Definition 5 (John [15, Definition 8]). *Let $L \subseteq \Sigma^*$ be a regular language and let $U, V \subseteq \Sigma^*$, $a \in \Sigma$. We call (U, a, V) a slice of L if and only if $U \neq \emptyset$ and $V \neq \emptyset$ and $UaV \subseteq L$. A slicing of L is a set of slices of L. Let S be the set of all slices of L. We define a partial order on S by considering $(U_1, a, V_1) \leq (U_2, a, V_2)$ if and only if $U_1 \subseteq U_2$ and $V_1 \subseteq V_2$. We define $S_{max} \subseteq S$, the set of maximal slices of L, by $S_{max} := \{(U, a, V) \in S \mid$ there is no $(U', a, V') \in S$ with $(U, a, V) < (U', a, V')\}$.*

Definition 6 (John [15, Definition 9]). *Assume $t_0 \notin S$ and $S_0 := S \cup \{t_0\}$. The follow-relation $\longrightarrow \subseteq S_0 \times S_0$ is defined for all slices (U_1, a, V_1) and $(U_2, b, V_2) \in S$:*

$$(U_1, a, V_1) \longrightarrow (U_2, b, V_2) :\Leftrightarrow U_1 a \subseteq U_2 \text{ and } bV_2 \subseteq V_1$$
$$t_0 \longrightarrow (U_2, b, V_2) :\Leftrightarrow \epsilon \in U_2$$
$$(U_1, a, V_1) \longrightarrow t_0 :\Leftrightarrow \epsilon \in V_1$$
$$t_0 \longrightarrow t_0 :\Leftrightarrow \epsilon \in L$$

Let $S' \subseteq S$ be a finite slicing of L. In order to read an automaton $A_{S'}$ out of S', each slice from S' is transformed into a transition of $A_{S'}$, and these transitions are connected via states and ϵ-transitions according to the follow-relation.

Theorem 2 (John [15, Theorem 2]). *The three following statements are equivalent for languages $L \subseteq \Sigma^*$ if the slicing S_{max} of L induces an unambiguous ϵ-NFA $A_{S_{max}}$:*

- *L is accepted by an ϵ-NFA*
- *$L = L(A_{S'})$ for some finite slicing $S' \subseteq S$*
- *S_{max} is finite*

Furthermore, $|S_{max}| \leq |S'| \leq |E_{\Sigma}|$.

Corollary 1 (John [15, Corollary 3]). *An unambiguous ϵ-NFA $A_{S_{max}}$ has the minimum number of non-ϵ-transitions.*

5   Bideterministic Automata Are Transition-Minimal ϵ-NFAs

It is well known that ϵ-NFAs can be more compact with respect to the number of transitions than NFAs. However, in the following, we will prove that bideterministic automata that we showed to be transition-minimal NFAs (see Section 3) are also transition-minimal ϵ-NFAs. This is a more general result and, actually, the transition minimality among NFAs can be obtained from this result as well.

Definition 7. *Let A be a trim ϵ-NFA. For each non-ϵ-transition t of A, we define the* transition slice *of t to be the slice $(U_t, l(t), V_t)$ of $L(A)$ where $U_t = \hat{\pi}(t)$ and $V_t = \hat{\varphi}(t)$.*

It is not difficult to see that the transition slice definition above is correct, that is, $(U_t, l(t), V_t)$ is a slice of $L(A)$. Indeed, clearly, $U_t = \hat{\pi}(t)$ and $V_t = \hat{\varphi}(t)$ are not empty sets and by Proposition 8, any string uav such that $u \in \hat{\pi}(t)$, $a = l(t)$ and $v \in \hat{\varphi}(t)$, is accepted by A.

Now, let $A = (Q, \Sigma, E, \{q_0\}, \{q_f\})$ be a trim bideterministic automaton. Then A has no ϵ-transitions. That means, $E = E_\Sigma$ and $E_\epsilon = \emptyset$. We will prove the following lemma:

Lemma 2. *For a bideterministic automaton A, let t_1 and t_2 be two different transitions of A, with the same label $l(t_1) = l(t_2) = a \in \Sigma$ and with the corresponding transition slices (U_{t_1}, a, V_{t_1}) and (U_{t_2}, a, V_{t_2}). Then $U_{t_1} \cap U_{t_2} = \emptyset$ and $V_{t_1} \cap V_{t_2} = \emptyset$.*

Proof. Let us suppose that $U_{t_1} \cap U_{t_2} \neq \emptyset$. Then there is a word $w \in U_{t_1}$ and $w \in U_{t_2}$. If $w = \epsilon$ then the initial state of A must have two out-transitions with the label a which is a contradiction since A is deterministic. If $|w| = k$, $k \geq 1$, then there exist two paths $\eta' = \eta'_1 \cdots \eta'_k$ and $\eta'' = \eta''_1 \cdots \eta''_k$ such that $l(\eta') = l(\eta'') = w$. Since A is deterministic, these paths have to coincide, that is, $\eta'_i = \eta''_i$, for $i = 1, \ldots, k$, and both paths end in the same state. Again, this state has two out-transitions with the label a, a contradiction. Therefore, $U_{t_1} \cap U_{t_2} = \emptyset$. Similarly, since the reversal of A is deterministic, it can be shown that $V_{t_1} \cap V_{t_2} = \emptyset$. $\square$

The following proposition is of central importance to obtain our result:

Proposition 9. *Each transition slice of a bideterministic automaton A is maximal.*

Proof. Let us suppose that there exists a transition t of A such that the corresponding transition slice $(U_t, l(t), V_t)$ is not maximal. Then, by Definition 5, there is some maximal slice (U, a, V) of $L(A)$ such that $l(t) = a$ and $(U_t, a, V_t) < (U, a, V)$. This implies that there is a string $uav \in L(A)$ such that $u \in U$ and $v \in V$ but either $u \notin U_t$ or $v \notin V_t$ (or both non-memberships hold). However, by Proposition 8, there must be some other transition t' of A with the transition slice $(U_{t'}, l(t'), V_{t'})$ such that $u \in U_{t'}$, $a = l(t')$, and $v \in V_{t'}$ and

$(U_{t'}, a, V_{t'}) \leq (U, a, V)$. Now, by Definition 5, we know that $U_t \subseteq U$ and $U_{t'} \subseteq U$, and therefore also $U_t \cup U_{t'} \subseteq U$. In the same way, $V_t \cup V_{t'} \subseteq V$.

Next, we can see that $(U_t \cup U_{t'}, a, V_t \cup V_{t'})$ is a slice of $L(A)$. Indeed, from $U_t \neq \emptyset$ and $U_{t'} \neq \emptyset$ we know that $U_t \cup U_{t'} \neq \emptyset$, and similarly, $V_t \cup V_{t'} \neq \emptyset$. Also, for every string $u_1 a v_1$ such that $u_1 \in U_t \cup U_{t'} \subseteq U$ and $v_1 \in V_t \cup V_{t'} \subseteq V$, we know that $u_1 a v_1 \in U a V \subseteq L(A)$, thus $(U_t \cup U_{t'})a(V_t \cup V_{t'}) \subseteq L(A)$, and we can conclude that $(U_t \cup U_{t'}, a, V_t \cup V_{t'})$ is a slice of $L(A)$.

Then there is a word $xay \in L(A)$ such that $x \in U_t$ and $y \in V_{t'}$. Since, by Lemma 2, there does not exist a transition t'' of A such that $x \in U_{t''}$, $a = l(t'')$ and $y \in V_{t''}$, we may conclude by Proposition 8 that $xay \notin L(A)$. We have obtained a contradiction. Thus, every transition slice of A is maximal.   □

Theorem 3. *A bideterministic automaton has the minimum number of transitions among all ϵ-NFAs accepting the same language.*

Proof. Let A be a bideterministic automaton. By Proposition 9, every transition slice of A is a maximal slice of $L(A)$. It is easy to see by Proposition 8 that there are no other maximal slices of $L(A)$. Thus, the set of maximal slices of $L(A)$ is given by $S_{max} := \{(U_t, l(t), V_t) \mid t \in E\}$. Note that $|S_{max}| = |E|$. The set S_{max} is used to form the ϵ-NFA $A_{S_{max}}$ by converting every slice from S_{max} into a transition of $A_{S_{max}}$ and connecting these transitions by ϵ-transitions according to the follow-relation of Definition 6. Because A is bideterministic, A is clearly unambiguous. It is not difficult to see that there is a one-to-one correspondence between the accepting paths of A and $A_{S_{max}}$. Thus, by Definition 3, $A_{S_{max}}$ is also unambiguous. By Theorem 2 and Corollary 1, $A_{S_{max}}$ has a minimum number of non-ϵ-transitions. Since the number of non-ϵ-transitions of $A_{S_{max}}$ is equal to the number of transitions of A, and there are no ϵ-transitions in A, we conclude that A is transition-minimal among all ϵ-NFAs accepting the given language.   □

Remark 1. Theorem 3 is a generalization of Proposition 7. Since NFAs form a subclass of ϵ-NFAs then from knowing that a bideterministic automaton is a transition-minimal ϵ-NFA, it can be concluded that it must be also a transition-minimal NFA.

References

1. Yu, S.: State complexity: recent results and open problems. Fundamenta Informaticae 64, 471–480 (2005)
2. Hromkovic, J.: Descriptional complexity of finite automata: concepts and open problems. Journal of Automata, Languages and Combinatorics 7, 519–531 (2002)
3. Moore, F.: On the bounds for state-set size in the proofs of equivalence between deterministic, nondeterministic, and two-way finite automata. IEEE Trans. Comput. C-20, 1211–1214 (1971)
4. Jiang, T., Ravikumar, B.: Minimal NFA problems are hard. SIAM J. Comput. 22, 1117–1141 (1993)
5. Tamm, H., Ukkonen, E.: Bideterministic automata and minimal representations of regular languages. Theoretical Computer Science 328, 135–149 (2004)

6. Angluin, D.: Inference of reversible languages. Journal of the Association for Computing Machinery 3, 741–765 (1982)
7. Pin, J.E.: On reversible automata. In: Simon, I. (ed.) LATIN 1992. LNCS, vol. 583, pp. 401–416. Springer, Heidelberg (1992)
8. Latteux, M., Roos, Y., Terlutte, A.: BiRFSA languages and minimal NFAs. Technical Report GRAPPA-0205, GRAPPA (2005)
9. Domaratzki, M., Salomaa, K.: Lower bounds for the transition complexity of NFAs. In: Královič, R., Urzyczyn, P. (eds.) MFCS 2006. LNCS, vol. 4162, pp. 315–326. Springer, Heidelberg (2006)
10. Gruber, H., Holzer, M.: Results on the average state and transition complexity of finite automata accepting finite languages. In: Proceedings of DCFS, Computer Science Technical Report, NMSU-CS-2006-001, New Mexico State University, pp. 267–275 (2006)
11. Hromkovic, J., Schnitger, G.: NFAs with and without ε-transitions. In: Caires, L., Italiano, G.F., Monteiro, L., Palamidessi, C., Yung, M. (eds.) ICALP 2005. LNCS, vol. 3580, pp. 385–396. Springer, Heidelberg (2005)
12. Arnold, A., Dicky, A., Nivat, M.: A note about minimal non-deterministic automata. Bull. EATCS 47, 166–169 (1992)
13. Sakarovitch, J.: Elements of Automata Theory (to appear)
14. John, S.: Minimal unambiguous ε-NFA. Technical Report TR-2003-22, Technical University Berlin (2003)
15. John, S.: Minimal unambiguous ε-NFA. In: Domaratzki, M., Okhotin, A., Salomaa, K., Yu, S. (eds.) CIAA 2004. LNCS, vol. 3317, pp. 190–201. Springer, Heidelberg (2005)
16. Shankar, P., Dasgupta, A., Deshmukh, K., Rajan, B.S.: On viewing block codes as finite automata. Theoretical Computer Science 290, 1775–1797 (2003)
17. McNaughton, R.: The loop complexity of pure-group events. Information and Control 11, 167–176 (1967)
18. Cohen, R.S.: Star height of certain families of regular events. J. Comput. Syst. Sci. 4, 281–297 (1970)
19. Hopcroft, J.E., Ullman, J.D.: Introduction to Automata Theory, Languages, and Computation. Addison-Wesley, Reading (1979)
20. Kameda, T., Weiner, P.: On the state minimization of nondeterministic automata. IEEE Trans. Comput. C-19, 617–627 (1970)
21. Tamm, H.: On minimality and size reduction of one-tape and multitape finite automata. PhD thesis, Department of Computer Science, University of Helsinki, Finland (2004)
22. Polak, L.: Minimalizations of NFA using the universal automaton. International Journal of Foundations of Computer Science 16, 999–1010 (2005)
23. Lombardy, S.: On the construction of reversible automata for reversible languages. In: Widmayer, P., Triguero, F., Morales, R., Hennessy, M., Eidenbenz, S., Conejo, R. (eds.) ICALP 2002. LNCS, vol. 2380, pp. 170–182. Springer, Heidelberg (2002)

Author Index

Lecture Notes in Computer Science

For information about Vols. 1–4467

please contact your bookseller or Springer

Vol. 4510: P. Van Hentenryck, L. Wolsey (Eds.), Integration of AI and OR Techniques in Constraint Programming for Combinatorial Optimization Problems. X, 391 pages. 2007.

Vol. 4509: Z. Kobti, D. Wu (Eds.), Advances in Artificial Intelligence. XII, 552 pages. 2007. (Sublibrary LNAI).

Vol. 4508: M.-Y. Kao, X.-Y. Li (Eds.), Algorithmic Aspects in Information and Management. VIII, 428 pages. 2007.

Vol. 4507: F. Sandoval, A. Prieto, J. Cabestany, M. Graña (Eds.), Computational and Ambient Intelligence. XXVI, 1167 pages. 2007.

Vol. 4506: D. Zeng, I. Gotham, K. Komatsu, C. Lynch, M. Thurmond, D. Madigan, B. Lober, J. Kvach, H. Chen (Eds.), Intelligence and Security Informatics: Biosurveillance. XI, 234 pages. 2007.

Vol. 4505: G. Dong, X. Lin, W. Wang, Y. Yang, J.X. Yu (Eds.), Advances in Data and Web Management. XXII, 896 pages. 2007.

Vol. 4504: J. Huang, R. Kowalczyk, Z. Maamar, D. Martin, I. Müller, S. Stoutenburg, K.P. Sycara (Eds.), Service-Oriented Computing: Agents, Semantics, and Engineering. X, 175 pages. 2007.

Vol. 4501: J. Marques-Silva, K.A. Sakallah (Eds.), Theory and Applications of Satisfiability Testing – SAT 2007. XI, 384 pages. 2007.

Vol. 4500: N. Streitz, A. Kameas, I. Mavrommati (Eds.), The Disappearing Computer. XVIII, 304 pages. 2007.

Vol. 4499: Y.Q. Shi (Ed.), Transactions on Data Hiding and Multimedia Security II. IX, 117 pages. 2007.

Vol. 4497: S.B. Cooper, B. Löwe, A. Sorbi (Eds.), Computation and Logic in the Real World. XVIII, 826 pages. 2007.

Vol. 4496: N.T. Nguyen, A. Grzech, R.J. Howlett, L.C. Jain (Eds.), Agent and Multi-Agent Systems: Technologies and Applications. XXI, 1046 pages. 2007. (Sublibrary LNAI).

Vol. 4495: J. Krogstie, A. Opdahl, G. Sindre (Eds.), Advanced Information Systems Engineering. XVI, 606 pages. 2007.

Vol. 4494: H. Jin, O.F. Rana, Y. Pan, V.K. Prasanna (Eds.), Algorithms and Architectures for Parallel Processing. XIV, 508 pages. 2007.

Vol. 4493: D. Liu, S. Fei, Z. Hou, H. Zhang, C. Sun (Eds.), Advances in Neural Networks – ISNN 2007, Part III. XXVI, 1215 pages. 2007.

Vol. 4492: D. Liu, S. Fei, Z. Hou, H. Zhang, C. Sun (Eds.), Advances in Neural Networks – ISNN 2007, Part II. XXVII, 1321 pages. 2007.

Vol. 4491: D. Liu, S. Fei, Z.-G. Hou, H. Zhang, C. Sun (Eds.), Advances in Neural Networks – ISNN 2007, Part I. LIV, 1365 pages. 2007.

Vol. 4490: Y. Shi, G.D. van Albada, J. Dongarra, P.M.A. Sloot (Eds.), Computational Science – ICCS 2007, Part IV. XXXVII, 1211 pages. 2007.

Vol. 4489: Y. Shi, G.D. van Albada, J. Dongarra, P.M.A. Sloot (Eds.), Computational Science – ICCS 2007, Part III. XXXVII, 1257 pages. 2007.

Vol. 4488: Y. Shi, G.D. van Albada, J. Dongarra, P.M.A. Sloot (Eds.), Computational Science – ICCS 2007, Part II. XXXV, 1251 pages. 2007.

Vol. 4487: Y. Shi, G.D. van Albada, J. Dongarra, P.M.A. Sloot (Eds.), Computational Science – ICCS 2007, Part I. LXXXI, 1275 pages. 2007.

Vol. 4486: M. Bernardo, J. Hillston (Eds.), Formal Methods for Performance Evaluation. VII, 469 pages. 2007.

Vol. 4485: F. Sgallari, A. Murli, N. Paragios (Eds.), Scale Space and Variational Methods in Computer Vision. XV, 931 pages. 2007.

Vol. 4484: J.-Y. Cai, S.B. Cooper, H. Zhu (Eds.), Theory and Applications of Models of Computation. XIII, 772 pages. 2007.

Vol. 4483: C. Baral, G. Brewka, J. Schlipf (Eds.), Logic Programming and Nonmonotonic Reasoning. IX, 327 pages. 2007. (Sublibrary LNAI).

Vol. 4482: A. An, J. Stefanowski, S. Ramanna, C.J. Butz, W. Pedrycz, G. Wang (Eds.), Rough Sets, Fuzzy Sets, Data Mining and Granular Computing. XIV, 585 pages. 2007. (Sublibrary LNAI).

Vol. 4481: J. Yao, P. Lingras, W.-Z. Wu, M. Szczuka, N.J. Cercone, D. Ślęzak (Eds.), Rough Sets and Knowledge Technology. XIV, 576 pages. 2007. (Sublibrary LNAI).

Vol. 4480: A. LaMarca, M. Langheinrich, K.N. Truong (Eds.), Pervasive Computing. XIII, 369 pages. 2007.

Vol. 4479: I.F. Akyildiz, R. Sivakumar, E. Ekici, J.C.d. Oliveira, J. McNair (Eds.), NETWORKING 2007. Ad Hoc and Sensor Networks, Wireless Networks, Next Generation Internet. XXVII, 1252 pages. 2007.

Vol. 4478: J. Martí, J.M. Benedí, A.M. Mendonça, J. Serrat (Eds.), Pattern Recognition and Image Analysis, Part II. XXVII, 657 pages. 2007.

Vol. 4477: J. Martí, J.M. Benedí, A.M. Mendonça, J. Serrat (Eds.), Pattern Recognition and Image Analysis, Part I. XXVII, 625 pages. 2007.

Vol. 4476: V. Gorodetsky, C. Zhang, V.A. Skormin, L. Cao (Eds.), Autonomous Intelligent Systems: Multi-Agents and Data Mining. XIII, 323 pages. 2007. (Sublibrary LNAI).

Vol. 4475: P. Crescenzi, G. Prencipe, G. Pucci (Eds.), Fun with Algorithms. X, 273 pages. 2007.

Vol. 4474: G. Prencipe, S. Zaks (Eds.), Structural Information and Communication Complexity. XI, 342 pages. 2007.

Vol. 4472: M. Haindl, J. Kittler, F. Roli (Eds.), Multiple Classifier Systems. XI, 524 pages. 2007.

Vol. 4471: P. Cesar, K. Chorianopoulos, J.F. Jensen (Eds.), Interactive TV: a Shared Experience. XIII, 236 pages. 2007.

Vol. 4470: Q. Wang, D. Pfahl, D.M. Raffo (Eds.), Software Process Dynamics and Agility. XI, 346 pages. 2007.

Vol. 4469: K.-c. Hui, Z. Pan, R.C.-k. Chung, C.C.L. Wang, X. Jin, S. Göbel, E.C.-L. Li (Eds.), Technologies for E-Learning and Digital Entertainment. XVIII, 974 pages. 2007.

Vol. 4468: M.M. Bonsangue, E.B. Johnsen (Eds.), Formal Methods for Open Object-Based Distributed Systems. X, 317 pages. 2007.